# INELASTIC ANALYSIS OF STRUCTURES UNDER VARIABLE LOADS

# SOLID MECHANICS AND ITS APPLICATIONS
Volume 83

*Series Editor:* G.M.L. GLADWELL
*Department of Civil Engineering*
*University of Waterloo*
*Waterloo, Ontario, Canada N2L 3GI*

*Aims and Scope of the Series*

The fundamental questions arising in mechanics are: *Why?*, *How?*, and *How much?* The aim of this series is to provide lucid accounts written bij authoritative researchers giving vision and insight in answering these questions on the subject of mechanics as it relates to solids.

The scope of the series covers the entire spectrum of solid mechanics. Thus it includes the foundation of mechanics; variational formulations; computational mechanics; statics, kinematics and dynamics of rigid and elastic bodies: vibrations of solids and structures; dynamical systems and chaos; the theories of elasticity, plasticity and viscoelasticity; composite materials; rods, beams, shells and membranes; structural control and stability; soils, rocks and geomechanics; fracture; tribology; experimental mechanics; biomechanics and machine design.

The median level of presentation is the first year graduate student. Some texts are monographs defining the current state of the field; others are accessible to final year undergraduates; but essentially the emphasis is on readability and clarity.

*For a list of related mechanics titles, see final pages.*

# Inelastic Analysis of Structures under Variable Loads

## Theory and Engineering Applications

Edited by

Dieter Weichert
*Institut für allgemeine Mechanik,*
*Rheinisch-Westfälische Technische Hochschule Aachen, Germany*

and

Giulio Maier
*Department of Structural Engineering,*
*Technical University (Politecnico) of Milan, Italy*

KLUWER ACADEMIC PUBLISHERS
DORDRECHT / BOSTON / LONDON

A C.I.P. Catalogue record for this book is available from the Library of Congress.

ISBN 0-7923-6645-X

---

Published by Kluwer Academic Publishers,
P.O. Box 17, 3300 AA Dordrecht, The Netherlands.

Sold and distributed in North, Central and South America
by Kluwer Academic Publishers,
101 Philip Drive, Norwell, MA 02061, U.S.A.

In all other countries, sold and distributed
by Kluwer Academic Publishers,
P.O. Box 322, 3300 AH Dordrecht, The Netherlands.

*Printed on acid-free paper*



Printed in the Netherlands.

## TABLE OF CONTENTS

## PREFACE

The question whether a structure or a machine component can carry the applied loads, and with which margin of safety, or whether it will become unserviceable due to collapse or excessive inelastic deformations, has always been a major concern for civil and mechanical engineers.

The development of methods to answer this technologically crucial question without analysing the evolution of the system under varying loads, has a long tradition that can be traced back even to the times of emerging mechanical sciences in the early 17th century. However, the scientific foundations of the theories underlying these methods, nowadays frequently called "direct", were established sporadically in the Thirties of the 20th century and systematically and rigorously in the Fifties.

Further motivations for the development of direct analysis techniques in applied mechanics of solids and structures arise from the circumstance that in many engineering situations the external actions fluctuate according to time histories not *a priori* known except for some essential features, e.g. variation intervals. In such situations the critical events (or "limit states") to consider, besides plastic collapse, are incremental collapse (or "ratchetting") and alternating plastic yielding, namely lack of "shakedown". Non evolutionary, direct methods for ultimate limit state analysis of structures subjected to variably-repeated external actions are the objectives of most papers collected in this book, which also contains a few contributions on related topics.

The early works on this subject were based upon simplifying assumptions allowing the solution of elementary practical problems without recourse to, at that time unavailable, effective computational tools. However, those assumptions turned out to be too restrictive in the light of the many recent advances in technological demands, computing capabilities and computational mechanics. Therefore, in the last few decades, many scientists have been involved in extending the basic theories, searching for alternative formulations and solving technologically key-problems in the spirit of direct approaches. In order to identify the directions of such modern developments one may classify them into the following groups:

- Adaptation of the general, theoretical achievements to specific types of structures such as trusses, frames, plates, shells, two- and three-dimensional solids and heterogeneous materials (such as composites) at the microscale.
- Generalisation of the basic theory to dynamics, i.e. to the time dependence due to inertia and damping forces.
- Reformulation of the fundamental theorems in the broader frame of geometrically non-linear theory of solid mechanics in order to account for configuration changes and their effects on equilibrium.
- Allowing for more sophisticated models of inelastic material behaviour, including non-linear hardening and softening, non-associated flow rules, viscous effects, multi-phase poro-plasticity and material damage.

- Development of computational procedures and specific algorithms by which direct methods can be efficiently used to solve large-scale industrial problems numerically.

The papers collected in this book discuss timely questions related to these topics and present research contributions consistent with the state-of-the-art in the field at the most advanced level and in line with the tradition of direct methods in structural and materials mechanics. They have been elaborated by their authors in the sequel of the European Mechanics (EUROMECH)-Colloquium 385 held at Aachen in October 1998 on *Inelastic Analysis of Structures under Variable Loads: Theory and Engineering Applications*. This conference was the latest in a sequence of meetings in the area of structural plasticity with more or less reference to direct methods for shakedown and limit analysis and related topics. Worth remembering here are the symposia *Foundations of Plasticity* (1972, Warsaw), *Engineering Plasticity by Mathematical Programming* (1977, Waterloo, Canada) and *Materiaux et Structures sous Chargement Cyclique* (1978), followed by the more specialised EUROMECH-Colloquia 174 *Inelastic Structures under Variable Loads* (1983, Palermo), 185 *Mathematical Programming Methods for the Plastic Analysis of Structures* (1984, London) and 298 *Inelastic Structures under Variable Loads* (1992, Warsaw). These antecedents to the Aachen meeting from which this volume evolved, inspire grateful memories of prominent contributors to them and leading researchers who have since passed away: J. Mandel, C. Massonnet, A. Sawczuk, J. König, Maria Duszek-Perzyna, J. Martin, P. Panagiotopoulos.

The editors thank both the participants in the EUROMECH-Colloquium 385 and the authors of the papers gathered in this book for their valuable contributions. Special thanks also to all those who helped to prepare the final version of the volume, in particular Dipl.-Math. Michael Ban.

Giulio Maier and Dieter Weichert

# AN EARLY UPPER BOUND METHOD FOR SHAKEDOWN OF BEAMS AND FRAMES

*A Historical Glance*

P.S. SYMONDS
*Brown University*
*Providence, Rhode Island 02912, U.S.A.*

## 1. Introduction

Placing this paper, which mainly describes a 50 year old and quite elementary approach to the shakedown problem, in the honored position of an introductory chapter in the Proceedings volume, serves at least to make evident the enormous increase in both the scope and the general sophistication of the more recent research efforts in this area of structural mechanics.

The writer had the good fortune to be a member of Professor Prager's group at Brown University when the "limit theorems" of plastic structural analysis were at a stage of intense development; they were undergoing birth pangs of emerging in their definitive forms. I am referring in particular to the forms they were given in the two papers of Drucker, Greenberg and Prager [5] and of Greenberg and Prager [6]. Perhaps it will be of interest to try to give some impressions of the mood and atmosphere of those times.

I joined the faculty of Brown University in September, 1947, with a joint appointment in the Division of Engineering and the newly created Graduate Division of Applied Mathematics. A Program of Advanced Instruction and Research in Mechanics had been running since 1941, when William Prager came from Istanbul to be its Director. This ended in 1946. Research was sponsored by military agencies such as the Office of Naval Research, among many others. Prager was adept at attracting research support with broadly defined objectives. One aimed at the establishment of theorems for the realistic analysis and efficient design of structures taking account of their ductility in the plastic range. Taking part in this effort were Daniel Drucker who had come to Brown shortly before I did and Herbert Greenberg, who had been a research associate in the Program of Advanced Instruction and Research in Mechanics, obtaining his Ph.D. under Prager. Among other professors were E.H. Lee and William Pell; research associates included Philip Hodge, Lawrence Malvern and Bernard Neal. Neal came in 1948. He was the first of a series of visitors from Cambridge University. The fundamental problem of structural plasticity was to find rigorous and general variational theorems governing a



*D. Weichert and G. Maier (eds.), Inelastic Analysis of Structures under Variable Loads,* 1–9.

structure as the load increases to its critical failure value. Greenberg, Drucker, Prager, Lee, Hodge and others dealt with approaches to this goal. Neal and and I worked together. We were intrigued by the problem of general loading, where only the limits are defined, and independent load components may be applied within them arbitrarily and repeatedly.

Prager took an interest in everything. In 1948 he published a paper on "Problem Types in the Theory of Perfectly Plastic Materials", in which he introduced a pictorial approach to display elastic-plastic structural behavior. The state of stress in a structure model (a 3-bar symmetric truss) was represented by points in a plane stress space. This simple structure is capable of displaying the Haar-Karman and Colonnetti principles for monotonically increasing loading, as well as Greenberg's generalization. It can also illustrate "shakedown": both the possibility of residual stress states such that no further plastic flow occurs if the load remains between its limits; and the possibility of failure by repeated positive and negative increments of plastic deformation when the limits are exceeded. Using the term "shakedown" in this context was Prager's invention. (He was a good linguist, and proud of his colloquial English; he liked to be called Bill by his peers.)

At that time we were very aware of the "competition", and of course of various detractors of what was called "limit design". Valuable contributions, especially with regard to experiments on beams and frames, had been made and were continuing at Lehigh and Cambridge universities. We were becoming aware of considerable earlier work, some quite astounding for its prescience. For example, Gabor Kazinczy, working as a young engineer for the city of Budapest, in 1913 and 1914 wrote analyses of tests on roof girders showing a clear grasp of the essence of the limit theorems. Both Friedrich Bleich ("Stahlhochbauten" Vol. 1, Berlin, 1932 Springer) and his son Hans Bleich (Der Bauingenieur, 1932, Heft 19/20) had clear ideas about both limit analysis and failure by repeated cyclic loading of continuous beams. In 1948 J.A. Van den Broek's book "Limit Design" (Wiley) appeared. The author says that his ideas on the subject originated from a lecture in 1917 of N.C. Kist at the Technical University in Delft. Although Van den Broek understood the essential phenomenon of equalization of critical bending moments in a symmetrically loaded clamped beam as the load approaches the failure magnitude, he took "failure" as occurring when the three moments reached their maximum values in the elastic range, not the fully plastic values; for I-beams, the former may be only some 10 percent smaller than the fully plastic value, and he chose it so as to be "slightly on the safe side".

Despite this conservative gesture, all was not exactly sweetness and light in the American solid mechanics community, with respect to people such as Drucker with new ideas about plasticity, and such as Van den Broek with unconventional notions of structural design. With regard to limit analysis, there were skeptics whose main objection was that any such method *must* be accompanied by "*means of determining the deformations corresponding to any given load up to the ultimate one*". (I am quoting a discusser of a paper on "Inelastic Bending" in the ASCE Transaction for 1948). He congratulates the author for his detailed methods for determining the deformations and for providing tables without which "the practical use of these methods is all but impossible", with special praise for his treatment of strain hardening. He agrees with the author that strain hardening is an essential property for the validity of the limit theorems. (Fortunately it is not!)

Since we were enjoying our collaboration, Neal persuaded Professor Baker, Head of the Engineering Department at Cambridge University, to invite me to spend the next year at Cambridge, England. Professor Prager generously acquiesced in this and I spent the next 18 months in Cambridge, where among other things we did experiments on cyclic loading of steel frames [18].

The approach to the "shakedown problem" that Bernard Neal and the writer published in 1951 [1, 2] followed directly from the method we had presented for the "plastic collapse" problem [3, 4]. The latter term refers to a structure subjected to a set of loads in fixed proportions to each other, and the former to the more realistic and difficult case where only the maximum and minimum values are assigned, each load being arbitrarily variable and repeatable within its limits. The fundamental theorems furnishing lower and upper bounds on plastic collapse loads on general bodies had then just been stated in modern terms [5, 6]. Melan's theorem [7], giving lower bounds in the shakedown problem, had been restudied [8, 9, 10]. Making the standard assumptions and idealizations for beams and frames, that plastic deformations occur only at "plastic hinges" at critical sections, where arbitrarily large rotations at constant moment magnitudes ("fully plastic bending moments") are allowed — ignoring deflection effects — we presented [3, 4] an efficient method for solving the plastic collapse problem of a continuous beam or plane frame. We summarize below our solution of this problem, as a preliminary to dealing with shakedown.

Figure 1 shows a simple frame problem. The two loads $V$ and $H$ are applied together at their maximum values, which we write as $P$ in both cases. There are five critical sections, as marked, where bending moments must be considered. The fully plastic moment is $M_p$ for all members. What is the limit load magnitude $P_L$? Upper bounds are easily computed by the mechanism method. This frame has three redundant moments and two independent equations of equilibrium. Hence there are two elementary mechanisms for vertical and horizontal forces shown in Fig. 2 as (a) and (b); a third mechanism (c) is obtained by combining those in (a) and (b) so as to eliminate the plastic hinge at section 2. Virtual rotations $\psi_i$ are shown in terms of the small angle $\theta$. The virtual work equations furnish a magnitude $P_c$ corresponding to each mechanism. Note that in computing the internal work the bending moment at each section always agrees in sign with the virtual hinge rotation; all products $M_{pi}\psi_i$ are positive. For example, using mechanism (c), we have

$$(H+V)l\theta = 2P_c l\theta = M_p(1+2+2+1)\theta = 6M_p\theta,$$

so that $P_c = 3M_p/l$, where $P_c \geq P_L$. Mechanisms (a) and (b) both furnish the result $P_c = 4M_p/l$. In this problem, having examined the three possible collapse mechanisms, we know that the smallest of them is the *actual* limit load magnitude $P_L$, i.e. with $P_c l = 3M_p$ the yield condition $|M_i| \leq M_p$ is satisfied at all the critical sections. In the paper [4], a less trivial (3-story, 2-bay) frame involving distributed loads was taken for the main example; a quite good preliminary solution could be obtained in about 20 minutes.

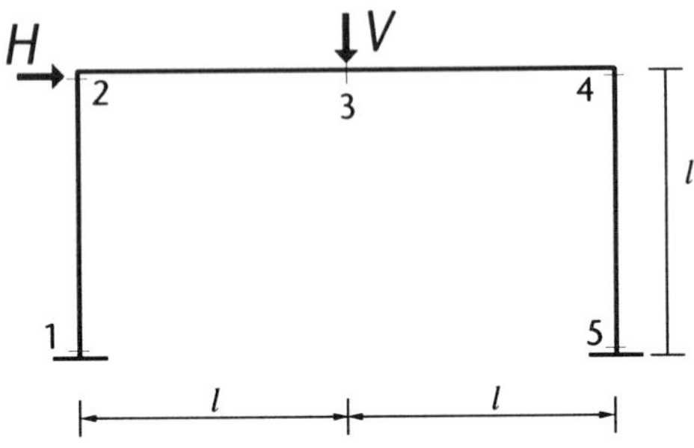


Figure 1. Simple frame problem used as example

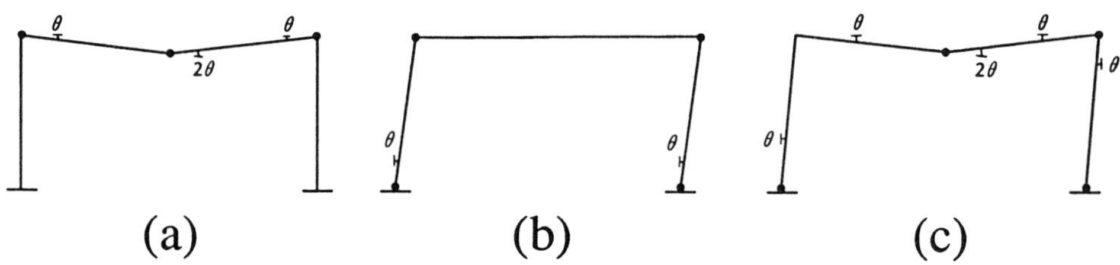


Figure 2. Mechanisms of simple plastic collapse and incremental collapse.

## 2. The Shakedown Problem

Instead of assuming $H = V = P$, we now prescribe only the extreme values:

$$-P \leq H(t) \leq P, \quad 0 \leq V(t) \leq P \tag{1a, 1b}$$

($t$ = time). We assume that $H(t)$ and $V(t)$ may have any values between these extremes, and that any program or cycle of loading may be repeated any number of times. What is the shakedown load magnitude $P_S$?

From Melan's theorem we knew that residual stresses and elastic stresses due to external loads on the stress-free structure play key roles in shakedown analysis. Also, from our experience with the plastic collapse problem we knew that collapse mechanisms and the device of combining mechanisms are convenient schemes for obtaining upper bounds, and often lead to *exact* solutions as well, satisfying lower as well as upper bound conditions. We therefore took as starting point an equation of virtual work involving concepts of both self-equilibrated residual stresses and kinematics of plastic collapse mechanisms, namely:

$$\sum m_i \psi_i = 0 \tag{2}$$

Here $m_i$ are residual moments at the critical sections, i.e. bending moments satisfying equations of equilibrium with zero values of the external loads; and $\psi_i$ are virtual rotations across plastic hinges at the critical sections in a particular collapse mechanism, written in terms of the small angle $\theta$. For example, for the mechanism in Fig. 2(c),

$$\psi_i = (\psi_1 \ \psi_2 \ \psi_3 \ \psi_4 \ \psi_5) = (\text{-1 } 0 \text{ } 2 \text{ -2 } 1)\theta \tag{3}$$

The sign convention, as already seen, attaches a plus sign to opening rotations, and a negative sign to closing rotations; the same convention is used for bending moments. The bending moments may be written in general as

$$M_i = m_i + M_i^e \tag{4}$$

where $M_i^e$ is the moment at section $i$ due to the external loads acting on the stress-free structure, assuming wholly elastic behavior. If all members have the same $EI$, the moment equations have the load coefficients shown in column (1) of Table 1. Their minimum and maximum values at the five critical sections, obtained by choosing the appropriate extreme values of $H$ and $V$, are shown in columns (2) and (3), respectively.

## 3. Upper Bound Calculation

Since the loads may vary in any way, we must look at the worst possibilities. The loading program may be a cycle that brings the moments at the critical sections *one by one* to the values required in one of the mechanisms of Fig. 2. The special value of $P$, say $P^*$, at which this happens corresponds to an *incipient plastic collapse.* This means that if $P^*$ is slightly exceeded, after the cycle has been completed the deflection in the chosen mechanism will have been increased by a small quantity. But the cycle may be repeated, and with each repetition the same increment of deflection will be added. The net effect is a process of *incremental collapse.* $P^*$ represents a borderline value for collapse in a particular mechanism. In the present problem there are three possible mechanisms. For each of them a value of $P^*$ may be computed by writing out Eq. 2 for that mechanism. The smallest of the three values is the minimum value of P at which incremental collapse may occur: $(P^*)_{min} = P_S$, the shakedown load.

If the loads are applied at a magnitude slightly smaller than this $P_S$, the moments at all the critical sections will be slightly smaller in magnitude than the $|M_i|$ required there for plastic collapse. The conditions of Melan's theorem will be satisfied and shakedown will occur. On the other hand, if this magnitude $P_S$ is exceeded, incremental collapse in some mechanism may occur, adding a finite deflection increment with each repetition of a cycle.

The calculation requires writing out Eq. 2 for each of the three mechanisms of Fig. 2. We give the details for the mechanism shown in Fig. 2(c). We write each $m_i$ from Eq. 4, putting $m_i = M_p - (M_i^e)_{max}$ when $M_i$ and $\psi_i$ are positive, and $m_i = -M_p - (M_i^e)_{min}$ when $M_i$ and $\psi_i$ are negative. Thus for mechanism 2(c), taking $\psi_i$ values from Eq. 3, we have

$$[-M_p - (M_1^e)_{min}](-\theta) + [M_p - (M_3^e)_{max}](2\theta) + [-M_p - (M_4^e)_{min}](-2\theta) + [M_p - (M_5^e)_{max}](\theta) = 0 \tag{5}$$

Inserting numerical values from columns (2) and (3) of Table 1:

$$[-M_p + .3125P^*l](-\theta) + [M_p - .3P^*l](2\theta) + [-M_p + .3875P^*l](-2\theta) + [M_p - .4125P^*l](\theta) = 0 \tag{6}$$

Hence

$$6M_p - 2.1P^*l = 0; \quad P^* = 2.857M_p/l \tag{7}$$

Similar calculations made for mechanisms (a) and (b) of Fig. 2 lead to larger values: $P^* = 2.909M_p/l$ and $3.077M_p/l$, respectively. We conclude that the shakedown load magnitude is $P_S = 2.857M_p/l$.

No mathematical proof seems necessary that the procedure sketched above furnishes an upper bound for each mechanism considered. If the chosen mechanism is wrong, there is a violation of the plastic moment condition at one or more of the critical sections elsewhere in the structure. But the sections where this occurs can be imagined to be *strengthened* so that the moment condition is satisfied, while its properties otherwise are unchanged. By Melan's theorem the load determined from the chosen mechanism then is the correct collapse load for the strengthened frame. But it can be taken as axiomatic that any strengthening of a structure either leaves its collapse load unchanged or increases it. The load corresponding to the chosen mechanism is therefore an upper bound on the collapse load of the original frame. The argument applies to incremental collapse as well as to simple plastic collapse under proportional loading.

TABLE 1. Col. (1) Elastic Coefficients;
(2)/(3) Minimum/Maximum Elastic Moments; (4) Residual Moments

| (1) | (2) | (3) | (4) |
|---|---|---|---|
| $M_1^e = -.3125Hl + .1Vl$ | $(M_1^e)_{min} = -.3125Pl$ | $(M_1^e)_{max} = .4125Pl$ | $m_1 = -.1071M_p$ |
| $M_2^e = .1875Hl - .2Vl$ | $(M_2^e)_{min} = -.3875Pl$ | $(M_2^e)_{max} = .1875Pl$ | $m_2 = .1787M_p$ |
| $M_3^e = 0Hl + .3Vl$ | $(M_3^e)_{min} = 0Pl$ | $(M_3^e)_{max} = .3Pl$ | $m_3 = .1429M_p$ |
| $M_4^e = -.1875Hl - .2Vl$ | $(M_4^e)_{min} = -.3875Pl$ | $(M_4^e)_{max} = .1875Pl$ | $m_4 = .1071M_p$ |
| $M_5^e = .3125Hl + .1Vl$ | $(M_5^e)_{min} = -.3125Pl$ | $(M_5^e)_{max} = .4125Pl$ | $m_5 = -.1786M_p$ |

Note that the quantities in square brackets in Eqs. 5 and 6 are the residual moments at sections 1, 3, 4, and 5; these are calculated at once using the value for the shake-down load $2.857M_p/l$. The moment at section 2 is readily obtained from either of the two mechanisms that include a hinge at that section; the other mechanism can be used as a check on the arithmetic. All the final values are given in column 4 of Table 1.

## 4. Alternating Plasticity

Variable and repeated loading may lead to failure not only by incremental collapse, but also by "alternating plasticity" when the external loads cause the range of the applied elastic bending moment at some section to exceed the elastic range of the material. Plastic strain increments alternating in sign during each repetition of a cycle of loading may lead to failure by fracture, often referred to as "low cycle fatigue" The load magnitude $P^{**}$ required for this type of failure is easily estimated. Let $\alpha$ denote the shape factor of a beam section, i.e. the ratio of the fully plastic moment to the yield moment. The elastic range may then be estimated as $2M_p/\alpha$. The algebraic differences between the maximum and minimum bending moments at the critical sections, obtained from Table 1, furnish the applied elastic moment range at each section for the load system considered. For example, at sections 1 and 5 the range of the elastic moment is $(.4125 + .3125)P^{**}l = .725P^{**}l$, and thus failure by alternating plasticity may occur if this exceeds $2M_p/\alpha$. Taking $\alpha = 1.20$, as a conservative value for I-sections, failure by alternating plasticity may occur if $P > P^{**} = 2.30M_p/l$. Hence this is the governing failure condition rather than incremental collapse, in this problem example, where the horizontal force is assumed to vary between $-P$ and $P$. However, if the horizontal force $H(t)$ does not change its direction, i.e. if Eq. (1a) is replaced by $0 \leq H(t) \leq P$, the situation is reversed: it is easily checked that the incremental collapse load $P_S$ is unchanged at $2.857M_p/l$, but failure by alternating plasticity then requires $P$ greater than about $4M_p/l$.

## 5. Comments

The first paper describing our new approach to shakedown was not [1], but [11], submitted to the First U.S. National Congress of Applied Mechanics in Chicago in 1951. The paper was presented orally but rejected for the Proceedings; the reviewer apparently did not believe it. Subsequent manuscripts met with ready acceptance. We were happy to have discovered a new approach to the shakedown problem by way of upper bounds, easier to compute than Melan's lower bounds. In 1956 Koiter [12] presented his "kinematic theorem" involving the plastic work done over a cycle of admissible plastic strains. This also furnishes upper bounds, a "second theorem" complementary to Melan's "first theorem" and presumably in some sense equivalent to our method. However, in his 1960 review paper [13], in a paragraph of "Historical Remarks" (p. 213), Koiter referred to our method as described in [1] as having *"some superficial resemblance to the second shakedown theorem, but it is in effect a clever method of applying the first theorem."* Here Koiter was wrong, strangely. Our method is unmistakably a kinematic approach. If one can be satisfied that all the possible collapse mechanisms have been examined for incremental collapse, the lowest bound of the finite set is the actual load at the shakedown limit. The only place where Melan's theorem need be applied is as a check, to verify that the stresses at all the critical sections satisfy all the $\pm M_p$ limits (as done in the example problem above).

Our method is described in several texts by American and British authors, e.g. Hodge [14] and Heyman [15, 16] (again, as a practical technique, not an application of a theorem). Martin [17] gives a careful review of the two shakedown theorems and illustrates the solution of both types of problem (plastic collapse and shakedown) by linear programming techniques, using essentially the present approach.

Finally, the significance of the limit load, computed either for collapse or shakedown, is that of a parameter of loading above which the deformations are regarded as essentially unlimited, according to the underlying assumptions (e.g. absence of strain hardening). The tests [18] that we perfomed in the Cambridge Engineering Laboratory in 1951 serve to illustrate this basic concept.

## References

1. Symonds, P.S. and Neal, B.G.: Recent progress in the plastic methods of structural analysis, Part II, *J. Franklin Inst.* **252** (6) (1951), 469–492.

2. Neal, B.G. and Symonds, P.S.: A method for calculating the failure load for a framed structure subjected to fluctuating loads, *J. Instn. Civil Engrs.* **35** (1951), 186–197.
3. Symonds, P.S. and Neal, B.G.: Recent progress in the plastic methods of structural analysis, Part I, *J. Franklin Inst.* **252** (5) (1951), 383–407.
4. Neal, B.G. and Symonds, P.S.: The rapid calculation of the plastic collapse load for a framed structure, Proc. Instn. Civil Engrs., Part III 1 (1952), 58–100 (with discussion).
5. Drucker, D.C., Prager, W., and Greenberg, H.J.: Extended limit design theorems for continuous media, *Quart. Appl. Math.* **9** (1951), 381–389.
6. Greenberg, H.J. and Prager, W.: Limit design of beams and frames, *Proc. ASCE* **77** (1951), Separate No. **59**; *Trans. ASCE* **117** (1952), 447–484 (with discussion).
7. Melan, E.: Der Spannungszustand eines "Hencky–Mises'chen" Kontinuums bei veraenderlicher Belastung, *Sitzungsber. Ak. Wiss. Wien IIa* **147** (1938), 73–87.
8. Symonds, P.S. and Prager, W.: Elastic–plastic analysis of structures subjected to loads varying arbitrarily between prescribed limits, *J.Appl. Mech.* **17** (1950), 315–323.
9. Symonds, P.S.: Shakedown in Continuous Media, *J. Appl. Mech.* **18** (1951), 85–89.
10. Neal, B.G.: The behaviour of framed structures under repeated loading, *Quart. J. Mech. Appl. Math.* **4** (1951), 78–84.
11. Symonds, P.S. and Neal, B.G.: New techniques for the computation of plastic failure loads of continuous frame structures, Report A11–62 (1951), 29 pages.
12. Koiter, W.T.: A new general theorem on shake–down of elastic–plastic structures, *Proc. Kon. Ned. Ak. Wet.* **B59** (1956), 24–34.
13. Koiter, W.T.: General theorems for elastic–plastic structures, in I.N. Sneddon and R. Hill (eds.) Progress in Solid Mechanics, North–Holland Publ. Co., Amsterdam, Ch. IV (1960), 165–221.
14. Hodge, P.G.: Plastic Analysis of Structures, McGraw–Hill, New York, NY; Ch. 5, 1959.
15. Heyman, J.: Plastic Design of Frames, Vol. 2 *Applications* Cambridge Univ. Press, Cambridge, England; Ch. 6., 1961.
16. Heyman, J.: Elements of the Theory of Structures, Cambridge Univ. Press, Cambridge, England; Ch. 10., 1966.
17. Martin, J.B.: Plasticity: Fundamentals and General Results, MIT Press, Cambridge MA; Chs. 18, 19, 1975.
18. Symonds, P.S.: Cyclic loading tests on small scale portal frames, Fourth Congress of Int. Assoc. for Bridge & Struct. Eng.; Final Report, Supplement (1953), 109-112.

# SHAKEDOWN LIMITS FOR A GENERAL YIELD CONDITION: IMPLEMENTATION AND APPLICATION FOR A VON MISES YIELD CONDITION

A. R. S. PONTER AND M. ENGELHARDT
*Department of Engineering,*
*University of Leicester,*
*Leicester, LE1 7RH,*
*United Kingdom*

## 1. Introduction

In previous papers [1,2], a procedure was described for the evaluation of limit loads and shakedown limits for a body subjected to cyclic loading. The procedure was based upon the "Elastic Compensation" method [9,10] where a sequence of linear problems are solved with spatially varying linear moduli. In [1] it was demonstrated that the method may be interpreted as a nonlinear programming method where the local gradient of the upper bound functional and the potential energy of the linear problem are matched at a current strain rate or during a strain rate history. This interpretation may be used to formulate a very general method for evaluating minimum upper bound solutions. Provided certain convexity conditions are satisfied, it is possible to define a sequence of linear problems where the functional monotonically reduces. The sequence then converges to the solution which corresponds to the absolute minimum of the functional, subject to constraints imposed by the class of strain rate histories under consideration. The theoretical basis for the method and convergence proofs are discussed in ref. [3,4].

In this paper we discuss the implementation of the method within a finite element code for limit load and shakedown solutions for a Von Mises yield condition. As the method calls upon procedures which form the basis of linear finite element analysis, it is possible to implement the method through the optional user procedures which are often included in commercial codes. For the solutions described in this paper the general code ABAQUS was used.

The paper consists of three main parts. Section (2) contains a summary of the method, based upon the theoretical structure of [3,4] but specialised to a Von Mises yield condition. Section (3) contains a convergence proof. Section (4) is concerned with the implementation of the method within a finite element code. This is followed, in Section



*D. Weichert and G. Maier (eds.), Inelastic Analysis of Structures under Variable Loads,* 11–29.

(5), by the solution of two shakedown problems involving variable load and variable temperature. Finally, in Section (6), the solution of an unconventional shakedown problem is discussed. The history of load is prescribed and the shakedown limit is required in terms of a minimum creep rupture stress for a maximum creep rupture life. This problems occurs in the life assessment method of British Energy, R5 [11], and demonstrates the flexibility of the method.

The ease of implementation, efficiency and reliability of the method indicate that it has considerable potential for application in design and life assessment methods where efficient methods are required for generating indicators of structural performance of structures.

## 2. Shakedown Limit for a Von Mises Yield Condition

Consider a body composed of an isotropic elastic-perfectly plastic solid which satisfies the Von Mises yield condition;

$$f(\sigma_{ij}) = \bar{\sigma} - \sigma_y \leq 0 \tag{2.1}$$

where $\bar{\sigma} = \left(\frac{3}{2}\sigma'_{ij}\,\sigma'_{ij}\right)^{1/2}$ denotes the Von Mises effective stress, $\sigma'_{ij} = \sigma_{ij} - \delta_{ij}\sigma_{kk}$ the deviatoric stress and $\sigma_y$ is a uniaxial yield stress. The plastic strain rate, $\dot{\varepsilon}^p_{ij}$, is given by the associated flow rule in the form of the Prandtl-Reuss relationship,

$$\dot{\varepsilon}^p_{ij} = \left(\frac{\dot{\bar{\varepsilon}}}{\sigma_y}\right)\sigma'_{ij} \quad \text{and} \quad \dot{\varepsilon}^p_{ij} = 0 \tag{2.2}$$

where

$$\dot{\bar{\varepsilon}} = \sqrt{2/3\,\dot{\varepsilon}^p_{ij}\dot{\varepsilon}^p_{ij}} \tag{2.3}$$

denotes the Von Mises effective strain rate.

Consider the following problem. A body of volume V and surface $S$ is subjected to a cyclic history of load $\lambda P_i(x_j,t)$ over $S_T$, part of $S$, and temperature $\lambda\theta(x_j,t)$ within V, where $\lambda$ is a load parameter. On the remainder of $S$, namely $S_U$, the displacement rate $\dot{u}_i = 0$. The linear elastic solution to the problem is denoted by $\lambda\hat{\sigma}_{ij}$. In the following we assume that the elastic solutions are chosen so that $\lambda \geq 0$.

We define a class of incompressible kinematically admissible strain rate histories $\dot{\varepsilon}^c_{ij}$ with a corresponding displacement increment fields $\Delta u^c_i$ and associated compatible strain increment,

$$\Delta\varepsilon_{ij}^{c} = \tfrac{1}{2}\left(\frac{\partial \Delta u_i^c}{\partial x_j} + \frac{\partial \Delta u_j^c}{\partial x_i}\right) \tag{2.4}$$

The strain rate history $\dot{\varepsilon}_{ij}^{c}$, which need not be compatible, satisfies the condition that,

$$\int_0^{\Delta t} \dot{\varepsilon}_{ij}^{c} dt = \Delta\varepsilon_{ij}^{c} \tag{2.5}$$

In terms of such a history of strain rate an upper bound on the shakedown limit is given by [2-5],

$$\int_V \int_0^{\Delta t} \sigma_y \bar{\dot{\varepsilon}}(\dot{\varepsilon}_{ij}^{c}) dt dV = \lambda_{UB}^{c} \int_V \int_0^{\Delta t} (\hat{\sigma}_{ij} \dot{\varepsilon}_{ij}^{c}) dt dV \tag{2.6}$$

where $\lambda_{UB}^{c} \geq \lambda_s$, with $\lambda_s$ the exact shakedown limit. In the following we describe a convergent method where a sequence of kinematically admissible strain increment fields, with associated strain rate histories, corresponds to a reducing sequence of upper bounds. The sequence converges to the shakedown limit $\lambda_s$ or the least upper bound associated with the class of displacement fields and strain rate histories chosen.

The iterative method relies upon the generation of a sequence of linear problems where the moduli of the linear problem are found by a matching process. For the Von Mises yield condition the appropriate class of strain rates chosen are incompressible so the linear problem is defined by a single shear modulus $\mu(t)$ which varies both spatially and during the cycle. Corresponding to an initial estimate of the strain rate history $\dot{\varepsilon}_{ij}^{i}$, a history of a shear modulus $\mu(x_i, t)$ may be defined by a matching condition,

$$\tfrac{3}{2}\mu \bar{\dot{\varepsilon}}^{i} = \sigma_y \tag{2.7}$$

A corresponding linear problem for a new kinematically admissible strain rate history, $\dot{\varepsilon}_{ij}^{f}$ and a time constant residual stress field $\bar{\rho}_{ij}^{f}$ may now be defined by

$$\dot{\varepsilon}_{ij}^{\prime f} = \frac{1}{\mu}\left(\bar{\rho}_{ij}^{\prime f} + \lambda \hat{\sigma}_{ij}^{\prime}\right) \tag{2.8}$$

and

$$\Delta\dot{\varepsilon}_{ij}^{\prime f} = \frac{1}{\bar{\mu}}\left(\bar{\rho}_{ij}^{\prime f} + \sigma_{ij}^{\prime in}\right) \quad \text{and} \quad \Delta\dot{\varepsilon}_{kk}^{\prime} = 0 \tag{2.9}$$

where $\lambda = \lambda_{UB}^{i}$ ,

$$\frac{1}{\bar{\mu}} = \int_0^{\Delta t} \frac{1}{\mu(t)} dt \quad \text{and} \quad \sigma_{ij}^{\prime in} = \bar{\mu}\left\{\int_0^{\Delta t} \frac{1}{\mu(t)} \lambda \hat{\sigma}_{ij}^{\prime} dt\right\} \tag{2.10}$$

The convergence proof , given below, then concludes that

$$\lambda_{UB}^{f} \leq \lambda_{UB}^{i} \tag{2.11}$$

where equality occurs if and only if $\dot{\varepsilon}_{ij}^{i} \equiv \dot{\varepsilon}_{ij}^{f}$ and $\lambda_{UB}^{f}$ is the upper bound corresponding to $\dot{\varepsilon}_{ij}^{c} = \dot{\varepsilon}_{ij}^{f}$. The repeated application of this procedure results in a monotonically reducing sequence of upper bounds which will converge to the least upper bound within a class of displacement fields and strain rate histories. A primary objective of the implementation of the technique is to ensure that the classes chosen ensures that the resulting minimum is sufficiently close to the absolute minimum to be of practical use as a substitute for an analytic solution.

The choice of the linear problem (2.7) to (2.10) has a simple physical interpretation. For the initial strain rate history $\dot{\varepsilon}_{ij}^{i}$, the shear modulus is chosen so that the rate of energy dissipation in the linear material is matched to that of the perfectly plastic material for the same strain rate history. At the same time the load parameter is adjusted so that the value corresponds to a global balance in energy dissipation through equality (2.6). In other words, the linear problem is adjusted so that it satisfies as many of the conditions of the plasticity problem as is possible. The fact that the resulting solution, when equilibrium is reasserted, is closer to the shakedown limit solution and produces a reduced upper bound should be no surprise. However, we need not rely upon such intuitive arguments as a formal proof of convergence may be constructed and this is given in the next section.

## 3. Convergence Proof for a Von Mises Yield Condition

The following convergence proof is based upon a more general proof in [4] for a general yield surface with a general class of linear problems. Here the form of proof is the same as that given in [4] but simplified and specialised to a Von Mises yield condition. Essentially, we need to demonstrate that, beginning with a strain rate history $\dot{\varepsilon}_{ij}^{i}$, the linear problem, equations (2.7) to (2.9), for the residual stress field $\overline{\rho}_{ij}^{f}$ and strain rate history $\dot{\varepsilon}_{ij}^{f}$ produces an upper bound on the shakedown limit which satisfies inequality (2.11) with equality only when $\dot{\varepsilon}_{ij}^{i} \equiv \dot{\varepsilon}_{ij}^{f}$. Convergence then results from the repeated application of the argument producing a monotonically reducing sequence of upper bounds.

We begin with some preliminary observations. The linear incompressible material defined by (2.7) possesses a strain rate potential $U(\dot{\varepsilon}_{ij})$ so that

$$\sigma_{ij}^{L} = \frac{\partial U(\dot{\varepsilon}_{ij})}{\partial \dot{\varepsilon}_{ij}}, \qquad U = \tfrac{3}{4}\mu \dot{\bar{\varepsilon}}^{2} \tag{3.1}$$

where $\sigma_{ij}^{L}$ is the stress generated by the linear material. The matching condition (2.7) may then be written in th more general form;

$$\sigma_{ij}^{pi} = \sigma_{ij}^{Li} = \frac{\partial U(\dot{\varepsilon}_{ij}^{i})}{\partial \dot{\varepsilon}_{ij}^{i}} \tag{3.2}$$

where $\sigma_{ij}^{pi}$ is the stress at yield associated with $\dot{\varepsilon}_{ij}^{i}$.

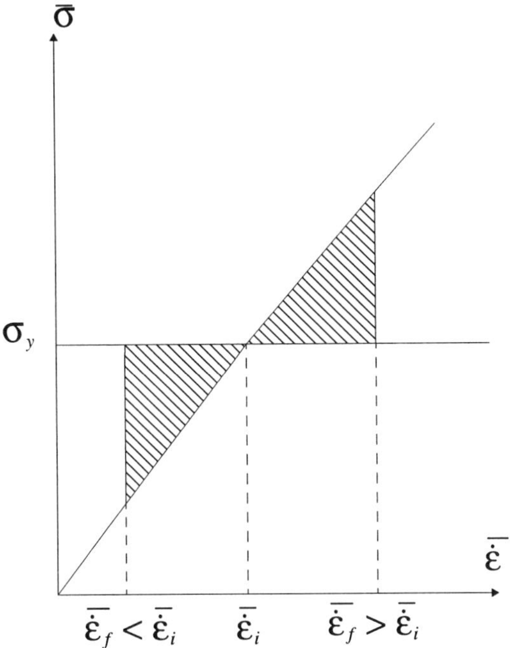


Figure 1. The inequality (3.4) for a Von Mises yield condition requires that the shaded regions have a positive area

The convergence proof requires a relationship between this linear material and the perfectly plastic material. This is provided by the following inequality which, in a more general context [3,4], may be regarded as a sufficient condition for convergence;

$$\sigma_{ij}^{pi}\dot{\varepsilon}_{ij}^{i} - \sigma_{ij}^{pf}\dot{\varepsilon}_{ij}^{f} \geq U(\dot{\varepsilon}_{ij}^{i}) - U(\dot{\varepsilon}_{ij}^{f}) \tag{3.3}$$

where $\dot{\varepsilon}_{ij}^{f}$ may be regarded as an arbitrary incompressible strain rate and where U is defined by the matching condition (3.2) and (2.7). For a Von Mises yield condition inequality (3.3) may be rewritten as;

$$\sigma_y(\bar{\dot{\varepsilon}}^{i} - \bar{\dot{\varepsilon}}^{f}) \geq U(\bar{\dot{\varepsilon}}^{i}) - U(\bar{\dot{\varepsilon}}^{f}) \tag{3.4}$$

This inequality is self evidently always satisfied, see Figure (1).

The convergence proof now commences by identifying $\dot{\varepsilon}_{ij}^{f}$ with the strain rate generated by the linear problem given by eqns (2.7) to (2.10). We first note that (3.1) and (2.8) may be written as;

$$\frac{\partial U(\dot{\varepsilon}_{ij}^{f})}{\partial \dot{\varepsilon}_{ij}^{f}} = \bar{\rho}_{ij}^{\prime f} + \lambda\hat{\sigma}_{ij}^{\prime} \tag{3.5}$$

On noting that U is a convex function of its argument and equation (3.5), the inequality (3.3) may be written in the extended form;

$$0 \le U(\dot{\varepsilon}_{ij}^{i}) - U(\dot{\varepsilon}_{ij}^{f}) - \frac{\partial U(\dot{\varepsilon}_{ij}^{f})}{\partial \dot{\varepsilon}_{ij}^{f}}(\dot{\varepsilon}_{ij}^{i} - \dot{\varepsilon}_{ij}^{f}) \le -(\sigma_{ij}^{pf} - \lambda\hat{\sigma}_{ij})\dot{\varepsilon}_{ij}^{f} + (\sigma_{ij}^{pi} - \lambda\hat{\sigma}_{ij})\dot{\varepsilon}_{ij}^{i}$$
$$+\bar{\rho}_{ij}^{f}(\dot{\varepsilon}_{ij}^{f} - \dot{\varepsilon}_{ij}^{i}) \qquad (3.6)$$

We now define $\lambda = \lambda_{UB}^{i}$ as the upper bound, defined by (2.6), corresponding to $\dot{\varepsilon}_{ij}^{c} = \dot{\varepsilon}_{ij}^{i}$. Integrating the right hand side of inequality (3.6) over the volume V and cycle, we obtain the following;

$$0 \le -\int_V \int_0^{\Delta t} (\sigma_{ij}^{pf} - \lambda_{UB}^{i}\hat{\sigma}_{ij})\dot{\varepsilon}_{ij}^{f}\,dtdV + \int_V \int_0^{\Delta t} \bar{\rho}_{ij}^{f}(\dot{\varepsilon}_{ij}^{f} - \dot{\varepsilon}_{ij}^{i})\,dtdV \qquad (3.7)$$

The contribution of the second term of the right hand side of (3.6) disappears due to the choice of $\lambda = \lambda_{UB}^{i}$. Note that the second term in (3.7) may be integrated to produce;

$$\int_V \int_0^{\Delta t} \bar{\rho}_{ij}^{f}(\dot{\varepsilon}_{ij}^{f} - \dot{\varepsilon}_{ij}^{i})\,dtdV = \int_V \bar{\rho}_{ij}^{f}(\Delta\varepsilon_{ij}^{f} - \Delta\varepsilon_{ij}^{i})\,dV \qquad (3.8)$$

If the linear problem is solved exactly the right hand side of (3.8) is zero from the virtual work theorem. But equally, if the linear problem is solved by a Rayleigh Ritz method where the rate version of the potential energy is minimized for a class of displacement fields, the right hand side of (3.8) is also zero as the residual stress field so generated is orthogonal to all compatible strain fields derived from the class of displacement fields. If we now return to inequality (3.7) and denote by $\lambda_{UB}^{f}$ the upper bound corresponding to $\dot{\varepsilon}_{ij}^{c} = \dot{\varepsilon}_{ij}^{f}$ in (2.6), we obtain;

$$0 \le (\lambda_{UB}^{i} - \lambda_{UB}^{f}) \int_V \int_0^{\Delta t} \hat{\sigma}_{ij}\dot{\varepsilon}_{ij}^{f}\,dtdV \qquad (3.9)$$

and hence $\lambda_{UB}^{f} \le \lambda_{UB}^{i}$ provided $\lambda$ is positive. Equality in (3.9) only occurs for $\dot{\varepsilon}_{ij}^{i} \equiv \dot{\varepsilon}_{ij}^{f}$ i.e when equality occurs in the first inequality in the extended inequality (3.6).

The repeated application of the procedure will result in a monotonically reducing sequence of upper bounds which will converge to a minimum when the difference between successive strain rate histories has a negligible effect upon the upper bound. In common with all other programming methods for shakedown limits, the absolute minimum upper bound will be reached for the chosen class of displacement fields in a finite element method only if it is ensured that the process allows access to all possible strain rate histories which contain the absolute minimum. For example, assumptions can be made in defining the initial strain rate history concerning the instants when plastic strains occur during the cycle. Such assumptions will affect the nature of all subsequent strain rate histories and result in convergence to a minimum subject to contraints. In the method described in [2] consistency arguments were applied at each iteration. A simpler approach is possible and this is discussed below.

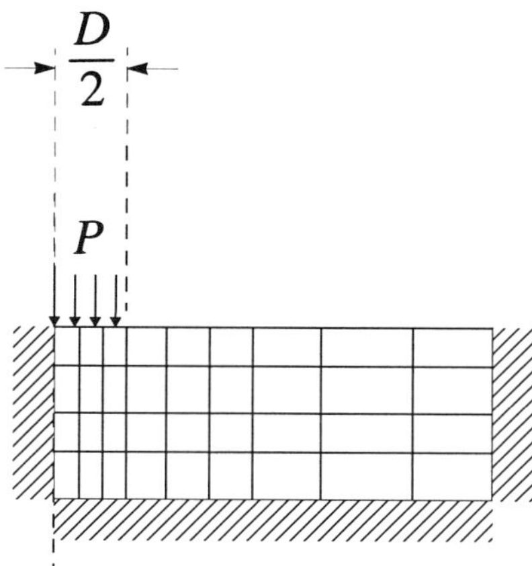


Figure 2. Finite element mesh for plane strain model of indentation of a half space. Eight noded quadrilateral elements

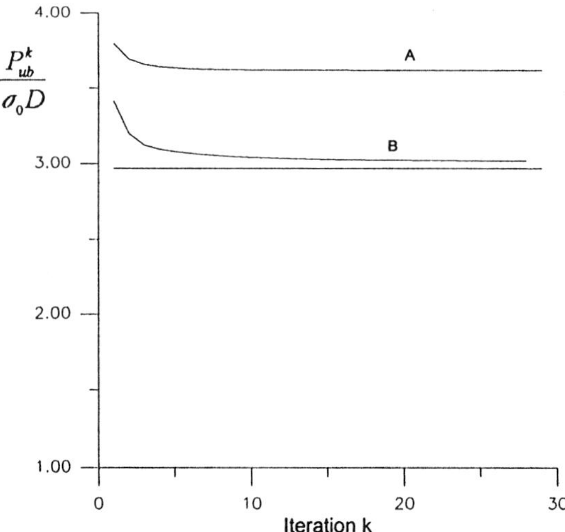


Figure 3. Convergence of the upper bound to the limit load for the indentation problem of Figure 1.. The lower curve corresponds to a mesh where each element in Figure 1. has been subdivided into 16 elements. The analytic solution (the Prandtl solution) is included for comparison.

## 4. Implementation of the Method - Limit Analysis

The method has been implemented in the commercial finite element code ABAQUS. The normal mode of operation of the code for non-linear analysis involves the solution of a sequence of linearised problems for incremental changes in stress, strain and displacement in time intervals corresponding to a predefined history of loading. At each increment, user routines allow a dynamic prescription of the Jacobian which defines the relationship between increments of stress and strain. The implementation involves carrying through a standard load history calculation for the body, but setting up the calculation sequence and Jacobian values so that each incremental solution provides the data for an iteration in the iterative process. Volume integral options evaluate the upper bound to the shakedown limit which is then provided to the user routines for the evaluation of the next iteration. In this way an exact implementation of the process may be achieved. The only source of error arises from the fact that ABAQUS uses Gaussian integration which is exact, for each element type, for a constant Jacobian within each element. The condition (2.7) is applied at each Gauss point and results in variations of the shear modulus $\mu(t)$ within each element. There is, therefore, a corresponding integration error but the effect of this on the upper bound is small. The primary advantages of this approach to implementation are practical. An implementation can be achieved which is:

1) easily transferable to other users of the code,
2) requires fairly minor additions to the basic routines of the code so that a reliable implementation can be achieved,
3) can be introduced for a wide range of element and problem types.

For the case of constant loads the formulation in the previous section reduces to the solution of the equation (2.8) or, equivalently, (2.9) for a shear modulus distribution defined by (2.7). In the upper bound (2.6) the time integration is not required. This formulation differs from the formulation given by Ponter and Carter [2] where each solution in the iterative process involves a stress field in equilibrium with an applied boundary load whereas in (2.7) the external loads are introduced through the linear elastic solution $\lambda\hat{\sigma}_{ij}$. The two formulations are entirely equivalent for linear elastic solutions which are solved by the same Galerkin finite element method as used in the iterative procedure. However, the sequence of calculations are not identical and, as was mentioned earlier, the Gaussian integration is not exact. In practice we find differences between the results of the two formulations which are negligible compared with the approximation errors of the finite element mesh.

Figures (2) shows the finite element mesh for the symmetric half of a plane strain indentation problem where a line load P is applied over a strip of width D. It may be recognised as one of the standard examples in the ABAQUS examples manual. The convergence of the upper bound is shown in Figure (3) for the mesh shown in Figure (2) and a refined mesh where each element has been subdivided into sixteen identical elements. The analytic solution ( the Prandtl solution) for the half space is also shown. The observed behaviour illustrates two points which are common to all the following solutions. As the method is a strict upper bound the solution converges to the analytic solution from above. The accuracy of the converged solution depends entirely on the ability of the class of displacement fields defined by the mesh geometry to represent the displacement field in the exact solution. Generally there is the need for a sufficient density of elements in the regions where the deformation field varies most rapidly. In this case the refining of the mesh geometry has a very significant effect on the accuracy of the converged solution.

The method has been used to solve a large number of problems involving structural components with cracks. Accurate limit load solutions for such problems are required for the application of life assessment methods in the power industry [11]. Empirical investigations of the convergence of upper bound to available analytic solutions have been carried out for a range of geometries. For a fixed element type, the error diminishes near linearly with a length scale which characterises the mean size of the elements. From such data it is possible to find empirical rules for the production of limit load solutions with an error of less than, say, 1%.

It is worth commenting on the sensitivity of solutions to the assumptions within the method. In the examples considered above, near incompressibility was achieved by using hybrid elements with a Poissons ratio of 0.49999 . In the convergence proof discussed in [3,4] a sufficient condition for convergence is provided by the requirement that the complementary energy surface for the linear material defined by the shear modulus $\mu$ which touches the yield surface at the "matching" point, giving rise to equation (3.1), otherwise must either coincide with the yield surface or lie outside it. For the Von Mises yield condition the complementary energy surface coincides with the yield surface when Poissons ratio $\nu = 0.5$, but for $\nu < 0.5$ it lies within the yield surface and contravenes the conditions of the convergence proof. As this is a sufficient condition for convergence, it is of some interest to observe the effect of introducing volumetric strains into the linear solutions by a fixed value of $\nu < 0.5$ throughout the process. Sequences of solutions for a range of values of $\nu$ have been generated. For the value used in the solutions discussed above further reduction in the difference from the ideal value of 0.5 has a negligible effect on the converged upper bound when compared with other sources of error. However, if a value of $\nu = 0.49$ is used for crack problems the upper bound converges to values which lies *below* the analytic solution for a sufficiently refined mesh. Hence the effect of adopting a value of $\nu$ significantly less than 0.5 will be to give results which may either be above or below the analytic solution and convergence cannot be assured.

## 5. Shakedown Solutions

The procedure described by equations (2.7) to (2.10) requires the definition of a shear modulus at each instant during the cycle. There are problems where the distribution of the strain rate history in time during the cycle is unknown in advance but there is an important class of problems where we know *ab initio* that plastic strains may only occur at certain instants within the cycle. If the loading history consists of a sequence of proportional changes in loads between a set of extreme points, as shown schematically in

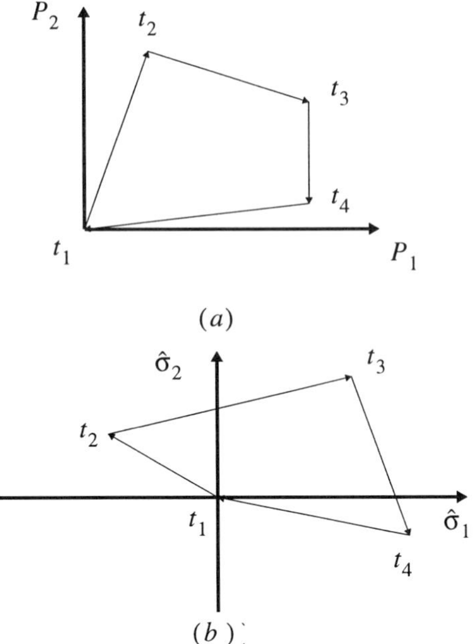


Figure 4. A history of load which describes a sequence of straight lines in load space, (a), produces a history of elastic stresses which describe straight lines in stress space, (b). As a result it is known *a priori* that plastic strains only occur at the r vertices during the cycle. r=4 in the Figure.

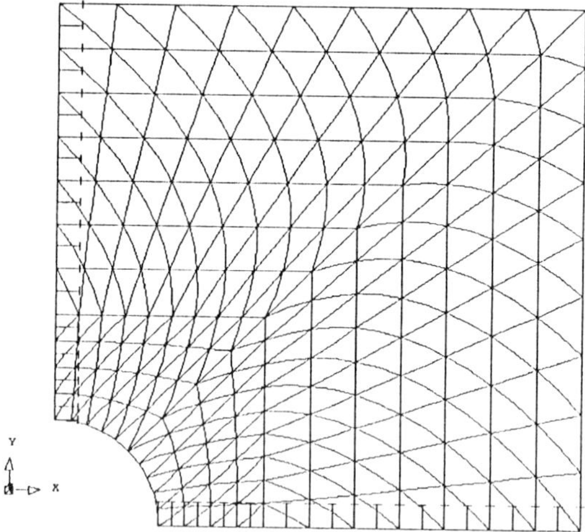


Figure 5. Finite element mesh and history of loading for plane stress problem subject to biaxial loading.

Figure (4) for a problem involving two loads $(P_1, P_2)$, then the linearly elastic stress history also describes a sequence of linear paths in stress space as shown. For a strictly convex yield condition, which includes the Von Mises yield condition in deviatoric stress space, the only instants when plastic strains can occur are at the vertices of the stress history, $\hat{\sigma}_{ij}(t_l)$, $l = 1$ to r. The strain rate history then becomes the sum of increments of plastic strain;

$$\Delta\varepsilon_{ij} = \sum_{l=1}^{r} \Delta\varepsilon_{ij}^{l} \tag{5.1}$$

and equations (2.8) to (2.10) become

$$\Delta\varepsilon_{ij}^{\prime f} = \frac{1}{\overline{\mu}}\left(\overline{\rho}_{ij}^{\prime f} + \sigma_{ij}^{\prime in}\right) \tag{5.2}$$

$$\sigma_{ij}^{\prime in} = \overline{\mu}\left\{\sum_{m=1}^{r} \frac{1}{\mu_m} \lambda \hat{\sigma}_{ij}^{\prime}(t_m)\right\} \tag{5.3}$$

where

$$\frac{1}{\overline{\mu}} = \sum_{l=1}^{r} \frac{1}{\mu_l} \quad \text{and} \quad \tfrac{3}{2}\mu_l \overline{\varepsilon}\left(\Delta\varepsilon_{ij}^{li}\right) = \sigma_y \tag{5.4}$$

The implementation of the method involves the following sequence of calculations: An initial solution assumes that plastic strains may occur at all r possible instants in the cycle. Initial, arbitary, values of the moduli $\mu_m = 1$ are chosen. As a result of this initial solution, the iterative method described in equations (5.2) to (5.4) is applied. The plastic strain components at instants where there is no strain in the converged solution then decline in relative magnitude until they make no contribution to the upper bound.

In the following solutions the iterative method was continued until there was no changed in the fifth significant figure in the computed upper bound for five consecutive iterations. The number of iterations required depended upon the nature of the optimal mechanism. For reverse plasticity mechanisms the number of iterations required could be quite high, in excess of 100, whereas for mechanisms where all the plastic strains occurred at a single instant at each point in the body (although not necessarily the same instant) the number of iterations required was significantly less and 50 iterations was a typical number. For a less exacting convergence criteria a significantly smaller number of iterations are required and variation of the upper bound with iteration numbers shown in Figure(3) is typical of both limit load and shakedown solutions.

Figure (5) shows the symmetric section of a finite element discretisation for a plane stress plate, with a circular hole, subjected to biaxial tension . The shakedown limit has been evaluated for the two histories of $(P_1(t), P_2(t))$ shown in Figure (6). The interaction diagram of the shakedown limit evaluated by the method are

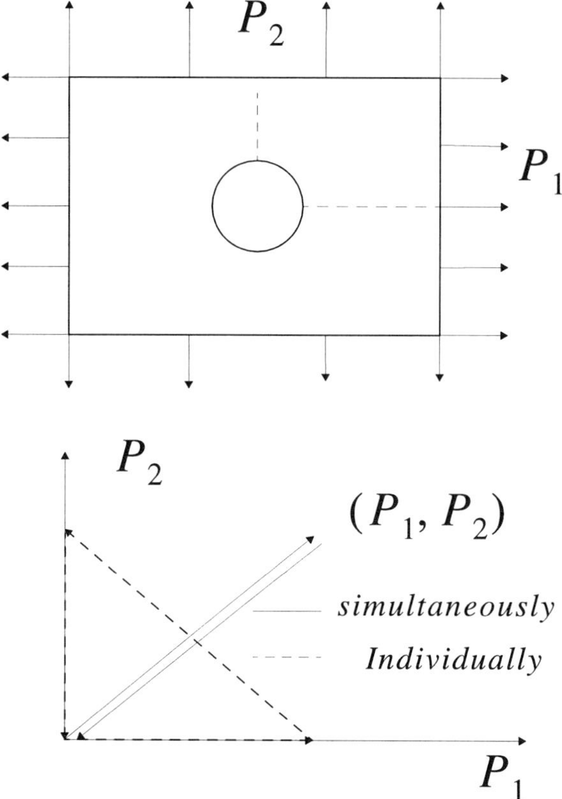


Figure 6. Loading histories for the two shakedown solutions shown in Figure 7.

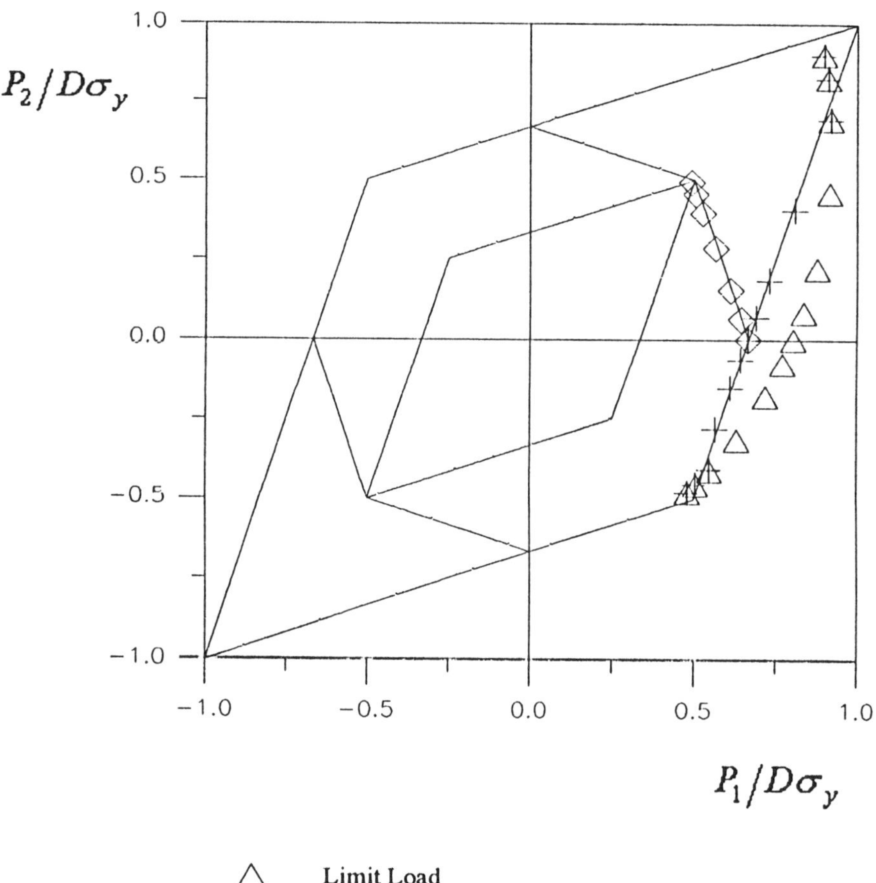


△ Limit Load

+ Shakedown Limit (Loads Applied Simultaneously)

◇ Shakedown Limit (Loads Applied Individually)

Figure 7. Limit load and shakedown limits for the geometry and histories of loading shown in Figure 6. and mesh shown in Figure 5. Note that the shakedown limit is identical to the least of the limit load or the reverse plasticity limit.

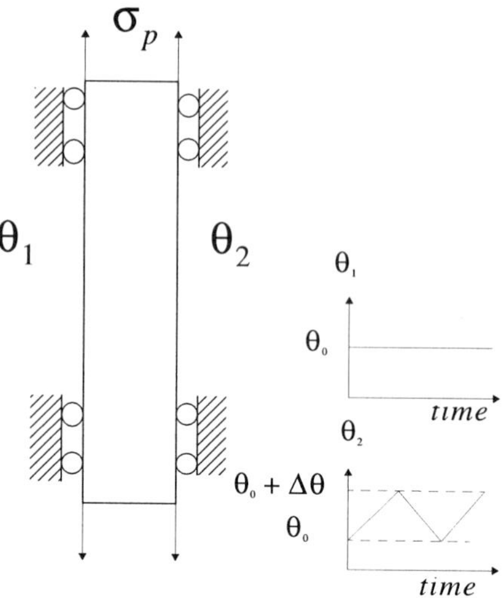


Figure 8. Bree problem - a plate or axisymmetric tube subjected to fluctuating temperature differences and an axial load.

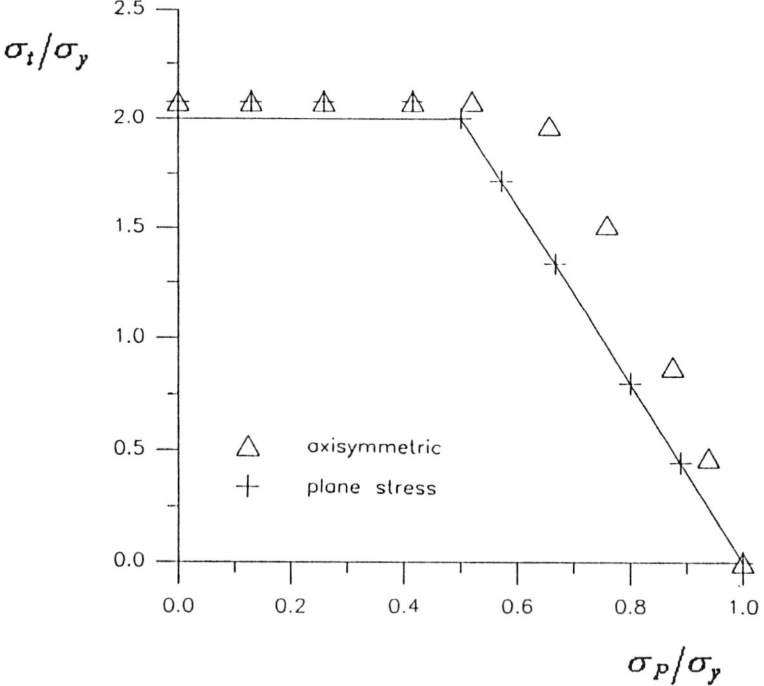


Figure 9. Shakedown limits for the Bree problem of Figure 8. modelled as both a plane stress and an axisymmetric thin walled tube. The solid line is the classic Bree solution for a Tresca yield condition.

shown in Figure (7) together with the limit load for monotonical increase in $(P_1, P_2)$. The elastic limit is also shown, i.e, the highest load levels for which the elastic solutions just lie within the elastic domain for the prescribed yield stress and also for a yield stress of $2\sigma_y$. It may be observed that in all cases the shakedown limit is given either by the limit load for the loads at some point in the cycle or at the elastic limit for $2\sigma_y$. As the initial loading point is zero load, this later condition corresponds to the variation of the elastic stresses lying within the yield surface if superimposed upon an arbitrary residual stress. This is a well known result and arises from the fact that the mechanism at the shakedown limit corresponds to a reverse plasticity condition at the point of stress concentration in the elastic solution, on either the major or minor axis of the hole surface.

Figure (8) shows the classic Bree problem where either a plate or a tube wall thickness is subjected to axial stress and a fluctuation temperature difference $\Delta\theta$ across the plate or through the wall thickness. The problem has been solved both as plane stress plate problem, where curviture of the plate due to thermal expansion is restrained, and as an axisymmetric cylinder. The two solutions for a temperature independent yield stress are both shown in Figure (9) in terms of $\sigma_t$, the maximum principal thermoelastic stress due to $\Delta\theta$. The plate solution coincides with the classic Bree solution for a Tresca yield condition (the problem is essentially one dimensional) whereas the solution for the axisymetric problem lies outside the classic Bree solution to a maximum extent of 15%, the maximum difference between the Tresca and Von-Mises yield condition. The reverse plasticity solution, which corresponds to $\sigma_t = 2\sigma_y$ in both cases, is overestimated by the both computed solutions. This is due to the way reverse plasticity limits are evaluated. The optimising strain rate history consists of increments of strain which result in a zero accumulation of strain over the cycle. The contribution of a single Gauss point (or in this problem a row of Gauss points) dominates over the contribution of all other Gauss points and the limit is governed by the variation of the elastic stress at that point. In Figure (9) we adopt for $\sigma_t$ the value of thermal stress at the surface whereas the reverse plasticity shakedown limit corresponds to the slightly smaller thermal stress at a neighbouring Gauss points.

## 6. An Extended Shakedown Limit

Consider the following problem. A body is subjected to a prescribed history of loading corresponding to load parameter $\lambda = 1$. The load history contains a significant component which derives from variations in temperature. The yield stress is assumed to vary with temperature and may have two values. At lower temperatures the yield stress equals a constant value $\sigma_y^{LT}$. At higher temperature it is replaced by a creep rupture stress $\sigma_c(t_f, \theta)$ which depends upon the time to creep rupture $t_f$, which is understood as a property of the structure as a whole, and the local temperature $\theta$. We require the largest creep rupture time $t_f$ for which the prescribed loads remain within the shakedown limit for this definition of the yield stress. This problem is posed within the methodology of the life assessment method R5 [11] as a means of assessing the remaining creep rupture life of the structure. Here we treat it as a novel shakedown problem where the parameter we wish to optimise, the creep rupture time $t_f$, is included in the definition of the shakedown problem through the definition of the yield stress at each point in the body and each instant during the cycle when plastic strains can occur.

Hence the yield stress at each point of the body at time $t_m$ is defined by;

$$\sigma_y(x_i,t_m)=\min\left\{\begin{array}{c}\sigma_y^{LT}\\ \sigma_c(t_f,\theta(x_i,t_m))\end{array}\right\} \tag{6.1}$$

We assume the following form for $\sigma_c(t_f,\theta)$;

$$\sigma_c(t_f,\theta)=\sigma_y^{LT}R\left(\frac{t_f}{t_0}\right)g\left(\frac{\theta}{\theta_0}\right) \tag{6.2}$$

and R(1)=g(1)=1 so that $\sigma_c(t_f,\theta)=\sigma_y^{LT}$ when $t_f=t_0$ and $\theta=\theta_0$. Hence we wish to compute the value of R for which the shakedown limit is given by $\lambda=1$.

For a prescribed mechanism of deformation at some stage in an iterative process with this definition of the yield stress there will be contributions to the volume integral of the plastic energy dissipation which originate from $\sigma_y^{LT}$ and $\sigma_c$. If we denote by $\dot{D}_p^{LT}$ and $\dot{D}_p^c$ the contributions to the total dissipation $\dot{D}_p$ given by those volumes and those times where the low temperature and creep stress operate;

$$\dot{D}_p=\dot{D}_p^{LT}+\dot{D}_p^C=\int_V\sum_{m=1}^{r}\left\{\sigma_y\dot{\varepsilon}\left(\Delta\varepsilon_{ij}^m\right)\right\}dV \tag{6.3}$$

then we can derive, from (6.2) and (6.3), the following relationships between small changes in $\lambda_{UB}$ and $R$ for a particular mechanism of deformation;

$$\Delta\lambda\dot{D}_p^E=\frac{\Delta R}{R}\dot{D}_p^c \tag{6.4}$$

where

$$\dot{D}_p^E=\int_V\sum_{m=1}^{r}(\hat{\sigma}_{ij}(t_m)\Delta\varepsilon_{ij}^m)dtdV \tag{6.5}$$

This relationship forms the basis for an iterative process which converges to the value of R and hence the rupture time $t_f$ corresponding to the shakedown for $\lambda=1$.

We begin by choosing an initial value of $R=R_0$ and $t_f$ so that the shakedown limit in the converged solution is expected to be $\lambda<1$. For fixed $R_0$ the iterative process is allowed to converge until the k-th iteration yields the first upper bound value of the load parameter which satisfies $\lambda_{UB}^{k} < 1$. The value of R is then changed according to equation (6.4) at each iteration so that $\lambda_{UB}$ returns to the preassigned value of $\lambda = 1$. i.e.,

$$\Delta R = R_o\left(1 - \lambda_{UB}^{k}\right)\left(\frac{\dot{D}_p^E}{\dot{D}_p^c}\right) > 0 \tag{6.6}$$

Hence

$$R_1 = R_0 + \Delta R > R_0 \tag{6.7}$$

and the process is repeated. At each iteration the value of R increases and converges, from below, to the value for which the shakedown limit is given by $\lambda = 1$.

In the following example we adopt the following simple form for the temperature dependence of $\sigma_c$;

$$g\left(\frac{\theta}{\theta_0}\right) = \frac{1}{\left(1 - \frac{\theta}{\theta_0}\right)} \tag{6.8}$$

In Figure (10) the solutions for the Bree problem are shown with $\theta_0 = 200^0 C$ and other material constants appropriate to a ferritic steel.

In Figure (10) the shakedown limit is shown for the three cases of R= 0.1, 0.4 and 1.5 . The contours shown were evaluated by converging to the value of the load parameter corresponding to the prescribed yield conditions given by equation (6.1) although it is worth noting that the dependency of the yield stress on temperature causes a change in the yield stress at each iteration. For the case of R=1.5, $\sigma_y = \sigma_y^{LT}$ throughout the volume. In Figure (11) we show the inverse problem where the load parameter $\lambda$ is prescribed and the value of R is evaluated corresponding to a shakedown state. The two solutions shown in Figure (11) corresponding to points A and B in Figure (10) for R=0.1. The two phases of the process can be seen where the initial value of $R_0 = 0.05$ is maintained constant for the first few iterations until $\lambda_{UB}^{k} < 1$, when R is allowed to increase according to the relationships (6.6) and (6.7) until convergence takes place. Convergence is slower in the case corresponding to point B in Figure (10), where, in the converged solution, a reverse plasticity mechanism operates.

This problem demonstrates the potential flexibility of the method. Traditionally, shakedown analysis has been seen as a method of defining a load parameter for a prescribed distribution of material properties and load history. It is clear from this example that the shakedown problem may be posed in other ways; in this particular problem the quantity optimised concerns a material property which enters the problem in only part of the volume and only during part of the load cycle. It is clearly possible, using the type of technique discussed in this section, to pose a variety of optimisation problems depending upon the needs of the problem. This introduces possibilities for shakedown analysis which have not previously been available.

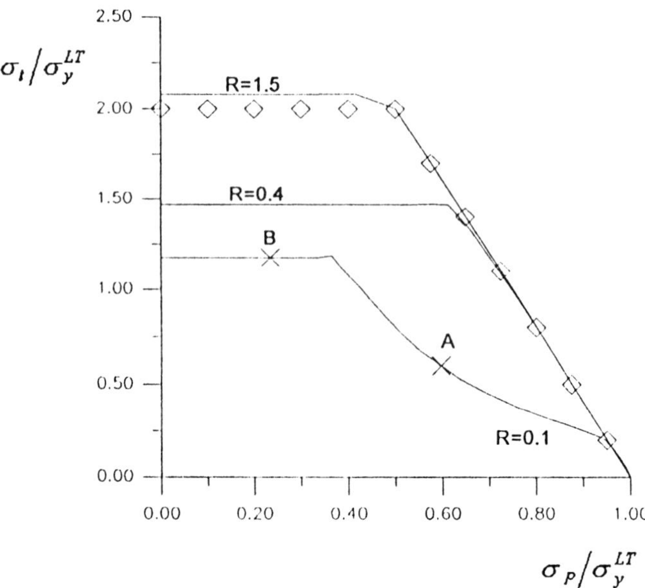


Figure 10. Shakedown limits for the problem discussed in Section 6 for prescribed values of R. The diamonds correspond to the Bree solution which coincides with the solution for R= 1.5. Points A and B refer to the solutions in Figure 11. for R=0.1.

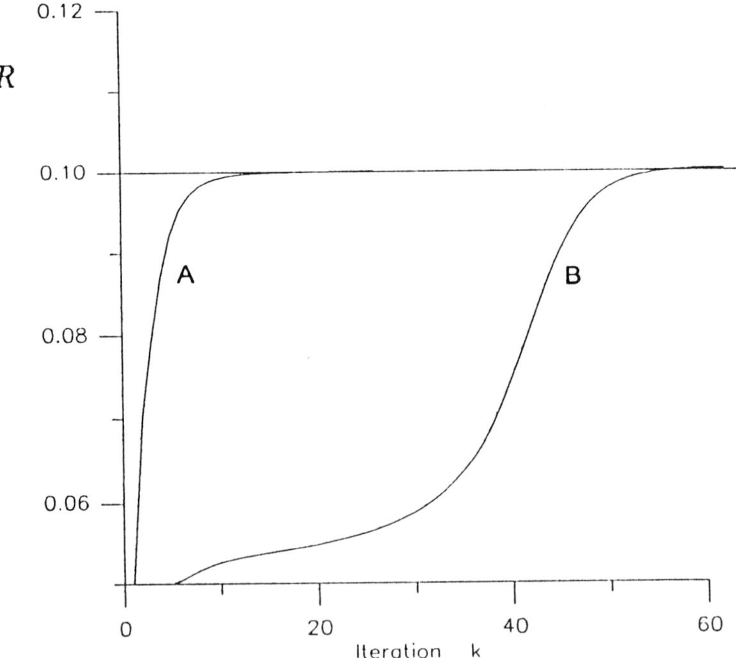


Figure 11. Convergence of R to 0.1 for the extended shakedown method discussed in Section 6. Curves labelled A and B correspond to the convergenec to the corresponding points in Figure 10. The slower convergence of case B is due to the dominance of a reverse plasticity mechanism.

## 7. Conclusions

The paper discusses a method of evaluating shakedown limits by a convergent non-linear programming method which has its origins in the "elastic compensation" method [9,10]. The existence of convergence proofs allows an implementation in a commercial finite element code which is both efficient and robust. The set of examples of both limit loads and shakedown limits given here demonstrate its numerical stability and the ability to approach the analytic solution from above through mesh refinement. The last example involves the optimisation of a creep rupture stress so that, for a prescribed load history, the body lies within a shakedown limit. Such problems occur in situations where both low temperature plasticity and high temperature creep properties limit the load capacity of the structure. It is clear from this example that convergent methods can be devised for such novel shakedown problems, thereby introducing a degree of flexibility into numerical methods for shakedown analysis which have not previously been available.

## 8. Acknowledgements

The work in this paper was supported by the University of Leicester and Nuclear Electric Ltd. The authors gratefully acknowledge this support.

## 9. References

1. Ponter A.R.S. and Carter K.F., "Limit state solutions, based upon linear elastic solutions with a statially varying elastic modulus". *Comput. Methods Appl. Mech. Engrg.*, Vol 140 (1997) pp 237-258.
2. Ponter A.R.S. and Carter K.F., "Shakedown State simulation techniques based on linear elastic solutions", *Comput. Methods Appl. Mech. Engrg.*, Vol 140 (1997) pp 259-279.
3. Ponter A. R. S., Fuschi P. and Engelhardt M., "Limit Analysis for a General Class of Yield Conditions", Department of Engineering, University of Leicester, Report - September 1999.
4. Ponter A. R. S. and Engelhardt, M. "Shakedown Limits for a General Yield Condition", Department of Engineering, University of Leicester, Report - September 1999.
5. Koiter, W.T., "General theorems of elastic-plastic solids", in *Progress in Solid Mechanics*, eds. J.N. Sneddon, R. Hill, Vol 1, pp167-221, 1960.
6. Gokhfeld, D. A. Cherniavsky, D.F., *"Limit Analysis of Structures at Thermal Cycling"*, Sijthoff\& Noordhoff. Alphen an Der Rijn, The Netherlands, 1980.
7. König, J.A., *"Shakedown of Elastic-Plastic Structures"*, PWN-Polish Scientific Publishers, Warsaw and Elsevier, Amsterdam, 1987.
8. Polizzotto, C., Borino, G., Caddemi, S and Fuschi, P., Shakedown Problems for materials with internal variables, *Eur. J. Mech. A/ Solids*, Vol 10, pp 621-639, 1991.
9. Seshadri R. and Fernando C.P.D., "Limit loads of mechanical components and structures based on linear elastic solutions using the GLOSS R-Node method", *Trans ASME PVP*, Vol 210-2, San Diego, pp125-134,1991.
10. Mackenzie D. and Boyle J.T., "A simple method of estimating shakedown loads for complex structures", *Proc. ASME PVP*, Denver, 1993.
11. Goodall, I. W., Goodman, A.M., Chell G. C., Ainsworth R. A., and Williams J. A., "R5: An Assessment Procedure for the High Temperature Response of Structures", Nuclear Electric Ltd., Report, Barnwood, Gloucester, 1991.

# SHAKEDOWN AT FINITE ELASTO-PLASTIC STRAINS

H. STUMPF
*Lehrstuhl für Allgemeine Mechanik, Ruhr-Universität Bochum*
*Universitätsstraße 150 IA3/126, D-44780 Bochum, Germany*
*e-mail: stumpf@am.bi.ruhr-uni-bochum.de*

B. SCHIECK
*FB Maschinenbau u. Wirtschaftsing., Fachhochschule Lübeck*
*Stephensonstraße 3, D-23562 Lübeck, Germany*
*e-mail: schieck@fh-luebeck.de*

## 1. Introduction

For structures undergoing small elastic-plastic deformations (i. e. small rotations and small strains), shakedown or non-shakedown can be determined by applying the well-known statical shakedown theorem of Melan [1], which leads to a lower bound of the load factor. Dual to the Melan's theorem Koiter [2] formulated a kinematical shakedown theorem yielding an upper bound of the load factor. While many publications deal with shakedown of structures at small deformations taking into account linear and non-linear hardening material behaviour (e.g. Stein et al. [3]) the search for generalisations of the shakedown theorem for large deformations was initiated by the paper of Maier [4]. Weichert [5, 6], Groß-Weege [7], Saczuk and Stumpf [8], Tritsch and Weichert [9] and Weichert and Hachemi [10] presented generalisations of the Melan's theorem for structures undergoing large deformations with large plastic strains and moderate or large rotations. The underlying idea is to determine a moderately or finitely deformed configuration of the structure as new reference and then to investigate the shakedown behaviour for superposed small deformations. Typical applications for these methods are thin plate and shell structures. While the shakedown analysis for arbitrary superposed small deformation histories can be performed with optimisation technique, the reference configuration has to be determined by using an appropriate shell finite element for moderate or large deformations. This type of shakedown analysis fails in the case of finite elastic-plastic strains.



*D. Weichert and G. Maier (eds.), Inelastic Analysis of Structures under Variable Loads,* 31–47.

Subject of the paper of Stumpf [11] was the shakedown analysis of structures undergoing finite elastic-plastic strains and rotations. There, an incremental shakedown analysis is proposed, where the shakedown of structures subject to arbitrary load histories within a given load domain can be determined by following the load cycle and its reverse one along the boundary of an admissible load domain. Of course, this procedure is very time-consuming and therefore not very satisfactory for general structures.

The aim of the present paper is to propose an appropriate shakedown analysis for structures at finite elasto-plastic strains and rotations without this deficiency. The underlying idea of our new shakedown method is to compute simultaneously all relevant load cases on the boundary of an admissible load domain, which are coupled by a common plastic strain field. As point of departure, one has to choose an admissible load domain that lies inside the elastic region, and then the admissible load domain has to be extended step by step up to its limit shakedown load size. During this extension the common plastic strain field evolves naturally. Numerical applications of the proposed shakedown method are considered, and the results are compared with those obtained by the reverse load cycle technique of [11].

The paper is organised as follows: In Section 2 we briefly recall the concept of finite elastoplasticity proposed in [12] and extensively applied numerically in [13]. In Section 3 a generalisation of the Melan's theorem for finite elasto-plastic strains and rotations is established. In Section 4 the incremental shakedown analysis for finite strains by the reverse load cycle technique is discussed and as numerical application the shakedown domains for a truss arch and a shallow cylindrical shell are determined. In Section 5 we present the new proposed shakedown method for finite strains by analysing simultaneously all relevant load cases with one common plastic strain field. This method is applied to a three-bar truss, for which the shakedown domains for small and finite strains are analysed, and to the truss arch of [11]. It is shown that the shakedown domain obtained with the proposed new shakedown method, is slightly larger than the one determined with the reverse load cycle technique.

All proofs of the proposed shakedown method for structures at finite elasto-plastic strains and rotations will be subject of a forthcoming paper (Schieck [14]). The special case of finite rotations but small strains is considered in detail in Polizzotto and Borino [15].

## 2. Constitutive model of finite strain elastoplasticity

In this section we recall some basic equations of the kinematic and constitutive modelling of finite strain elastoplasticity proposed in [12] and [13]. This theory is formulated in symmetric elastic and plastic strain and stretch measures, respectively, and is independent of the plastic rotation. Therefore, a constitutive equation for the plastic spin is superfluous. Point of departure of the kinematical concept is the multiplicative decomposition of the total deformation gradient $\mathbf{F}$ into elastic and plastic constituents,

$$\mathbf{F} = \mathbf{F}^e \, \mathbf{F}^p , \tag{1}$$

where the index $^e$ refers to the elastic part of the deformation and $^p$ to the plastic part. Applying the polar decomposition theorem to $\mathbf{F}^e$ and $\mathbf{F}^p$, the decomposition

$$\mathbf{F} = \mathbf{R}^e \, \bar{\mathbf{U}}^e \, \mathbf{R}^p \, \mathbf{U}^p \tag{2}$$

is obtained, where $\mathbf{R}$ denotes proper rotation tensor and $\mathbf{U}$ stretch tensor. The overbar on $\bar{\mathbf{U}}^e$ indicates that $\bar{\mathbf{U}}^e$ is directionally coupled with the intermediate configuration given by $\mathbf{F}^p$. Since in a macroscopic model of finite elasto-plasticity of polycrystalline materials the plastic rotation $\mathbf{R}^p$ cannot be determined uniquely, we follow [12] and introduce the back-rotated right elastic stretch tensor

$$\mathbf{U}^e := \mathbf{R}^{pT} \, \bar{\mathbf{U}}^e \, \mathbf{R}^p \ . \tag{3}$$

This leads to the following unique right and left elasto-plastic decomposition of the total deformation gradient,

$$\mathbf{F} = \mathbf{Q} \, \mathbf{U}^e \, \mathbf{U}^p = \mathbf{V}^e \, \mathbf{V}^p \, \mathbf{Q} , \tag{4}$$

where

$$\mathbf{V}^e = \mathbf{Q} \, \mathbf{U}^e \mathbf{Q}^T \ , \ \ \mathbf{V}^p = \mathbf{Q} \, \mathbf{U}^p \mathbf{Q}^T \tag{5}$$

are the forward-rotated elastic and plastic stretch tensors. It is easy to prove that $\mathbf{U}^e$ and $\mathbf{U}^p$ are Lagrangean invariant and $\mathbf{V}^e$ and $\mathbf{V}^p$ Eulerian invariant. Analogously to eqn (5) we define the back-rotated Kirchhoff stress tensor $\boldsymbol{\tau}_0$ and the back-rotated Kirchhoff-type back-stress tensor $\boldsymbol{\alpha}_0$,

$$\boldsymbol{\tau}_0 = \mathbf{Q}^T \boldsymbol{\tau} \, \mathbf{Q} \ , \quad \boldsymbol{\alpha}_0 = \mathbf{Q}^T \boldsymbol{\alpha} \, \mathbf{Q} . \tag{6}$$

The tensors $\boldsymbol{\tau}_0$ and $\boldsymbol{\alpha}_0$ are Lagrangean invariant. Since the appropriate rate of the Lagrangean objective quantities $\mathbf{U}^e$, $\mathbf{U}^p$, $\boldsymbol{\tau}_0$ and $\boldsymbol{\alpha}_0$ is the material time derivative, the appropriate corotational rate of $\mathbf{V}^e$, $\mathbf{V}^p$, $\boldsymbol{\tau}$ and $\boldsymbol{\alpha}$ is given by

$$\overset{\nabla}{(\,.\,)} = \mathbf{Q} \frac{\partial}{\partial t}\left(\mathbf{Q}^T (\,.\,) \mathbf{Q}\right) \mathbf{Q}^T = \frac{\partial}{\partial t}(\,.\,) - \boldsymbol{\Omega}(\,.\,) + (\,.\,)\boldsymbol{\Omega} , \tag{7}$$

where

$$\begin{aligned}\mathbf{\Omega} = \dot{\mathbf{Q}}\,\mathbf{Q}^{\mathrm{T}} &= \left(\dot{\mathbf{F}}\,\mathbf{F}^{-1} - \mathbf{Q}\,\dot{\mathbf{U}}^{\mathrm{e}}\,\mathbf{U}^{\mathrm{e}-1}\mathbf{Q}^{\mathrm{T}} - \mathbf{Q}\,\mathbf{U}^{\mathrm{e}}\,\dot{\mathbf{U}}^{\mathrm{p}}\,\mathbf{U}^{\mathrm{p}-1}\,\mathbf{U}^{\mathrm{e}-1}\mathbf{Q}^{\mathrm{T}}\right)_{\mathrm{skw}} \\ &= \left(\dot{\mathbf{F}}\,\mathbf{F}^{-1} - \overset{\nabla}{\mathbf{V}}{}^{\mathrm{e}}\,\mathbf{V}^{\mathrm{e}-1} - \mathbf{V}^{\mathrm{e}}\,\overset{\nabla}{\mathbf{V}}{}^{\mathrm{p}}\,\mathbf{V}^{\mathrm{p}-1}\,\mathbf{V}^{\mathrm{e}-1}\right)_{\mathrm{skw}}\end{aligned} \tag{8}$$

is the spin of the rotation tensor $\mathbf{Q}$. The co-rotational rates $\overset{\nabla}{\mathbf{V}}{}^{\mathrm{e}}$ and $\overset{\nabla}{\mathbf{V}}{}^{\mathrm{p}}$, which are needed to determine $\mathbf{\Omega}$, can be obtained by solving numerically the equations for the elastic and plastic deformation rates

$$\mathbf{d}^{\mathrm{e}} = \left(\overset{\nabla}{\mathbf{V}}{}^{\mathrm{e}}\,\mathbf{V}^{\mathrm{e}-1}\right)_{\mathrm{sym}} \tag{9}$$

and

$$\mathbf{d}^{\mathrm{p}} = \left(\mathbf{V}^{\mathrm{e}}\,\overset{\nabla}{\mathbf{V}}{}^{\mathrm{p}}\,\mathbf{V}^{\mathrm{p}-1}\,\mathbf{V}^{\mathrm{e}-1}\right)_{\mathrm{sym}} \tag{10}$$

for $\overset{\nabla}{\mathbf{V}}{}^{\mathrm{e}}$ and $\overset{\nabla}{\mathbf{V}}{}^{\mathrm{p}}$. Taking into account the Clausius-Duhem inequality and the principle of maximum plastic dissipation (see $[13]_2$), the incremental constitutive equation for the Kirchhoff stress tensor is obtained as

$$\overset{\nabla}{\boldsymbol{\tau}} = \mathbb{C}^{\mathrm{ep}} \bullet \mathbf{d}\,, \tag{11}$$

where

$$\mathbf{d} = \left(\dot{\mathbf{F}}\,\mathbf{F}^{-1}\right)_{\mathrm{sym}} = \mathbf{d}^{\mathrm{e}} + \mathbf{d}^{\mathrm{p}} \tag{12}$$

is the deformation rate and

$$\mathbb{C}^{\mathrm{ep}} = \mathbb{C}^{\mathrm{e}} - \frac{\left(\mathbb{C}^{\mathrm{e}} \bullet \dfrac{\partial\phi}{\partial\boldsymbol{\tau}}\right) \otimes \left(\dfrac{\partial\phi}{\partial\boldsymbol{\tau}} \bullet \mathbb{C}^{\mathrm{e}}\right)}{\dfrac{\partial\phi}{\partial\boldsymbol{\tau}} \bullet \mathbb{C}^{\mathrm{e}} \bullet \dfrac{\partial\phi}{\partial\boldsymbol{\tau}} - \dfrac{\partial\phi}{\partial\mathbf{V}^{\mathrm{p}}} \bullet \dfrac{\partial\overset{\nabla}{\mathbf{V}}{}^{\mathrm{p}}}{\partial\mathbf{d}^{\mathrm{p}}} \bullet \dfrac{\partial\phi}{\partial\boldsymbol{\tau}}} \tag{13}$$

the consistent elasto-plastic tangent material tensor. In (13), $\phi = \phi(\boldsymbol{\tau}, \boldsymbol{\alpha}, k)$ denotes the yield condition with $k = k(\mathbf{V}^{\mathrm{p}})$ the set of internal variables, and $\mathbb{C}^{\mathrm{e}}$ denotes the tangent elasticity tensor.

Alternatively to (11) and (13) the incremental constitutive equation can also be formulated in back-rotated form as

$$\dot{\boldsymbol{\tau}}_{\mathrm{o}} = \mathbb{C}^{\mathrm{ep}}_{\mathrm{o}} \bullet \mathbf{d}_{\mathrm{o}} \tag{14}$$

with the back-rotated deformation rate

$$\mathbf{d}_o = \mathbf{Q}^T \mathbf{d}\,\mathbf{Q} = \mathbf{U}^{e-1}\,\mathbf{U}^{p-1}\left(\dot{\mathbf{F}}^T\,\mathbf{F}\right)_{sym}\,\mathbf{U}^{p-1}\,\mathbf{U}^{e-1} \tag{15}$$

and the back-rotated consistent elasto-plastic material tangent operator

$$\mathbb{C}_o^{ep} = \mathbb{C}_o^e - \frac{\left(\mathbb{C}_o^e \bullet \frac{\partial\phi}{\partial\boldsymbol{\tau}_o}\right)\otimes\left(\frac{\partial\phi}{\partial\boldsymbol{\tau}_o}\bullet\mathbb{C}_o^e\right)}{\frac{\partial\phi}{\partial\boldsymbol{\tau}_o}\bullet\mathbb{C}_o^e\bullet\frac{\partial\phi}{\partial\boldsymbol{\tau}_o} - \frac{\partial\phi}{\partial\mathbf{U}^p}\bullet\frac{\partial\dot{\mathbf{U}}^p}{\partial\mathbf{d}_o^p}\bullet\frac{\partial\phi}{\partial\boldsymbol{\tau}_o}}, \tag{16}$$

where $\mathbb{C}_o^e$ denotes the back-rotated tangent elasticity tensor. The back-rotated elastic and plastic deformation rates are obtained analogously to eqn (15) as

$$\mathbf{d}_o^e = \mathbf{Q}^T\,\mathbf{d}^e\,\mathbf{Q} = \left(\dot{\mathbf{U}}^e\,\mathbf{U}^{e-1}\right)_{sym} \tag{17}$$

and

$$\mathbf{d}_o^p = \mathbf{Q}^T\,\mathbf{d}^p\,\mathbf{Q} = \left(\mathbf{U}^e\,\dot{\mathbf{U}}^p\,\mathbf{U}^{p-1}\,\mathbf{U}^{e-1}\right)_{sym}. \tag{18}$$

## 3. Extension of Melan's theorem for finite elasto-plastic strains and rotations

For an arbitrary load history within a given load domain, shakedown of the structure is characterised by limited plastic flow. It means, that after a finite number of load cycles the structure reacts purely elastically such that no further plastic flow occurs. In the classical geometrically linear shakedown analysis one has to determine a fictitious time-independent self stress field such, that the sum of the self stress and of the stresses due to the external loads does not violate the yield limit. But in the case of geometrical non-linearity the self stress field is not unique and not time-independent, e.g. for snap through – snap back problems. Therefore, the fictitious time-independent self stress field must be replaced by a fictitious time-independent plastic strain field. Then the geometrical non-linear extension of Melan's theorem can be proposed as follows:

*Generalisation of Melan's theorem for finite strains and rotations:*
For a given elasto-plastic structure subject to loads varying arbitrarily within a given load domain, a necessary condition for shakedown is the existence of a time-independent plastic stretch field $\hat{\mathbf{U}}^p$ such that the fictitious elastic stress response $\hat{\boldsymbol{\tau}}$ to the loads and the plastic stretch $\hat{\mathbf{U}}^p$ do not violate the yield limit.

Polizzotto and Borino [15] proposed and proved this extended Melan theorem for finite deformations but small strains. In Schieck [14] the proof will be given for finite deformations and large strains. Due to the complexity of this proof it can't be presented here.

## 4. Incremental finite strain shakedown analysis with the reverse load cycle technique

### 4.1. FUNDAMENTALS

Since the classical optimisation method to determine the fictitious time-independent self stress field for the largest possible admissible load domain cannot be applied for large strains, Stumpf [11] proposed to compute incrementally reverse load cycles along the boundary of the load domain in order to determine a common plastic strain field. For large strains we have to take into account that the response of the structure is path- and history-dependent. This means, if after a sufficiently large number of cycles there is no additional plastification at the next cycle, it may happen that by reversing the direction of the load cycle additional plastification occurs. Therefore, one has to analyse all possible extremal load paths along the boundary of the load domain in both directions, and if there is no additional plastification for both directions, shakedown is ensured. In this case a common plastic strain field for all loads within and on the boundary of the load domain exists and has been found. By this procedure the shakedown analysis of arbitrary structures undergoing finite strains and rotations can be performed, if a finite element for large elasto-plastic strains and rotations is available (as it is the case at the institute of the first author).

The convergence of the procedure can be controlled by determining the plastic dissipation for each load cycle. If the increase of dissipation vanishes, shakedown is obtained. However, the convergence can be rather weak. In this case it is reasonable to compute from time to time load cycles with a small overload. If after several load cycles no additional dissipation is obtained, one has to compute the set of all extremal load paths along the boundaries of the domain in both directions. If also there no additional plastification occurs, shakedown of the structure is obtained. If a weak convergence of the dissipation increase is observed, it may happen that due to restricted computational effort this method leads to a smaller shakedown domain than the maximally possible one.

## 4.2. NUMERICAL APPLICATIONS

### 4.2.1. *Shallow truss arch*

For the shallow truss arch shown in fig. 1, the shakedown domain was determined in [11] applying the reverse load cycle method. The structure carries a vertical load varying between zero and a maximum, and a horizontal load alternating between plus/minus of an extremal value. It develops large rotations and shows buckling effects in many load paths although the strains remain small. All necessary data are given in fig. 1, where a, l describe the geometry, A is the cross section of the bars, and E, $E_T$, $\tau_y$ denote Young's modulus, kinematic hardening modulus and initial yield stress, respectively.

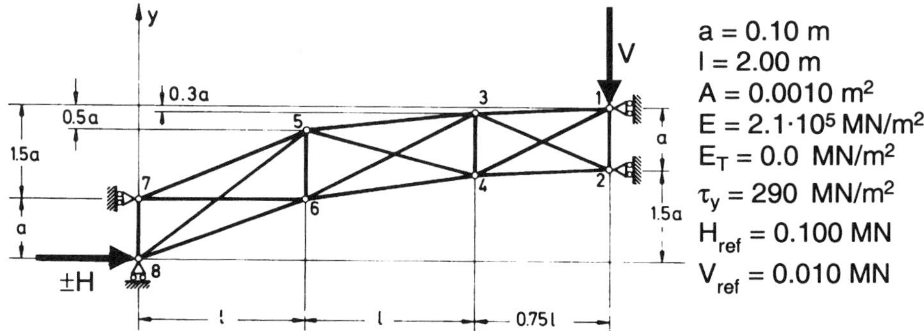


*Figure 1.* Shakedown of a shallow truss arch: problem definition

The shakedown domain obtained in [11] is represented in fig. 2 by the dashed bold line.

### 4.2.2. *Shallow cylindrical shell*

Fig. 3 shows a shallow cylindrical shell panel that is designed similar to the truss arch of fig. 1. This similarity concerns geometry and stiffness. The parameter a describes the geometry, and E, $E_T$, $\tau_y$ denote Young's modulus, kinematic hardening modulus and initial yield stress, respectively. The shakedown domain of this shell problem was analysed by the reverse load cycle method and is shown in fig. 4.

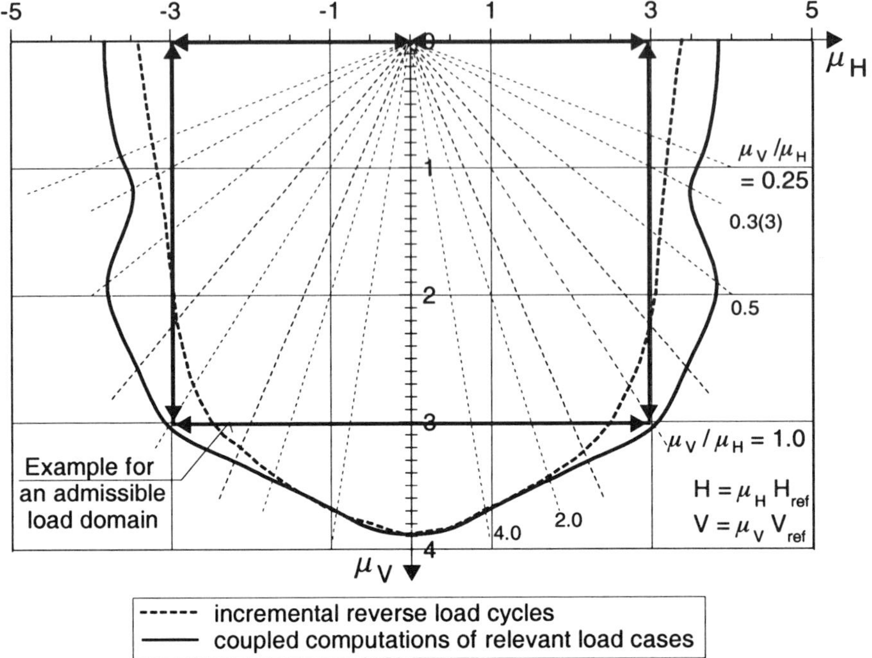


*Figure 2.* Shakedown of a shallow truss arch: shakedown domain

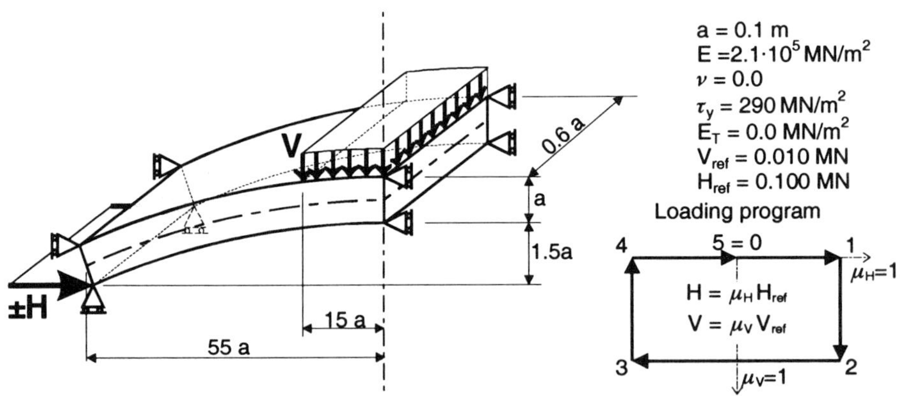


*Figure 3.* Shakedown of a shallow cylindrical shell panel: problem definition

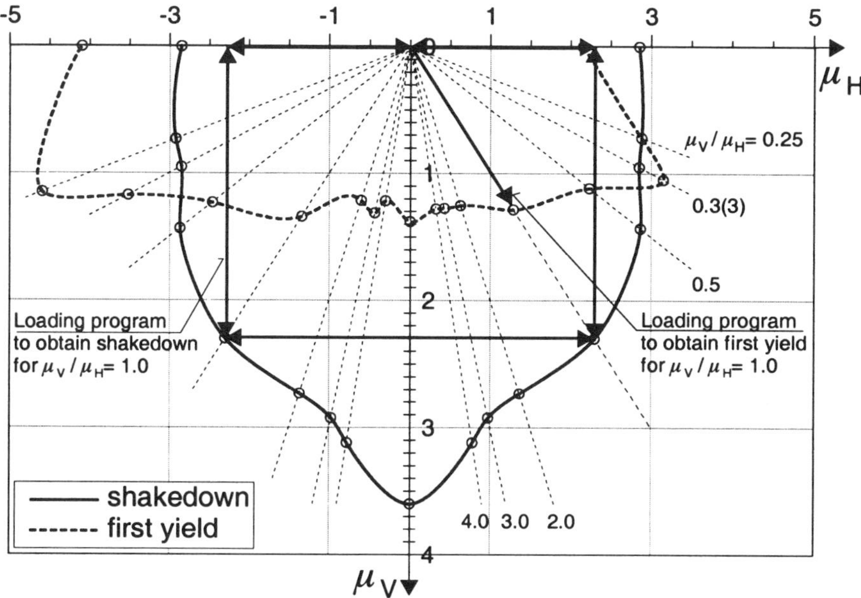


*Figure 4.* Shakedown of a shallow cylindrical shell panel: shakedown domain

## 5. Coupled simultaneous computation of all relevant load cases with one common plastic strain field

The basic idea of the new procedure proposed in this paper is to compute all relevant load cases simultaneously, where the corresponding deformations are coupled by one common plastic strain field $\mathbf{E}^{\mathrm{p}}$. By increasing the loads the common plastic strain field evolves. When the loads reach the limit, where a common plastic strain field $\mathbf{E}^{\mathrm{p}}$ cannot be found any more, $\lim \mathbf{E}^{\mathrm{p}}$ determines the fictitious time-independent plastic stretch field $\hat{\mathbf{U}}^{\mathrm{p}} = \sqrt{\mathbf{1} + 2 \lim \mathbf{E}^{\mathrm{p}}}$, at which all superposed admissible loads can be carried without additional plastification.

### 5.1. THE CONCEPT OF RELEVANT LOAD CASES

Now the question arises, whether for each admissible load domain really all admissible load cases on the boundary must be considered, or if there exists only a small number of relevant load cases. These relevant load cases must be identified in a way that they determine the plastic response of the structure for the whole admissible load domain.

Let us assume that the structure subject to variable loads shows a monotonous behaviour in the load-displacement relation within the admissible load domain, or more precisely, monotonous relations between stresses and loads as they are defined by eqns (20) and (21). Then, if the load domain is bounded by a polygon, only the load cases at the corners of the admissible load domain are relevant load cases (see fig. 5), because the yield condition is convex.

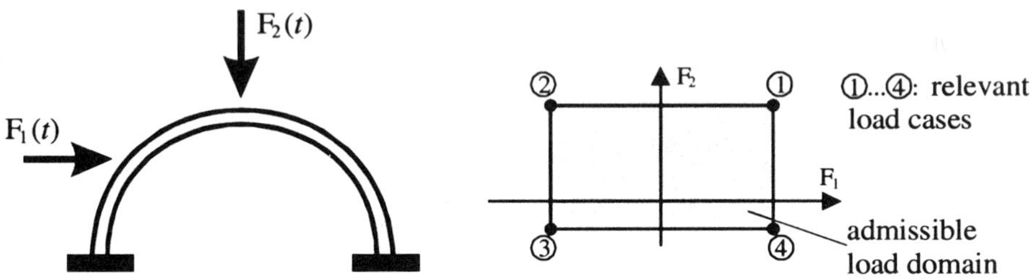


*Figure 5.* Structure with variable loads, admissible load domain and relevant load cases

To prove this, let us consider two corner points of the admissible load domain and name them $i=1$ and $i=2$. At each of the corner points the v. Mises yield condition is assumed to hold identically:

$$\phi_i = \phi(\boldsymbol{\tau}_{oi}, \boldsymbol{\alpha}_o, k) = \sqrt{\left[\boldsymbol{\tau}_{oi} - \boldsymbol{\alpha}_o\right]_{\text{dev}} \cdot \left[\boldsymbol{\tau}_{oi} - \boldsymbol{\alpha}_o\right]_{\text{dev}}} - k = 0. \tag{19}$$

Since the plastic strain field is the same for all load cases, the back stress $\boldsymbol{\alpha}_o$ and the internal variable $k$ are also identical. Monotonous relations between the stresses and the loads shall be defined as follows: during the time increase from $t_1$ to $t_2$ and with the loads varying linearly from corner 1 to corner 2 of the admissible load domain and with stress changes according to

$$\boldsymbol{\tau}_o(t) = \boldsymbol{\tau}_{o1} + \int_{t_1}^{t} \dot{\boldsymbol{\tau}}_o \, \mathrm{d}t \,, \qquad \boldsymbol{\tau}_{o2} = \boldsymbol{\tau}_{o1} + \int_{t_1}^{t_2} \dot{\boldsymbol{\tau}}_o \, \mathrm{d}t \,, \tag{20}$$

the conditions

$$\operatorname{sgn}\left(\left[\boldsymbol{\tau}_{o1} - \boldsymbol{\alpha}_o\right]_{\text{dev}} \cdot \dot{\boldsymbol{\tau}}_o\right) = \textit{const.} \quad \text{and} \quad \operatorname{sgn}\left(\left[\boldsymbol{\tau}_{o2} - \boldsymbol{\alpha}_o\right]_{\text{dev}} \cdot \dot{\boldsymbol{\tau}}_o\right) = \textit{const.} \tag{21}$$

must hold. For geometrically linear problems this condition is always satisfied. In case of strong geometrical non-linear behaviour like snap through eqn (21) is violated. In this case the snap through – snap back loads must be considered as additional relevant load cases.

Introducing eqn $(20)_2$ into eqn (19) yields

$$\left[\boldsymbol{\tau}_{o1} - \boldsymbol{\alpha}_o + \int_{t_1}^{t_2} \dot{\boldsymbol{\tau}}_o \, dt\right]_{dev} \bullet \left[\boldsymbol{\tau}_{o1} - \boldsymbol{\alpha}_o + \int_{t_1}^{t_2} \dot{\boldsymbol{\tau}}_o \, dt\right]_{dev} - k^2$$
$$= 2\left[\boldsymbol{\tau}_{o1} - \boldsymbol{\alpha}_o\right]_{dev} \bullet \int_{t_1}^{t_2} \dot{\boldsymbol{\tau}}_o \, dt + \int_{t_1}^{t_2} \left[\dot{\boldsymbol{\tau}}_o\right]_{dev} dt \bullet \int_{t_1}^{t_2} \left[\dot{\boldsymbol{\tau}}_o\right]_{dev} dt = 0, \tag{22}$$

from which

$$\left[\boldsymbol{\tau}_{o1} - \boldsymbol{\alpha}_o\right]_{dev} \bullet \int_{t_1}^{t_2} \dot{\boldsymbol{\tau}}_o \, dt < 0 \tag{23}$$

is obtained. Then, it follows from eqn (21) that

$$\left[\boldsymbol{\tau}_{o1} - \boldsymbol{\alpha}_o\right]_{dev} \bullet \dot{\boldsymbol{\tau}}_o < 0. \tag{24}$$

This allows to estimate

$$\left[\boldsymbol{\tau}_o(t) - \boldsymbol{\alpha}_o\right]_{dev} \bullet \left[\boldsymbol{\tau}_o(t) - \boldsymbol{\alpha}_o\right]_{dev} - k^2$$
$$= \left[\boldsymbol{\tau}_{o1} - \boldsymbol{\alpha}_o + \int_{t_1}^{t} \dot{\boldsymbol{\tau}}_o \, dt\right]_{dev} \bullet \left[\boldsymbol{\tau}_{o1} - \boldsymbol{\alpha}_o + \int_{t_1}^{t} \dot{\boldsymbol{\tau}}_o \, dt\right]_{dev} - k^2$$
$$= \underbrace{2\left[\boldsymbol{\tau}_{o1} - \boldsymbol{\alpha}_o\right]_{dev} \bullet \int_{t_1}^{t} \dot{\boldsymbol{\tau}}_o \, dt}_{O(t-t_1) < 0} + \underbrace{\int_{t_1}^{t} \left[\dot{\boldsymbol{\tau}}_o\right]_{dev} dt \bullet \int_{t_1}^{t} \left[\dot{\boldsymbol{\tau}}_o\right]_{dev} dt}_{O(t-t_1)^2 > 0}, \tag{25}$$
$$\leq 0 \quad \text{for} \quad t_1 \leq t \leq t_2 .$$

Thus, under the assumption of monotonous behaviour due to eqns (20) and (21) we have proved that the v. Mises yield condition

$$\phi = \phi(\boldsymbol{\tau}_o(t), \boldsymbol{\alpha}_o, k) = \sqrt{\left[\boldsymbol{\tau}_o(t) - \boldsymbol{\alpha}_o\right]_{dev} \bullet \left[\boldsymbol{\tau}_o(t) - \boldsymbol{\alpha}_o\right]_{dev}} - k \leq 0 \tag{26}$$

is satisfied for the whole stress path, while the load changes from one corner of the admissible load domain to the other.

However, the assumption of monotonous load-strain-stress relations requires structural stability. If structural stability is not assured, one has to divide the admissible load domain into piecewise monotonous regions. Then, the load cases at the intersections between the border lines of the monotonous regions and the boundary of the admissible load domain are additional relevant load cases, which have to be taken into account.

## 5.2. BASIC EQUATIONS

According to the proposed procedure one starts with an admissible load domain that lies fully in the elastic region. By increasing the loads of the relevant load cases the admissible load domain is expanded step by step until it reaches its limiting shakedown size. During this procedure the common plastic stretch field $\hat{\mathbf{U}}^{\mathrm{p}}$ defined at the beginning of this section develops according to

$$\dot{\hat{\mathbf{U}}}^{\mathrm{p}} = \sum_i \dot{\mathbf{U}}_i^{\mathrm{p}} , \tag{27}$$

where the plastic stretch increment $\dot{\mathbf{U}}_i^{\mathrm{p}}$ due to the relevant load case $i$ is obtained from eqn (18) by solving it numerically for $\mathbf{d}_{\mathrm{o}i}^{\mathrm{p}}$,

$$\dot{\mathbf{U}}_i^{\mathrm{p}} = \left( \frac{\partial \mathbf{d}_{\mathrm{o}i}^{\mathrm{p}}}{\partial \dot{\mathbf{U}}_i^{\mathrm{p}}} \right)^{-1} \bullet \mathbf{d}_{\mathrm{o}i}^{\mathrm{p}} = \left( \frac{\partial \left( \mathbf{U}_i^{\mathrm{e}} \, \dot{\mathbf{U}}_i^{\mathrm{p}} \, \hat{\mathbf{U}}^{\mathrm{p}-1} \, \mathbf{U}_i^{\mathrm{e}-1} \right)_{\mathrm{sym}}}{\partial \dot{\mathbf{U}}_i^{\mathrm{p}}} \right)^{-1} \bullet \mathbf{d}_{\mathrm{o}i}^{\mathrm{p}} . \tag{28}$$

The plastic deformation increments $\mathbf{d}_{\mathrm{o}i}^{\mathrm{p}}$ of the relevant load cases $i$ with plastic flow are determined by the associated flow rule

$$\mathbf{d}_{\mathrm{o}i}^{\mathrm{p}} = \dot{\lambda}_i \frac{\partial \phi_i}{\partial \boldsymbol{\tau}_{\mathrm{o}i}} , \quad \phi_i := \phi(\boldsymbol{\tau}_{\mathrm{o}i}, \boldsymbol{\alpha}_{\mathrm{o}}, k) , \quad \dot{\lambda}_i \, \phi_i = 0 \tag{29}$$

and the consistency condition

$$\begin{aligned} \dot{\phi}_i &= \frac{\partial \phi_i}{\partial \boldsymbol{\tau}_{\mathrm{o}i}} \bullet \mathbb{C}_{\mathrm{o}}^{\mathrm{e}} \bullet \left( \mathbf{d}_{\mathrm{o}i} - \sum_j \mathbf{d}_{\mathrm{o}j}^{\mathrm{p}} \right) + \sum_j \left( \frac{\partial \phi_j}{\partial \boldsymbol{\alpha}_{\mathrm{o}}} \bullet \frac{\partial \dot{\boldsymbol{\alpha}}_{\mathrm{o}}}{\partial \mathbf{d}_{\mathrm{o}}^{\mathrm{p}}} + \frac{\partial \phi_j}{\partial k} \bullet \frac{\partial \dot{k}}{\partial \mathbf{d}_{\mathrm{o}}^{\mathrm{p}}} \right) \bullet \mathbf{d}_{\mathrm{o}j}^{\mathrm{p}} \\ &= \frac{\partial \phi}{\partial \boldsymbol{\tau}_{\mathrm{o}i}} \bullet \mathbb{C}_{\mathrm{o}}^{\mathrm{e}} \bullet \left( \mathbf{d}_{\mathrm{o}i} - \sum_j \dot{\lambda}_j \frac{\partial \phi_j}{\partial \boldsymbol{\tau}_{\mathrm{o}j}} \right) \\ &\quad + \sum_j \dot{\lambda}_j \left( \frac{\partial \phi_j}{\partial \boldsymbol{\alpha}_{\mathrm{o}}} \bullet \frac{\partial \dot{\boldsymbol{\alpha}}_{\mathrm{o}}}{\partial \mathbf{d}_{\mathrm{o}}^{\mathrm{p}}} + \frac{\partial \phi_j}{\partial k} \bullet \frac{\partial \dot{k}}{\partial \mathbf{d}_{\mathrm{o}}^{\mathrm{p}}} \right) \bullet \frac{\partial \phi_j}{\partial \boldsymbol{\tau}_{\mathrm{o}j}} \end{aligned}$$

$$
\begin{aligned}
&= \frac{\partial \phi}{\partial \boldsymbol{\tau}_{oi}} \bullet \mathbb{C}_o^e \bullet \mathbf{d}_{oi} \\
&\quad -\sum_j \dot{\lambda}_j \left[ \frac{\partial \phi_j}{\partial \boldsymbol{\tau}_{oj}} \bullet \mathbb{C}_o^e \bullet \frac{\partial \phi_j}{\partial \boldsymbol{\tau}_{oj}} - \left( \frac{\partial \phi_j}{\partial \boldsymbol{\alpha}_o} \bullet \frac{\partial \dot{\boldsymbol{\alpha}}_o}{\partial \mathbf{d}_o^p} + \frac{\partial \phi_j}{\partial k} \bullet \frac{\partial \dot{k}}{\partial \mathbf{d}_o^p} \right) \bullet \frac{\partial \phi_j}{\partial \boldsymbol{\tau}_{oj}} \right] \qquad (30) \\
&= 0
\end{aligned}
$$

The latter one forms a system of linear equations for $\dot{\lambda}_i$, which can be solved as long as alternating plastification does not appear, i. e.

$$
\sum_i \dot{\lambda}_i \frac{\partial \phi(\boldsymbol{\tau}_{oi}, \boldsymbol{\alpha}_o, k)}{\partial \boldsymbol{\tau}_{oi}} = 0 \quad \Leftrightarrow \quad \dot{\lambda}_i = 0 \qquad (31)
$$

must hold. With the solution of eqn (30) for $\dot{\lambda}_i$, which can formally written as

$$
\dot{\lambda}_i = \sum_k \mathbf{A}_{ik} \bullet \mathbf{d}_{ok} \qquad (32)
$$

where $\mathbf{A}_{ik}$ forms a matrix of tensors, we can derive the plastically coupled tangent material tensor as follows,

$$
\begin{aligned}
\dot{\boldsymbol{\tau}}_{oi} &= \mathbb{C}_o^e \bullet \left( \mathbf{d}_{oi} - \sum_j \dot{\lambda}_j \frac{\partial \phi_j}{\partial \boldsymbol{\tau}_{oj}} \right) \\
&= \sum_k \left( \delta_{ik}\, \mathbb{C}_o^e - \mathbb{C}_o^e \bullet \sum_j \frac{\partial \phi_j}{\partial \boldsymbol{\tau}_{oj}} \otimes \mathbf{A}_{ik} \right) \bullet \mathbf{d}_{ok} \qquad (33) \\
&=: \sum_k \mathbb{C}_{o\ ik}^{ep} \bullet \mathbf{d}_{ok} \, ,
\end{aligned}
$$

where $\delta_{ik}$ denotes the Kronecker symbol. In components we obtain

$$
\left[ \dot{\boldsymbol{\tau}}_{oi} \right]^{lm} = \left[ \mathbb{C}_{o\ ik}^{ep} \right]^{lmpq} \left[ \mathbf{d}_{ok} \right]_{pq} , \qquad (34)
$$

where $\mathbb{C}_{o\ ik}^{ep}$ is the coupled elastic-plastic tangent material tensor of 6th order.

Now, one recognises easily, that in the corresponding finite element procedure we obtain coupling between the nodal degrees of freedom of all load cases under consideration. Therefore, the nodes of the corresponding finite elements need separated degrees of freedom for each relevant load case. However, there exists only one common plastic strain field and one common "plastic history" for all relevant load cases. The shakedown computations are finished and the maximum size of the admissible load domain is determined, if either alternating plastification occurs, i. e. eqn (31) is violated, or if in a specified load case so many plastic zones (plastic hinges) have developed that the structure becomes a mechanism (kinematic chain).

### 5.3. NUMERICAL APPLICATIONS

The method for a geometrical non-linear shakedown analysis proposed in this paper is tested by two numerical examples. The first one is a very simple structure of three bars (fig. 6) and serves to check the procedure for large strains and large rotations. The second example is the shallow truss arch of fig. 1, which shows buckling effects with large rotations in many load paths although the strains remain small.

#### 5.3.1. *Three bar truss*

The truss shown in fig. 6 is a suitable and most simple example to prove the numerical applicability of the proposed method. All bars are of the same material and have the same cross section $EA = const.$ The constitutive equation for the true (back-rotated) bar force $S$ is assumed as

$$S = EA \ln \frac{l}{l_p} = EA \left( \ln \frac{l}{L} - \ln \frac{l_p}{L} \right), \tag{35}$$

where $l$ is the actual length, $L$ the initial length and $l_p$ is the relaxed plastically deformed length. This constitutive equation matches the multiplicative decomposition of the deformation gradient (eqn (4)), because for truss structures the multiplicative decomposition leads always to the additive decomposition of the Lagrangean logarithmic (Hencky) strains. Moreover, the logarithmic elastic material law is a good approximation for materials undergoing large elastic strains. The true plastic bar force $S_{\text{plast}}$ is assumed to be constant, i. e. plastic hardening or softening just compensates the change of the true cross section. This simple material law is chosen, because in this paper the new shakedown computational procedure and not the geometrically non-linear model of finite elastoplasticity (see [13]) is subject of our main interest.

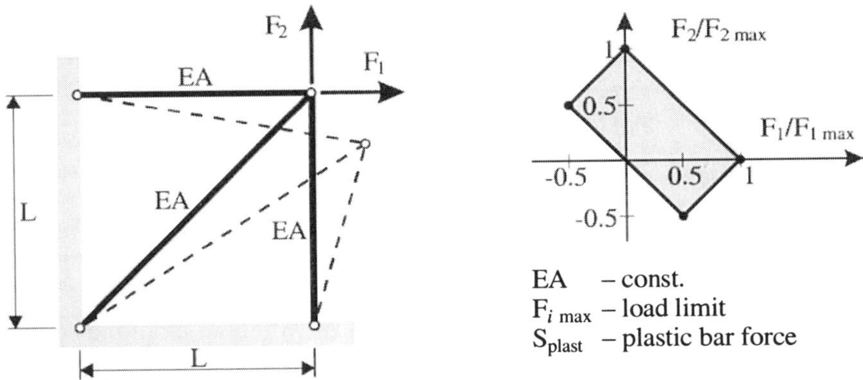


*Figure 6.* Three bar truss and the admissible load domain

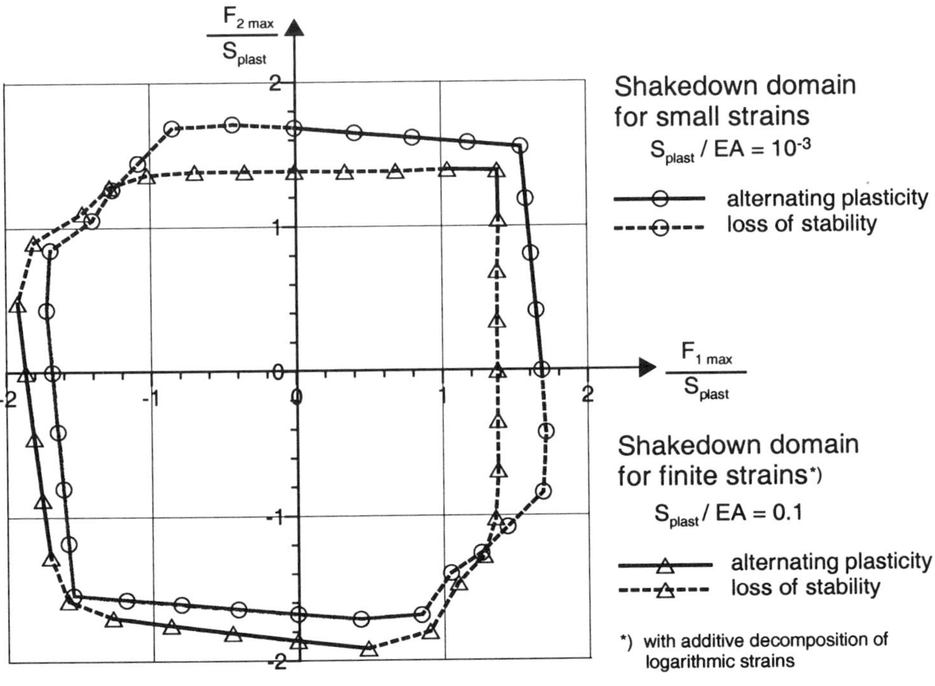


*Figure 7.* Shakedown domains for the 3-bar truss at small and large strains

The shape of the admissible load domain for the loads $F_1$ and $F_2$ is sketched in fig. 6. There $F_{1\,\max}$ and $F_{2\,\max}$ are the limiting values for the loads. During the shakedown analysis $F_{1\,\max}$ and $F_{2\,\max}$ are increased stepwise until either alternating plasticity occurs or the structure becomes a plastic mechanism (plastic kinematic chain) and looses its stability. The set of all possible combinations of $F_{1\,max}$ and $F_{2\,max}$ is then the shakedown domain, which is shown in fig. 7. There, two shakedown domains are presented, one obtained for small elastic strains, $S_{plast}/EA = 10^{-3}$, and one for moderately large elastic strains, $S_{plast}/EA = 10^{-1}$. In the latter case also the plastic strains become large of about $\pm 20\%$ and more. One such deformed configuration is depicted in fig. 6 with dashed lines. In fig. 7, one recognises that the geometrical non-linearities have significant influence on the shakedown behaviour. Such an example with high geometrical non-linearity cannot be analysed sufficiently by the established Melan's optimisation procedures. The only alternative to our new proposed method is the incremental reverse load cycle technique described in Section 4.

5.3.2. *Shallow truss arch*

The shallow truss arch of fig. 1 was analysed also with the method of a coupled computation of the relevant load cases. The shakedown domain obtained in this way is shown in fig. 2 and is represented by the bold line. It is seen that this method leads to a slightly larger shakedown domain compared with the one determined by applying the reverse load cycle technique.

For the interpretation of the shakedown domains, it is necessary to mention here that they are obtained under the assumption that the corners of the investigated admissible load domains are reached in any case. Therefore, due to the geometrical non-linearities the shakedown domains need not be convex. The reason is that the plastic deformations may give the structure a better shape for carrying the applied loads. However, if one looks for an admissible load domain, where the corner points need not be attained in any case, one must restrict the search to the largest convex subset of the shakedown domain.

## Acknowledgement

The authors gratefully acknowledge financial support of the Deutsche Forschungsgemeinschaft (DFG) within the SFB-Project 398/A7.

## References

1. Melan, E.: *Theorie statisch unbestimmter Systeme aus ideal plastischem Baustoff*, Sitzber. Akad. Wiss., Wien, IIa, 145 (1936), 195 pp.
2. Koiter, W.T.: General theorems for elastic-plastic solids. In: Sneddon, I.N. and Hill, R. (eds). *Progress in solid mechanics*, North-Holland, Amsterdam (1960), 165-221.
3. Stein, E., Zhang, G., Huang, Y.: Modeling and computation of shakedown problems for nonlinear hardening materials, *Comp. Meth. Appl. Mech. Engng.* **1-2** (1993), 247-272.
4. Maier, G.: A shakedown matrix theory allowing for work-hardening and second-order geometric effects. In: Sawczuk, A. (ed.). *Foundations of plasticity*, North-Holland, Amsterdam (1973), 417-433.
5. Weichert, D.: Shakedown at finite displacements; a note on Melan's theorem, *Mech. Res. Comm.* **11** (1984), 121-127.
6. Weichert, D.: On the influence of geometrical nonlinearities on the shakedown of elastic-plastic structures, *Int. J. Plasticity* **2.2** (1986), 135-148.
7. Groß-Weege, J.: A unified formulation of statical shakedown criteria for geometrical nonlinear problems, *Int. J. Plasticity* **6.4** (1990), 433-447.
8. Saczuk, J. and Stumpf, H.: On statical shakedown theorems for nonlinear problems. *Mitt. Inst. Mechanik* **74**, Ruhr-Univ. Bochum, 1990.
9. Tritsch, J.-B. and Weichert, D.: Case studies on the influence of geometric effects on the shakedown of structures, *SMIA* **36** (1995), 309-320, Mróz et al. (Eds.), Kluver Academic Publishers Dordrecht, Boston, London, 1995.
10. Weichert, D. and Hachemi, A.: Influence of geometrical nonlinearities on the shakedown of damaged structures, *Int. J. Plasticity* **14.9** (1998), 891-907
11. Stumpf, H.: Theoretical and computational aspects in the shakedown analysis of finite elastoplasticity, *Int. J. Plasticity* **9** (1993), 583-602.
12. Schieck, B. and Stumpf, H.: The appropriate corotational rate, exact formula for the plastic spin and constitutive model for finite elastoplasticity. *Int. J. Solids Structures* **32.24** (1995), 3643-3667.
13. Schieck, B., Smoleński, W.M., Stumpf, H.: A shell finite element for large strain elastoplasticity with anisotropies - Part I: Shell theory and variational principle - Part II: Constitutive equations and numerical applications. *Int. J. Solids Structures* (1999, in print).
14. Schieck, B.: Melan's theorem at large strains, in preparation (1999).
15. Polizzotto, C. and Borino, G.: Shakedown and steady-state responses of elastic-plastic solids in large displacements, *Int. J. Solids Structures* **33.23** (1996), 3415-3437.

# SHAPE OPTIMIZATION UNDER SHAKEDOWN CONSTRAINTS

K. WIECHMANN, F.–J. BARTHOLD, E. STEIN
*Institute of Structural Mechanics and Computational Mechanics, University of Hannover, Appelstr. 9 A, D-30167 Hannover, Germany*

## Abstract

Based on MELAN's static shakedown theorem for linear unlimited kinematic hardening material behaviour, we formulate an integrated approach for all necessary variations within direct analysis, shakedown analysis and variational design sensitivity analysis based on convected coordinates. Using a special formulation of the optimization problem of shakedown analysis, we easily derive the necessary variations of residuals, objectives and constraints. Subsequent discretizations w.r.t. displacements and geometry using e.g. an isoparametric finite element method yield the well known *tangent stiffness matrix* and *tangent sensitivity matrix*, as well as the corresponding matrices for the variation of the LAGRANGE-functional. Thus, all expressions on the element level are dependent only on the nodal values of the displacements and the coordinates but not on a single design variable or the corresponding design velocity field. Remarks on the computer implementation and a numerical example show the efficiency of the proposed formulation.

## 1. Introduction

Elastic shakedown of elasto-plastic systems subjected to variable loading occurs if, after initial yielding, plastification subsides and the system behaves elastically. This is due to the fact that a stationary residual stress field is formed and the total dissipated energy becomes stationary. Elastic shakedown (or simply shakedown) of a system is regarded as a safe state. It is important to know whether a system under given variable loading shakes down or not. But besides this, other important tasks have to be considered: the maximal possible enlargement of a load domain while the geometry of a given structure is kept fixed; the optimization (shape design or material parameters) of the structure for a fixed load domain; the investigation of the sensitivity of the load factor with respect to changes in geometry,



*D. Weichert and G. Maier (eds.), Inelastic Analysis of Structures under Variable Loads,* 49–68.

e.g. in order to take into account geometrical imperfections. All these problems require an efficient strategy for deriving sensitivity information, where special attention must be paid to an adequate numerical formulation and implementation to prevent the computations from becoming very time-consuming. We will limit our attention to the investigation of the structural optimization of systems subjected to shakedown conditions. Classical PRANDTL-REUSS elasto-plasticity with linear unlimited kinematic hardening with VON MISES yield criterion is seen to be a relevant model problem with practical importance.

## 2. State of the Art

In 1932 BLEICH [7] was the first to formulate a shakedown theorem for simple hyperstatic systems consisting of elastic, perfectly plastic materials. This theorem was generalized by MELAN [17, 18] in 1938 to continua with elastic, perfectly plastic and linear unlimited kinematic hardening behaviour. KOITER [12] introduced a kinematic shakedown theorem for an elastic, perfectly plastic material in 1956 which was dual to MELAN'S static shakedown theorem. Since then extensions of these theorems for applications of thermoloadings, dynamic loadings, geometrically nonlinear effects, internal variables and nonlinear kinematic hardening have been carried out by different authors, see e.g. [8, 14, 16, 21, 30, 20, 27]. Several papers have been published concerning especially 2-D and 3-D problems, see [11, 13, 22, 23, 15]. The shakedown investigation of these problems leads to difficult mathematical problems. Thus, in most of these papers, approximate solutions based on the kinematic shakedown theorem of KOITER or on the assumption of a special failure form, were derived. But these solutions often lost their bounding character due to the fact that simplifying flow rules or wrong failure forms were estimated. Thus, the use of the finite element method was beneficial for the numerical treatment of shakedown problems, see [6, 9, 10, 24, 27].

Structural optimization essentially needs an efficient strategy for performing the sensitivity analysis, i.e. for calculating the design variations of functionals modeling the objective and the constraints of the optimization problem. These demands are addressed within the so-called design sensitivity analysis which has been discussed in the literature for about the last three decades and especially for shape design sensitivity within the past fifteen years.

Two basic methods, i.e. the *material derivative* approach and the *domain parametrization* approach, have been used to derive the shape design sensitivity expressions. The *material derivative* approach dates back to 1981 and was later extended to several different viewpoints. The *domain parametrization* method, also called *control volume* approach, was briefly introduced by CÉA in 1981, but did not gain popularity until HABER published a modified formulation in 1987. For a concise description and further hints to literature see [29, 1].

In [2] BARTHOLD ET AL. presented a systematic reformulation of these two equivalent approaches called the *separation* approach taking advantage of their merits but avoiding their drawbacks. He formulated an integrated approach of variations with respect to geometry and displacements within direct analysis and variational design sensitivity analysis based on convected coordinates. This rigorous formulation allows one to separate the influence of geometry and displacement and thus permits an efficient linearization with respect to one of these quantities at time. This approach was outlined for general hyperelastic material behaviour with large strains. It was then extended to problems with linear and finite elasto-plastic material behaviour for single load-cases, see [32, 31]. However, an integrated formulation for elasto-plastic problems with multiple load-cases under shakedown constraints has not yet been published.

## 3. Shakedown analysis of structures with unlimited linear kinematic hardening behaviour

Table I presents the constitutive equations of the classical PRANDTL-REUSS elasto-plasticity with linear unlimited kinematic hardening and VON MISES yield criterion.

The necessary shakedown condition for this material is that there exist at least one residual stress field $\bar{\boldsymbol{\rho}}(\mathbf{X})$ and one backstress field $\bar{\boldsymbol{\gamma}}(\mathbf{X})$, such that

$$\Phi[\boldsymbol{\sigma}^E,\bar{\boldsymbol{\rho}},\bar{\boldsymbol{\gamma}}] = \|\operatorname{dev}[\boldsymbol{\sigma}^E(\mathbf{X},t)+\bar{\boldsymbol{\rho}}(\mathbf{X})-\frac{2}{3}\bar{\boldsymbol{\gamma}}(\mathbf{X}))]\| - \sqrt{\frac{2}{3}}Y_0 \leq 0 \quad \forall \mathbf{X}\in\Omega_o \quad (1)$$

is satisfied for all possible loads $\mathbf{P}(t)$ within the given load domain $\mathcal{M}$.

The following static shakedown theorem due to MELAN states, that the necessary shakedown condition eq.(1) in the sharper form eq.(2) is also sufficient.

THEOREM 1. *If there exist a time-independent residual stress field* $\bar{\boldsymbol{\rho}}(\mathbf{X})$*, a time-independent backstress field* $\bar{\boldsymbol{\gamma}}(\mathbf{X})$ *and a factor* $m>1$*, such that the condition*

$$\Phi[m\boldsymbol{\sigma}^E(\mathbf{X},t), m\bar{\boldsymbol{\rho}}(\mathbf{X}), m\bar{\boldsymbol{\gamma}}(\mathbf{X}))] \leq 0 \quad (2)$$

*holds for all* $\mathbf{P}(t)$ *in* $\mathcal{M}$ *and for all* $\mathbf{X}$ *in* $\Omega_o$*, then the system will shake down.*

Based upon this static shakedown theorem, an optimization problem can be formulated. We will investigate for a given system and a given load domain $\mathcal{M}$ how much the load domain $\mathcal{M}$ can be increased or must be decreased, respectively, such that the system will still shake down. The investigation of this optimization problem will be called *shakedown analysis* in the sequel. Introducing a load factor $\beta$, we can reformulate the problem as follows: calculate the maximal possible load factor $\beta$ such that the system will shake down for the increased or decreased

TABLE I. Set of constitutive equations

| | | | | | | | |
|---|---|---|---|---|---|---|---|
| Kinematics | $\boldsymbol{\varepsilon}^e$ | $=$ | $\nabla^s \mathbf{u} - \boldsymbol{\varepsilon}^p$ | | | | |
| | $e^e$ | $=$ | $\boldsymbol{\varepsilon}^e : \mathbf{1}$, | $\tilde{\boldsymbol{\varepsilon}}^e$ | $=$ dev$[\boldsymbol{\varepsilon}^e]$ | | |
| 1. Free energy | $\Psi$ | $=$ | $\hat{\Psi}_{\text{vol}}(e^e)$ | $+$ | $\hat{\Psi}_{\text{iso}}(\tilde{\boldsymbol{\varepsilon}}^e)$ | $+$ | $\hat{K}(\boldsymbol{\alpha})$ |
| | | $=$ | $\frac{1}{2}\kappa(e^e)^2$ | $+$ | $\mu \text{tr}[\tilde{\boldsymbol{\varepsilon}}^e \tilde{\boldsymbol{\varepsilon}}^e]$ | $+$ | $\frac{1}{2} H \boldsymbol{\alpha} : \boldsymbol{\alpha}$ |
| 2. Macro stresses | $\boldsymbol{\sigma}$ | $=$ | $\partial_{e^e} \hat{\Psi}_{\text{vol}}(e^e) \cdot \mathbf{1}$ | $+$ | $\partial_{\tilde{\boldsymbol{\varepsilon}}^e} \hat{\Psi}_{\text{iso}}(\tilde{\boldsymbol{\varepsilon}}^e)$ | | |
| | | $=$ | $\kappa e^e \cdot \mathbf{1}$ | $+$ | $2\mu \tilde{\boldsymbol{\varepsilon}}^e$ | | |
| 3. Micro stresses | $\boldsymbol{\gamma}$ | $=$ | $\partial_{\boldsymbol{\alpha}} \hat{K}(\boldsymbol{\alpha})$ | | | | |
| | | $=$ | $H \boldsymbol{\alpha}$ | | | | |
| 4. Plastic dissipation | $\mathcal{D}^p$ | $=$ | $\boldsymbol{\sigma} : \dot{\boldsymbol{\varepsilon}}^p - \boldsymbol{\gamma} \dot{\boldsymbol{\alpha}} \geq 0$ | | | | |
| 5. Yield criterion | $\Phi$ | $=$ | $\hat{\Phi}(\boldsymbol{\sigma}, \boldsymbol{\gamma})$ | | | | |
| | | $=$ | $\Vert \text{dev}[\boldsymbol{\sigma}] - \frac{2}{3}\boldsymbol{\gamma} \Vert - \sqrt{2/3}\, Y_o$ | | | | |
| 6. Flow rule | $\dot{\boldsymbol{\varepsilon}}^p$ | $=$ | $\partial_{\boldsymbol{\sigma}} \hat{\Phi}(\boldsymbol{\sigma}, \boldsymbol{\gamma})$ | | | | |
| | | $=$ | $\lambda \left[ \text{dev}[\boldsymbol{\sigma}] - \frac{2}{3}\boldsymbol{\gamma} \right] / \Vert \text{dev}[\boldsymbol{\sigma}] - \frac{2}{3}\boldsymbol{\gamma} \Vert$ | $=$ | $\lambda \mathbf{n}$ | | |
| 7. Evolution | $\dot{\boldsymbol{\alpha}}$ | $=$ | $-\partial_{\boldsymbol{\gamma}} \hat{\Phi}(\boldsymbol{\sigma}, \boldsymbol{\gamma})$ | | | | |
| | | $=$ | $\frac{2}{3} \lambda \mathbf{n}$ | | | | |
| 8. Loading/unloading | $\lambda$ | $\geq$ | $0$, | $\Phi \leq 0$, | $\lambda \Phi = 0$ | | |

$\mu$: shear modulus, $\kappa$: bulk modulus, $H$: hardening parameter, $Y_0$: yield stress

load domain $\mathcal{M}_\beta$, respectively. This optimization problem consists of the scalar valued objective $\beta$, the equilibrium conditions for the residual stresses $\bar{\boldsymbol{\rho}}$ and for the elastic stresses $\boldsymbol{\sigma}^E$, i.e.

$$\mathcal{G}^\rho = \int_{\Omega_o} \text{Grad}_X \boldsymbol{\eta} : \bar{\boldsymbol{\rho}}(\mathbf{X}) \, dV_{\Omega_o}, \tag{3a}$$

$$\mathcal{G}^{\sigma^E} = \int_{\Omega_o} \text{Grad}_X \boldsymbol{\eta} : \boldsymbol{\sigma}^E(\mathbf{X}, t) \, dV_{\Omega_o} - \int_{\Omega_o} \boldsymbol{\eta} : \mathbf{b}(\mathbf{X}, t) \, dV_{\Omega_o} - \int_{\partial\Omega_o} \boldsymbol{\eta} : \mathbf{p}(\mathbf{X}, t) \, dA_{\partial\Omega_0} \tag{3b}$$

($\boldsymbol{\eta}$ denotes a test function and $\mathbf{b}$ and $\mathbf{p}$ are the applied loads) and the shakedown conditions, i.e. eq.(2). Thus the optimization problem can be represented as

$$\begin{array}{ll} \text{Objective} & \beta \rightarrow max \\ \text{Constraints} & \mathcal{G}^\rho = 0 \\ & \mathcal{G}^{\sigma^E} = 0 \\ & \Phi \leq 0 \quad \forall \quad \mathbf{X} \in \Omega_o. \end{array} \tag{4a-d}$$

Due to the fact that the shakedown conditions must be fulfilled for any time $t > 0$ the number of constraints of this optimization problem is infinite. To overcome this problem we assume that the load domain has the form of a convex $n$-dimensional polyhedron with $M$ load vertices, see e.g. Figure 1.

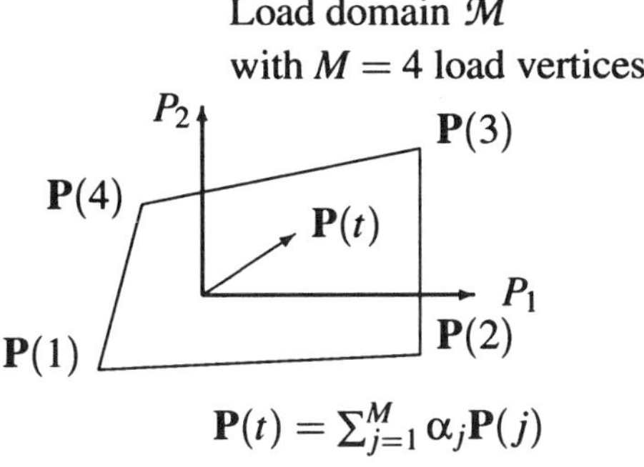


*Figure 1.* Load domain

Thus any point within the load domain can be described by a convex combination of the load vertices

$$\mathbf{P}(t) = \sum_{j=1}^{M} \alpha_j(t)\mathbf{P}(j), \text{ where } \sum_{j=1}^{M} \alpha_j(t) = 1. \tag{5}$$

The time-dependent elastic stresses $\boldsymbol{\sigma}^E$ can then be represented as

$$\boldsymbol{\sigma}^E(t) = \sum_{j=1}^{M} \alpha_j(t)\boldsymbol{\sigma}^E(j). \tag{6}$$

As a result of this formulation the number of shakedown conditions is limited to the number of load vertices of the load domain.

Taking advantage of this formulation, STEIN AND ZHANG [25] introduced the following special optimization problem for unlimited linear kinematic hardening material behaviour:

$$\begin{array}{lll} \text{Objective} & \beta \rightarrow max & \\ \text{Constraints} & \mathcal{G}_j(\mathbf{X}, \mathbf{u}_j^o) = 0, & j \in [1, \dots, M] \\ & \Phi_j(\mathbf{X}, \mathbf{u}_j^o, \beta, \mathbf{y}) \leq 0, & j \in [1, \dots, M] \quad \forall \quad \mathbf{X} \in \Omega_o, \end{array} \tag{7a-c}$$

where $\bar{\boldsymbol{\rho}}$ denotes the residual stresses, $\bar{\boldsymbol{\gamma}}$ denotes the backstresses and $\mathbf{y}$ is the difference of residual stresses and backstresses, i.e. $\mathbf{y} := \bar{\boldsymbol{\rho}} - \bar{\boldsymbol{\gamma}}$, and will be denoted as *internal stresses* in the sequel. Note, that the internal stresses $\mathbf{y}$ are not constrained and thus the equality constraints for the backstresses $\mathcal{G}^{\rho}$ can be eliminated from the formulation. Furthermore no time-dependency occurs, whereas the optimization problem now depends on the displacements corresponding to the loading of the $j$-th load vertex. $\mathcal{G}_j$ now denotes the weak form of equilibrium corresponding to the $j$-th load vertex where the superscript $E$ is dropped for convenience. In order to simplify the notation, this optimization problem is replaced by a LAGRANGE-functional

$$L(\mathbf{X},\mathbf{u}_j^o,\mathbf{z},\boldsymbol{\mu},\boldsymbol{\lambda}) = -\beta + \mu_j \mathcal{G}_j(\mathbf{X},\mathbf{u}_j^o) + \lambda_j \Phi_j(\mathbf{X},\mathbf{u}_j^o,\mathbf{z}) \rightarrow \text{stat}, \tag{8}$$

where $\mu_j$ and $\lambda_j$ are LAGRANGE-multipliers for the equality and inequality constraints, respectively, and the vector $\mathbf{z}$ is the solution of the optimization problem and consists of the load factor $\beta$ and the internal stresses $\mathbf{y}$, i.e. $\mathbf{z}^T = [\beta, \mathbf{y}^T]$. The KUHN-TUCKER conditions which are necessary for an optimal solution are

$$\begin{aligned} \delta_z L =: \mathcal{L}(\mathbf{X},\mathbf{u}_j^o,\mathbf{z},\boldsymbol{\mu},\boldsymbol{\lambda}) &= 0, \\ \mathcal{G}_j(\mathbf{X},\mathbf{u}_j^o) &= 0, \quad \mu_j \neq 0, \quad \mu_j \mathcal{G}_j = 0, \\ \Phi_j(\mathbf{X},\mathbf{u}_j^o,\mathbf{z}) &\leq 0, \quad \lambda_j \geq 0, \quad \lambda_j \Phi_j = 0. \end{aligned} \tag{9a-g}$$

A detailed presentation of these equations for the chosen model problem is given in eq.(24a,b). This general formulation of the optimization problem is the basis for a sensitivity analysis described in the next section.

## 4. Variational design sensitivity analysis

### 4.1. GENERAL REMARKS

The sensitivity analysis of the objective function or the constraint functions under consideration can be performed with different methods. Our approach is based upon the variational design sensitivity analysis of the investigated functionals. This means the variations of the continuous formulation are calculated and then discretized in a subsequent step in order to get computable expressions. The main advantage of this methodology is that the sensitivity analysis can be formulated analogous to and consistent with the structural analysis and the shakedown analysis, where the weak form of equilibrium and the LAGRANGE-functional presented above are formulated for the continuous structure and are then discretized. Thus, many formulations that were derived for and are being used for the direct analysis (structural analysis and shakedown analysis) can also be applied to sensitivity analysis.

The separation approach to continuum mechanics based on convected coordinates [2] yields a decomposition of the deformation mapping $\mathbf{x} = \boldsymbol{\varphi}(\mathbf{X},t)$ into an independent geometry mapping $\mathbf{X} = \tilde{\boldsymbol{\psi}}(\boldsymbol{\Theta})$ and a displacement mapping $\mathbf{u} = \tilde{\mathbf{v}}(\boldsymbol{\Theta},t)$, see Figure 2. Different gradient operators can be defined, i.e. $\text{grad}_x$, $\text{Grad}_X$ and $\text{GRAD}_\Theta$ , see Figure 3, corresponding to the independent variables $\mathbf{x},\mathbf{X}$ and $\boldsymbol{\Theta}$ of the considered domains $\Omega_t, \Omega_o$ and $\mathcal{T}_\Theta$, respectively. Furthermore, the convected basis vectors are $\tilde{\mathbf{E}}_i$ on $\mathcal{T}_\Theta$, $\mathbf{G}_i := \partial\mathbf{X}/\partial\Theta^i$ on $\Omega_\circ$ and $\mathbf{g}_i := \partial\mathbf{x}/\partial\Theta^i$ on $\Omega_t$.

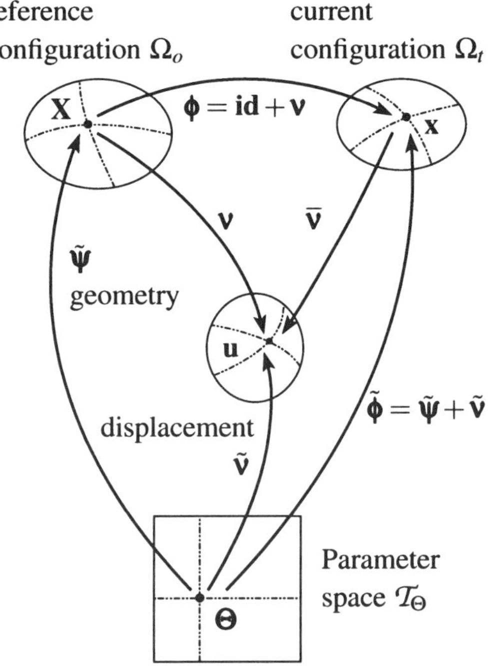


*Figure 2.* Configurations and mappings in continuum mechanics

Thus, the material displacement gradient can be decomposed as follows

$$\mathbf{H} = \operatorname{Grad}\mathbf{v}(\mathbf{X}) = \operatorname{GRAD}\tilde{\mathbf{v}}(\boldsymbol{\Theta})\ [\operatorname{GRAD}\tilde{\boldsymbol{\psi}}(\boldsymbol{\Theta})]^{-1}. \tag{10}$$

The gradients $\operatorname{GRAD}\tilde{\mathbf{v}}(\boldsymbol{\Theta})$ and $\operatorname{GRAD}\tilde{\boldsymbol{\psi}}(\boldsymbol{\Theta})$ are used to decompose continuum mechanical tangent mappings and to perform pull back and push forward transformations between $\Omega_o$ or $\Omega_t$ and the parameter space $\mathcal{T}_\Theta$.

## 4.2. SENSITIVITY ANALYSIS OF SHAKEDOWN ANALYSIS PROBLEMS

### 4.2.1. *General formulation*

Sensitivity analysis of problems with shakedown conditions should be based on the chosen numerical implementation of the direct problem in order to obtain correct and consistent expressions and numerically efficient implementations. The key points of our approach are:

1. The variation of the geometry mapping $\delta\mathbf{X} = \delta\tilde{\psi}(\boldsymbol{\Theta})$ is assumed to be given.
2. The equilibrium condition $\mathcal{G}_j = 0$ of the $j$-th load vertex is solved to update the displacement mapping $\mathbf{u}_j^o = \tilde{\mathbf{v}}_j^o(\boldsymbol{\Theta})$, i.e.

$$\tilde{\mathcal{G}}_j(\boldsymbol{\Theta}, \tilde{\psi}(\boldsymbol{\Theta}), \tilde{\mathbf{v}}_j^o(\boldsymbol{\Theta})) = 0. \tag{11}$$

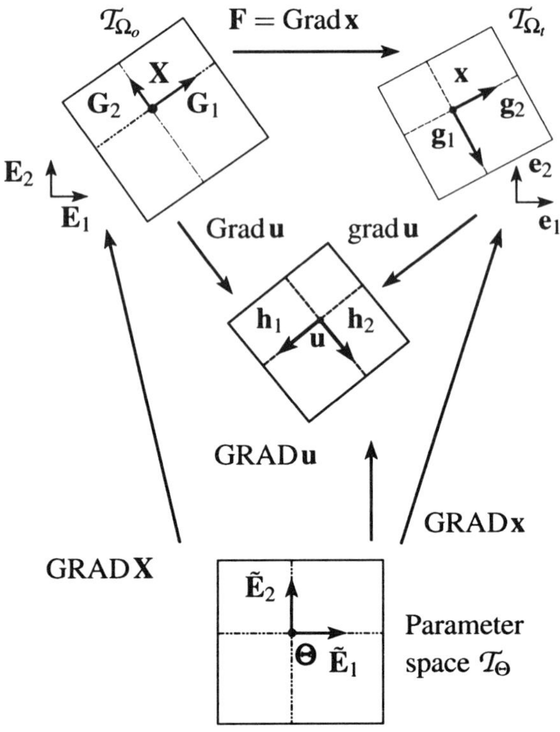


*Figure 3.* Tangent mappings and transformation operators

3. The optimality condition $\mathcal{L} = 0$ is solved to calculate the solution of the shakedown analysis problem $\mathbf{z}$, where $\mathbf{z}^T = [\beta, \mathbf{y}^T]$, i.e.

$$\mathcal{L}(\boldsymbol{\Theta}, \tilde{\psi}(\boldsymbol{\Theta}), \tilde{\mathbf{v}}_j^o(\boldsymbol{\Theta}), \mathbf{z}, \boldsymbol{\mu}, \boldsymbol{\lambda}) = 0. \tag{12}$$

4. The variation of the equilibrium condition $\mathcal{G}_j = 0$ yields

$$\begin{aligned} 0 &= \delta_s \mathcal{G}_j + \delta_{u_j^o} \mathcal{G}_j \\ &= s_j(\boldsymbol{\eta}, \delta\mathbf{X}) + k_j(\boldsymbol{\eta}, \delta\mathbf{u}_j^o) \end{aligned} \tag{13}$$

where $s_j(\boldsymbol{\eta}, \delta\mathbf{X})$ denotes the *tangent sensitivity* of the $j$-th load vertex

$$s_j(\boldsymbol{\eta}, \delta\mathbf{X}) = \frac{\partial \mathcal{G}_j}{\partial \mathbf{X}} \delta\mathbf{X} \tag{14}$$

and $k_j(\boldsymbol{\eta}, \delta\mathbf{u}_j^o)$ denotes the *tangent stiffness* corresponding to the $j$-th load vertex

$$k_j(\boldsymbol{\eta}, \delta\mathbf{u}_j^o) = \frac{\partial \mathcal{G}_j}{\partial \mathbf{u}_j^o} \delta\mathbf{u}_j^o. \tag{15}$$

5. Thus, the variation of the displacement mapping $\delta \mathbf{u}_j^o$ is implicitly defined by eq.(13) and can be represented as

$$\frac{\partial G_j}{\partial \mathbf{u}_j^o} \delta \mathbf{u}_j^o = -\frac{\partial G_j}{\partial \mathbf{X}} \delta \mathbf{X}. \tag{16}$$

6. The variation of the optimality condition $\mathcal{L} = 0$, see eq.(12), yields

$$\begin{aligned} 0 &= \quad \delta_z \mathcal{L} \quad + \quad \delta_s \mathcal{L} \quad + \quad \delta_{u_j^o} \mathcal{L} \\ &= l_z(\delta \bar{\mathbf{z}}, \delta \mathbf{z}) + l_s(\delta \bar{\mathbf{z}}, \delta \mathbf{X}) + l_{u_j^o}(\delta \bar{\mathbf{z}}, \delta \mathbf{u}_j^o) \end{aligned} \tag{17}$$

where three bilinear forms are introduced. Here $l_z(\delta \bar{\mathbf{z}}, \delta \mathbf{z})$ is well known from shakedown analysis; it denotes the variation of the optimality condition $\mathcal{L}$ w.r.t. $\mathbf{z}$. As the optimality condition itself is the variation of the LAGRANGE-functional $L$ w.r.t. $\mathbf{z}$ it is necessary to distinguish between first and second variations. Later we will denote the first variation as a variation w.r.t. $\delta \bar{\mathbf{z}}^T = [\delta \bar{\beta}, \delta \bar{\mathbf{y}}^T]$ and the second as a variation w.r.t. $\delta \mathbf{z}^T = [\delta \beta, \delta \mathbf{y}^T]$, i.e.

$$l_z(\delta \bar{\mathbf{z}}, \delta \mathbf{z}) = \frac{\partial \mathcal{L}}{\partial \mathbf{z}} \delta \mathbf{z}. \tag{18}$$

The term $l_s(\delta \bar{\mathbf{z}}, \delta \mathbf{X})$ denotes the variation of the optimality condition $\mathcal{L}$ w.r.t. geometry, and can be represented as

$$l_s(\delta \bar{\mathbf{z}}, \delta \mathbf{X}) = \frac{\partial \mathcal{L}}{\partial \mathbf{X}} \delta \mathbf{X}. \tag{19}$$

And finally $l_{u_j^o}(\delta \bar{\mathbf{z}}, \delta \mathbf{u}_j^o)$ denotes the variation of the optimality condition w.r.t. the displacements of the $j$-th load vertex, i.e.

$$l_{u_j}(\delta \bar{\mathbf{z}}, \delta \mathbf{u}_j^o) = \frac{\partial \mathcal{L}}{\partial \mathbf{u}_j^o} \delta \mathbf{u}_j^o. \tag{20}$$

7. Thus, the variation of the load factor $\delta \beta$ and the internal stresses $\delta \mathbf{y}$ of the shakedown analysis problem is implicitly defined by eq.(17) and can be represented as

$$\frac{\partial \mathcal{L}}{\partial \mathbf{z}} \delta \mathbf{z} = -\frac{\partial \mathcal{L}}{\partial \mathbf{X}} \delta \mathbf{X} - \frac{\partial \mathcal{L}}{\partial \mathbf{u}_j^o} \delta \mathbf{u}_j^o. \tag{21}$$

Using eq.(16) we can express the variation of the displacement field $\delta \mathbf{u}_j^o$ in terms of the variation of the geometry mapping $\delta \mathbf{X}$ and thus, the variation of the solution of the shakedown analysis problem $\delta \mathbf{z}^T = [\delta \beta, \delta \mathbf{y}^T]$; this means the variation of the load factor and the internal stresses, can also be expressed in terms of $\delta \mathbf{X}$.

4.2.2. *Special formulation for the chosen model problem*
To derive the variational formulation presented above, the total variation of the weak form of equilibrium and of the optimality condition must be supplied, see eq.(11) and (17).

*Variation of the weak form of equilibrium.* For notational convenience we limit our attention to the internal part of the weak form of equilibrium

$$\mathcal{G}_j^{int} = \int_{\Omega_o} \mathrm{Grad}_X \boldsymbol{\eta} : \boldsymbol{\sigma}_j^E \, dV_{\Omega_o}. \tag{22}$$

Its total variation consists of partial variations w.r.t. displacements and w.r.t. geometry and it is computed by a *pull-back - variation - push forward*–scheme, i.e.

$$\begin{aligned}
\delta \mathcal{G}_j^{int} =& \delta \left[ \int_{\mathcal{T}_\Theta} \mathrm{Grad}_X \boldsymbol{\eta} : \boldsymbol{\sigma}_j^E \, J_\Psi \, dV_{\mathcal{T}_\Theta} \right] \\
=& \int_{\mathcal{T}_\Theta} \mathrm{Grad}_X \boldsymbol{\eta} : \delta[\boldsymbol{\sigma}_j^E] \, J_\Psi \, dV_{\mathcal{T}_\Theta} \\
&+ \int_{\mathcal{T}_\Theta} \boldsymbol{\sigma}_j^E : \delta[\mathrm{Grad}_X \boldsymbol{\eta}] \, J_\Psi \, dV_{\mathcal{T}_\Theta} \\
&+ \int_{\mathcal{T}_\Theta} \boldsymbol{\sigma}_j^E : \mathrm{Grad}_X \boldsymbol{\eta} \, \delta[J_\Psi] \, dV_{\mathcal{T}_\Theta} \\
=& \int_{\Omega_o} \mathrm{Grad}_X \boldsymbol{\eta} : \mathbb{C}^E : \mathrm{Grad}_X \delta \mathbf{u}_j^o \, dV_{\Omega_o} \\
&- \int_{\Omega_o} \mathrm{Grad}_X \boldsymbol{\eta} : \mathbb{C}^E : \mathrm{Grad}_X \mathbf{u}_j^o \, \mathrm{Grad}_X \delta \mathbf{X} \, dV_{\Omega_o} \\
&- \int_{\Omega_o} \boldsymbol{\sigma}_j^E : \mathrm{Grad}_X \boldsymbol{\eta} \, \mathrm{Grad}_X \delta \mathbf{X} \, dV_{\Omega_o} \\
&+ \int_{\Omega_o} \boldsymbol{\sigma}_j^E : \mathrm{Grad}_X \boldsymbol{\eta} \, \mathrm{Div} \, \delta \mathbf{X} \, dV_{\Omega_o}.
\end{aligned} \tag{23}$$

The first part of this total variation corresponds to the partial variation of the elastic stresses $\boldsymbol{\sigma}^E$ w.r.t. the displacements $\mathbf{u}_j^o$ of the $j$-th load vertex. This part is well known from finite element analysis. The last three terms correspond to the variation of the elastic stresses $\boldsymbol{\sigma}^E$, the material gradient of the test-functions $\mathrm{Grad}_X \boldsymbol{\eta}$ and the JACOBI-determinant of the transformation onto the parameter space $J_\Psi$ w.r.t. geometry $\mathbf{X}$.

*Variation of the optimality condition.* This variation consists of the partial variations w.r.t. **z**, **X** and $\mathbf{u}_j^o$ and is equal to zero. The optimality condition $\mathcal{L}$ of the LAGRANGE-functional $L$ consists of partial variations w.r.t the load factor denoted as $\mathcal{L}_\beta$ and w.r.t. the internal stresses denoted as $\mathcal{L}_y$, i.e. $\mathcal{L} = \mathcal{L}_\beta + \mathcal{L}_y$. For the chosen model problem these terms read

$$\begin{aligned} \mathcal{L}_\beta &= -\delta\beta + \lambda_j[2\boldsymbol{\sigma}_j : \mathbf{P} : \boldsymbol{\sigma}_j^E \delta\beta], && (24a)\\ \mathcal{L}_y &= \phantom{-\delta\beta} + \lambda_j[2\boldsymbol{\sigma}_j : \mathbf{P} : \delta\mathbf{y}]. && (24b) \end{aligned}$$

Thus, the total variation of the optimality condition $\delta\mathcal{L}$ can be split additively into two parts: the total variation of $\mathcal{L}_\beta$, which is the partial variation of the LAGRANGE-functional $L$ w.r.t. the load factor $\beta$; and the total variation of $\mathcal{L}_y$, which is the partial variation of the LAGRANGE-functional $L$ w.r.t. the internal stresses $\mathbf{y}$. Each of these total variations $\delta\mathcal{L}_\beta$ and $\delta\mathcal{L}_y$ consists of partial variations w.r.t. the load factor $\beta$, the internal stresses $\mathbf{y}$, the displacements of the $j$-th load vertex $\mathbf{u}_j^o$ and the geometry $\mathbf{X}$, as indicated in the following equations.

$$\begin{array}{rlllllllll|l} \delta\mathcal{L}_\beta & = & 2\lambda_j & \delta\bar{\beta} & \boldsymbol{\sigma}_j^E & : \mathbf{P} : & \mathbb{I} & : & \boldsymbol{\sigma}_j^E\,\delta\beta & & \delta\beta \\ & + & 2\lambda_j & \delta\bar{\beta} & \boldsymbol{\sigma}_j^E & : \mathbf{P} : & \mathbb{I} & : & \delta\mathbf{y} & & \delta\mathbf{y} \\ & + & 2\lambda_j & \delta\bar{\beta} & \boldsymbol{\sigma}_j & : \mathbf{P} : & \mathbb{C}^E & : & \mathrm{Grad}_X\,\delta\mathbf{u}_j^o & & \\ & + & 2\lambda_j & \delta\bar{\beta} & \boldsymbol{\sigma}_j^E & : \mathbf{P} : & \beta\mathbb{C}^E & : & \mathrm{Grad}_X\,\delta\mathbf{u}_j^o & & \delta\mathbf{u}_j^o \\ & - & 2\lambda_j & \delta\bar{\beta} & \boldsymbol{\sigma}_j & : \mathbf{P} : & \mathbb{C}^E & : & \mathbf{H}_j\,\mathrm{Grad}_X\,\delta\mathbf{X} & & \\ & - & 2\lambda_j & \delta\bar{\beta} & \boldsymbol{\sigma}_j^E & : \mathbf{P} : & \beta\mathbb{C}^E & : & \mathbf{H}_j\,\mathrm{Grad}_X\,\delta\mathbf{X} & & \delta\mathbf{X} \end{array} \qquad (25a)$$

$$\begin{array}{rllllllll|l} \delta\mathcal{L}_y & = & 2\lambda_j & \delta\bar{\mathbf{y}} & : \mathbf{P} : & \mathbb{I} & : & \boldsymbol{\sigma}_j^E\,\delta\beta & & \delta\beta \\ & + & 2\lambda_j & \delta\bar{\mathbf{y}} & : \mathbf{P} : & \mathbb{I} & : & \delta\mathbf{y} & & \delta\mathbf{y} \\ & + & 2\lambda_j & \delta\bar{\mathbf{y}} & : \mathbf{P} : & \beta\mathbb{C}^E & : & \mathrm{Grad}_X\,\delta\mathbf{u}_j^o & & \delta\mathbf{u}_j^o \\ & - & 2\lambda_j & \delta\bar{\mathbf{y}} & : \mathbf{P} : & \beta\mathbb{C}^E & : & \mathbf{H}_j\,\mathrm{Grad}_X\,\delta\mathbf{X} & & \delta\mathbf{X} \end{array} \qquad (25b)$$

where the following notation is used

$\mathbb{C}^E$ : linear elastic tangent operator,
$\mathbb{I}$ : fourth-order unity tensor,
$\mathbf{P}$ : fourth-order deviatoric projection tensor,
$\mathbf{H}_j$ : displacement gradient for the $j$-th load vertex.

Note that the structures of the derived variations are almost identical. Furthermore the partial variations of the functionals $\mathcal{L}_\beta$ and $\mathcal{L}_y$ w.r.t. the load factor $\beta$ and w.r.t. the internal stresses $\mathbf{y}$ are required for solving the shakedown analysis problem and are already known. The only additional effort needed for the sensitivity analysis is the computation of the partial variations w.r.t. the displacements $\mathbf{u}_j$ and w.r.t. geometry $\mathbf{X}$.

## 5. Numerical solution of shakedown analysis problems with unlimited linear kinematic hardening using the finite element method

### 5.1. GENERAL CONSIDERATIONS

The numerically efficient implementation of shakedown analysis problems is a very important task due to the fact that normally a discretized optimization problem with very many constraints has to be analysed. This is because, for perfect plasticity or limited kinematic hardening, the shakedown constraints must be calculated for every load vertex in every Gaussian point of the discretized structure. Thus numerical solution procedures have been formulated like a reduced bases technique or a special SQP-algorithm, see [26, 28]. Nevertheless it was shown by STEIN AND ZHANG [25] that for unlimited linear kinematic hardening the formulation of the optimization problem can be simplified because of the local nature of the failure of problems with this kind of material behaviour. The following strategy allows one to take advantage of the special structure of the shakedown optimization problem:

- In a first step the solution of the equilibrium conditions $\mathcal{G}_j = 0$ for any load vertex $j$ is computed.
- Then the vectors of elastic stresses $\underline{\sigma}_j^E(i)$ in any Gaussian point $i$ of the discretized structure are calculated.
- In a final step the solution of the global discretized shakedown optimization problem is calculated by solving local optimization problems at every Gaussian point based upon the following lemma, see [25].

LEMMA 1. *The global maximal load factor* $\beta$ *of a shakedown analysis problem with unlimited linear kinematic hardening material behaviour is given by*

$$\beta = \bar{\beta} \quad \textit{with} \quad \bar{\beta} = \min_{i=1,NGP} \beta_i, \tag{26}$$

*where* $\beta_i$ *is the solution of the local sub-problem defined at the i-th Gaussian point*

$$\begin{array}{ll} \textit{Objective} & \beta_i \rightarrow \max \\ \textit{Constraint} & \Phi[\beta_i \underline{\sigma}_j^E(i) + \underline{y}(i)] \leq 0 \quad \forall \quad j = [1, \dots, M]. \end{array} \tag{27a-b}$$

Note, that the structure of the local optimization problems is very simple because the number of constraints is equal to the number of load vertices. We adopt this solution strategy, and formulate the sensitivity analysis consistent with this formulation, see Figure 4. First of all the sensitivities of the elastic stresses are computed, and after the shakedown analysis problem has been solved its sensitivities are calculated.

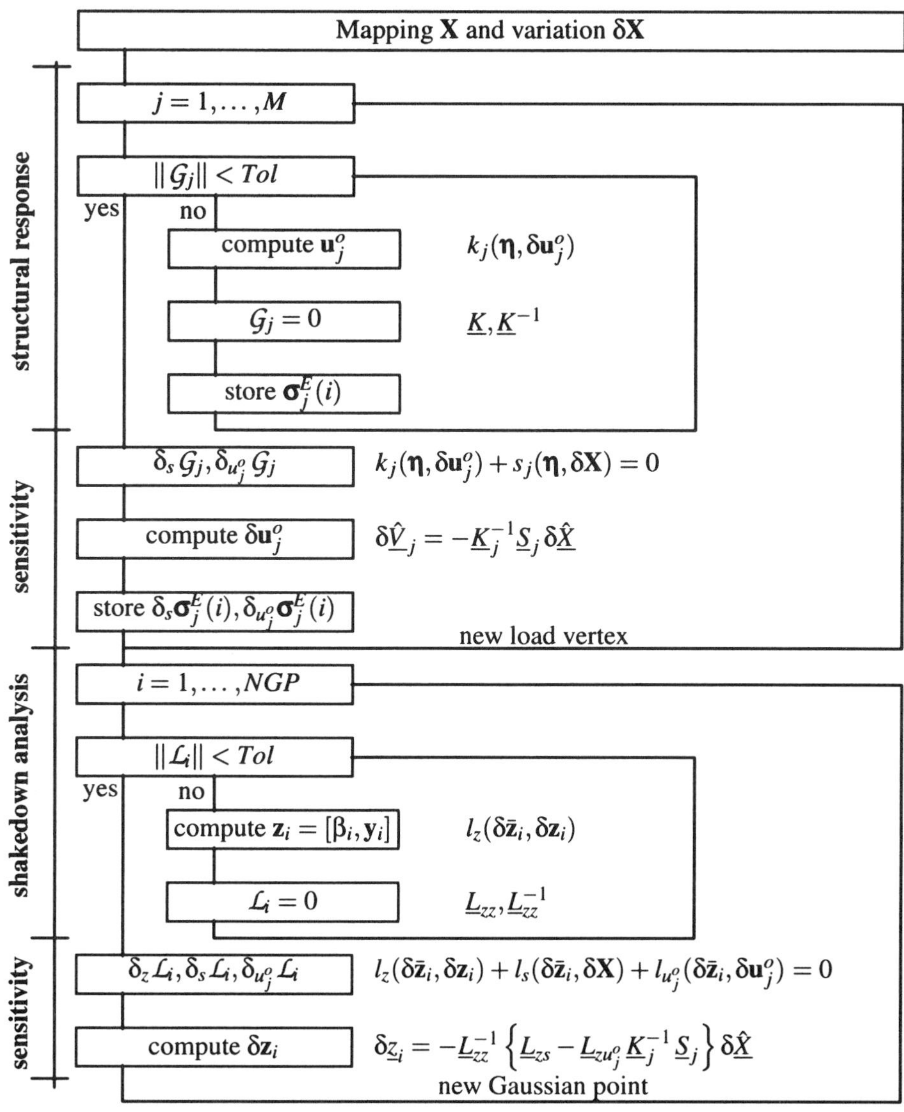


*Figure 4.* Flow chart

## 5.2. FINITE ELEMENT DISCRETIZATION

The continuous formulation derived in Section 4.2 is now discretized by using a standard displacement formulation. Two expressions, the total variation of the weak form of equilibrium and the total variation of the optimality condition of the shakedown analysis problem, must be investigated.

The total variation of the weak form with respect to design using the introduced bilinear forms $s_j(\boldsymbol{\eta}, \delta\mathbf{X})$ and $k_j(\boldsymbol{\eta}, \delta\mathbf{u}_j^o)$ is, see eq.(13),

$$\begin{aligned} \delta \mathcal{G}_j &= \delta_s \mathcal{G}_j + \delta_{u_j^o} \mathcal{G}_j \\ &= s_j(\boldsymbol{\eta}, \delta\mathbf{X}) + k_j(\boldsymbol{\eta}, \delta\mathbf{u}_j^o) = 0. \end{aligned} \tag{28}$$

The FEM-discretization with $\boldsymbol{\eta}_h, \delta\mathbf{u}_h$ and $\delta\mathbf{X}_h$ on each element domain leads to an approximation of the variation of the considered functionals by the derivative with respect to scalar valued design variables

$$\begin{aligned} \delta \mathcal{G}_j = 0 \longrightarrow \frac{d\mathcal{G}_j^h}{ds} &= \bigcup_{e=1}^{E} \left\{ s_j^{e,h}(\boldsymbol{\eta}^h, \delta\mathbf{X}^h) + k_j^{e,h}(\boldsymbol{\eta}^h, \delta\mathbf{u}_j^{o,h}) \right\} \\ &= \underline{\hat{\eta}}^T \, \underline{S}_j \, \delta\underline{\hat{X}} + \underline{\hat{\eta}}^T \, \underline{K}_j \, \delta\underline{\hat{u}}_j^o \\ &= \underline{\hat{\eta}}^T \, \{ \underline{S}_j \, \delta\underline{\hat{X}} + \underline{K}_j \, \delta\underline{\hat{u}}_j^o \} = 0. \end{aligned} \tag{29}$$

Thus, we obtain for each virtual node coordinate vector $\delta\underline{\hat{X}} \in \mathbb{R}^k$ the induced virtual node displacement vector $\delta\underline{\hat{u}}_j^o \in \mathbb{R}^n$

$$\delta\underline{\hat{u}}_j^o = -\underline{K}_j^{-1} \, \underline{S}_j \, \delta\underline{\hat{X}}, \tag{30}$$

where $\underline{K}_j$ is the global tangential stiffness matrix of order $(n \times n)$, $\underline{S}_j$ is the global tangential sensitivity matrix of order $(n \times k)$, $n$ is the overall number of degrees of freedom and $k$ is the overall number of coordinates, see [3].

The total variation of an optimality condition for any local sub-problem with respect to design using the introduced bilinear forms $l_z(\delta\bar{\mathbf{z}}_i, \delta\mathbf{z}_i)$, $l_s(\delta\bar{\mathbf{z}}_i, \delta\mathbf{X})$ and $l_{u_j^o}(\delta\bar{\mathbf{z}}_i, \delta\mathbf{u}_j^o)$ is, see eq. (17),

$$\begin{aligned} \delta \mathcal{L}_i &= \delta_z \mathcal{L}_i + \delta_s \mathcal{L}_i + \delta_{u_j^o} \mathcal{L}_i \\ &= l_z(\delta\bar{\mathbf{z}}_i, \delta\mathbf{z}_i) + l_s(\delta\bar{\mathbf{z}}_i, \delta\mathbf{X}) + l_{u_j^o}(\delta\bar{\mathbf{z}}_i, \delta\mathbf{u}_j^o) = 0. \end{aligned} \tag{31}$$

The FEM-discretization on each element domain leads to an approximation of the variation of the considered functionals by the derivative with respect to scalar valued design variables

$$\begin{aligned} \delta \mathcal{L}_i = 0 \longrightarrow \frac{d\mathcal{L}_i^h}{ds} &= \left\{ l_z^{e,h}(\delta\bar{\mathbf{z}}_i^h, \delta\mathbf{z}_i^h) + l_s^{e,h}(\delta\bar{\mathbf{z}}_i^h, \delta\mathbf{X}^h) + l_{u_j^o}^{e,h}(\delta\bar{\mathbf{z}}_i^h, \delta\mathbf{u}_j^{o,h}) \right\} \\ &= \delta\underline{z}_i^T \, \underline{L}_{zz} \, \delta\underline{\hat{z}}_i + \delta\underline{z}_i^T \, \underline{L}_{zs} \, \delta\underline{\hat{X}} + \delta\underline{z}_i^T \, \underline{L}_{zu_j^o} \, \delta\underline{\hat{u}}_j^o \\ &= \delta\underline{z}_i^T \left\{ \underline{L}_{zz} \, \delta\underline{\hat{z}}_i + \underline{L}_{zs} \, \delta\underline{\hat{X}} + \underline{L}_{zu_j^o} \, \delta\underline{\hat{u}}_j^o \right\} = 0. \end{aligned} \tag{32}$$

Thus, by using eq.(30) we obtain for each virtual node coordinate vector $\delta\underline{\hat{X}} \in \mathbb{R}^k$ the induced virtual vector $\delta\underline{\hat{z}} \in \mathbb{R}^l$

$$\delta\underline{\hat{z}} = -\underline{L}_{zz}^{-1} \left\{ \underline{L}_{zs} - \underline{L}_{zu_j^o} \, \underline{K}_j^{-1} \, \underline{S}_j \right\} \delta\underline{\hat{X}} \tag{33}$$

where $\underline{L}_{zz}$ is the local Hessian matrix of order $(l \times l)$, $\underline{L}_{zs}$ is the local shakedown sensitivity matrix of order $(l \times k)$, $\underline{L}_{zu_j^o}$ is the local shakedown displacement matrix of order $(l \times n)$ and $l$ is the number of unknowns of the local shakedown problem.

### 5.3. STORAGE REQUIREMENTS

Two different methods are suitable for the implementation of the shakedown analysis and its variation.

One way is to compute the global displacement vectors $\underline{u}_j^o$ and the elastic stress vectors $\underline{\sigma}_j^E(i)$ at each Gaussian point $i$ for each load vertex. These stress vectors $\underline{\sigma}_j^E(i)$ as well as their total variations $\delta\underline{\sigma}_j^E(i)$ must be stored to solve the shakedown analysis problem and the sensitivity analysis problem, respectively.

In order to minimize the storage required to perform shakedown analysis and its sensitivity analysis we use the following strategy. The vectors of the global displacements $\underline{u}_j^o$ are stored for each load vertex $j = 1, .., M$. The vectors of the elastic stresses $\underline{\sigma}_j^E(i)$ at each Gaussian point are recomputed in order to solve the local optimization problem. Additionally, in order to perform the sensitivity analysis, the total variation of the global displacement vectors $\delta\underline{u}_j^o$ is stored for each load vertex $j = 1, ..., M$ and for each scalar valued design variable $s = 1, ..., NDV$. The vectors of the total variation of the elastic stresses $\delta\underline{\sigma}_j^E(i)$ at each Gaussian point are then recomputed in order to solve the sensitivity analysis problem. Thus, implementing the sensitivity analysis for a shakedown analysis problem requires additional storage that is equal to that needed to implement the shakedown analysis multiplied with the number of design variables of the shape optimization problem.

## 6. Numerical Example

The formulation of sensitivity analysis of geometrically linear elasto-plastic materials as described above was implemented into a finite element research tool for (IN)elastic (A)nalysis and (OPT)imization (INA-OPT).

### 6.1. SQUARE PLATE WITH A CENTRAL CIRCULAR HOLE

The system depicted in Figure 5 is a square plate with a central circular hole. The dimension of the whole structure is $20cm \times 20cm$. The diameter of the central

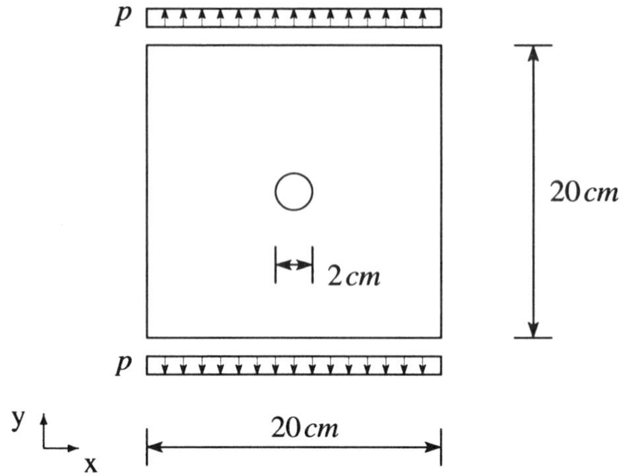


*Figure 5.* Geometry and loading conditions

hole is $2cm$. This structure is loaded by a uniformly distributed load $p$ in the $y$-direction. Due to symmetry conditions only one quarter of the structure was investigated, see Figure 6A. It was discretized with 112 standard two-dimensional isoparametric $q1$-displacement elements. The geometry of the circular hole was modeled by a Bézier curve with five control points, see Figure 6B.

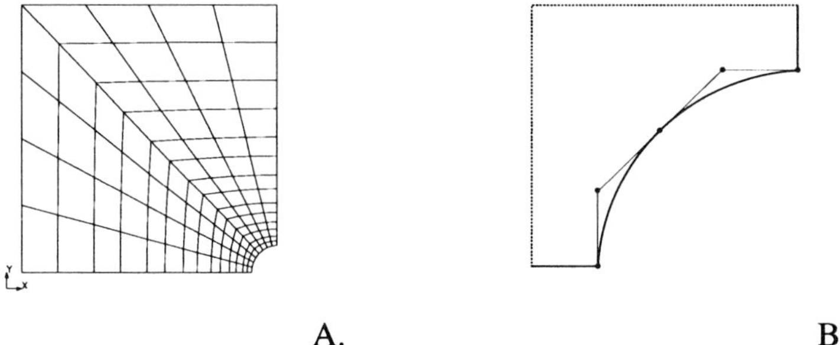


*Figure 6.* A) Finite element discretization. B) Geometry model of the circular hole

The material parameters that have been adopted in the numerical analysis are shown in Table II.

With this initial geometry and material formulation five different optimization problems have been stated, see Table III. For all of these optimization problems a weight minimization was performed. In the first of these formulations only elastic material behaviour was considered and a static load $p = 3.0kN$ was applied. Displacements as well as stresses were constrained in this formulation. The same loading conditions and constraints were used for the second optimization problem but in this case elasto-plastic material behaviour was considered. For the third and

TABLE II. Material parameter

| | | | | |
|---|---|---|---|---|
| Young's modulus: | $E$ | = | 206.90 | $kN/mm^2$ |
| Poisson's ratio: | $\nu$ | = | 0.290 | |
| Bulk modulus: | $\kappa$ | = | 164.206 | $kN/mm^2$ |
| Shear modulus: | $\mu$ | = | 80.1938 | $kN/mm^2$ |
| Initial yield stress: | $Y_0$ | = | 0.45 | $kN/mm^2$ |
| Linear Hardening: | $H$ | = | .12924 | $kN/mm^2$ |

the fourth of the optimization problems a dynamic loading $p = 3.0\sin\left(\frac{2\pi}{0.1}t\right)kN$ was applied. Elastic material behaviour was considered in problem three, whereas elasto-plastic material behaviour was used in problem four. Again displacements and stresses were controlled. In the fifth optimization problem shakedown constraints were applied and a load domain was investigated with initial load vertices $-1.5 \leq p \leq 3.0kN$. For details on the formulation of shape optimization of problems with linear elasto-plastic material behaviour see [32] and of problems with dynamic loading see [19].

TABLE III. Formulation of the five different optimization problems

| | Material behaviour | Loading | active Constraints |
|---|---|---|---|
| 1 | elastic | $p = 3.0kN$ | $\boldsymbol{\sigma}$ |
| 2 | elasto-plastic | $p = 3.0kN$ | $\mathbf{u}$ |
| 3 | elastic | $p(t) = 3.0\sin\left(\frac{2\pi}{0.1}t\right)kN$ | $\boldsymbol{\sigma}$ |
| 4 | elasto-plastic | $p(t) = 3.0\sin\left(\frac{2\pi}{0.1}t\right)kN$ | $\mathbf{u}$ |
| 5 | elasto-plastic (shakedown) | $-1.5 \leq p \leq 3.0kN$ | $\beta \rightarrow$ max |

Table III shows that in optimization problems 1 and 3 (elastic material behaviour) the constraints for the stresses were active, whereas in optimization problems 2 and 4 (elasto-plastic material behaviour) the constraints for the displacements were active. In optimization problem 5 neither stresses nor displacements were limited, here the maximization of the global load factor $\beta$ was adopted as a constraint.

The maximal reduction of weight is gained by optimization problem 2 (elasto-plastic material behaviour, single static load case). The least saving of weight is obtained in optimization 5 (shakedown constraints), see Table IV.

Figure 7 shows A-E, the optimized structures for optimization problems 1-5. It can be seen that the general shapes of the structures are similar. During the optimization, all structures tend to open in loading direction in order to reduce weight, whereas the hole is kept small in the direction perpendicular to the loading direction. Due to plastic yielding, savings are larger in elasto-plastic problems (Fig. 7 B, D) than in elastic problems (Fig. 7 A, C). On the other hand, savings are smaller in dynamic problems (Fig. 7 C, D) than in statically loaded problems (Fig. 7 A, B). In problem 5 (Fig. 7 E) weight is minimized and the load factor is maximized in a nested optimization; the outer boundary is not changed and the savings are least.

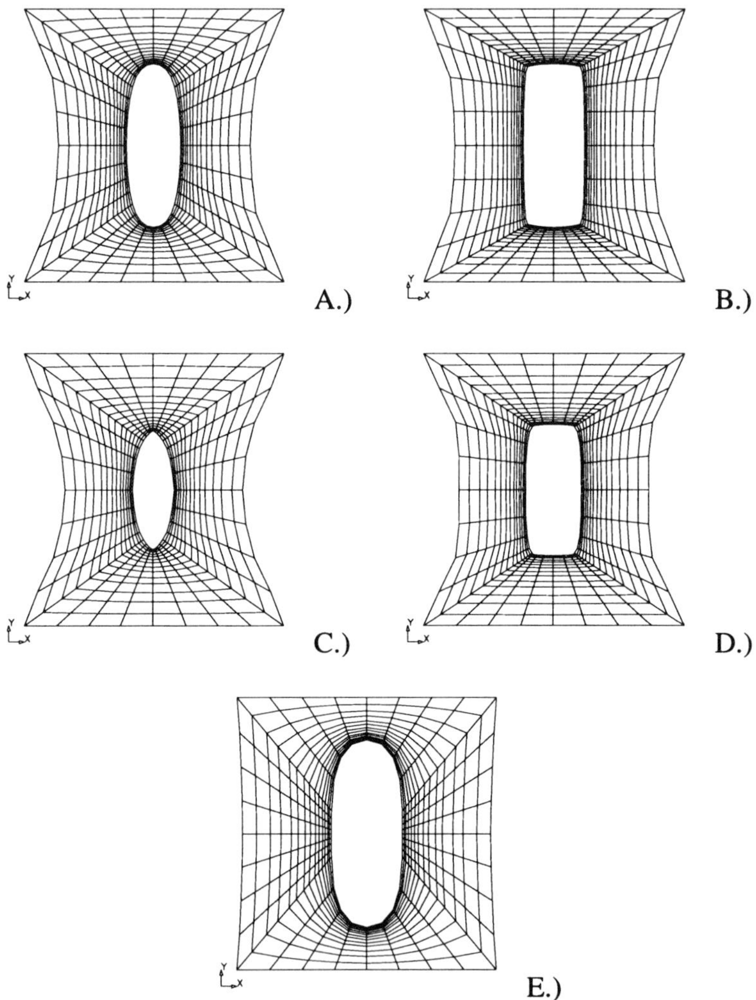


*Figure 7.* Optimized structures. A.) Static loading, elastic material behaviour B.) Static loading, elasto-plastic material behaviour C.) Dynamic loading, elastic material behaviour D.) Dynamic loading, elasto-plastic material behaviour E.) Shakedown constraints, elasto-plastic material behaviour

TABLE IV. Weight reduction of the optimized structures

| | | Reduction of weight |
|---|---|---|
| 1 | static elastic optimization | 26.6 % |
| 2 | static elasto-plastic optimization | 28.8 % |
| 3 | dynamic elastic optimization | 19.0 % |
| 4 | dynamic elasto-plastic optimization | 25.3 % |
| 5 | optimization under shakedown conditions | 18.3 % |

## 7. Conclusions

The proposed representation of variational design sensitivity describes an integrated treatment of all necessary linearizations in structural analysis and sensitivity analysis of shakedown problems. It is directed towards an easily applicable variational design sensitivity analysis and efficient numerical algorithms.

The proposed methodology and the investigated problems are in line with our research on variational design sensitivity analysis described in [2, 3, 4, 5, 32].

## References

1. Arora, J. S.: 1993, 'An exposition of the material derivative approach for structural shape sensitivity analysis'. *Comp. Meth. Appl. Mech. Engng.* **105**, 41–62.
2. Barthold, F.-J. and E. Stein: 1996, 'A continuum mechanical based formulation of the variational sensitivity analysis in structural optimization. Part I: Analysis'. *Structural Optimization* **11**, 29–42.
3. Barthold, F.-J. and E. Stein: 1997, 'Implementing Variational Design Sensitivity Analysis'. In: *Second World Congress of Structural and Multidisciplinary Optimization*. pp. 181–186.
4. Barthold, F.-J. and K. Wiechmann: 1997, 'Variational design sensitivity for inelastic deformation'. In: *Proc. COMPLAS 5*. pp. 792–797.
5. Barthold, F.-J. and K. Wiechmann: 1999, 'Concepts for variational design sensitivity analysis of inelastic materials'. *Structural Optimization*, to be submitted.
6. Belytschko, T.: 1972, 'Plane Stress Shakedown Analysis by Finite Elements'. *Int. J. Mech. Sci.* **14**, 619–625.
7. Bleich, H.: 1932, 'Über die Bemessung statisch unbestimmter Stahlwerke unter der Berücksichtigung des elastisch-plastischen Verhaltens der Baustoffe'. *Bauingenieur* **13**, 261–267.
8. Corradi, L. and G. Maier: 1973, 'Inadaptation Theorems in the Dynamics of Elastic-Workhardening Structures'. *Ing.-Arch.* **43**, 44–57.
9. Corradi, L. and I. Zavelani: 1974, 'A Linear Programming Approach to Shakedown Analysis of Structures'. *Comp. Math. Appl. Mech. Eng.* **3**, 37–53.
10. Dang, H. N. and P. Morelle: 1981, 'Numerical Shakedown Analysis of Plates and Shells of Revolution'. In: *Proceedings of 3rd World Congress and Exhibition on FEMs*. Beverly Hills.

11. Gokhfeld, D. A. and O. F. Cherniavsky: 1980, *Limit Analysis of Structures at Thermal Cycling*. Sijthoff & Noordhoff.
12. Koiter, W. T.: 1956, 'A New General Theorem on Shakedown of Elastic-Plastic Structures'. *Proc. Koninkl. Acad. Wet.* **B 59**, 24–34.
13. König, J. A.: 1966, 'Theory of Shakedown of Elastic-Plastic Structures'. *Arch. Mech. Stos.* **18**, 227–238.
14. König, J. A.: 1969, 'A Shakedown Theorem for Temperature Dependent Elastic Moduli'. *Bull. Acad. Polon. Sci. Ser. Sci. Tech.* **17**, 161–165.
15. Leckie, F. A.: 1965, 'Shakedown Pressure for Flush Cylinder-Sphere Shell Interaction'. *J. Mech. Eng. Sci.* **7**, 367–371.
16. Maier, G.: 1972, 'A Shakedown Matrix Theory Allowing for Workhardening and Second-Order Geometric Effects'. In: *Proc. Symp. Foundations of Plasticity*. Warsaw.
17. Melan, E.: 1938a, 'Der Spannungszustand eines Mises-Henckyschen Kontinuums bei veränderlicher Belastung'. *Sitzber. Akad. Wiss. Wien IIa* **147**, 73–78.
18. Melan, E.: 1938b, 'Zur Plastizität des räumlichen Kontinuums'. *Ing.-Arch.* **8**, 116–126.
19. Meyer, L., F.-J. Barthold, and E. Stein: 1997, 'Remarks on shape optimization of dynamically loaded structures'. In: *Second World Congress of Structural and Multidisciplinary Optimization*. pp. 607–612.
20. Polizzotto, C., G. Borino, S. Caddemi, and P. Fuschi: 1991, 'Shakedown Problems for Material Models with Internal Variables'. *Eur. J. Mech. A/Solids* **10**, 787–801.
21. Prager, W.: 1956, 'Shakedown in Elastic-Plastic Media Subjected to Cycles of Load and Temperature'. In: *Proc. Symp. Plasticita nella Scienza delle Costruzioni*. Bologna.
22. Sawczuk, A.: 1969a, 'Evaluation of Upper Bounds to Shakedown Loads of Shells'. *J. Mech. Phys. Solids* **17**, 291–301.
23. Sawczuk, A.: 1969b, 'On Incremental Collapse of Shells under Cyclic Loading'. In: *Second IUTAM Symp. on Theory of Thin Shells*. Kopenhagen.
24. Shen, W. P.: 1986, 'Traglast- und Anpassungsanalyse von Konstruktionen aus elastisch, ideal plastischem Material'. Ph.D. thesis, Inst. für Computeranwendungen, Universität Stuttgart.
25. Stein, E. and G. Zhang: 1992, 'Theoretical and Numerical Shakedown Analysis for Kinematic Hardening Materials'. In: *3rd Conf. on Computational Methods in Mechanics and Engineering*. Barcelona.
26. Stein, E., G. Zhang, and J. A. König: 1992, 'Shakedown with Nonlinear Hardening including Structural Computation using Finite Element Method'. *Int. J. Plasticity* **8**, 1–31.
27. Stein, E., G. Zhang, and R. Mahnken: 1993, 'Shake-down Analysis for Perfectly Plastic and Kinematic Hardening Materials'. In: *Progress in Computational Analysis of Inelastic Structures*. Springer Verlag, pp. 175–244.
28. Stein, E., G. Zhang, R. Mahnken, and J. A. König: 1990, 'Micromechanical Modeling and Computation of Shakedown with Nonlinear Kinematic Hardening including Examples for 2-D Problems'. In: *Proc. CSME Mechanical Engineering Forum*. Toronto, pp. 425–430.
29. Tortorelli, D. A. and Z. Wang: 1993, 'A systematic approach to shape sensitivity analysis'. *Solids and Structures* **30**(9), 1181–1212.
30. Weichert, D.: 1986, 'On the Influence of Geometrical Nonlinearities on the Shakedown of Elastic-Plastic Structures'. *Int. J. Plasticity* **2**, 135–148.
31. Wiechmann, K. and F.-J. Barthold: 1998, 'Remarks on variational design sensitivity analysis of structures with large elasto-plastic deformations'. In: *Proc. 7th AIAA/USAF/NASA/ISMO Symposium on Multidisciplinary Analysis and Optimization*. pp. 349–358.
32. Wiechmann, K., F.-J. Barthold, and E. Stein: 1997, 'Optimization of elasto-plastic structures using the finite element method'. In: *Second World Congress of Structural and Multidisciplinary Optimization*. pp. 1013–1018.

# NUMERICAL SIMULATIONS OF THERMO–VISCOPLASTIC FLOW PROCESSES UNDER CYCLIC DYNAMIC LOADINGS

W. DORNOWSKI
*Military University of Technology*
*Kaliskiego 2, 01–489 Warsaw, Poland*

P. PERZYNA
*Institute of Fundamental Technological Research*
*Polish Academy of Sciences,*
*Świętokrzyska 21, 00–049 Warsaw, Poland*

Abstract. The main objective of the paper is the description of the behaviour and fatigue damage of inelastic solids in plastic flow processes under dynamic cyclic loadings.

A general constitutive model of elasto–viscoplastic damaged polycrystalline solids is developed within the thermodynamic framework of the rate type covariance structure with a finite set of the internal state variables. A set of the internal state variables is assumed and interpreted such that the theory developed takes account of the effects as follows: (i) plastic non–normality; (ii) plastic strain induced anisotropy (kinematic hardening); (iii) softening generated by microdamage mechanisms; (iv) thermomechanical coupling (thermal plastic softening and thermal expansion); (v) rate sensitivity.

To describe suitably the time and temperature dependent effects observed experimentally and the accumulation of the plastic deformation and damage during dynamic cyclic loading process the kinetics of microdamage and the kinematic hardening law have been modified. The relaxation time is used as a regularization parameter. The viscoplastic regularization procedure assures the stable integration algorithm by using the finite difference method. Particular attention is focused on the well–posedness of the evolution problem (the initial–boundary value problem) as well as on its numerical solutions. Convergence, consistency, and stability of the discretised problem are discussed. The Lax–Richtmyer equivalence theorem is formulated and conditions under which this theory is valid are examined. Utilizing the finite difference method for regularized elasto–viscoplastic model, the numerical investigation of the three–dimensional dynamic adiabatic deformation in a particular body under cyclic loading condition is presented. Particular examples have been considered, namely a dynamic, adiabatic and isothermal, cyclic loading processes for a smooth cylindrical tensile bar. The problem is assumed as axisymmetrical. The accumulation of damage and equivalent plastic deformation on each considered cycle has been obtained. It has been found that this accumulation of microdamage distinctly depends on the wave shape of the assumed loading cycle.

## 1. Introduction

A number of plasticity models have been recently proposed for cyclic loadings, cf. Auricchio, Taylor and Lubliner [5], Auricchio and Taylor [4], Chaboche [6],



*D. Weichert and G. Maier (eds.), Inelastic Analysis of Structures under Variable Loads,* 69–94.

Dafalias and Popov [9], Duszek and Perzyna [12], Mróz [21], Ristinmaa [35], Van der Giessen [41,42] and Wang and Ohno [43].

However none of these theories is able to describe properly the mechanism of fatigue damage when time and strain rate effects are important. Such effects have been observed by Sidey and Coffin [38]. They investigated oxygen–free high conductivity (OFHC) copper at 673 K using unequal strain rates to produce the wave shape. It has been observed that the intrinsic micro–damage process does very much depend on the strain rate effects, cf. chapter 2.

The main objective of the present paper is the development of the thermo–elasto–viscoplasticity theory of damaged polycrystalline solids which takes into account the time and temperature dependent effects observed experimentally and the accumulation of the plastic deformation and damage during dynamic cyclic loading processes.

To describe these effects we intend to modify a constitutive model of thermo–elasto–vicsoplastic damaged polycrystalline solids developed by Duszek–Perzyna and Perzyna [14]. The main modification concerns the kinematic of microdamage and the kinematic hardening law. In chapter 3 a general constitutive model of elasto–viscoplastic damaged polycrystalline solids is developed with the thermodynamic framework of the rate type covariance structure with a finite set of the internal state variables. A set of the internal state variables consists of the equivalent plastic deformation, volume fraction porosity and the residual stress (the back stress). The theory developed takes account of the effects as follows: (i) plastic non–normality; (ii) plastic induced anisotropy (kinematic hardening); (iii) softening generated by microdamage mechanisms; (iv) thermomechanical coupling (thermal plastic softening and thermal expansion); (v) rate sensitivity.

The relaxation time is used as a regularization parameter.

In chapter 4 the numerical solution of the initial–boundary value problem (evolution problem) is discussed. Particular attention has been focused on the viscoplastic regularization procedure for the solution of the dynamical initial–boundary value problems under cyclic loadings. Convergence, consistency and stability of the discretised problem by means of the finite difference method are examined.

Chapter 5 is devoted to the numerical investigation of dynamic adiabatic and isothermal, cyclic loading processes for a smooth cylindrical tensile bar and in chapter 6 final comments are presented.

## 2. Experimental and physical motivations

Sidey and Coffin [38] investigated the mechanisms of fatigue as they bear on fatigue damage at elevated temperature when time and strain rate effects are important. This regime is often referred to as that of time–dependent fatigue.

Test on oxygen-free high conductivity (OFHC) copper at 673 K are examined using unequal strain rates to produce the wave shape. Some typical wave shape were considered, include equal–equal, slow–fast and fast–slow. Specimens with a gage length of 12.70 mm and diameter of 6.35 mm were used. Strain–controlled fatigue tests were carried out at 673 K in air with a total strain range of 1.0 percent. In each test the cycle time was kept constant at 600 s but the tensile and compressive strain rates were varied so that a study of wave–shape effects could be made. In the case of the unbalanced loops, the ratio of fast to slow strain rates was fixed at 100 to 1 with the slow strain rate being $1.7 \times 10^{-5}$ $s^{-1}$. A strain rate of $3.3 \times 10^{-5}$ $s^{-1}$ was used in equal ramp rate tests. After failure, the gage–length was sectioned longitudinally for metallographic observations.

TABLE I. Effect of wave–shape on the number of cycles to failure of OFHC copper at 673 K

| Cycle | Slow–Fast | Equal | Fast–Slow |
|---|---|---|---|
| $N_f$ | 104 | 380 | 1138 |
| $t_f(h)$ | 17 | 63 | 190 |

Table I shows the number of cycles to failure and the time to failure under the various testing conditions. It can be seen that when the total cycle time is kept constant the number of cycles to failure decreases as the tensile–going strain rate decreases.

In the fast–slow test the crack path is transgranular and has started from the surface. Many transgranular surface cracks had initiated and grown for depths up to 0.5 mm but presumably these had ceased growing when one crack became dominant. No internal intergranular cracks were observed.

In contrast, failure in the slow–fast test was intergranular. Near the fracture edge, extensive intergranular cracks can be seen which have been opened out by the final failure. Many of these cracks were wedge type, but at higher magnifications linked cavities could be seen. It was noted that cavitation was present throughout the gage length. Thus the failure was typical of that for creep fracture with most of the cracks being orientated at right angles to the applied stress direction. The fracture path in the equal strain rate test was intergranular. Also internal and short surface intergranular cracks were observed. The surface cracks were about one grain in depth and less numerous than in the slow–fast specimen. There was evidence of cavities either near the fracture or in the bulk of the specimen.

The slow strain rate tension test failed by the propagation of an intergranular crack. Microscopically the specimen was very similar to the equal ramp test with intergranular wedge cracks being situated near the fracture surface and no cracking in the bulk region.

Thus, metallography of the specimen indicated that the decrease in fatigue life was associated with a change in the fracture mode from transgranular to intergranular cracking. Cavity damage occurred only when the tensile–going strain rate was less than the compressive–going strain rate.

## 3. Constitutive modelling for dynamic cyclic loadings

### 3.1. BASIC ASSUMPTIONS AND DEFINITIONS

Let us assume that a continuum body is an open bounded set $\mathcal{B} \subset I\!R^3$, and let $\phi : \mathcal{B} \to \mathcal{S}$ be a $C^1$ configuration of $\mathcal{B}$ in $\mathcal{S}$. The tangent of $\phi$ is denoted $\mathbf{F} = T\phi$ and is called the deformation gradient of $\phi$.

Let $\{X^A\}$ and $\{x^a\}$ denote coordinate systems on $\mathcal{B}$ and $\mathcal{S}$ respectively. Then we refer to $\mathcal{B} \subset I\!R^3$ as the reference configuration of a continuum body with particles $X \in \mathcal{B}$ and to $\mathcal{S} = \phi(\mathcal{B})$ as the current configuration with points $\mathbf{x} \in \mathcal{S}$. The matrix $\mathbf{F}(\mathbf{X}, t) = \partial\phi(\mathbf{X}, t)/\partial\mathbf{X}$ with respect to the coordinate bases $\mathbf{E}_A(\mathbf{X})$ and $\mathbf{e}_a(\mathbf{x})$ is given by

$$F^a_A(\mathbf{X}, t) = \frac{\partial\phi^a}{\partial X^A}(\mathbf{X}, t), \tag{1}$$

where a mapping $\mathbf{x} = \phi(\mathbf{X}, t)$ represents a motion of a body $\mathcal{B}$.

We consider the local multiplicative decomposition

$$\mathbf{F} = \mathbf{F}^e \cdot \mathbf{F}^p, \tag{2}$$

where $(\mathbf{F}^e)^{-1}$ is the deformation gradient that releases elastically the stress on the neighbourhood $\phi(\mathcal{N}(\mathbf{X}))$ in the current configuration.

Let us define the total and elastic Finger deformation tensors

$$\mathbf{b} = \mathbf{F} \cdot \mathbf{F}^T, \quad \mathbf{b}^e = \mathbf{F}^e \cdot \mathbf{F}^{e^T}, \tag{3}$$

respectively, and the Eulerian strain tensors as follows

$$\mathbf{e} = \frac{1}{2}(\mathbf{g} - \mathbf{b}^{-1}), \quad \mathbf{e}^e = \frac{1}{2}(\mathbf{g} - \mathbf{b}^{e^{-1}}), \tag{4}$$

where $\mathbf{g}$ denotes the metric tensor in the current configuration.

By definition[1]

$$\mathbf{e}^p = \mathbf{e} - \mathbf{e}^e = \frac{1}{2}(\mathbf{b}^{e^{-1}} - \mathbf{b}^{-1}) \tag{5}$$

we introduce the plastic Eulerian strain tensor.

To define objective rates for vectors and tensors we use the Lie derivative[2]. Let us define the Lie derivative of a spatial tensor field $\mathbf{t}$ with respect to the velocity field $\boldsymbol{\upsilon}$ as

$$\mathrm{L}_{\boldsymbol{\upsilon}}\mathbf{t} = \phi_* \frac{\partial}{\partial t}(\phi^* \mathbf{t}), \tag{6}$$

where $\phi^*$ and $\phi_*$ denote the pull–back and push–forward operations, respectively.

The rate of deformation is defined as follows

$$\mathbf{d}^\flat = \mathrm{L}_{\boldsymbol{\upsilon}}\mathbf{e}^\flat = \frac{1}{2}\mathrm{L}_{\boldsymbol{\upsilon}}\mathbf{g} = \frac{1}{2}\left(g_{ac}\upsilon^c\,|_b + g_{cb}\upsilon^c\,|_a\right)\mathbf{e}^a \otimes \mathbf{e}^b, \tag{7}$$

[1] For precise definition of the finite elasto–plastic deformation see Perzyna [29].

[2] The algebraic and dynamic interpretations of the Lie derivative have been presented by Abraham et al. [1], cf. also Marsden and Hughes [20].

where the symbol $\flat$ denotes the index lowering operator and $\otimes$ the tensor product,

$$\upsilon^a \mid_b = \frac{\partial \upsilon^a}{\partial x^b} + \gamma^a_{bc}\upsilon^c, \tag{8}$$

and $\gamma^a_{bc}$ denotes the Christoffel symbol for the general coordinate systems $\{x^a\}$. The components of the spin $\boldsymbol{\omega}$ are given by

$$\omega_{ab} = \frac{1}{2}\left(g_{ac}\upsilon^c \mid_b - g_{cb}\upsilon^c \mid_a\right) = \frac{1}{2}\left(\frac{\partial \upsilon_a}{\partial x^b} - \frac{\partial \upsilon_b}{\partial x^a}\right). \tag{9}$$

Similarly

$$\mathbf{d}^{e^\flat} = \mathrm{L}_{\boldsymbol{\upsilon}}\mathbf{e}^{e^\flat}, \qquad \mathbf{d}^{p^\flat} = \mathrm{L}_{\boldsymbol{\upsilon}}\mathbf{e}^{p^\flat}, \tag{10}$$

and

$$\mathbf{d} = \mathbf{d}^e + \mathbf{d}^p. \tag{11}$$

Let $\boldsymbol{\tau}$ denote the Kirchhoff stress tensor related to the Cauchy stress tensor $\boldsymbol{\sigma}$ by

$$\boldsymbol{\tau} = J\boldsymbol{\sigma} = \frac{\rho_{Ref}}{\rho}\boldsymbol{\sigma}, \tag{12}$$

where the Jacobian $J$ is the determinant of the linear transformation $\mathbf{F}(\mathbf{X},t) = (\partial/\partial X)\phi(\mathbf{X},t)$, $\rho_{Ref}(\mathbf{X})$ and $\rho(\mathbf{x},t)$ denote the mass density in the reference and current configuration, respectively.

The Lie derivative of the Kirchhoff stress tensor $\boldsymbol{\tau} \in \mathbf{T}^2(\mathcal{S})$ (elements of $\mathbf{T}^2(\mathcal{S})$ are called tensors on $\mathcal{S}$, contravariant of order 2) gives

$$\begin{aligned}
\mathrm{L}_{\boldsymbol{\upsilon}}\boldsymbol{\tau} &= \phi_* \frac{\partial}{\partial t}(\phi^* \boldsymbol{\tau}) \\
&= \left\{\mathbf{F} \cdot \frac{\partial}{\partial t}[\mathbf{F}^{-1} \cdot (\boldsymbol{\tau} \circ \phi) \cdot \mathbf{F}^{-1^T}] \cdot \mathbf{F}^T\right\} \circ \phi^{-1} \\
&= \dot{\boldsymbol{\tau}} - (\mathbf{d} + \boldsymbol{\omega}) \cdot \boldsymbol{\tau} - \boldsymbol{\tau} \cdot (\mathbf{d} + \boldsymbol{\omega})^T,
\end{aligned} \tag{13}$$

where $\circ$ denotes the composition of mappings. In the coordinate systems (13) reads

$$\begin{aligned}
(\mathrm{L}_{\boldsymbol{\upsilon}}\boldsymbol{\tau})^{ab} &= F^a_A \frac{\partial}{\partial t}(F_c^{-1^A} \tau^{cd} F_d^{-1^B}) F^b_B \\
&= \frac{\partial \tau^{ab}}{\partial t} + \frac{\partial \tau^{ab}}{\partial x^c}\upsilon^c - \tau^{cb}\frac{\partial \upsilon^a}{\partial x^c} - \tau^{ac}\frac{\partial \upsilon^b}{\partial x^c}.
\end{aligned} \tag{14}$$

Equation (14) defines the Oldroyd rate of the Kirchhoff stress tensor $\boldsymbol{\tau}$ (cf. Oldroyd, 1950).

### 3.2. CONSTITUTIVE POSTULATES

Let us assume that: (i) conservation of mass, (ii) balance of momentum, (iii) balance of moment of momentum, (iv) balance of energy, (v) entropy production inequality hold.

We introduce the four fundamental postulates:

(i) Existence of the free energy function. It is assumed that the free energy function is given by

$$\psi = \hat{\psi}(\mathbf{e}, \mathbf{F}, \vartheta; \boldsymbol{\mu}), \tag{15}$$

where $\mathbf{e}$ denotes the Eulerian strain tensor, $\mathbf{F}$ is deformation gradient, $\vartheta$ temperature and $\boldsymbol{\mu}$ denotes a set of the internal state variables.
To extend the domain of the description of the material properties and particularly to take into consideration diferent dissipation effects we have to introduce the internal state variables represented by the vector $\boldsymbol{\mu}$.

(ii) Axiom of objectivity (spatial covariance). The constitutive structure should be invariant with respect to any diffeomorphism (any motion) $\boldsymbol{\xi} : \mathcal{S} \to \mathcal{S}$ (Marsden and Hughes [20]). Assuming that $\boldsymbol{\xi} : \mathcal{S} \to \mathcal{S}$ is a regular, orientation preserving map transforming $\mathbf{x}$ into $\mathbf{x}'$ and $T\boldsymbol{\xi}$ is an isometry from $T_{\mathbf{x}}\mathcal{S}$ to $T_{\mathbf{x}'}\mathcal{S}$, we obtain the axiom of material frame indifference (cf. Truesdell and Noll [40]).

(iii) The axiom of the entropy production. For any regular motion of a body $\mathcal{B}$ the constitutive functions are assumed to satisfy the reduced dissipation inequality

$$\frac{1}{\rho_{Ref}}\boldsymbol{\tau} : \mathbf{d} - (\eta\dot{\vartheta} + \dot{\psi}) - \frac{1}{\rho\vartheta}\mathbf{q} \cdot \operatorname{grad}\vartheta \geq 0, \tag{16}$$

where $\rho_{Ref}$ and $\rho$ denote the mass density in the reference and actual configuration, respectively, $\boldsymbol{\tau}$ is the Kirchhoff stress tensor, $\mathbf{d}$ the rate of deformation, $\eta$ is the specific (per unit mass) entropy, and $\mathbf{q}$ denotes the heat flow vector field. Marsden and Hughes [20] proved that the reduced dissipation inequality (16) is equivalent to the entropy production inequality first introduced by Coleman and Noll [7] in the form of the Clausius–Duhem inequality. In fact the Clausius–Duhem inequality gives a statement of the second law of thermodynamics within the framework of mechanics of continuous media, cf. Duszek and Perzyna [13].

(iv) The evolution equation for the internal state variable vector $\boldsymbol{\mu}$ is assumed in the form as follows

$$\mathrm{L}_{\boldsymbol{\upsilon}}\boldsymbol{\mu} = \hat{\mathbf{m}}(\mathbf{e}, \mathbf{F}, \vartheta, \boldsymbol{\mu}), \tag{17}$$

where the evolution function $\hat{\mathbf{m}}$ has to be determined based on careful physical interpretation of a set of the internal state variables and analysis of available experimental observations.
The determination of the evolution function $\hat{\mathbf{m}}$ (in practice a finite set of the evolution functions) appears to be the main problem of the modern constitutive modelling.

### 3.3. FUNDAMENTAL ASSUMPTIONS

The main objective is to develop the rate type constitutive structure for an elastic–viscoplastic material in which the effects of the plastic non–normality, plastic strain induced anisotropy (kinematic hardening), micro–damaged mechanism and thermomechanical coupling are taken into consideration. To do this it is sufficient to assume a finite set of the internal state variables. For our practical purposes it is sufficient to assume that the internal state vector $\boldsymbol{\mu}$ has the form

$$\boldsymbol{\mu} = (\in^p, \xi, \boldsymbol{\alpha}), \tag{18}$$

where $\in^p$ is the equivalent viscoplastic deformation, i.e.

$$\in^p = \int_0^t \left(\frac{2}{3}\mathbf{d}^p : \mathbf{d}^p\right)^{\frac{1}{2}} dt, \tag{19}$$

$\xi$ is volume fraction porosity and takes account for micro–damaged effects and $\boldsymbol{\alpha}$ denotes the residual stress (the back stress) and aims at the description of the kinematic hardening effects.

Let us introduce the plastic potential function $f = f(\tilde{J}_1, \tilde{J}_2, \vartheta, \boldsymbol{\mu})$, where $\tilde{J}_1$, $\tilde{J}_2$ denote the first two invariants of the stress tensor $\tilde{\boldsymbol{\tau}} = \boldsymbol{\tau} - \boldsymbol{\alpha}$.

Let us postulate the evolution equations as follows

$$\mathbf{d}^p = \Lambda \mathbf{P}, \qquad \dot{\xi} = \Xi, \qquad \mathrm{L}_{\boldsymbol{\upsilon}} \boldsymbol{\alpha} = \mathbf{A}, \tag{20}$$

where for elasto–viscoplastic model of a material we assume (cf. Perzyna [24,25,29])

$$\Lambda = \frac{1}{T_m} \langle \Phi(\frac{f}{\kappa} - 1) \rangle, \tag{21}$$

$T_m$ denotes the relaxation time for mechanical disturbances, the isotropic work–hardening–softening function $\kappa$ is

$$\kappa = \hat{\kappa}(\in^p, \vartheta, \xi), \tag{22}$$

$\Phi$ is the empirical overstress function, the bracket $\langle\cdot\rangle$ defines the ramp function,

$$\mathbf{P} = \left.\frac{\partial f}{\partial \boldsymbol{\tau}}\right|_{\xi=const} \left(\left\|\frac{\partial f}{\partial \boldsymbol{\tau}}\right\|\right)^{-1}, \tag{23}$$

$\Xi$ and $\mathbf{A}$ denote the evolution functions which have to be determined.

### 3.4. INTRINSIC MICRO–DAMAGE PROCESS

An analysis of the experimental observations for cycle fatigue damage mechanisms at high temperature of metals performed by Sidey and Coffin [38] suggests that the intrinsic micro–damage process does very much depend on the strain rate effects as well as on the wave shape effects. In the tests in which duration of extension stress was larger than duration of compression stress (in single cycle) decreasing of the fatigue lifetime was observed and fracture mode changed from a transgranular fracture for the fast–slow wave shape to an intergranular single–crack fracture for equal ramp rates to interior cavitation for the slow–fast test.

To take into consideration these observed time dependent effects it is advantageous to use the proposition of the description of the intrinsic micro–damage process presented by Perzyna [27,28] and Duszek–Perzyna and Perzyna [14].

Let us assume that the intrinsic micro–damage process consists of the nucleation and growth mechanism[3].

Physical considerations (cf. Curran et al. [8] and Perzyna [27]) have shown that the nucleation of microvoids in dynamic loading processes which are characterized by very short time duration is governed by the thermally–activated mechanism. Based on this heuristic suggestion and taking into account the influence of the stress triaxiality on the nucleation mechanism we postulate for rate dependent plastic flow

$$\left(\dot{\xi}\right)_{nucl} = \frac{1}{T_m} h^*(\xi, \vartheta) \left[ \exp \frac{m^*(\vartheta) \mid \tilde{I}_n - \tau_n(\xi, \vartheta, \epsilon^p) \mid}{k\vartheta} - 1 \right], \tag{24}$$

where $k$ denotes the Boltzmann constant, $h^*(\xi, \vartheta)$ represents a void nucleation material function which is introduced to take account of the effect of microvoid interaction, $m^*(\vartheta)$ is a temperature dependent coefficient, $\tau_n(\xi, \vartheta, \epsilon^p)$ is the porosity, temperature and equivalent plastic strain dependent threshold stress for microvoid nucleation,

$$\tilde{I}_n = a_1 \tilde{J}_1 + a_2 \sqrt{\tilde{J}_2'} + a_3 \left(\tilde{J}_3'\right)^{\frac{1}{3}} \tag{25}$$

defines the stress intensity invariant for nucleation, $a_i$ $(i = 1, 2, 3)$ are the material constants, $\tilde{J}_1$ denotes the first invariant of the stress tensor $\tilde{\boldsymbol{\tau}} = \boldsymbol{\tau} - \boldsymbol{\alpha}$, $\tilde{J}_2'$ and $\tilde{J}_3'$ are the second and third invariants of the stress deviator $\tilde{\boldsymbol{\tau}}' = (\boldsymbol{\tau} - \boldsymbol{\alpha})'$.

For the growth mechanism we postulate (cf. Johnson [18]; Perzyna [27,28]; Perzyna and Drabik [30,31])

$$\left(\dot{\xi}\right)_{grow} = \frac{1}{T_m} \frac{g^*(\xi, \vartheta)}{\sqrt{\kappa_0}} \left[ \tilde{I}_g - \tau_{eq}(\xi, \vartheta, \epsilon^p) \right], \tag{26}$$

where $T_m \sqrt{\kappa_0}$ denotes the dynamic viscosity of a material, $g^*(\xi, \vartheta)$ represents a void growth material function and takes account for void interaction, $\tau_{eq}(\xi, \vartheta, \epsilon^p)$ is the porosity, temperature and equivalent plastic strain dependent void growth threshold stress,

$$\tilde{I}_g = b_1 \tilde{J}_1 + b_2 \sqrt{\tilde{J}_2'} + b_3 \left(\tilde{J}_3'\right)^{\frac{1}{3}}, \tag{27}$$

defines the stress intensity invariant for growth and $b_i$ $(i = 1, 2, 3)$ are the material constants.

---

[3] Recent experimental observation results (cf. Shockey et al. [37]) have shown that coalescence mechanism can be treated as nucleation and growth process on a smaller scale. This conjecture simplifies very much the description of the intrinsic micro–damage process by taking account only of the nucleation and growth mechanisms.

Finally the evolution equation for the porosity $\xi$ has the form

$$\begin{aligned}\dot{\xi} &= \frac{h^*(\xi,\vartheta)}{T_m}\left[\exp\frac{m^*(\vartheta)\mid \tilde{I}_n - \tau_n(\xi,\vartheta,\in^p)\mid}{k\vartheta} - 1\right]\\ &+\frac{g^*(\xi,\vartheta)}{T_m\sqrt{\kappa_0}}\left[\tilde{I}_g - \tau_{eq}(\xi,\vartheta,\in^p)\right]. \end{aligned} \tag{28}$$

This determines the evolution function $\Xi$.

### 3.5. KINEMATIC HARDENING

For a constitutive model describing the behaviour of a material under cyclic loading processes the crucial role plays the evolution equation for the back stress $\boldsymbol{\alpha}$, which is responsible for the description of the induced plastic strain anisotropy effects.

We shall follow some fundamental results obtained by Duszek and Perzyna [12]. Let us postulate

$$\mathrm{L}_{\boldsymbol{\upsilon}}\boldsymbol{\alpha} = \mathbf{A}(\mathbf{d}^p, \tilde{\boldsymbol{\tau}}, \vartheta, \xi). \tag{29}$$

Making use of the tensorial representation of the function $A$ and taking into account that there is no change of $\boldsymbol{\alpha}$ when $\tilde{\boldsymbol{\tau}} = 0$ and $\mathbf{d}^p = 0$ the evolution law (29) can be written in the form (cf. Truesdell and Noll [40])

$$\begin{aligned}\mathrm{L}_{\boldsymbol{\upsilon}}\boldsymbol{\alpha} &= \eta_1\mathbf{d}^p + \eta_2\tilde{\boldsymbol{\tau}} + \eta_3\mathbf{d}^{p^2} + \eta_4\tilde{\boldsymbol{\tau}}^2 + \eta_5\left(\mathbf{d}^p\cdot\tilde{\boldsymbol{\tau}} + \tilde{\boldsymbol{\tau}}\cdot\mathbf{d}^p\right)\\ &+\eta_6\left(\mathbf{d}^{p^2}\cdot\tilde{\boldsymbol{\tau}} + \tilde{\boldsymbol{\tau}}\cdot\mathbf{d}^{p^2}\right) + \eta_7\left(\mathbf{d}^p\cdot\tilde{\boldsymbol{\tau}}^2 + \tilde{\boldsymbol{\tau}}^2\cdot\mathbf{d}^p\right)\\ &+\eta_8\left(\mathbf{d}^{p^2}\cdot\tilde{\boldsymbol{\tau}}^2 + \tilde{\boldsymbol{\tau}}^2\cdot\mathbf{d}^{p^2}\right), \end{aligned} \tag{30}$$

where $\eta_1, \ldots, \eta_8$ are functions of the basic invariants of $\mathbf{d}^p$ and $\tilde{\boldsymbol{\tau}}$, the porosity parameter $\xi$ and temperature $\vartheta$.

A linear approximation of the general evolution law (30) leads to the result

$$\mathrm{L}_{\boldsymbol{\upsilon}}\boldsymbol{\alpha} = \eta_1\mathbf{d}^p + \eta_2\tilde{\boldsymbol{\tau}}. \tag{31}$$

This kinetic law represents the linear combination of the Prager and Ziegler kinematic hardening rules (cf. Prager [32] and Ziegler [44]).

To determine the connection between the material functions $\eta_1$ and $\eta_2$ we take advantage of the geometrical relation (cf. Duszek and Perzyna [12])

$$(\mathrm{L}_{\boldsymbol{\upsilon}}\boldsymbol{\alpha} - r\mathbf{d}^p) : \mathbf{Q} = 0, \tag{32}$$

where

$$\mathbf{Q} = \left[\frac{\partial f}{\partial \boldsymbol{\tau}} + \left(\frac{\partial f}{\partial \xi} - \frac{\partial \kappa}{\partial \xi}\right)\frac{\partial \xi}{\partial \boldsymbol{\tau}}\right]\left\|\frac{\partial f}{\partial \boldsymbol{\tau}} + \left(\frac{\partial f}{\partial \xi} - \frac{\partial \kappa}{\partial \xi}\right)\frac{\partial \xi}{\partial \boldsymbol{\tau}}\right\|^{-1}, \tag{33}$$

and $r$ denotes the new material function.

The relation (32) leads to the result

$$\eta_2 = \frac{1}{T_m}\langle\Phi(\frac{f}{\kappa} - 1)\rangle\,[r(\xi,\vartheta) - \eta_1]\,\frac{\mathbf{P}:\mathbf{Q}}{\tilde{\boldsymbol{\tau}}:\mathbf{Q}}. \tag{34}$$

Finally the kinematic hardening evolution law takes the form

$$\mathrm{L}_{\boldsymbol{\upsilon}}\boldsymbol{\alpha} = \frac{1}{T_m}\langle\Phi(\frac{f}{\kappa} - 1)\rangle\left[r_1(\xi,\vartheta)\mathbf{P} + r_2(\xi,\vartheta)\frac{\mathbf{P}:\mathbf{Q}}{\tilde{\boldsymbol{\tau}}:\mathbf{Q}}\tilde{\boldsymbol{\tau}}\right], \tag{35}$$

where

$$r_1(\xi,\vartheta) = \eta_1, \qquad r_2(\xi,\vartheta) = r - \eta_1. \tag{36}$$

It is noteworthy to add that the developed procedure can be used as general approach for obtaining various particular kinematic hardening laws. As an example let us assume that the evolution function $\mathbf{A}$ in (29) instead of $\mathbf{d}^p$ and $\tilde{\boldsymbol{\tau}}$ depends on $\mathbf{d}^p$ and $\boldsymbol{\alpha}$ only (cf. Agah–Tehrani et al. [2]). Then instead of (35) we obtain

$$\mathrm{L}_{\boldsymbol{\upsilon}}\boldsymbol{\alpha} = \frac{1}{T_m}\langle\Phi(\frac{f}{\kappa} - 1)\rangle\,[\zeta_1(\xi,\vartheta)\mathbf{P} - \zeta_2(\xi,\vartheta)\boldsymbol{\alpha}], \tag{37}$$

where

$$\zeta_1 = r_1, \qquad \zeta_2 = -r_2(\xi,\vartheta)\frac{\mathbf{P}:\mathbf{Q}}{\boldsymbol{\alpha}:\mathbf{Q}}. \tag{38}$$

When the infinitesimal deformations and rate independent response of a material are assumed and the intrinsic micro–damage effects are neglected then the kinematic hardening law (37) reduces to that proposed by Armstrong and Frederick [3].

The kinematic hardening law (37) leads to the nonlinear stress–strain relation with the characteristic saturation effect. The material function $\zeta_1(\xi,\vartheta)$ for $\xi = \xi_0$ and $\vartheta = \vartheta_0$ can be interpreted as an initial value of the kinematic hardening modulus while the material function $\zeta_2(\xi,\vartheta)$ determines the character of the nonlinearity of kinematic hardening. The particular forms of the functions $\zeta_1$ and $\zeta_2$ have to take into account the degradation nature of the influence of the intrinsic micro–damage process on the evolution of anisotropic hardening.

### 3.6. THERMODYNAMIC RESTRICTIONS AND RATE TYPE CONSTITUTIVE RELATIONS

Suppose the axiom of the entropy production holds. Then the constitutive assumption (15) and the evolution equations (20) lead to the results as follows

$$\begin{gathered}\boldsymbol{\tau} = \rho_{Ref}\frac{\partial\hat{\psi}}{\partial\mathbf{e}}, \qquad \eta = -\frac{\partial\hat{\psi}}{\partial\vartheta}, \\ -\frac{\partial\hat{\psi}}{\partial\boldsymbol{\mu}}\cdot\mathrm{L}_{\boldsymbol{\upsilon}}\boldsymbol{\mu} - \frac{1}{\rho\vartheta}\mathbf{q}\cdot\mathrm{grad}\vartheta \geq 0.\end{gathered} \tag{39}$$

The rate of internal dissipation is determined by

$$\begin{aligned}\vartheta\hat{i} &= -\frac{\partial\hat{\psi}}{\partial\boldsymbol{\mu}}\cdot \mathrm{L}_{\boldsymbol{\upsilon}}\boldsymbol{\mu}\\ &= -\left[\frac{\partial\hat{\psi}}{\partial\in^p}\sqrt{\frac{2}{3}}+\frac{\partial\hat{\psi}}{\partial\boldsymbol{\alpha}}:\left(r_1\mathbf{P}+r_2\frac{\mathbf{P}:\mathbf{Q}}{\tilde{\boldsymbol{\tau}}:\mathbf{Q}}\tilde{\boldsymbol{\tau}}\right)\right]\Lambda-\frac{\partial\hat{\psi}}{\partial\xi}\Xi. \qquad (40)\end{aligned}$$

Operating on the stress relation $(39)_1$ with the Lie derivative and keeping the internal state vector constant, we obtain (cf. Duszek–Perzyna and Perzyna [14])

$$\mathrm{L}_{\boldsymbol{\upsilon}}\boldsymbol{\tau}=\mathcal{L}^e:\mathbf{d}-\mathcal{L}^{th}\dot{\vartheta}-[(\mathcal{L}^e+\mathbf{g}\boldsymbol{\tau}+\boldsymbol{\tau}\mathbf{g}):\mathbf{P}]\frac{1}{T_m}\langle\Phi(\frac{f}{\kappa}-1)\rangle, \qquad (41)$$

where

$$\begin{aligned}\mathcal{L}^e &= \rho_{Ref}\frac{\partial^2\hat{\psi}}{\partial\mathbf{e}^2},\\ \mathcal{L}^{th} &= -\rho_{Ref}\frac{\partial^2\hat{\psi}}{\partial\mathbf{e}\partial\vartheta}. \qquad (42)\end{aligned}$$

Substituting $\dot{\psi}$ into the energy balance equation and taking into account the results $(39)_3$ and (40) gives

$$\rho\vartheta\dot{\eta}=-\mathrm{div}\mathbf{q}+\rho\vartheta\hat{i}. \qquad (43)$$

Operating on the entropy relation $(39)_2$ with the Lie derivative and substituting the result into (43) we obtain

$$\rho c_p\dot{\vartheta}=-\mathrm{div}\mathbf{q}+\vartheta\frac{\rho}{\rho_{Ref}}\frac{\partial\boldsymbol{\tau}}{\partial\vartheta}:\mathbf{d}+\rho\chi^*\boldsymbol{\tau}:\mathbf{d}^p+\rho\chi^{**}\dot{\xi}, \qquad (44)$$

where the specific heat

$$c_p=-\vartheta\frac{\partial^2\hat{\psi}}{\partial\vartheta^2} \qquad (45)$$

and the irreversibility coefficients $\chi^*$ and $\chi^{**}$ are determined by

$$\begin{aligned}\chi^* &= -\left[\left(\frac{\partial\hat{\psi}}{\partial\in^p}-\vartheta\frac{\partial^2\hat{\psi}}{\partial\vartheta\partial\in^p}\right)\sqrt{\frac{2}{3}}\right.\\ &\quad\left.+\left(\frac{\partial\hat{\psi}}{\partial\boldsymbol{\alpha}}-\vartheta\frac{\partial^2\hat{\psi}}{\partial\vartheta\partial\boldsymbol{\alpha}}\right):\left(r_1\mathbf{P}+r_2\frac{\mathbf{P}:\mathbf{Q}}{\tilde{\boldsymbol{\tau}}:\mathbf{Q}}\tilde{\boldsymbol{\tau}}\right)\right]\frac{1}{\boldsymbol{\tau}:\mathbf{P}}, \qquad (46)\\ \chi^{**} &= -\left(\frac{\partial\hat{\psi}}{\partial\xi}-\vartheta\frac{\partial^2\hat{\psi}}{\partial\vartheta\partial\xi}\right).\end{aligned}$$

## 4. Numerical solution of the initial–boundary value problem (evolution problem)

### 4.1. FORMULATION OF THE EVOLUTION PROBLEM

Find $\varphi$ as function of $t$ and $\mathbf{x}$ satisfying[4]

$$\left.\begin{array}{ll}\text{(i)} & \dot{\varphi} = \mathcal{A}(t,\varphi)\varphi + \mathbf{f}(t,\varphi);\\ \text{(ii)} & \varphi(0) = \varphi^0(\mathbf{x});\\ \text{(iii)} & \text{The boundary conditions.}\end{array}\right\} \tag{47}$$

The evolution problem (47) describes an adiabatic inelastic flow process provided

$$\varphi = \begin{bmatrix}\phi\\ \boldsymbol{v}\\ \rho\\ \boldsymbol{\tau}\\ \boldsymbol{\alpha}\\ \xi\\ \vartheta\end{bmatrix}, \quad \mathbf{f} = \begin{bmatrix} \boldsymbol{v}\\ 0\\ 0\\ -\frac{1}{T_m}\langle\Phi(\frac{f}{\kappa}-1)\rangle\left[\left(\mathcal{L}^e + \frac{\chi^*}{\rho c_p}\mathcal{L}^{th}\boldsymbol{\tau} + \mathbf{g}\boldsymbol{\tau} + \boldsymbol{\tau}\mathbf{g}\right):\mathbf{P}\right] - \frac{\chi^{**}\Xi}{\rho c_p}\mathcal{L}^{th}\\ \frac{1}{T_m}\langle\Phi(\frac{f}{\kappa}-1)\rangle\left[r_1(\xi,\vartheta)\mathbf{P} + r_2(\xi,\vartheta)\frac{\mathbf{P}:\mathbf{Q}}{\tilde{\boldsymbol{\tau}}:\mathbf{Q}}\tilde{\boldsymbol{\tau}}\right]\\ \Xi\\ \frac{1}{T_m}\langle\Phi(\frac{f}{\kappa}-1)\rangle\frac{\chi^*}{\rho c_p}\boldsymbol{\tau}:\mathbf{P} + \frac{\chi^{**}}{\rho c_p}\Xi \end{bmatrix},$$

$$\mathcal{A} = \begin{bmatrix} 0 & 0 & 0 & 0 & 0\ 0\ 0\\ 0 & 0 & \frac{\boldsymbol{\tau}}{\rho_{Ref}\rho}\text{grad} & \frac{1}{\rho_{Ref}}\text{div} & 0\ 0\ 0\\ 0 & -\rho\text{div} & 0 & 0 & 0\ 0\ 0\\ 0 & \mathbb{E}:\text{sym}\frac{\partial}{\partial\mathbf{x}} + 2\text{sym}(\boldsymbol{\tau}:\frac{\partial}{\partial\mathbf{x}}) & 0 & 0 & 0\ 0\ 0\\ 0 & 2\text{sym}(\boldsymbol{\alpha}:\frac{\partial}{\partial\mathbf{x}}) & 0 & 0 & 0\ 0\ 0\\ 0 & 0 & 0 & 0 & 0\ 0\ 0\\ 0 & \frac{\vartheta}{c_p\rho_{Ref}}\frac{\partial\boldsymbol{\tau}}{\partial\vartheta}:\text{sym}\frac{\partial}{\partial\mathbf{x}} & 0 & 0 & 0\ 0\ 0 \end{bmatrix}, \tag{48}$$

where

$$\mathbb{E} = \mathcal{L}^e - \frac{\vartheta}{c_p\rho_{Ref}}\mathcal{L}^{th}\frac{\partial\boldsymbol{\tau}}{\partial\vartheta}. \tag{49}$$

It is noteworthy that the spatial operator $\mathcal{A}$ has the same form as in thermo–elastodynamics while all dissipative effects generated by viscoplastic flow phenomena influence the process through the nonlinear function $\mathbf{f}$.

A strict solution of (47) with $\mathbf{f}(t,\varphi) \equiv 0$ (i.e. the homogeneous evolution problem) is defined as a function $\varphi(t) \in \mathrm{E}$ (a Banach space) such that

$$\varphi(t) \in \mathcal{D}(\mathcal{A}), \quad \text{for all} \quad t \in [0, t_f], \tag{50}$$

$$\lim_{\Delta t\to 0}\left\|\frac{\varphi(t+\Delta t) - \varphi(t)}{\Delta t} - \mathcal{A}\varphi(t)\right\|_{\mathrm{E}} = 0 \quad \text{for all} \quad t \in [0, t_f].$$

[4] We shall follow here some fundamental results which have been discussed in Richtmyer and Morton [34], Strang and Fix [39], Richtmyer [33], Dautray and Lions [10], Gustafsson, Kreiss and Oliger [15] and Łodygowski and Perzyna [19].

The boundary conditions are taken care of by restricting the domain $\mathcal{D}(\mathcal{A})$ to elements of E that satisfy those conditions; they are assumed to be linear and homogeneous, so that the set $\mathbf{S}$ of all $\varphi$ that satisfy them is a linear manifold; $\mathcal{D}(\mathcal{A})$ is assumed to be contained in $\mathbf{S}$.

The choice of the Banach space E, as well as the domain of $\mathcal{A}$, is an essential part of the formulation of the evolution problem.

### 4.2. WELL-POSEDNESS OF THE EVOLUTION PROBLEM

The homogeneous evolution problem (i.e. for $\mathbf{f} \equiv 0$) is called well posed (in the sense of Hadamard) if it has the following properties (cf. Richtmyer [33] and Hughes et al. [16]):

(i) The strict solutions are uniquely determined by their initial elements;
(ii) The set $Y$ of all initial elements of strict solutions is dense in the Banach space E;
(iii) For any finite interval $[0, t_0]$, $t_0 \in [0, t_f]$ there is a constant $K = K(t_0)$ such that every strict solution satisfies the inequality

$$||\varphi(t)|| \leq K||\varphi^0||, \quad \text{for} \quad 0 \leq t \leq t_0. \tag{51}$$

The inhomogeneous evolution problem (47) will be called well posed if it has a unique solution for all reasonable choices of $\varphi^0$ and $\mathbf{f}(t, \varphi)$ and if the solution depends continuously, in some sense, on those choices.

It is evident that any solution is unique, because of the uniqueness of the solutions of the homogeneous evolution problem. Namely, the difference of two solutions, for given $\varphi^0$ and given $\mathbf{f}(\cdot)$, is a solution on the homogeneous problem with zero as initial element, hence must be zero for all $t$.

It is possible to show (cf. Richtmyer [33]) that strict solutions exist for sets of $\varphi^0$ and $\mathbf{f}(\cdot)$ that are dense in E and $E_1$ (a new Banach space), respectively.

Let $\{\mathbb{F}_t^*; t \geq 0\}$ be a semi-group generated by the operator $\mathcal{A} + \mathbf{f}(\cdot)$ and $\{\mathbb{F}_t; t \geq 0\}$ be a semi-group generated by the operator $\mathcal{A}$.

Then we can write the generalized solution of the nonhomogenous evolution problem (47) in alternative forms

$$\begin{aligned} \varphi(t, \mathbf{x}) &= \mathbb{F}^*(t)\varphi^0(\mathbf{x}) \\ &= \mathbb{F}(t)\varphi^0(\mathbf{x}) + \int_0^t \mathbb{F}(t-s)\mathbf{f}(s, \varphi(s))\mathrm{d}s. \end{aligned} \tag{52}$$

The generalized solution of the nonhomogeneous evolution problem (47) in the form $(52)_2$ is the integral equation.

The successive approximations for $(52)_2$ are defined to be the functions $\varphi_0, \varphi_1, \ldots$, given by the formulas

$$\begin{aligned} \varphi_0(t) &= \varphi^0, \\ \varphi_{k+1}(t) &= \mathbb{F}(t)\varphi^0 + \int_0^t \mathbb{F}(t-s)\mathbf{f}(s, \varphi_k(s))\mathrm{d}s, \\ & \qquad k = 0, 1, 2, \ldots; \qquad t \in [0, t_f]. \end{aligned} \tag{53}$$

It is possible to show that these functions actually exist on $t \in [0, t_f]$ if the continuous function $\mathbf{f}$ is Lipschitz continuous with respect to the second argument uniformly with respect to $t \in [0, t_f]$. Then $(52)_2$ has unique solution (cf. Ionescu and Sofonea [17]).

### 4.3. APPLICATION OF THE FINITE DIFFERENCE METHOD

Let us consider the evolution problem in the form of (47). The actual configuration of the considered body at time $t$ is $\phi_t(\mathcal{B} \cup \partial\mathcal{B})$. Let us introduce in the configuration a regular difference net of nodes $(i, j, k)$ with coordinates $x_i^1 = i\Delta x^1$, $x_j^2 = j\Delta x^2$ and $x_k^3 = k\Delta x^3$, $i, j, k \in N$, where $N$ is a set of natural numbers. Of course, some of the nodes belong to $\partial\phi_t(\mathcal{B})$ and are used to approximate the boundary conditions. Time is approximated by a discrete sequence of moments $t_n = n\Delta t$, where $\Delta t$ is time step, $n \in N$.

For all functions $\varphi = \hat{\varphi}(\mathbf{x}, t)$ of the analysed problem we postulate the following approximation in the domain $\Delta\mathcal{S} = \Delta x^1 \times \Delta x^2 \times \Delta x^3$ of a difference mesh (cf. Fig. 1):

$$\begin{aligned} \varphi(\mathbf{x}, t) \cong \varphi_h(\mathbf{x}, t) &= \mathbf{a}_1(t) + \mathbf{a}_2(t)x^1 + \mathbf{a}_3(t)x^2 + \mathbf{a}_4(t)x^3 \\ &\quad \mathbf{a}_5(t)x^1x^2 + \mathbf{a}_6(t)x^1x^3 + \mathbf{a}_7(t)x^2x^3 + \mathbf{a}_8(t)x^1x^2x^3, \\ &\quad \mathbf{x} \in \Delta\mathcal{S}. \end{aligned} \tag{54}$$

The functions $\mathbf{a}_1(t), \ldots, \mathbf{a}_8(t)$ depend only on time, are determined by the value of the function $\varphi_w(t) = [\varphi_1(t), \ldots, \varphi_8(t)]^T$ in the node points of difference mesh, (cf. Fig. 1). Hence the approximation functions (54) can be written in the form

$$\varphi_h(\mathbf{x}, t) = \mathbf{N}(\mathbf{x})\varphi_w(t), \qquad \mathbf{x} \in \Delta\mathcal{S}, \tag{55}$$

where

$$N_1(\mathbf{x}) = q\left(-\Delta x^1 + 2x^1\right)\left(-\Delta x^2 + 2x^2\right)\left(\Delta x^3 - 2x^3\right),$$

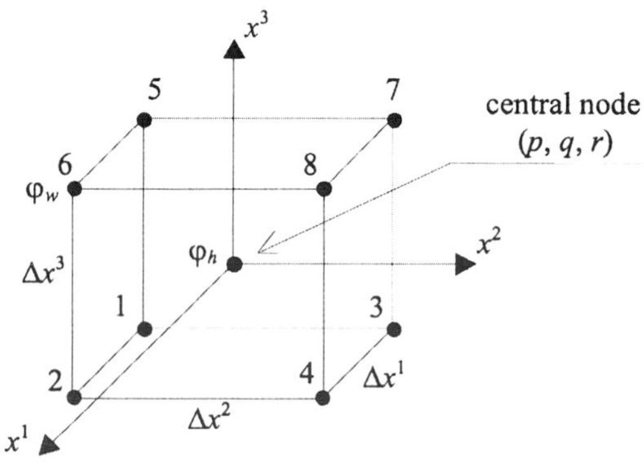


*Figure 1.* Fundamental finite difference mesh.

$$
\begin{aligned}
N_2(\mathbf{x}) &= q\left(\Delta x^1+2x^1\right)\left(-\Delta x^2+2x^2\right)\left(-\Delta x^3+2x^3\right),\\
N_3(\mathbf{x}) &= q\left(-\Delta x^1+2x^1\right)\left(\Delta x^2+2x^2\right)\left(-\Delta x^3+2x^3\right),\\
N_4(\mathbf{x}) &= q\left(\Delta x^1+2x^1\right)\left(\Delta x^2+2x^2\right)\left(\Delta x^3-2x^3\right),\\
N_5(\mathbf{x}) &= q\left(-\Delta x^1+2x^1\right)\left(-\Delta x^2+2x^2\right)\left(\Delta x^3+2x^3\right),\\
N_6(\mathbf{x}) &= q\left(\Delta x^1+2x^1\right)\left(\Delta x^2-2x^2\right)\left(\Delta x^3+2x^3\right),\\
N_7(\mathbf{x}) &= q\left(\Delta x^1-2x^1\right)\left(\Delta x^2+2x^2\right)\left(\Delta x^3+2x^3\right),\\
N_8(\mathbf{x}) &= q\left(\Delta x^1+2x^1\right)\left(\Delta x^2+2x^2\right)\left(\Delta x^3+2x^3\right),\\
q &= \frac{1}{8\Delta x^1\Delta x^2\Delta x^3}.
\end{aligned}
\tag{56}
$$

Equations (56) allow to determine values of the function $\varphi_h(\mathbf{x},t)$ in any point of the difference mesh, $\mathbf{x}\in\Delta\mathcal{S}$. For the central point $\mathbf{x}=\mathbf{x}_0$, $N_1=\cdots=N_8=\frac{1}{8}$ and $\varphi_h(t)=[\varphi_1(t)+\ldots+\varphi_8(t)]\frac{1}{8}$.

By using (55) we can determine the matrix of the difference operators which approximate the first partial derivatives of the function $\varphi(\mathbf{x},t)$ for $\mathbf{x}\in\Delta\mathcal{S}$. We have

$$
\frac{\partial}{\partial\mathbf{x}}\varphi(\mathbf{x},t)\cong\frac{\partial}{\partial\mathbf{x}}\varphi_h(\mathbf{x},t)=\frac{\partial}{\partial\mathbf{x}}\mathbf{N}(\mathbf{x})\varphi_w(t)=\mathbf{R}(\mathbf{x})\varphi_w(t). \tag{57}
$$

The matrix of the difference operator $\mathbf{R}(\mathbf{x})$ for the central point takes the form

$$
\begin{aligned}
\mathbf{R}(\mathbf{x}=\mathbf{x}_0) &= \left.\frac{\partial}{\partial\mathbf{x}}\mathbf{N}(\mathbf{x})\right|_{\mathbf{x}=\mathbf{x}_0}\\
&= \begin{bmatrix}
\frac{-1}{\Delta x^1} & \frac{1}{\Delta x^1} & \frac{-1}{\Delta x^1} & \frac{1}{\Delta x^1} & \frac{-1}{\Delta x^1} & \frac{1}{\Delta x^1} & \frac{-1}{\Delta x^1} & \frac{1}{\Delta x^1}\\
\frac{-1}{\Delta x^2} & \frac{-1}{\Delta x^2} & \frac{1}{\Delta x^2} & \frac{1}{\Delta x^2} & \frac{-1}{\Delta x^2} & \frac{-1}{\Delta x^2} & \frac{1}{\Delta x^2} & \frac{1}{\Delta x^2}\\
\frac{-1}{\Delta x^3} & \frac{-1}{\Delta x^3} & \frac{-1}{\Delta x^3} & \frac{-1}{\Delta x^3} & \frac{1}{\Delta x^3} & \frac{1}{\Delta x^3} & \frac{1}{\Delta x^3} & \frac{1}{\Delta x^3}
\end{bmatrix}.
\end{aligned}
\tag{58}
$$

In similar way we can find the difference form of the spatial difference operator $\mathcal{A}[\varphi(\mathbf{x},t),t]$ of the considered evolution problem (47)

$$
\mathcal{A}[\varphi(\mathbf{x},t),t]\varphi(\mathbf{x},t)\cong\mathcal{A}[\varphi_h(\mathbf{x},t),t]\mathbf{N}(\mathbf{x})\varphi_w(t)=\mathcal{R}[\varphi_h(\mathbf{x},t),\mathbf{x},t]\varphi_w(t), \tag{59}
$$

hence

$$
\mathcal{R}[\varphi_h(\mathbf{x},t),\mathbf{x},t]=\mathcal{A}[\varphi_h(\mathbf{x},t),t]\mathbf{N}(\mathbf{x})\quad\text{for}\quad\mathbf{x}\in\Delta\mathcal{S}. \tag{60}
$$

For the central node, $\mathbf{x}=\mathbf{x}_0$ the difference operator (60) depends only on time.

As a result of the proposed approximation of the evolution problem (47) with respect to the spatial variables we obtain a set of differential equations with respect to time and difference equations with respect to spatial variables

$$
\frac{\mathrm{d}}{\mathrm{d}t}\varphi_h(t)=\mathcal{R}[\varphi_h(t),t]\varphi_w(t)+\mathbf{f}[\varphi_h(t),t]. \tag{61}
$$

For the approximation of (61) with respect to time we use the evident scheme of the first order in the form

$$
\frac{\mathrm{d}}{\mathrm{d}t}\varphi_h(t)\cong\frac{\varphi_h^{n+1}-\varphi_h^n}{\Delta t}=\mathcal{R}(\varphi_h^n,t_n)\varphi^n+\mathbf{f}(\varphi_h^n,t_n). \tag{62}
$$

The solution of (62) is reduced to the realization of the recurrence relation

$$\varphi_h^{n+1} = \mathbf{C}_h^n \varphi_w^n + \Delta t \mathbf{f}_h^n. \tag{63}$$

The difference operator

$$\mathbf{C}_h^n = (\Delta t \mathcal{R}_h^n + \mathbf{N}) \tag{64}$$

couple dependent variables and various points of difference mesh.

In explicit finite difference scheme for a set of the partial differential equations (47)(i) of the hiperbolic type the condition of stability is the criterion of Courant–Friedrichs–Lewy

$$\Delta t_{n,n+1} \leq \min \left( \frac{\Delta L_{p,q,r}^n}{\left| c_{p,q,r}^n \right|} \right), \tag{65}$$
$$p = 1, 2, 3, \ldots, P; \quad q = 1, 2, 3, \ldots, Q; \quad r = 1, 2, 3, \ldots, R,$$

where $\Delta t_{n,n+1}$ denotes time step, $c_{p,q,r}^n$ denotes the velocity of the propagation of the disturbances in the vicinity of the central node $(p, q, r)$, $\Delta L_{p,q,r}^n$ is the minimum distance between the mesh nodes which are in the vicinity of the node (cf. Fig. 1).

### 4.4. THE LAX–RICHTMYER EQUIVALENCE THEOREM

We can now state the Lax–Richtmyer equivalence theorem (cf. Richtmyer and Morton [34], Strang and Fix [39], Dautray and Lions [10] and Gustafsson, Kreiss and Oliger [15]).

**Theorem.** Suppose that the evolution problem (47) is well–posed for $t \in [0, t_0]$ and that it is approximated by the scheme (63), which we assume consistent. Then the scheme is convergent if and only if it is stable.

The proof of the Lax–Richtmyer equivalence theorem for the case when the partial differential operator $\mathcal{A}$ in (47) is independent of $\varphi$ can be found in Dautray and Lions [10].

Remark. Let us consider the evolution problem (47) with

$$\mathbf{f}(t, \varphi) \neq 0 \tag{66}$$

and $\varphi^0 = 0$, and also the corresponding approximation (63). We have

$$\varphi_h^{n+1} = \Delta t \sum_{j=1}^{n} \left[ C_h(\Delta t) \right]^{n-j} \mathbf{f}_h^j. \tag{67}$$

If $\mathcal{A}$ is the infinitesimal generator of a semigroup $\{\mathbb{F}(t)\}$ we can write

$$\varphi(t) = \int_0^t \mathbb{F}(t-s)\mathbf{f}(s)\mathrm{d}s. \tag{68}$$

Under suitable hypotheses on the convergence of $\mathbf{f}_h^j$ to $\mathbf{f}(j\Delta t)$ we can show that expression (67) converges to (68) if the scheme is stable and consistent.

## 5. Particular examples

### 5.1. THE IDENTIFICATION PROCEDURE OF THE MATERIAL FUNCTIONS AND CONSTANTS

The plastic potential function $f$ is assumed in the form (cf. Perzyna [26] and Shima and Oyane [36])

$$f = \left\{ \tilde{J}_2' + [n_1(\vartheta) + n_2(\vartheta)\xi] \tilde{J}_1^2 \right\}^{\frac{1}{2}} \tag{69}$$

where

$$n_1(\vartheta) = 0, \qquad n_2(\vartheta) = \text{const.} \tag{70}$$

The isotropic work–hardening–softening function $\kappa$ is postulated as (cf. Perzyna [27,28] and Nemes and Eftis [22])

$$\begin{aligned} \kappa &= \hat{\kappa}(\in^p, \vartheta, \xi) \\ &= \{\kappa_s(\vartheta) - [\kappa_s(\vartheta) - \kappa_0(\vartheta)] \exp[-\delta(\vartheta) \in^p]\} \left[ 1 - \left( \frac{\xi}{\xi_F} \right)^{\beta(\vartheta)} \right], \end{aligned} \tag{71}$$

where

$$\begin{gathered} \kappa_s(\vartheta) = \kappa_s^* - \kappa_s^{**}\overline{\vartheta}, \qquad \kappa_0(\vartheta) = \kappa_0^* - \kappa_0^{**}\overline{\vartheta}, \\ \delta(\vartheta) = \delta^* - \delta^{**}\overline{\vartheta}, \qquad \beta(\vartheta) = \beta^* - \beta^{**}\overline{\vartheta}, \qquad \overline{\vartheta} = \frac{\vartheta - \vartheta_0}{\vartheta_0}. \end{gathered} \tag{72}$$

The overstress function $\Phi\left(\frac{f}{\kappa} - 1\right)$ is assumed in the form

$$\Phi\left(\frac{f}{\kappa} - 1\right) = \left(\frac{f}{\kappa} - 1\right)^m. \tag{73}$$

The evolution equations for the kinematic hardening parameter $\boldsymbol{\alpha}$ are assumed in the form (37) with

$$\zeta_1(\xi, \vartheta) = \zeta_1^* - \zeta_1^{**}\overline{\vartheta}, \qquad \zeta_2(\xi, \vartheta) = \zeta_2^* - \zeta_2^{**}\overline{\vartheta}. \tag{74}$$

The evolution equation for the porosity $\xi$ is postulated as

$$\dot{\xi} = \dot{\xi}_{grow} = \frac{g^*(\xi, \vartheta)}{T_m \sqrt{\kappa_0(\vartheta)}} \left[ \tilde{I}_g - \tau_{eq}(\xi, \vartheta, \in^p) \right] \tag{75}$$

where (cf. Dornowski [11])

$$\begin{aligned} g^*(\xi, \vartheta) &= c_1(\vartheta) \frac{\sqrt{\kappa_0(\vartheta)}}{\kappa_0(\vartheta)} \frac{\xi}{1 - \xi}, \\ \tilde{I}_g &= b_1 \tilde{J}_1 + b_2 \sqrt{\tilde{J}_2}, \end{aligned} \tag{76}$$

$$\tau_{eq}(\xi,\vartheta,\in^p) = c_2(\vartheta)(1-\xi)\ln\frac{1}{\xi}\left\{2\kappa_s(\vartheta) - [\kappa_s(\vartheta)-\kappa_0(\vartheta)]\,F(\xi_0,\xi,\vartheta)\right\},$$
$$c_1(\vartheta) = \text{const}, \qquad c_2(\vartheta) = \text{const},$$
$$F(\xi_0,\xi,\vartheta) = \left(\frac{\xi_0}{1-\xi_0}\frac{1-\xi}{\xi}\right)^{\frac{2}{3}\delta} + \left(\frac{1-\xi}{1-\xi_0}\right)^{\frac{2}{3}\delta}.$$

As in the infinitesimal theory of elasticity we assume linear properties of the material, i.e.

$$\mathcal{L}^e = 2\mu\mathbf{I} + \lambda(\mathbf{g}\otimes\mathbf{g}) \tag{77}$$

where $\mu$ and $\lambda$ denote the Lamé constants, and the thermal expansion matrix is postulated as

$$\mathcal{L}^{th} = (2\mu + 3\lambda)\theta\mathbf{g}, \tag{78}$$

where $\theta$ is the thermal expansion constant.

TABLE II. Material constants for OFHC copper

| | | | |
|---|---|---|---|
| $\kappa_s^* = 41.6$ MPa, | $\kappa_s^{**} = 12.1$ MPa, | $\kappa_0^* = 23.1$ MPa, | $\kappa_0^{**} = 6.9$ MPa, |
| $\delta^* = 28.0$, | $\delta^{**} = 8.1$, | $\beta^* = 5$, | $\beta^{**} = 1.5$, |
| $\vartheta_0 = 293$ K, | $\xi_F = 0.32$, | $\rho_{Ref} = 8930$ kg/m$^3$, | $\mu = 48.5$ GPa, |
| $\lambda = 138.2$ GPa, | $\theta = 18\cdot 10^{-6}$ K$^{-1}$, | $T_m = 4.6\ \mu$s, | $m = 2$, |
| $\zeta_1^* = 30$ GPa, | $\zeta_1^{**} = 8.7$ GPa | $\zeta_2^* = 500$ | $\zeta_2^{**} = 145$, |
| $c_1 = 1.5\cdot 10^{-3}$, | $c_2 = 6.2\cdot 10^{-2}$, | $b_1 = 0.50$ | $b_2 = 0.86$, |
| $\xi_0 = 3\cdot 10^{-4}$, | $\chi^* = 0.85$, | $\chi^{**} = 0$ | $c_p = 381$ J/kg K. |

## 5.2. DYNAMIC ADIABATIC AND ISOTHERMAL, CYCLIC LOADING PROCESSES FOR A CYLINDRICAL TENSILE BAR

A subject of the numerical analysis is a smooth cylindrical tensile bar. Adiabatic and isothermal, dynamic cyclic loading processes are considered. The material of a bar is OFHC copper, cf. Table II. We consider similar specimen as has been tested by Sidey and Coffin [38], cf. Fig. 2.

We consider the cyclic displacement constraint in the form of the three different loading characteristics (adiabatic and isothermal) in time namely slow–fast, equal–equal and fast–slow. All three types of constraint are represented by positive cycles, pulsating from zero and having the same amplitude $\overline{V}_{max} = 0.14$ mm and period $T_0 = 600\ \mu$s. They have different time for tensile deformation $T_t$ and compress deformation $T_c$ in a cycle. The tensile and compress constraints are described by the sine functions. The numerical results obtained by means of the finite difference method (cf. geometry and finite difference discretization of the specimen in Fig. 2) are presented in Figs. 3 and 4.

Stress–strain ($\sigma^{ZZ} - e_{ZZ}$) response in the point ($R = 0,\ \ Z = 49.7$) for various loading processes is presented in Fig. 3.

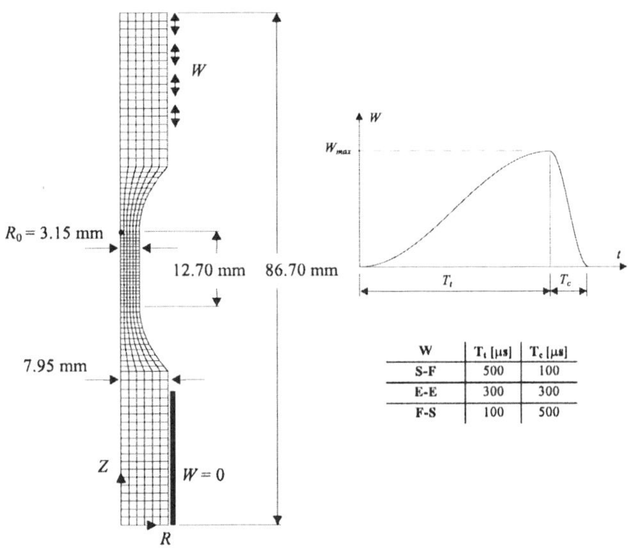


| W | $T_t$ [μs] | $T_c$ [μs] |
|---|---|---|
| S-F | 500 | 100 |
| E-E | 300 | 300 |
| F-S | 100 | 500 |

*Figure 2.* Geometry, finite difference discretization and kinematic constraints of the specimen.

The maximum stress $\sigma^{ZZ}$ versus number of cycles in the point $(R = 0,\ \ Z = 49.7)$ is plotted in Fig. 4. Different strain rate in a cycle for different kind of adiabatic process influences the character of changes of the considered stress.

Difference in amplitude of tensile stress for S–F and F–S loadings are implied by viscosity of the material and microdamage mechanism. For all cases of loadings the effect of plastic hardening is characteristic. The saturation of hardening is first observed for S–F loading process (after 15 cycles), cf. Fig. 4. Next the amplitude is decreasing due to softening generated by microdamage mechanism. The conclusion can be drown that cycles with longer time of applied tensile stress lead sooner to the softening of the material.

Evolution of damage in the point $(R = 0,\ \ Z = 49.7)$ for varoius loading processes is presented in Fig. 5. This evolution of demage is very different for various kind of loadings.

In Fig. 6 the evolution of the damage $\xi_g$ in the point $(R = 0,\ \ Z = 49.7)$ for different forms of the stress intensity invariant (adiabatic process) and for various loading processes are showed.

Distributions of equivalent plastic deformation, damage $\xi_{grow}$ and temperature along the gauge section of the specimen after 50 cycles of the slow–fast adiabatic process are presented in Fig. 7.

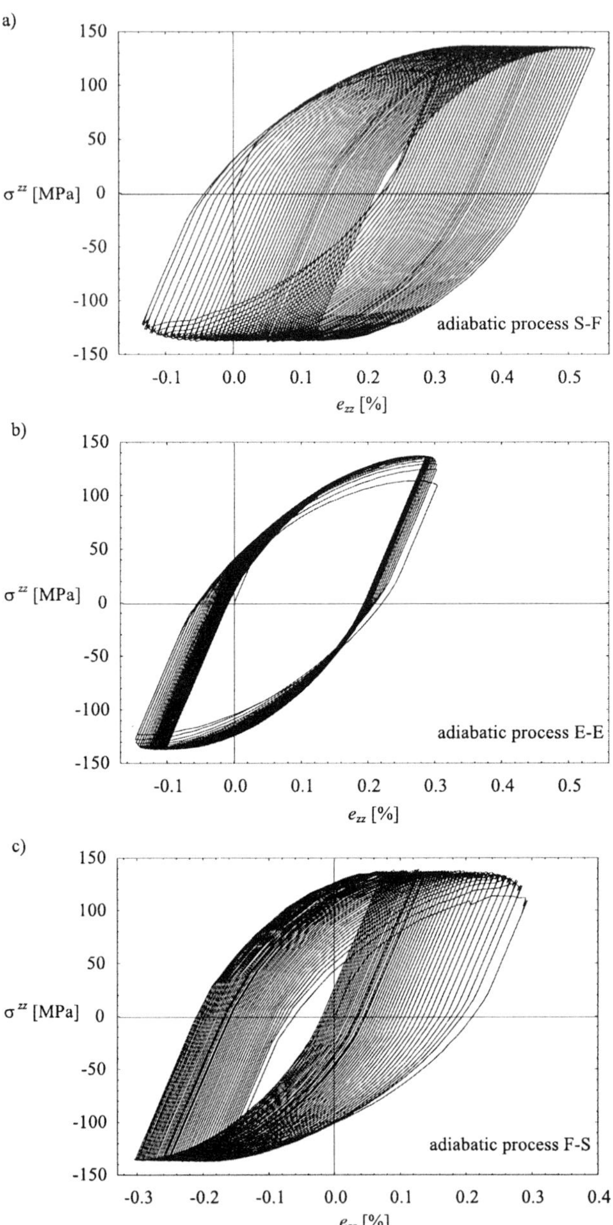


*Figure 3.* Stress-strain ($\sigma^{ZZ} - e_{ZZ}$) response for various loading processes ($R = 0$, $Z = 49.7$); a) S–F, b) E–E, c) F–S.

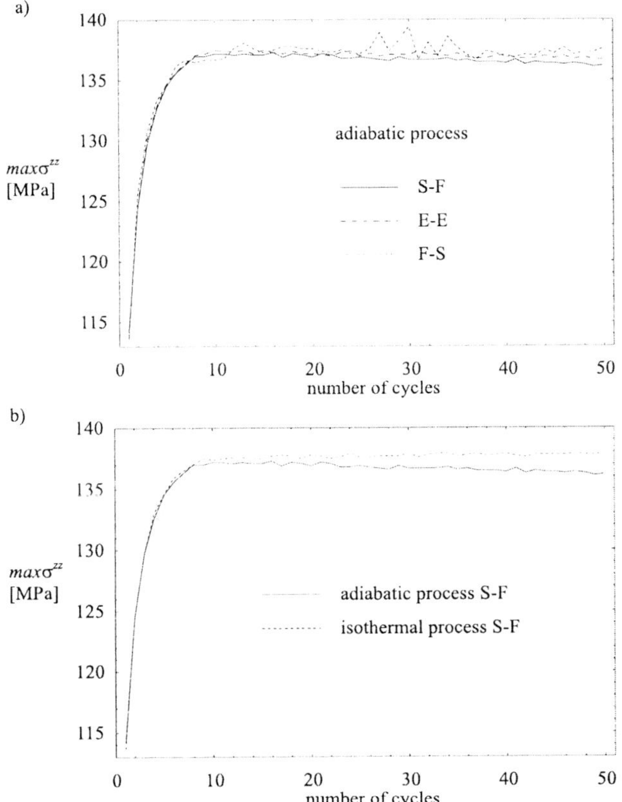


*Figure 4.* Maximum stress $\sigma^{ZZ}$ per cycle as a function of number of cycles ($R = 0$, $Z = 49.7$); a) effect of the wave shape (adiabatic process), b) effect of the temperature in the slow–fast process.

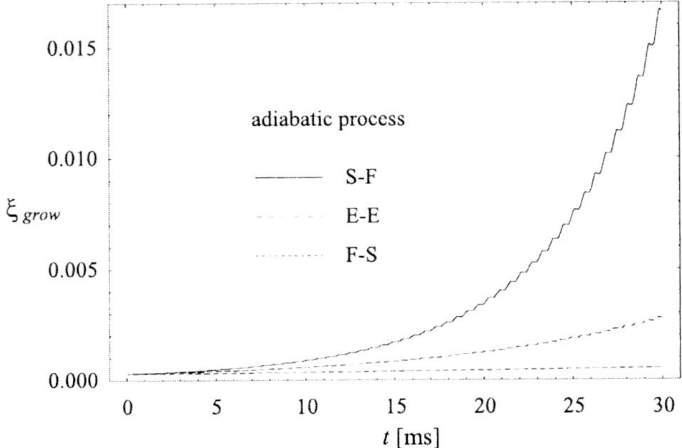


*Figure 5.* Evolution of damage $\xi_{grow}$, ($R = 0$, $Z = 49.7$).

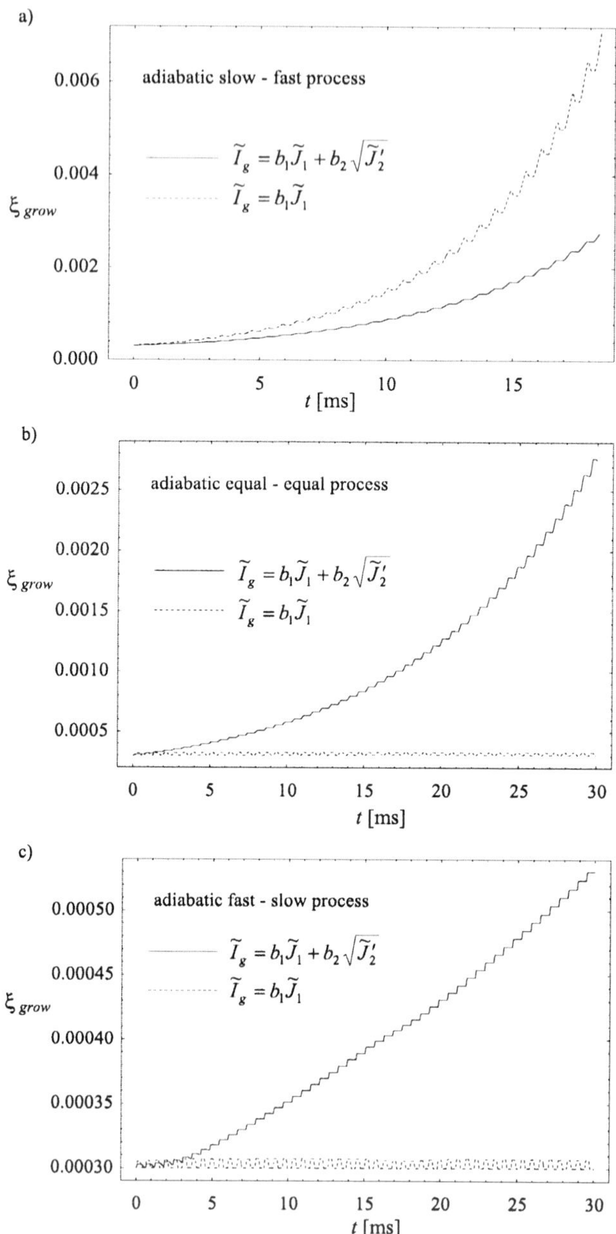


*Figure 6.* Evolution of the damage $\xi_g$ ($R = 0$, $Z = 49.7$) for different forms of the stress intensity invariant (adiabatic process) for various loading processes.

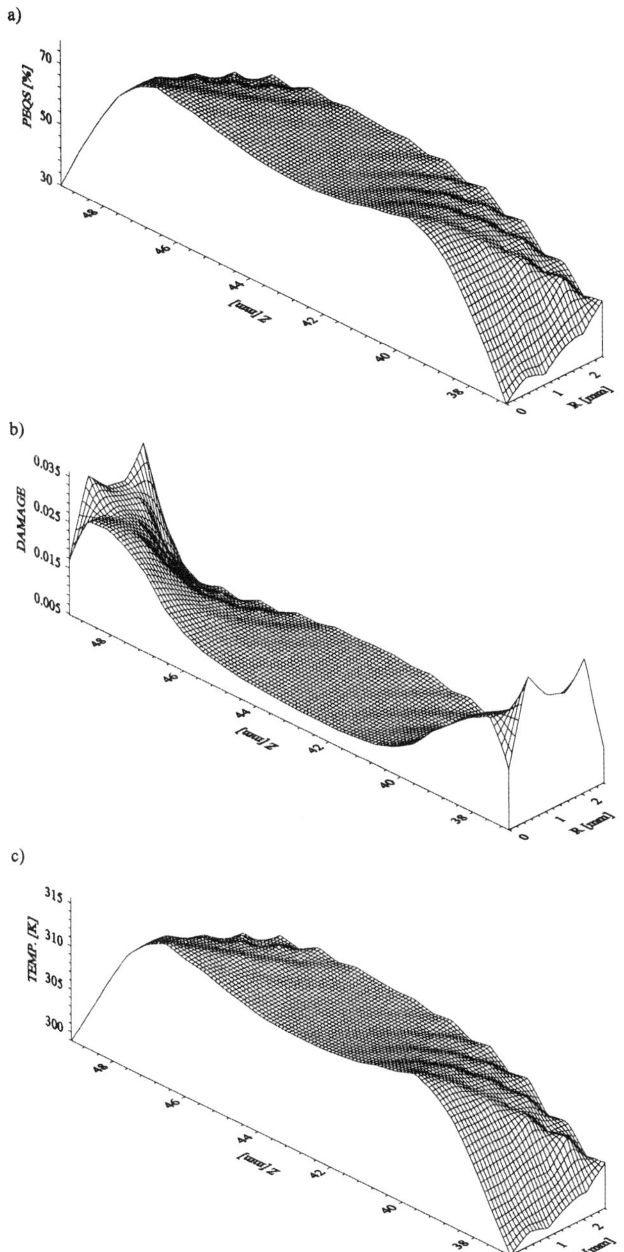


*Figure 7.* Distributions along the gauge section, after 50 cycles of the slow–fast adiabatic process; a) plastic equivalent strain, b) damage $\xi_{grow}$, c) temperature.

## 6. Final comments

It is noteworthy that the theory of thermo–elasto–viscoplasticity developed in this paper has been inspired by the brilliant experimental investigations presented by Sidey and Coffin [38]. This experimental work has brought deep understanding of the intrinsic microdamage mechanism during dynamic cyclic loading process and clearly has shown that cycle fatigue damage mechanism at high temperature of metals does very much depend on the strain rate and wave shape effects. We hope that a new constitutive model proposed is sufficiently simple in its nature that it can be applicable to the numerical solution of initial–boundary value problems under cyclic loadings.

The crucial idea in this theory is the very efficient interpretation of a finite set of the internal state variables as the equivalent plastic deformation, volume fraction porosity and the residual stress (the back stress). To describe suitably the time and temperature dependent effects observed experimentally and the accumulation of the plastic deformation and damage during dynamic cyclic loading process the kinetics of microdamage and the kinematic hardening law have been modified. The performed numerical simulations of the dynamic, cyclic loading process have proven the usefulness of the thermo–elasto–viscoplastic theory. Since the rate independent plastic response can be obtained as the limit case when the relaxation time is equal to zero hence the theory of viscoplasticity offers the regularization procedure for the solution of the dynamical initial–boundary problems under cyclic loadings. Due to that we have possibility to investigate numerically the fatigue damage. The viscoplastic regularization procedure assures the stable integration algorithm by using the finite difference method. Convergence, consistency, and stability of the discretised problem are discussed. The Lax–Richtmyer equivalence theorem is formulated and conditions under which this theorem is valid are examined. The accumulation of damage and equivalent plastic deformation on each considered cycle has been obtained. It has been numerically found that accumulation of microdamage distinctly depends on the wave shape of the assumed loading cycle.

*Acknowledgment*
The paper has been partly prepared within the research programme sponsored by the Committee of Scientific Research under Grant 7 T07 A013 10.

## REFERENCES

1. Abraham, R., Marsden, J.E. and Ratiu, T. (1988) *Manifolds, Tensor Analysis and Applications*, Springer, Berlin.
2. Agah–Tehrani, A., Lee, E.H., Malett, R.L. and Onat, E.T. (1987) The theory of elastic–plastic deformation at finite strain with induced anisotropy modelled isotropic–kinematic hardening, *J. Mech. Phys. Solids* **35**, 43–60.

3. Armstrong, P.J. and Frederick, C.O. (1966) A mathematical representation of the multiaxial Baushinger effect, *CEGB Report RD/B/N731*, Central Electricity Generating Board.
4. Auricchio, F. and Taylor, R.L. (1995) Two material models for cyclic plasticity models: Nonlinear kinematic hardening and generalized plasticity, *Int. J. Plasticity* **11**, 65-98.
5. Auricchio, F., Taylor, R.L. and Lubliner, J. (1992) Application of a return map algorithm to plasticity models, in D.R.J.Owen and E.Onate (eds.), *COMPLAS Computational Plasticity: Fundamentals and Applications*, Barcelona, pp. 2229–2248.
6. Chaboche, J.L. (1986) Time – independent constitutive theories for cyclic plasticity, *Int. J. Plasticity* **2**, 149–188.
7. Coleman, B.D. and Noll, W. (1963) The thermodynamics of elastic materials with heat conduction and viscosity, *Arch. Rational Mech. Anal.* **13**, 167–178.
8. Curran, D.R., Seaman, L. and Shockey, D.A. (1987) Dynamic failure of solids, *Physics Reports* **147**, 253–388.
9. Dafalias, Y.F. and Popov, E.P. (1976) Plastic internal variable formalism of cyclic plasticity, *J. Appl. Mech.* **43**, 645–651.
10. Dautray, R. and Lions, J.-L. (1993) *Mathematical Analysis and Numerical Methods for Science and Technology*, Vol. 6. *Evolution Problems II*, Springer, Berlin.
11. Dornowski, W. (1998) The influence of finite deformation on the mechanism of microdamage in structural metals, *Biuletyn WAT* (in print).
12. Duszek, M.K. and Perzyna, P. (1991a) On combined isotropic and kinematic hardening effects in plastic flow processes, *Int. J. Plasticity* **7**, 351–363.
13. Duszek, M.K. and Perzyna, P. (1991b) The localization of plastic deformation in thermoplastic solids, *Int. J. Solids Structures* **27**, 1419–1443.
14. Duszek-Perzyna, M.K. and Perzyna, P. (1994) Analysis of the influence of different effects on criteria for adiabatic shear band localization in inelastic solids, in R.C.Batra and H.M.Zbib (eds.), *Material Instabilities: Theory and Applications*, ASME Congress, Chicago, 9–11 November 1994, AMD–Vol. 183/MD–Vol.50, ASME, New York, pp. 59–85.
15. Gustafsson, B., Kreiss, H.O. and Oliger, J. (1995) *Time Dependent Problems and Difference Methods*, John Wiley, New York.
16. Hughes T.J.R., Kato T. and Marsden J.E. (1977) Well-posed quasilinear second order hyperbolic system with application to nonlinear elastodynamics and general relativity, *Arch. Rat. Mech. Anal.* **63**, 273–294.
17. Ionescu, I.R. and Sofonea, M. (1993) *Functional and Numerical Methods in Viscoplasticity*, Oxford.
18. Johnson, J.N. (1981) Dynamic fracture and spallation in ductile solids, *J. Appl. Phys.* **52**, 2812–2825.
19. Łodygowski, T. and Perzyna, P. (1997) Localized fracture of inelastic polycrystalline solids under dynamic loading processes, *Int. J. Damage Mechanics* **6**, 364–407.
20. Marsden, J.E. and Hughes, T.J.R. (1983) *Mathematical Foundations of Elasticity*, Prentice–Hall, Englewood Cliffs, New York.
21. Mróz, Z. (1967) On the description of anisotropic workhardening, *J. Mech. Phys. Solids* **15**, 163–175.
22. Nemes, J. A. and Eftis, J. (1993) Constitutive modelling on the dynamic fracture of smooth tensile bars, *Int. J. Plasticity* **9**, 243–270.
23. Oldroyd, J. (1950) On the formulation of rheological equations of state, *Proc. Roy. Soc. (London)* **A 200**, 523–541.

24. Perzyna, P. (1963) The constitutive equations for rate sensitive plastic materials, *Quart. Appl. Math.*, **20**, 321–332.
25. Perzyna, P. (1971) Thermodynamic theory of viscoplasticity, *Advances in Applied Mechanics* **11**, 313–354.
26. Perzyna, P. (1984) Constitutive modelling of dissipative solids for postcritical behaviour and fracture, *ASME J. Eng. Materials and Technology* **106**, 410-419.
27. Perzyna, P. (1986a) Internal state variable description of dynamic fracture of ductile solids, *Int. J. Solids Structures* **22**, 797-818.
28. Perzyna, P. (1986b) Constitutive modelling for brittle dynamic fracture in dissipative solids, *Arch. Mechanics* **38**, 725–738.
29. Perzyna, P. (1995) Interactions of elastic–viscoplastic waves and localization phenomena in solids, in L.J. Wegner and F.R. Norwood (eds.), *Nonlinear Waves in Solids*, Proc. IUTAM Symposium, August 15–20, 1993, Victoria, Canada, ASME Book No AMR 137, pp. 114–121.
30. Perzyna, P. and Drabik, A. (1989) Description of micro–damage process by porosity parameter for nonlinear viscoplasticity, *Arch. Mechanics* **41**, 895–908.
31. Perzyna, P. and Drabik, A. (1999) Micro–damage mechanism in adiabatic processes, *Int. J. Plasticity* (submitted for publication).
32. Prager, W. (1955) The theory of plasticity: A survey of recent achievements, (J. Clayton Lecture), *Proc. Inst. Mech. Eng.* **169**, 41–57.
33. Richtmyer, R.D. (1978) *Principles of Advance Mathematical Physics*, Vol. I, Springer, New York.
34. Richtmyer, R.D. and Morton, K.W. (1967) *Difference Methods for Initial–Value Problems*, John Wiley, New York.
35. Ristinmaa, M. (1995) Cyclic plasticity model using one yield surface only, *Int. J. Plasticity* **11**, 163–181.
36. Shima, S. and Oyane, M. (1976) Plasticity for porous solids, *Int. J. Mech. Sci.* **18**, 285–291.
37. Shockey, D.A., Seaman, L. and Curran, D.R. (1985) The microstatistical fracture mechanics approach to dynamic fracture problem, *Int. J. Fracture* **27**, 145–157.
38. Sidey, D. and Coffin, L.F. (1979) Low–cycle fatigue damage mechemism at high temperature, in J.T.Fong (ed.), *Fatigue Mechanism*, Proc. ASTM STP 675 Symposium, Kansas City, Mo., May 1978, Baltimore, pp. 528–568.
39. Strang, G. and Fix, G.J. (1973) *An Analysis of the Finite Element Method*, Prentice–Hall, Englewood Cliffs.
40. Truesdell C. and Noll W. (1965) *The nonlinear field theories*, Handbuch der Physik, Band III/3, Springer, Berlin, pp. 1–579.
41. Van der Giessen, E. (1989) Continuum models of large deformation plasticity, Part I: Large deformation plasticity and the concept of a natural reference state, *Eur. J. Mech., A/Solids* **8**, 15.
42. Van der Giessen, E. (1989) Continuum models of large deformation plasticity, Part II: A kinematic hardening model and the concept of a plastically induced orientational structure, *Eur. J. Mech., A/Solids* **8**, 89.
43. Wang, J.–D. and Ohno, N. (1991) Two equivalent forms of nonlinear kinematic hardening: application to nonisothermal plasticity, *Int. J. Plasticity* **7**, 637–650.
44. Ziegler, H. (1959) A modification of Prager's hardening rule, *Quart. Appl. Math.* **17**, 55–65.

# SUBDOMAIN BOUNDING TECHNIQUE FOR SHAKEDOWN ANALYSIS OF STRUCTURES

Y. LIU, Z. CEN and Y. XU
*Department of Engineering Mechanics, Tsinghua University, Beijing, 100084, China*

## Abstract

A bounding technique is proposed to handle non-linear mathematical programming for shakedown analysis of structures. The idea is to divide the original structure into several substructures, each of which can be dealt with respectively. Upper and lower bounds of the limit load factor of shakedown for the original structure can be obtained from similar load factors for each of the substructures. The Melan's theorem and hybrid stress finite element are used in the programming formulaton and numerical discretization. An elastic-perfectly plastic material is considered here. Numerical applications illustrate the validity of the present technique.

## Notation

| | |
|---|---|
| $\lambda_s$ | limit load factor of shakedown for the original structure |
| $\varphi$ | yield function, $\varphi(\sigma_{ij})=\frac{1}{2}(\sigma_{ij}-\frac{1}{3}\sigma_{kk}\delta_{ij})(\sigma_{ij}-\frac{1}{3}\sigma_{kk}\delta_{ij})-\sigma_s^2$ |
| $\sigma_{ij}^0$ | elastic stress field with load factor $\lambda=1$ |
| $\rho_{ij}$ | self-equilibrium stress field |
| $\omega(U)$ | the set of first kind of self-equilibrium stress field in $U$ |
| $\Omega(U)$ | the set of second kind of self-equilibrium stress field in $U$ |
| $V_k$ | the $K$-th subdomain or substructure of the body $V$ |
| $\lambda_k$ | the optimum load factor of shakedown for the $K$-th subdomain $V_k$ when $\rho_{ij}$ is restricted to $\omega(V_k)$ |
| $\mu_k$ | the optimum load factor of shakedown for the $K$-th subdomain $V_k$ when $\rho_{ij}$ is restricted to $\Omega(V_k)$ |
| $u_i$, $\delta u_i$ | the vector of nodal displacement and its virtual variation |



*D. Weichert and G. Maier (eds.), Inelastic Analysis of Structures under Variable Loads*, 95–105.

## 1. Introduction

In recent years, shakedown analysis of elastic-plastic structures has increasingly gained importance in certain engineering situations concerning civil and mechanical engineering, nuclear power plants, the chemical industry, energy conversion and future fusion technologies. Many structures in engineering problems are subjected to variable mechanical loads, temperature field and other sources varying between certain prescribed limits. The shakedown is a necessary condition for safety of such kinds of structures. If after every possible loading path inside the given load domain the plastic deformation will cease to develop and a structure will exhibit a pure elastic response, then the structure will shake down. If not, the structure will fail by one or both of two failure modes, namely, alternating plasticity and incremental collapse. The foundations of shakedown theories were given by Melan (1938) and Koiter (1960), who derived sufficient criteria for shakedown and non-shakedown, respectively, of elastic-perfectly plastic bodies.

One purpose of shakedown analysis is to find the limit range of cyclic loading within which an elastic-plastic body shakes down to an elastic state, i.e. no further plastic deformation takes place after several cycles of loading. In the context of computational mechanics such a shakedown problem can be treated as an optimization problem by using FEM. Generally, the dimension of the optimization problem via FEM is very large.

Now we consider that a body is subjected to a single parameter load $P_i(x,t)=\lambda(t)P_i^0(x)$. In terms of Melan's theorem (1938), a mathematical programming problem can be formulated as:

$$\begin{aligned} &\text{Maximize} && \lambda_s \\ &\text{Subject to} && \varphi[\lambda\sigma_{ij}^0(x)+\rho_{ij}(x)]\le 0, \forall\lambda\in[0,\lambda_s], x\in V \end{aligned} \tag{1}$$

where $\sigma_{ij}^0(x)$ is the elastic solution for $\lambda(t)=1$, and $\rho_{ij}(x)$ belongs to the set of self-equilibrium stress fields in the body $V$ and plays the role of variables in the mathematical programming (MP) formulation.

The main reason why numerical applications of the classical shakedown theory (direct approach) are relatively rare, is the fact that these approaches suffer from problems of computational efficiency. The direct approach leads to a problem of mathematical optimization, which requires a large amount of computer memory for complex structures. Furthermore, for nonlinear yield conditions, the solution of the respective nonlinear optimization problem often requires highly iterative procedures and is therefore very time-consuming. This results from the fact, that the nonlinear programming approaches imply adopting solution schemes based more on purely mathematical considerations than on the physics of the problem, thus often becoming unnecessarily expensive. Belytschko (1972) solved the shakedown problem by means of equilibrated plane stress finite elements and nonlinear mathematical programming techniques. Corradi and Zavelani (1974) found two static formulations in linear programming, from which the relevant dual kinematic versions are obtained via duality properties. Stein and Zhang (1992) used a reduced base technique to solve the static shakedown problem. O.Debordes (1978) first used subdomain techniques in shakedown analysis. The aim of the present study is to propose a subdomain bounding technique to deal with the mathematical programming problems for shakedown analysis of complex structures.

## 2. Subdomain Bounding Theorem

**Definition 1:** A stress field $\rho_{ij}(x)$ in domain $U$ belongs to the first kind of self-equilibrium stress field in domain $U$ and is denoted as: $\rho_{ij}(x) \in \omega(U)$, if:

$$\text{i)} \quad \rho_{ij,j} = 0 \quad \text{in} \quad U \tag{2}$$

$$\text{ii)} \quad \rho_{ij} n_j = 0 \quad \text{on} \quad \partial U \tag{3}$$

**Definition 2:** A stress field $\rho_{ij}(x)$ in domain $U$ belongs to the second kind of self-equilibrium stress field in domain $U$ and is denoted as: $\rho_{ij}(x) \in \Omega(U)$, if:

$$\text{i)} \quad \rho_{ij,j} = 0 \quad \text{in} \quad U \tag{4}$$

$$\text{ii)} \quad \iint_{\partial U} \rho_{ij} n_j ds = 0 \tag{5}$$

**Theorem:** If $\lambda_s$ is the shakedown limit load factor of the domain $V$, as defined by the optimum objective value of (1), and the domain $V$ is divided into $n$ subdomains $V_1$, $V_2$, $\cdots$, $V_n$, then:

$$\mathrm{Min}\{\lambda_1, \lambda_2, \cdots, \lambda_n\} \le \lambda_s \le \mathrm{Min}\{\mu_1, \mu_2, \cdots, \mu_n\} \tag{6}$$

where $\lambda_1, \lambda_2, \cdots, \lambda_n$ and $\mu_1, \mu_2, \cdots, \mu_n$ are defined as the optimum objective values of the following MP problems which are formulated in each subdomain independently:

$$\begin{aligned} &\underset{\rho_{ij} \in \omega(V_k)}{\text{Maximize}} \quad \lambda_k \\ &\text{subject to} \quad \varphi[\lambda \sigma_{ij}^0(x) + \rho_{ij}(x)] \le 0, \forall \lambda \in [0, \lambda_k], x \in V_k \end{aligned} \tag{7}$$

$$\begin{aligned} &\underset{\rho_{ij} \in \Omega(V_k)}{\text{Maximize}} \quad \mu_k \\ &\text{subject to} \quad \varphi[\mu \sigma_{ij}^0(x) + \rho_{ij}(x)] \le 0, \forall \mu \in [0, \mu_k], x \in V_k \end{aligned} \tag{8}$$

**Proof**: Let $\rho_{ij}(x)$ in each subdomain be the solution field of problem (7) for $x$ belonging to that domain, then there is

$$\varphi\left[\lambda\sigma_{ij}^{o}(x)+\rho_{ij}(x)\right]\leq 0 \quad \text{for} \quad x\in V, \ \lambda\in[0,\lambda_m], \tag{9}$$

where

$$\lambda_m = \text{Min}\{\lambda_1,\lambda_2,\ldots,\lambda_n\}, \tag{10}$$

thus we have $\lambda_s \geq \lambda_m$.

If the body $V$ can shake down and the corresponding self-equilibrium stress field is $\tau_{ij}(x)$ which is denoted by $\tau_{ij}^{(k)}(x)$ in the subdomain $V_k(k=1, 2, ..., n)$, it is obvious that

$$\iint_{\partial V_k} \tau_{ij}^{(k)} n_j ds = 0\,, \tag{11}$$

thus

$$\tau_{ij}^{(k)}(x)\in\Omega(V_k), \tag{12}$$

it follows from the definition that

$$\mu_k \geq \lambda_s (k=1,2,\ldots,n), \tag{13}$$

thus

$$\lambda_s \leq \mu_m = \text{Min}\{\mu_1,\mu_2,\ldots,\mu_n\}. \tag{14}$$

Q. E. D.

## 3. The Hybrid Finite Element Discretization

The fields $\sigma_{ij}^{0}(x)$ and $\rho_{ij}(x)$ in formulation (1) are constructed here by the hybrid stress finite element discretization . $\sigma_{ij}^{0}(x)$ is obtained by the well-known procedure of elastic analysis. $\rho_{ij}(x)$ is constructed by the same stress mode in each element as that for $\sigma_{ij}^{0}(x)$ , that is:

$$\rho = P\alpha \tag{15}$$

where $\boldsymbol{\alpha}$ is the vector of self-equilibrium stress parameters for each element and $\boldsymbol{P}$ is the interpolation function matrix.

The self-equilibrium requirement (3) for the first kind of fields and (5) for the second kind of fields are satisfied in a variational sense here. From the principle of virtual work we have the following equation for the self-equilibrium field $\rho_{ij}(x)$ :

$$\iiint_V \rho_{ij} \frac{1}{2} \delta[u_{i,j} + u_{j,i}] dV = 0$$

which can be transformed into: $$\iint_{\partial V} \rho_{ij} n_j \delta u_i ds = 0$$

or in a discretized form: $$\sum_{k=1}^{n_e} \iint_{\partial V_k} \rho_{ij} n_j \delta u_i ds = 0 \tag{16}$$

If $\delta u_i$ is interpolated by its values in the nodes of the FE mesh and if $\delta u_i$ for each node is independent, then Eqs. (15) and (16) provide the first kind of self-equilibrium fields: $\rho_{ij}(x) \in \omega(V)$.

If $\delta u_i$ is independent in each inner node of the domain and constitutes a virtual rigid motion on boundary, then Eqs. (15) and (16) provide the second kind of self-equilibrium fields: $\rho_{ij}(x) \in \Omega(V)$. This is illustrated by an axisymmetric problem as follows.

If $\delta u_i$ is independent in each node, we will obtain a linear constraint equation from Eq. (16)

$$\begin{bmatrix} a_{11} & \cdots & a_{1n} \\ a_{21} & \cdots & a_{2n} \\ \vdots & & \vdots \\ a_{m1} & \cdots & a_{mn} \end{bmatrix} \begin{Bmatrix} \beta_1 \\ \beta_2 \\ \vdots \\ \beta_n \end{Bmatrix} = \begin{Bmatrix} 0 \\ 0 \\ \vdots \\ 0 \end{Bmatrix} \tag{17}$$

Now, if we replace all the rows corresponding to the virtual displacements of boundary nodes in $z$ direction in Eq. (16) by one row which is the sum of these rows, then a linear constraint equation follows from Eq. (17)

$$\begin{bmatrix} b_{11} & \cdots & b_{1n} \\ b_{21} & \cdots & b_{2n} \\ \vdots & & \vdots \\ b_{11} & \cdots & b_{1n} \end{bmatrix} \begin{Bmatrix} \beta_1 \\ \beta_2 \\ \vdots \\ \beta_n \end{Bmatrix} = \begin{Bmatrix} 0 \\ 0 \\ \vdots \\ 0 \end{Bmatrix} \tag{18}$$

Eqs. (15) and (18) will give a self-equilibrium field of the second kind, because the unique degree of freedom of rigid motion in axisymmetric problems is the dislocation along the $z$ direction.

If formulation (1) is not a large scale mathematical programming, we can solve it directly by the available MP algorithm. For the prescribed numerical examples, we have adopted the axisymmetric hybrid stress element AXH9C (Spilker, 1981). The yield function $\varphi$ is averaged at each element, like in Hung (1976). A general nonlinear programming code SCDD (the Synthesized Constrained Dual-Descent method, Wan, 1983) is applied here.

## 4. Numerical Examples

In this section we apply the above subdomain bounding technique to solve some examples and if possible, compare the calculated results with those obtained by other methods.

(1) A storage energy vessel under variable internal pressure(0 -- P)

The pressure vessel is shown in Fig.1. The section under consideration for numerical analysis is divided into four subdomains, with 39 elements in all, as shown in Fig.2. The Young's modulus is $E = 2.1\times10^5$ MPa, the Poisson's ratio is $\nu$=0.28 and the yield stress of material is $\sigma_s$=900.00MPa. In order to calculate the elastic stress for $\lambda(t)=1, P=P^0$, we set the reference load $P^0$=0.032MPa, then we obtain $\lambda_4 = 2404.5$, $\mu_4 = 2513.5$, and we find that, with the load factor $\lambda(t) = \mu_4$, the subdomains $V_1, V_2$ and $V_3$ are still in the elastic range:

$$\varphi\left[\lambda\sigma_{ij}^0(x)\right] \le 0$$
$$\forall\lambda \in [0, \mu_4]$$
$$\mathbf{x} \in [V_1, \ V_2, \ V_3]$$

Thus, we have

$$\mu_1 \ge \lambda_1 \ge \mu_4$$
$$\mu_2 \ge \lambda_2 \ge \mu_4$$
$$\mu_3 \ge \lambda_3 \ge \mu_4$$
$$\mathrm{Min}\{\lambda_1, \lambda_2, \lambda_3, \lambda_4\} = \lambda_4$$
$$\mathrm{Min}\{\mu_1, \mu_2, \mu_3, \mu_4\} = \mu_4$$

Let the shakedown limit load factor $\lambda_s = \frac{1}{2}(\lambda_4 + \mu_4)$. We obtain the shakedown limit load $P_s = \lambda_s P^0$=78.68MPa, and the error bound due to the uncertainty is less than

$$\mathrm{Max}\left\{\left|\frac{\lambda_s - \lambda_4}{\lambda_4}\right|, \left|\frac{\lambda_s - \mu_4}{\mu_4}\right|\right\} = 2.26\%$$

Thus, the bounding technique is successful for this problem.

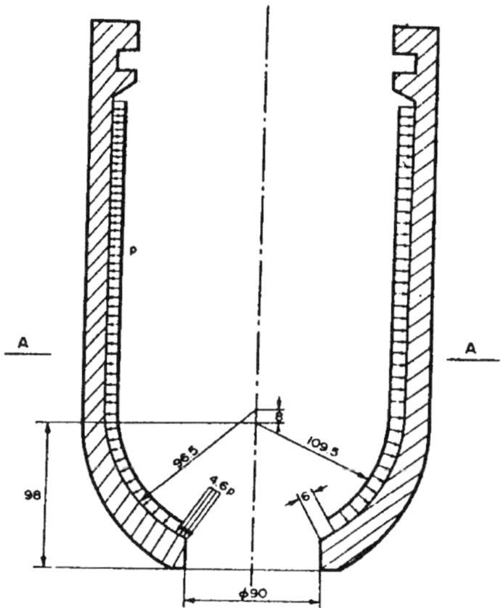


Fig.1 The geometry of storage energy vessel under internal pressure

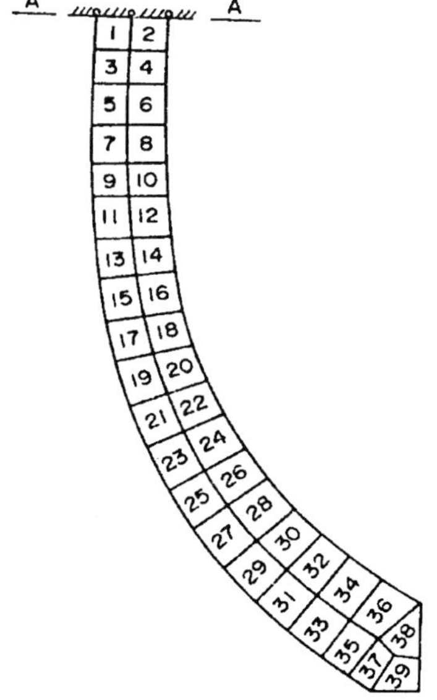


Fig.2 The finite element mesh

(2) A tube subjected to variable internal pressure and temperature field

A free-end tube with a linear temperature distribution across the thickness of wall is considered here. $\theta$ denotes the temperature difference between the inside and outside of the tube. The following parameters are adopted in the numerical analysis: $R/h$=10, $T_0=2(1-\nu)\sigma_s/E\alpha$, $P_0=\sigma_s h/R$. Here $P_0$ and $T_0$ are the theoretical collapse internal pressure and shakedown limit of pulsating temperature difference respectively, when the shell is subjected to only one category of loads. Two loading cases are considered:

(1) The internal pressure is constant whereas the temperature $T$ varies between 0 and $T^*$;
(2) Internal pressure $P$ and temperature $T$ vary independently within the bounds $0 \le P \le P^*$ and $0 \le T \le T^*$.

The corresponding shakedown load domains are shown in Fig.3. Loading case 1 was treated by Bree (1967) and Hyde (1985). It should be noted that the different forms of failure (alternating plasticity and rachetting) are distinguished by the turning point of the bounding curves of loading domains. It can be seen that the present results are in good agreement with those of Bree(1967).

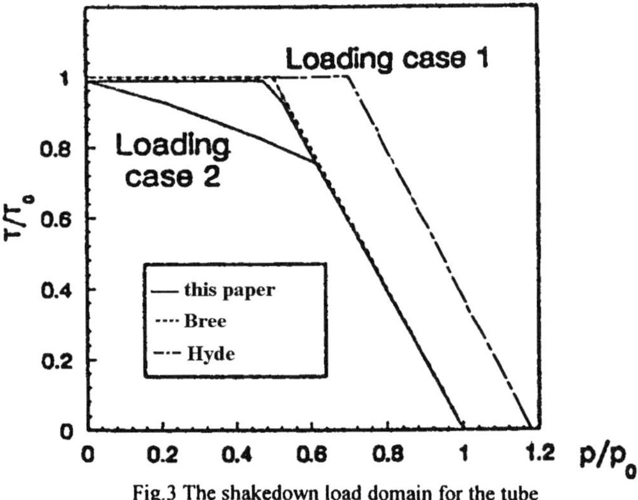

Fig.3 The shakedown load domain for the tube

(3) A complete spherical shell under variable internal pressure and temperature field

It is assumed that the action of both variable internal pressure and temperature is slow enough and the temperature is distributed across the thickness of shell according to Gohkfeld and Cherniavsky (1980):

$$T(r) = T_b + (T_a - T_b)\frac{a(b-r)}{r(b-a)}$$

where *a, b, r* are the inner, outer and current radii, respectively. Let $T_b$=0℃ and $b/a$=1.02. The loading cases like in example (2) are adopted. The present numerical results indicate a good agreement with the analytical solutions of Gohkfeld and Cherniavsky (1980), as shown in Fig.4.

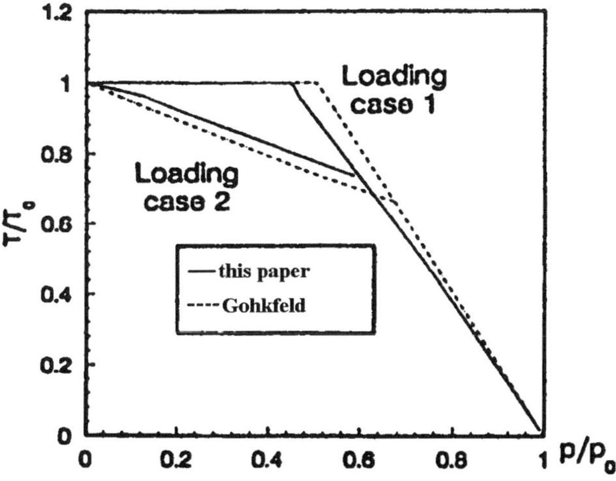


Fig. 4 The shakedown load domain for the spherical shell

(4) A spherical shell with a part-through slot under variable internal pressure and temperature field

A spherical shell with a part-through slot is considered here. The geometric dimensions of spherical shell are the same as those in example (3), and the dimension of part-through slot, as shown in Fig.5, is $r/H$=1.0, $c/H$=0.5. It is assumed that the temperature of the outside of spherical shell is 0℃. Two loading cases are considered like in example (2). The calculated results of loading domain, as shown in Fig.6, indicate that the local slot affects the shakedown load domain greatly under the combinations of thermal and mechanical loads.

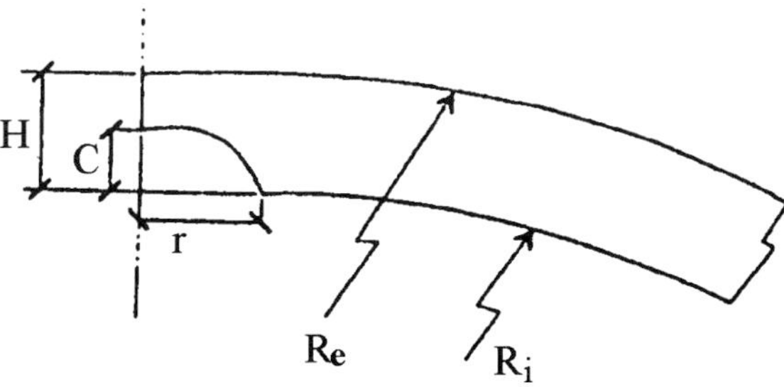


Fig. 5 The geometry of spherical shell with a part-through slot

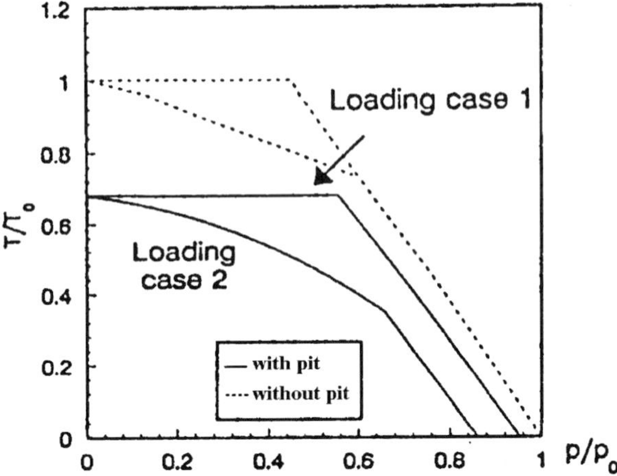


Fig. 6 The shakedown load domain for the spherical shell with a part-through slot

## 5. Concluding Remarks

Optimization problems derived from shakedown theory by using finite element discretizations are generally of large dimensions. To treat these problems effectively, special numerical approaches must be developed. The bounding technique proposed in this paper provides a possible means to deal with shakedown analysis of large scale and complex structures. It overcomes the difficulty of numerical calculation and so the shakedown analysis of structures under various combinations of steady and variable mechanical and thermal loading can more readily be performed in practice. The numerical results obtained here are satisfactory. Generally, the denser the adopted finite element meshes are, the smaller the discrepancy between upper and lower bounds of shakedown limit factor is. However, more practice is needed before this technique can be successfully applied to other types of structures besides the shell-like structures where the self-equilibrium stress field $\rho_{ij}(x)$ seems to interact weakly among the connected subdomains. The self-equilibrium stress field is constructed here by Eq. (15), together with the constraint (16), and it only approximately meets the requirement of completeness of self-equilibrium stress. Therefore, the numerical results tend to be conservative.

It should be noted here that the subdomain bounding relation (6) is valid analytically, irrespective of the numerical method used to obtain $\lambda_1, \lambda_2, \cdots, \lambda_n$, $\mu_1, \mu_2, \cdots, \mu_n$ and $\lambda_s$, so long as $\lambda_s$ is obtained by the same numerical method rather than by the subdomain technique.

### Acknowledgement

The authors would like to express their sincere gratitude to Prof. H. Yang for his help in computations.

## References


1. Melan, E. (1938) *Theorie statisch unbestimmter Tragwerke aus idealplastischem Baustoff*, Sitzungsbericht der Akademie der Wissenschaften (Wien) Abt. IIA, **195**, 145-195.
2. Koiter, W. T. (1960) Genernal theorems for elastic-plastic solids, in Sneddon, I. N. and Hill, R.(eds.), *Progress in Solid Mechanics*, North-Holland, Amsterdam, pp.165-221.
3. Belytschko, T. (1972) Plane stress shakedown analysis by finite elements, *Int. J. Mech. Sci.* **14**, 619-625.
4. Corradi, L. and Zavelani, A. (1974) A linear programming approach to shakedown analysis of structures, *Comp. Meth. App. Mech. Engng.* **3**, 37-53.
5. Stein, E., Zhang, G. and König, J. A. (1992) Shakedown with nonlinear hardening including structural computation using finite element method, *Int. J. Plasticity* **8**, 1-31.
6. O.Debordes (1978) Sur le calcul à l'adaptation des structures élastoplastiques parfaites,*Annales des Ponts et Chaussées – 1.Trim*,7-21.
7. Spilker, R.L.(1981) Improved hybrid stress axisymmetric elements including behavior for nearly incompressible materials, *Int. J. Numer. Meth. Engng.* **17**, 483-501.
8. Hung, N.D. and König, J.A. (1976) A finite element formulation for shakedown problems using a yield eriteria of the mean, *Comp. Meth. Appl. Mech. Engng.* **8**, 179-192.
9. Wan, Y. et al. (1983) *A Collection of the Programs in Common Use for Optimization,* Worker Press, Beijing, China. (in Chinese)
10. Prager, W. and Hodge, P. G. (1951) *Theory of Perfectly Plastic Solids*, John Wiley, New York.
11. Gohkfeld, P.A. and Cherniavsky, O.F. (1980) *Limit Analysis of Structure at Thermal Cycling*, Sijthoff and Noordhoff Press, The Netherlands.
12. Bree, J.(1967) Elastic-plastic behavior of thin tubes subjected to internal pressure and intermittent high-heat fluxes with application to fast-nuclear-reactor fuel elements, *Journal of Strain Analysis* **2**, 226-238.
13. Hyde, T. H., Sahari, B. B. and Webster, J. J. (1985) The effect of axial loading and axial restraint on the thermal ratchetting of thin tubes, *Int. J. Mech. Sci.* **27**, 679-692.

# FAILURE INVESTIGATION OF FIBER-REINFORCED COMPOSITE MATERIALS BY SHAKEDOWN ANALYSIS

A. HACHEMI, F. SCHWABE, D. WEICHERT
*Institute of General Mechanics, RWTH Aachen*
*Templergraben 64, 52056 Aachen, Germany*

**Abstract**

A methodology is proposed to investigate failure of composite materials under thermo-mechanical variable loads. By using the homogenization technique of periodic media, the plastically admissible ranges of macroscopic stresses can be found from the shakedown analysis of representative volume elements of the considered composite.

## 1. Introduction

Modern engineering relies increasingly on the use of materials especially designed for the specific functional requirements on the mechanical components made from these materials. Composites have turned out to be particularly interesting in this respect, if their different constituents are chosen in such a way that their combined properties surpass those of competing conventional materials. As an example we quote metal-matrix composites (MMC's), where the ductility of the metal-matrix combined with the hardness of embedded brittle ceramic particles allows one to produce materials that exhibit longer lifetime than conventional materials under repeated shock loading, without losing hardness, as is required, e.g., for cold forging tools [1].

Due to the microstructural heterogeneity, local damage in such materials is in general caused by a combination of different effects depending upon the mechanical properties, shape and volume fraction of the individual constituents of the composite and their interaction on the local level. In fact, in MMC's e.g., local accumulation of plastic deformations can lead to material damage in the ductile matrix material, eventually generating micro-cracks. Later in the loading process these may propagate and initiate failure of the considered structural element. To predict this, it is important



*D. Weichert and G. Maier (eds.), Inelastic Analysis of Structures under Variable Loads,* 107–119.

to understand and to model properly the mechanical processes on the microstructural level and to link them to the macroscopic material properties.

In this paper, we apply the static shakedown theorems to the assessment and design of such composites undergoing thermal and/or mechanical loads. Similar studies using the kinematical approach have been carried out in [2-3] and [23]. For the application of this method, no precise knowledge of the evolution of loads is required and it is sufficient to know the loading domain in the space of generalized loads embracing the total loading history. For engineering purposes, such a situation is in general much more realistic than to assume the loading history to be precisely given, as it is the case if incremental simulation techniques are used.

The two principal theoretical ingredients of the proposed methodology are shakedown theory [18] applied to the analysis of unit cells on the micro-scale level, and the theory of homogenization [5, 25]. Under the assumption of the periodicity of the composite, the results from micro-scale are linked to the macro-scale material properties. From this point of view, the present methodology can be regarded as an extension of that proposed by Suquet for limit analysis [25]. In the present paper, we do not review the achievements in these two fields of research; the reader may find the fundamentals of shakedown theory e.g. in [6, 13-15, 17-19] and on more recent achievements in [20-22, 26-27].

## 2. Material model and constitutive relations

The composite material is assumed to consist of continuous elastic fibers embedded according to a regular pattern in an elastic-plastic metal matrix as shown in Figure 1. A perfect bond is assumed to exist at the matrix-fiber interface (see e.g. [29]). The assumed independence of the problem with respect to the fiber axis leads to the plane strain hypothesis. Thus, the problem is reduced to a two-dimensional case with periodicity in the fiber section plane. We introduce, as is customary in homogenization theory, a representative volume element (unit cell), which describes the heterogeneities at the microscopic scale. We call $\mathbf{x}$ the macroscopic scale for which the inhomogeneities have very small dimensions, and $\mathbf{y}$ the microscopic scale.

For the theoretical formulation, linear kinematical hardening is taken into account by using internal parameters according to the concept of *Generalized Standard Material Model* (GSMM) [9]. For this, generalized total, elastic, plastic and thermal strains and generalized stresses are introduced defined on the micro-level by the sets

$$\mathbf{e} = [\boldsymbol{\varepsilon}, \mathbf{0}],\ \mathbf{e}^e = [\boldsymbol{\varepsilon}^e, \boldsymbol{\omega}],\ \mathbf{e}^p = [\boldsymbol{\varepsilon}^p, \boldsymbol{\kappa}],\ \mathbf{e}^\vartheta = [\boldsymbol{\varepsilon}^\vartheta, \mathbf{0}],\ \mathbf{s} = [\boldsymbol{\sigma}, \boldsymbol{\pi}]. \tag{1}$$

Here, $\boldsymbol{\varepsilon}^e$, $\boldsymbol{\varepsilon}^p$ and $\boldsymbol{\varepsilon}^\vartheta$ are respectively the observed elastic, plastic and thermally induced parts of the total strain tensor $\boldsymbol{\varepsilon}$. The observable stresses are represented by the stress tensor $\boldsymbol{\sigma}$, and the quantities $\boldsymbol{\omega}$, $\boldsymbol{\kappa}$ and $\boldsymbol{\pi}$ are the $r$-dimensional vectors of internal elastic and plastic parameters and back-stresses, respectively. The dimension $r$ depends upon the particular choice of hardening model.

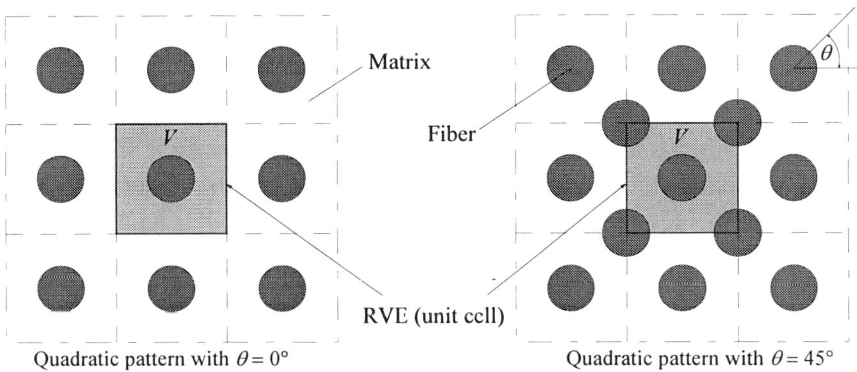


*Figure 1.* A periodic fiber-reinforced composite.

The elastic-plastic damage behavior of materials is introduced through the concept of *effective stress* [12]. This means that the behavior of the damaged material can be represented by the constitutive equations of the virgin material where the usual generalized stresses on the micro-level are replaced by the effective generalized stresses defined by

$$\tilde{\mathbf{s}} = \frac{\mathbf{s}}{1 - D} \cdot \qquad (2)$$

Here, the value $D = 0$ corresponds to the undamaged state, $D \in (0, D_c)$ corresponds to a partly damaged state and $D = D_c$ defines the complete local rupture ($D_c \in [0, 1]$). A superposed tilda indicates quantities related to the damaged state of the material.

According to the restriction to geometrically linear theory, the total generalized strains $\mathbf{e}$ can be split into purely elastic, purely plastic and thermally induced parts $\mathbf{e}^e$, $\mathbf{e}^p$ and $\mathbf{e}^\vartheta$, respectively

$$\mathbf{e} = \mathbf{e}^e + \mathbf{e}^p + \mathbf{e}^\vartheta \qquad (3)$$

with

$$\boldsymbol{\varepsilon} = \boldsymbol{\varepsilon}^e + \boldsymbol{\varepsilon}^p + \boldsymbol{\varepsilon}^\vartheta \qquad (4)$$

$$\mathbf{0} = \boldsymbol{\omega} + \boldsymbol{\kappa}. \qquad (5)$$

In order to introduce the constitutive equations in the formulation of the shakedown theorem, we consider the thermodynamic potential $\Psi$, assumed to be convex function of all observable and internal variables (cf. [11, 16])

$$\Psi(\boldsymbol{\varepsilon}^e, \boldsymbol{\kappa}, D, T) = \Psi_e(\boldsymbol{\varepsilon}^e, D, T) + \Psi_p(\boldsymbol{\kappa}, D) \tag{6}$$

with

$$\rho \Psi_e = \frac{1}{2}(1-D)\,\mathbf{L}:(\boldsymbol{\varepsilon}^e - \alpha_\vartheta \vartheta \mathbf{I}):(\boldsymbol{\varepsilon}^e - \alpha_\vartheta \vartheta \mathbf{I}) + C_\varepsilon \vartheta^2 \tag{7}$$

$$\rho \Psi_p = \frac{1}{2}(1-D)\,\mathbf{Z}.\boldsymbol{\kappa}.\boldsymbol{\kappa} \tag{8}$$

where $\rho$ is the mass density, $C_\varepsilon$ is the specific heat at constant strain, $\alpha_\vartheta$ the coefficient of isotropic thermal expansion, $\vartheta$ the difference between the absolute temperature ($T$) and the reference temperature ($T_0$), $\mathbf{L}$ and $\mathbf{Z}$ are the tensors of elasticity of observable elastic strains and internal elastic parameters and $\mathbf{I}$ is the identity tensor of second rank. The operators ($\cdot$) and (:) stands for simple and double tensor contraction, respectively. Then, the material constitutive equations are

$$\boldsymbol{\sigma} = \rho \frac{\partial \Psi}{\partial \boldsymbol{\varepsilon}^e} = (1-D)\mathbf{L}:(\boldsymbol{\varepsilon}^e - \alpha_\vartheta \vartheta \mathbf{I}) \tag{9}$$

$$\boldsymbol{\pi} = -\rho \frac{\partial \Psi}{\partial \boldsymbol{\kappa}} = -(1-D)\,\mathbf{Z}.\boldsymbol{\kappa} \tag{10}$$

$$G = -\rho \frac{\partial \Psi}{\partial D} = \frac{1}{2}\mathbf{L}:(\boldsymbol{\varepsilon}^e - \alpha_\vartheta \vartheta \mathbf{I}):(\boldsymbol{\varepsilon}^e - \alpha_\vartheta \vartheta \mathbf{I}) + \frac{1}{2}\mathbf{Z}.\boldsymbol{\kappa}.\boldsymbol{\kappa}. \tag{11}$$

The thermodynamic force $G$ conjugate to the damage variable $D$ is the energy function of the undamaged material [11, 16].

The generalized stresses $\boldsymbol{\Sigma}(\mathbf{x})$ and strains $\mathbf{E}(\mathbf{x})$ on the macro-level are linked to generalized stresses $\mathbf{s}(\mathbf{y})$ and strains $\mathbf{e}(\mathbf{y})$ on the micro-level through [10]

$$\boldsymbol{\Sigma}(\mathbf{x}) = \langle \mathbf{s}(\mathbf{y}) \rangle = \frac{1}{V}\int_{(V)} \mathbf{s}(\mathbf{y})\,\mathrm{d}V \tag{12}$$

$$\mathbf{E}(\mathbf{x}) = \langle \mathbf{e}(\mathbf{y}) \rangle = \frac{1}{V}\int_{(V)} \mathbf{e}(\mathbf{y})\,\mathrm{d}V \tag{13}$$

where $V$ denotes the volume of the (periodic) representative volume element (RVE) as shown in Figure 1.

We assume the validity of the normality rule for plastic flow, such that

$$\dot{\mathbf{e}}^p \in \delta\varphi(\mathbf{s}) \tag{14}$$

where $\delta\varphi(\mathbf{s})$ denotes the sub-gradients of the plastic potential $\varphi(\mathbf{s})$ [9] which is the indicator function of a convex generalized elastic domain P(**s**, **y**) of all plastically admissible stress states

$$\mathbf{s}(\mathbf{y}) \in \mathrm{P}(\mathbf{s}, \mathbf{y}), \quad \forall \mathbf{y} \in V. \tag{15}$$

P(**s**, **y**) is defined by means of a yield function $\mathscr{F}(\tilde{\mathbf{s}}, \mathbf{y})$

$$\mathrm{P}(\mathbf{s}, \mathbf{y}) = \{\mathbf{s} \,/\, \mathscr{F}(\tilde{\mathbf{s}}, \mathbf{y}) \leq 0, \ \forall \mathbf{y} \in V\}. \tag{16}$$

Here, it is assumed that the yield function $\mathscr{F}(\tilde{\mathbf{s}}, \mathbf{y})$ is of von Mises type

$$\mathscr{F}(\tilde{\mathbf{s}}, \mathbf{y}) = \left[\frac{3}{2}\left(\frac{\boldsymbol{\sigma}^D}{1-D} - \frac{\boldsymbol{\pi}}{1-D}\right) : \left(\frac{\boldsymbol{\sigma}^D}{1-D} - \frac{\boldsymbol{\pi}}{1-D}\right)\right]^{1/2} - \sigma_F(\mathbf{y}) \tag{17}$$

where $\sigma_F(\mathbf{y})$ denotes the yield stress and $\boldsymbol{\sigma}^D$ the deviatoric part of $\boldsymbol{\sigma}$. The convexity of $\mathscr{F}(\tilde{\mathbf{s}}, \mathbf{y})$ and the validity of the normality rule can be expressed by the generalized maximum plastic work inequality

$$<(\mathbf{s} - \mathbf{s}^{(s)}) : \dot{\mathbf{e}}^p> \geq 0, \quad \forall \mathbf{s}^{(s)}(\mathbf{y}) \in \bar{\mathrm{P}}(\mathbf{s}^{(s)}, \mathbf{y}) \tag{18}$$

where $\mathbf{s}^{(s)} = [\boldsymbol{\sigma}^{(s)}, \boldsymbol{\pi}^{(s)}]$ is any safe state of generalized stresses defined by

$$\bar{\mathrm{P}}(\mathbf{s}^{(s)}, \mathbf{y}) = \{\mathbf{s}^{(s)} \,/\, \mathscr{F}(\tilde{\mathbf{s}}^{(s)}, \mathbf{y}) < 0, \ \forall \mathbf{y} \in V\}. \tag{19}$$

## 3. Formulation of shakedown theorems

It follows from the adopted concepts of *effective stress* and the *generalized standard material model* that the formulation of the statical shakedown theorem for damaged linear kinematical hardening materials is formally the same as presented in [28] for the case of undamaged elastic-perfectly plastic materials.

Then, the convenient macroscopic admissible domain $P'''(\mathbf{\Sigma}, \mathbf{x})$ of composite material may be defined as the set of macroscopic states of generalized stress $\mathbf{\Sigma}(\mathbf{x})$ for all microscopic states of plastically admissible generalized stress $\mathbf{s}(\mathbf{y})$

$$P'''(\mathbf{\Sigma}, \mathbf{x}) = \{\mathbf{\Sigma} \,/\, \mathbf{\Sigma} = <\mathbf{s}> \,,\ \mathbf{s}(\mathbf{y}) \in P(\mathbf{s}, \mathbf{y}),\ \forall \mathbf{y} \in V\}. \tag{20}$$

The computation of the macroscopic admissible domain will give a reliable prediction of the failure of composite materials. To determine $P'''(\mathbf{\Sigma}, \mathbf{x})$, the shakedown analysis is carried out on the micro-level. For this, we introduce the notion of a *reference representative volume element* ($RVE^{(c)}$) differing from the actual one only by the fact that the material is supposed to behave purely elastically. All quantities related to this reference representative volume element are indicated by superscript *(c)*. The internal parameters to describe the state of hardening and damage in the material vanish naturally for the $RVE^{(c)}$, so that the generalized strains and stresses are given by

$$\mathbf{e}^{(c)} = \mathbf{e}^{e(c)} = [\boldsymbol{\varepsilon}^{(c)}, \mathbf{0}],\ \mathbf{e}^{p(c)} = [\mathbf{0}, \mathbf{0}],\ \mathbf{e}^{\vartheta(c)} = [\boldsymbol{\varepsilon}^{\vartheta(c)}, \mathbf{0}],\ \mathbf{s}^{(c)} = [\boldsymbol{\sigma}^{(c)}, \mathbf{0}]. \tag{21}$$

The statical shakedown theorem states the following: if there exists a safety factor $\alpha > 1$, a time-independent field of generalized stresses $\bar{\mathbf{s}}^r = [\bar{\boldsymbol{\sigma}}^r, \bar{\boldsymbol{\pi}}]$ and a *Sanctuary of Elasticity* [21]

$$\bar{P}'''(\mathbf{\Sigma}^{(s)}, \mathbf{x}) \subset P'''(\mathbf{\Sigma}, \mathbf{x}) \tag{22}$$

with

$$\bar{P}'''(\mathbf{\Sigma}^{(s)}, \mathbf{x}) = \{\mathbf{\Sigma}^{(s)} \,/\, \mathbf{\Sigma}^{(s)} = <\mathbf{s}^{(s)}> \,,\ \mathbf{s}^{(s)}(\mathbf{y}) \in \bar{P}(\mathbf{s}^{(s)}, \mathbf{y}),\ \forall \mathbf{y} \in V\} \tag{23}$$

then the periodic composite material shakes down. Here, the safe state of generalized stresses $\mathbf{s}^{(s)}$ is defined as usual (see e.g. [17])

$$\mathbf{s}^{(s)} = \alpha\, \mathbf{s}^{(c)} + \bar{\mathbf{s}}^r \tag{24}$$

where $\mathbf{s}^{(c)} = [\boldsymbol{\sigma}^{(c)}, \mathbf{0}]$ is the generalized stress field which would occur in the $RVE^{(c)}$ under the same boundary conditions as the actual RVE such that the following relations hold for given macroscopic strains $\mathbf{E}$

$$\mathrm{Div}\, \boldsymbol{\sigma}^{(c)} = \mathbf{0} \quad \text{in } V \tag{25}$$

$$\mathbf{u}^{(c)} = \mathbf{E}.\mathbf{y} \quad \text{on } \partial V \tag{26}$$

$$\boldsymbol{\varepsilon}^{(c)} = \nabla_s(\mathbf{u}^{(c)}) \tag{27}$$

$$\boldsymbol{\sigma}^{(c)} = \mathbf{L}:\boldsymbol{\varepsilon}^{(c)} \tag{28}$$

However, the field of the residual stresses $\bar{\boldsymbol{\sigma}}^r$ satisfies

$$\text{Div}\, \bar{\boldsymbol{\sigma}}^r = \mathbf{0} \qquad \text{in } V \tag{29}$$

$$\bar{\boldsymbol{\sigma}}^r.\mathbf{n} = \mathbf{0} \qquad \text{on } \partial V \tag{30}$$

where $\mathbf{n}$ is the outward normal vector to $\partial V$, Div the divergence operator and $\nabla_s$ the symmetric gradient-operator.

The domain of external loads $\mathscr{D}$ can be represented by an $n$-dimensional polyhedron, defined by

$$\mathscr{D} = \left\{ \mathscr{P} \,/\, \mathscr{P} = \sum_{i=1}^{n} \mu_i \mathscr{P}_i^0,\ \ \mu_i \in [\mu_i^-, \mu_i^+] \right\} \tag{31}$$

where $\mathscr{P}$ is the vector of generalized loads, $\mu_i$ are scalar multipliers with lower and upper bounds $\mu_i^-$ and $\mu_i^+$, respectively. $\mathscr{P}_i^0$ are $n$ fixed and independent generalized loads (macroscopic stresses, macroscopic strains, temperature changes or combinations). Then, the static shakedown theorem for the determination of the macroscopic admissible domain $\bar{\mathrm{P}}'''$ against failure due to inadmissible damage or unlimited accumulation of plastic deformations can be expressed by the following optimization problem

$$\alpha_{SD} = \max_{\bar{\boldsymbol{\sigma}}^r,\, \bar{\boldsymbol{\pi}},\, D} \alpha \tag{32}$$

with the subsidiary conditions (29)-(30) and

$$D < D_c \tag{33}$$

$$\mathscr{F}_I\left(\alpha \frac{\boldsymbol{\sigma}^{(c)}}{1-D}(\mathscr{P}) + \frac{\bar{\boldsymbol{\sigma}}^r}{1-D} - \frac{\bar{\boldsymbol{\pi}}}{1-D},\ \sigma_F\right) < 0 \tag{34}$$

$$\mathscr{F}_L\left(\alpha \frac{\boldsymbol{\sigma}^{(c)}}{1-D}(\mathscr{P}) + \frac{\bar{\boldsymbol{\sigma}}^r}{1-D},\ \sigma_S\right) < 0 \tag{35}$$

such that

$$\boldsymbol{\Sigma}^{(s)} = \alpha < \boldsymbol{\sigma}^{(c)} > + < \bar{\boldsymbol{\sigma}}^r >. \tag{36}$$

## 4. Discretization of shakedown problems

The present formulation of shakedown theorems is translated into a numerical simulation program by using displacement-based finite elements for the discretization of the problem. To calculate the elastic stresses $\boldsymbol{\sigma}^{(c)}$ in the reference RVE$^{(c)}$, we use the virtual work principle

$$\frac{1}{V}\int_{(V)} \{\boldsymbol{\sigma}^{(c)}\}\{\delta\boldsymbol{\varepsilon}^{(c)}\}\, \mathrm{d}V = \{\boldsymbol{\Sigma}\}\{\delta\mathbf{E}\}. \tag{37}$$

Analogously, the field of residual stresses is defined by

$$\int_{(V)} \{\bar{\boldsymbol{\sigma}}^{r}(\mathbf{y})\}\{\delta\boldsymbol{\varepsilon}(\mathbf{y})\}\, \mathrm{d}V = 0. \tag{38}$$

By using the well-known Gauss-Legendre technique, this integral can be calculated for each finite element. The integration has to be carried out over all Gaussian points *NGE* with their weighting factors $w_i$ in the considered element "$e$"

$$\int_{(V^e)} \{\bar{\boldsymbol{\sigma}}^{r}(\mathbf{y})\}\{\delta\boldsymbol{\varepsilon}(\mathbf{y})\}\, \mathrm{d}V = \sum_{i=1}^{NGE}\sum_{k=1}^{NK} w_i\, |\mathbf{J}_i|\, [\, [\mathbf{B}_k(\mathbf{y}_i)]\, \{\delta\mathbf{u}_k^e\}]^T \{\bar{\boldsymbol{\sigma}}^{r}(\mathbf{y}_i)\}. \tag{39}$$

Here, $[\mathbf{B}_k(\mathbf{y})]$ and $\{\delta\mathbf{u}_k^e\}$ denote the $k$-th matrix of the derivatives of the shape functions and the vector of virtual displacement of the $k$-th node of the element "$e$", respectively. *NK* denotes the total number of nodes of each element and $|\mathbf{J}_i|$ the determinant of the Jacobian matrix. By summation of the contributions of all elements and by variation of the virtual node-displacements with regard to the boundary conditions, one finally gets the linear system of equations (see e.g. [7, 24])

$$\sum_{i=1}^{NG} [\mathbf{C}_i]\, \{\bar{\boldsymbol{\sigma}}_i^r\} = [\mathbf{C}]\, \{\bar{\boldsymbol{\sigma}}^r\} = \{\mathbf{0}\}. \tag{40}$$

The total number of Gaussian points of the RVE$^{(c)}$ is denoted by *NG*, $[\mathbf{C}]$ is a constant matrix, uniquely defined by the discretized system and the boundary conditions and $\{\bar{\boldsymbol{\sigma}}^r\}$ is the global residual stress vector of the discretized RVE$^{(c)}$.

Then, the discrete formulation of shakedown problems for the determination of admissible domain of macroscopic stresses against failure due to inadmissible damage or accumulated plastic deformations is given by the solution of the following non-linear optimization problems

$$\alpha_{SD} = \max_{\bar{\boldsymbol{\sigma}}_i^r,\, \bar{\boldsymbol{\pi}}_i,\, D_i} \alpha \tag{41}$$

with the subsidiary conditions

$$[\mathbf{C}]\ \{\bar{\boldsymbol{\sigma}}^r\} = \{\mathbf{0}\} \tag{42}$$

$$D_i < D_c \tag{43}$$

$$\mathscr{F}_I\left(\alpha \frac{\boldsymbol{\sigma}_i^{(c)}}{1 - D_i}(\mathscr{P}_j) + \frac{\bar{\boldsymbol{\sigma}}_i^r}{1 - D_i} - \frac{\bar{\boldsymbol{\pi}}_i}{1 - D_i},\ \sigma_F\right) < 0 \tag{44}$$

$$\mathscr{F}_L\left(\alpha \frac{\boldsymbol{\sigma}_i^{(c)}}{1 - D_i}(\mathscr{P}_j) + \frac{\bar{\boldsymbol{\sigma}}_i^r}{1 - D_i},\ \sigma_S\right) < 0 \tag{45}$$

$$\forall i \in [1, NG],\ \ \forall j \in [1, 2^n]$$

such that

$$\boldsymbol{\Sigma}^{(s)} = \alpha < \boldsymbol{\sigma}^{(c)} > + < \bar{\boldsymbol{\sigma}}^r >. \tag{46}$$

with $\alpha_{SD}$ as objective function to be optimized with respect to $\bar{\boldsymbol{\sigma}}_i^r$, $\bar{\boldsymbol{\pi}}_i$ and $D_i$ and with relations (42) and (43-45) as linear and non-linear constraints, respectively. The condition (43) assures structural safety against failure due to material damage. The condition (45) assures that safe states of stresses, according to definition (24), are never outside the loading surface $\mathscr{F}_L$ and so guarantees implicitly the boundedness of the back-stresses $\bar{\boldsymbol{\pi}}_i$ (see e.g. [8]). The damage parameter and the yield criteria have to be fulfilled in $NG$ points and for $2^n$ combinations of loads. The solution of this optimization problem is carried out by using the code LANCELOT which is based on an augmented Lagrangian method (for more detail about the algorithm of optimization, refer to [4]).

## 5. Numerical examples

To illustrate the method, we consider typical problems for $Al/Al_2O_3$ fiber-reinforced composite materials, with the following properties of the individual phases:

TABLE 1. Material properties of the individual phases of the composite

| Phases | $E$ (GPa) | $\nu$ | $\sigma_S$ (MPa) | $\alpha_\vartheta$ ($10^{-6}\,K^{-1}$) |
|---|---|---|---|---|
| Matrix | 70 | 0.3 | 80 | 22 |
| Fibers | 370 | 0.3 | ---- | 8 |

In all the numerical examples, we ignore material damage and hardening effects, and we assume perfect bonding between fibers and matrix. The matrix is assumed to be elastic-perfectly plastic, and fibers are supposed to be purely elastic with circular cross-section. The calculation are made on a unit cell for quadratic arrangement with $\theta = 0°$ and $\theta = 45°$ which is discretized by two-dimensional finite elements under the assumption of plane strains. The specific type of loading is restrictive in that sense that only principal states of stresses are considered. Triangular isoparametric elements with six nodes are used. For symmetry reasons, only a quarter of the unit cell is considered; this involves 216 elements with 473 nodes for the case of $\theta = 0°$, and 210 element with 461 nodes for the case of $\theta = 45°$ (Figure 2). The unit cell is subjected to bi-axial uniform displacement $u_1$ and $u_2$ at the edges ($u_1 = \mu_1\, u_1^0$ ; $0 \leq \mu_1 \leq \mu_1^+$ and $u_2 = \mu_2\, u_2^0$ ; $0 \leq \mu_2 \leq \mu_2^+$) and to constant distribution of temperature $\Delta\vartheta$.

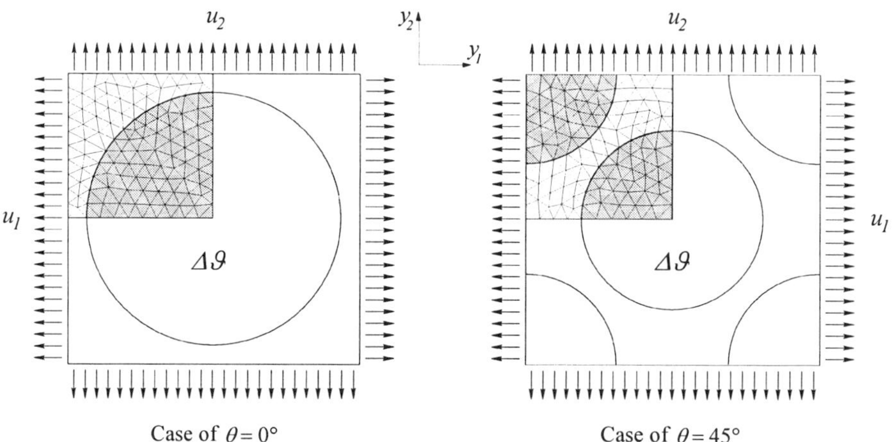


*Figure 2.* Finite element discretization.

For given regular quadratic patterns of periodicity of the reinforcing elastic fibers in a ductile matrix as illustrated in Figure 2, given material properties of the fibers and the matrix (Table 1) and given volume fraction of fibers (here 60%), one has to determine the safe domains in the space of given generalized loads. The shakedown domains are shown in Figure 3 for both $\theta = 0°$ and $\theta = 45°$. As expected, one observes that under the given loading conditions these domains are quite different for the different angles $\theta$.

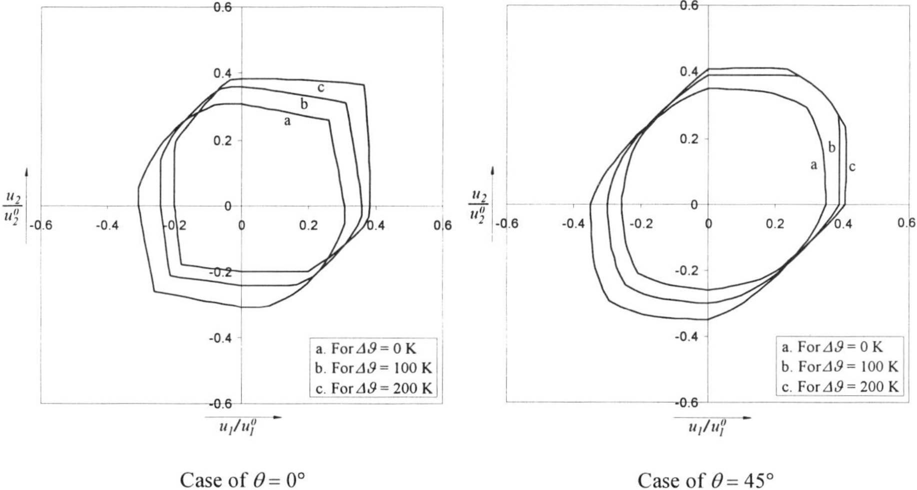


*Figure 3.* Shakedown domains.

The results show the sensitivity of the composite to prescribed thermal loads. It can be seen that the shakedown domains increase (decrease) if the composite is under traction (compression). In Figure 4, the results show the variation of shakedown domains for the composite under compression with the temperature amplitude normalized by $\Delta\vartheta_0 = 100$ K for $\theta = 0°$ and $\theta = 45°$. The results show that the thermal residual stresses induced by the thermal loads strongly affect the size and the shape of the shakedown domains. For $\theta = 0°$, the influence of the thermal loads is much stronger than for $\theta = 45°$.

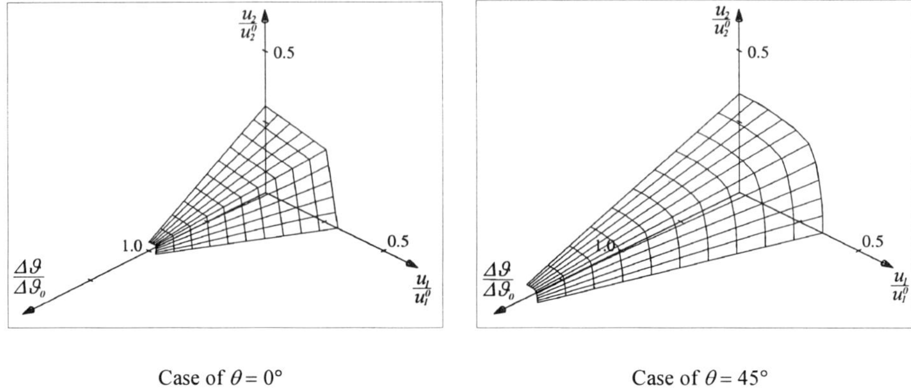


Case of $\theta = 0°$ Case of $\theta = 45°$

*Figure 4.* Variation of shakedown domains with amplitude of temperature.

## 6. Conclusions

In this paper we propose a methodology for assessing certain types of composites exposed to constant thermal and variable mechanical loads. Damaging physical effects, plasticity and material damage are taken into account in the theoretical formulation. Although the methodology is general for the considered class of materials, the numerical examples have been restricted to plane strain and to bi-axial mechanical loading along the principal directions of macroscopic stresses.

## 7. References


1. Berns, H., Melander, A., Weichert, D., Asnafi, N., Broeckmann, C. and Gross-Weege, A.: A new material for cold forging tools, *Comput. Mat. Sci.* **11** (1998), 166-180.
2. Carvelli, V., Maier, G. and Taliercio, A.: Shakedown analysis of periodic heterogeneous materials by a kinematic approach, *J. Mech. Engng. (Strojnícky Casopis)* **50** (1999), 229-240.
3. Carvelli, V., Maier, G. and Taliercio, A.: Kinematic limit analysis of periodic heterogeneous media, *Comput. Model. Engng. Sci.* **1** (2000), 15-26.
4. Conn, A.R., Gould, N.I.M. and Toint, Ph.L.: *LANCELOT: A Fortran Package for Large-Scale Nonlinear Optimization (Release A)*, Springer-Verlag, Berlin, 1992.
5. De Buhan, P. and Taliercio, A.: A homogenization approach to the yield strength of composite materials, *Eur. J. Mech. A/Solids* **10** (1991), 129-154.
6. Gokhfeld, D.A. and Cherniavsky, O.F.: *Limit Analysis of Structures at Thermal Cycling*, Leyden, Sijthoff and Noordhoff, 1980.

7. Gross-Weege, J.: On the numerical assessment of the safety factor of elastic-plastic structures under variable loading, *Int. J. Mech. Sci.* **39** (1997), 417-433.
8. Hachemi, A. and Weichert, D.: Application of shakedown theory to damaging inelastic material under mechanical and thermal loads, *Int. J. Mech. Sci.* **39** (1997), 1067-1076.
9. Halphen, B. and Nguyen, Q.S.: Sur les matériaux standards généralisés, *J. Méc.* **14** (1975), 39-63.
10. Hill, R.: Elastic properties of reinforced solids: some theoretical principles, *J. Mech. Phys. Solids* **11** (1963), 357-372.
11. Ju, J.W.: On energy-based coupled elastoplastic damage theories: constitutive modelling and computational aspects, *Int. J. Solids Struct.* **25** (1989), 803-833.
12. Kachanov, L.M.: Time of the rupture process under creep conditions, *Izv. Akad. Nauk, S.S.R., Otd. Tech. Nauk* **8** (1958), 26-31.
13. Koiter, W.T.: General theorems for elastic-plastic solids. In: Sneddon, I.N. and Hill, R. (eds.), *Progress in Solid Mechanics*, Amsterdam, North-Holland, pp. 165-221, 1960.
14. König, J.A. and Maier, G.: Shakedown analysis of elastic-plastic structures: a review of recent developments, *Nucl. Engng. Design* **6** (1981), 81-95.
15. König, J.A.: *Shakedown of Elastic-Plastic Structures*, Elsevier, Amsterdam, 1987.
16. Lemaitre, J. and Chaboche, J.L.: *Mécanique des Matériaux Solides*, Dunod, Paris, 1985.
17. Mandel, J.: Adaptation d'une structure plastique écrouissable et approximations, *Mech. Res. Comm.* **3** (1976), 483-488.
18. Melan, E.: Theorie statisch unbestimmter Systeme aus ideal-plastischem Baustoff, *Sitber. Akad. Wiss. Wien, Abt. IIa* **145** (1936), 195-218.
19. Melan, E.: Zur Plastizität des räumlichen Kontinuums, *Ing. Arch.* **9** (1938), 116-126.
20. Mróz, Z., Weichert, D. and Dorosz, S.: *Inelastic Behaviour of Structures under Variable Loads*, Kluwer Academic Publishers, Dordrecht, 1995.
21. Nayroles, B. and Weichert, D.: La notion de sanctuaire d'élasticité et l'adaptation des structures, *C.R. Acad. Sci.* **316** (1993), 1493-1498.
22. Polizzotto, C., Borino, G., Caddemi, S. and Fuschi, P.: Shakedown problems for material models with internal variables, *Eur. J. Mech. A/Solids* **10** (1991), 621-639.
23. Ponter, A.R.S. and Leckie, F.A.: On the behaviour of metal matrix composites subjected to cyclic thermal loading, *J. Mech. Phys. Solids*, **46** (1998), 2183-2199.
24. Stein, E., Zhang, G. and Huang, Y.: Modeling and computation of shakedown problems for nonlinear hardening materials, *Comput. Methods Appl. Mech. Engng.* **103** (1993), 247-272.
25. Suquet, P.: Analyse limite et homogénéisation, *C.R. Acad. Sci.* **296** (1983), 1355-1358.
26. Weichert, D. and Gross-Weege, J.: The numerical assessment of elastic-plastic sheets under variable mechanical and thermal loads using a simplified two-surface yield condition, *Int. J. Mech. Sci.* **30** (1988), 757-767.
27. Weichert, D. and Hachemi, A.: Influence of geometrical nonlinearities on the shakedown of damaged structures, *Int. J. Plasticity* **14** (1998), 891-907.
28. Weichert, D., Hachemi, A. and Schwabe, F.: Shakedown analysis of composites, *Mech. Res. Comm.* **26** (1999), 309-318.
29. Zahl, D.B., Schmauder, S. and McMeeking, R.M.: Transverse strength of metal matrix composites reinforced with strongly bonded continuous fibers in regular arrangements, *Acta metall. Mater.* **42** (1994), 2983-2997.

# ANALYSIS OF MASONRY STRUCTURES SUBJECT TO VARIABLE LOADS: A NUMERICAL APPROACH BASED ON DAMAGE MECHANICS

A. CALLERIO*, E. PAPA* AND A. NAPPI**

** Department of Structural Engineering, Politecnico of Milano*
*Piazza Leonardo da Vinci, 32 - 20133 Milano (Italy)*
*** Department of Civil Engineering, University of Trieste*
*Piazzale Europa, 1 - 34127 Trieste (Italy)*

## Abstract

An elastic-plastic model that takes into account damage effects is developed and applied to masonry structures. Masonry is described as an orthotropic material and inelastic strains are governed by an associated flow rule based upon a piecewise-linear yield surface. They are split into a non-reversible (plastic) part and a recoverable part. Reversible strains are related to damage, whose effects are quantified by updating the material stiffness matrix. The model has been checked by comparing numerical results and test data concerned with masonry walls (with and without openings) subjected to plane stress states.

## 1. Introduction

Since masonry is an aggregate of bricks with interposed mortar joints, its mechanical properties, which may be assumed as macroscopically orthotropic, depend on the characteristics of each phase (brick and mortar). Starting from this consideration, models for masonry can be divided in two categories: heterogeneous models and homogeneous models. In the first set of models, bricks and mortar are separately discretised. The second group requires the definition of an equivalent homogeneous material, namely a fictitious material whose mechanical properties are equivalent to the overall, average properties of the given, non-homogeneous material.
Within each category, several numerical models based on damage mechanics concepts were developed in the last decades, using a discrete or smeared crack approach (see e.g. [6, 9, 11]). Alternatively, in the discrete approach, crack growth induces a material discontinuity: to properly model the process of crack closure and opening due to cyclic loads, a new structure geometry has to be described and a unilateral constraint has to be introduced. A drawback of these numerical models comes from the large computational effort necessary to analyse real masonry structures. The smeared crack



*D. Weichert and G. Maier (eds.), Inelastic Analysis of Structures under Variable Loads,* 121–134.

approach considers areas characterised by continuous displacement fields and uniform mechanical properties, that imply a reduced stiffness (when compared with the initial one). Thus, cracks are not exactly localised but distributed throughout a certain region. Consequently, this approach can be used with satisfactory results when the material degradation is due to a set of cracks spread over a defined zone.

Homogeneous models, which require a lower computational effort, are based on the definition of an equivalent continuum. Its properties can be obtained either by introducing constitutive laws able to simulate the directional, experimental behaviour of masonry, or by starting from the properties of the masonry components, *i.e.*, mortar and bricks, and using a suitable homogenisation process.

Several macro-models were developed that do not require any homogenisation procedure. They were either specifically conceived for masonry (see e.g. [5, 20, 23]) or derived from models for concrete by introducing slight modifications (see e.g. [4, 8, 24]).

On the other hand, several authors proposed different homogenisation processes by interpreting masonry as a composite material. Encouraging results were obtained (see e.g. [7, 15, 21, 22]), in spite of limitations due to highly different Young's moduli [10]. The computational effort of these models depends upon the use of the homogenisation procedure. Such effort is greater if the process is continually applied as the material is subjected to damage phenomena.

On the contrary, the analysis is simpler if the homogenisation procedure is applied only to determine the initial elastic properties, while inelastic effects are considered for the equivalent, homogeneous material. In this case, however, some care is needed to realise if failure takes place in mortar or bricks.

The elastic-plastic constitutive law with damage presented here belongs to the above class of macro-models for which the homogenisation process is applied just once, when the material is in its initial, virgin state. It was conceived to describe the behaviour of masonry subjected to cyclic in-plane excitation. The constitutive law is derived by assuming an orthotropic equivalent material, whose average physical and mechanical properties are obtained from experimental results on brickwork. Starting from test data on masonry walls subjected to multi-axial loads, a piecewise-linear surface that bounds the elastic domain is introduced. Its evolution law is defined by considering piecewise linear functions, that provide hardening and softening rules.

During the non-linear finite element analysis, the increment of the plastic strain vector is obtained by solving a Linear Complementarity Problem (LCP). At the end of each time step, this vector is split into a plastic (non-reversible) and a reversible part. The material damage is taken into account by considering the change of the stiffness matrix as a function of this latter fraction of the inelastic deformation.

## 2. An elastic-plastic damage model for masonry

Masonry will be assumed as an orthotropic continuum subjected to plane stress conditions. Accordingly, the stiffness matrix **D** of masonry has the following pattern

$$\mathbf{D} = \begin{bmatrix} E_{11}^2 / Q & E_{11}E_{22}\nu_{12} / Q & 0 \\ E_{11}E_{22}\nu_{12} / Q & E_{11}E_{22} / Q & 0 \\ 0 & 0 & G_{12} \end{bmatrix}$$

with $Q = \left(E_{11} - E_{22}\nu_{12}^2\right)$. The parameter $E_{11}$ ($E_{22}$) is Young's modulus in direction 1 (2), $G_{12}$ is the shear modulus and $\nu_{12}$ Poisson's ratio. The symmetry of the stiffness matrix is obtained imposing that $-\nu_{21}/E_{22} = -\nu_{12}/E_{11}$ (hyperelastic model). These parameters may be determined through a homogenisation technique starting from the elastic constants of the components (mortar and brick), whose behaviour is supposed to be homogeneous and isotropic.

The elastic domain is bounded by a piecewise-linear surface (fully analogous to the yield surface in plasticity; [13, 14]), defined in the stress space $\sigma_{11}$-$\sigma_{22}$-$\sigma_{12}$ (with $\sigma_{22}$ normal to mortar beds). The entire surface is made of eighteen planes and is symmetrical with respect to the $\sigma_{11}$-$\sigma_{22}$ plane [18]. The part of this surface concerned with non negative values of $\sigma_{12}$ is shown in Fig. 1. The piecewise linear curve on the plane $\sigma_{12}$=0 is an approximation of traditional curves that give the limit normal stresses for brickwork. Planes schematically shown by white quadrangles are due to the typical decrease of limit normal stresses associated to increments of shear. Finally, dark areas imply the limitation of shear stresses found through experimental tests. Their projections on the $\sigma_{22}$-$\sigma_{12}$ plane are the segments $\alpha$ (for the grey plane in Fig. 1) and $\beta$ (for the black surface) in Fig. 2. Plane $\alpha$ is introduced in order to limit the maximum shear stress in mortar joints [2] and to obtain a better agreement with experimental data. The compression strength is obviously greater than the tensile one.

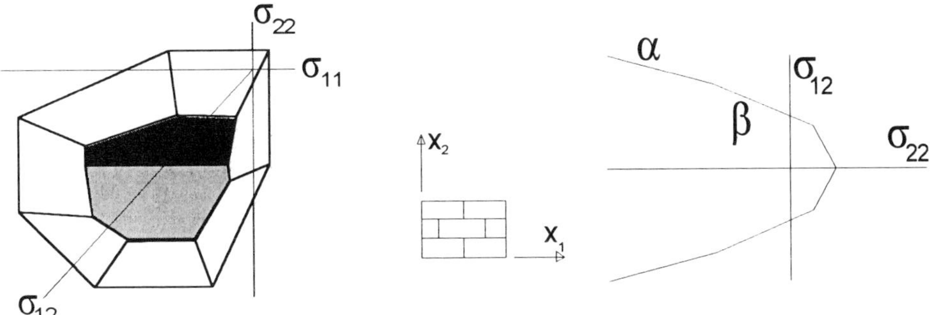


*Figure 1.* Piecewise-linear surface

*Figure 2.* Section in the plane $\sigma_{11}$=0

When the model is used to solve a structural problem, the load history is divided into a convenient number of time steps during which increments of the external actions are applied. The response to each step is found by means of the following iterative scheme:

a) stresses are predicted by assuming a linear elastic response to the strain increments;

b) no inelastic strains are assumed if the predicted stress point belongs to the current elastic domain

c) if the predicted stress point is beyond the current piecewise linear surface, inelastic strain increments are computed on the basis of an associated flow rule by using the same theoretical background which is typical of classical plasticity. First, they are determined as traditional plastic strains. However, at the end of each time step they are split in two contributions [19]: a non recoverable (plastic) strain and a deformation related to the damage processes, which are quantified by updating the stiffness matrix (as discussed below).

The yield surface is updated by means of hardening/softening rules governed by functions which are also piecewise linear. Fig. 3 gives a typical plot where the distance of the $k$-th plane from the origin of the stress space ($\mathbf{r}^k$) is shown as a function of the strain component ($\mathbf{e}^k$) normal to the same plane. Parameters $d_1$ (such as $d_2$ in Fig. 3) refer to the slopes of the different linear branches. After yielding, each branch (*i.e.*, each plastic mode) is associated to one plastic multiplier that represents one contribution to the inelastic strain component normal to the *k-th* plane. Thus, inelastic strain increments are obtained by properly combining plastic multipliers:

$$\Delta\varepsilon^i = \mathbf{n}\,\Delta\mu \qquad with \qquad \mathbf{n} = [\mathbf{n}^1 \mid \mathbf{n}^2 \mid \ldots \mid \mathbf{n}^Y] \qquad (1a,b)$$

where $\mathbf{n}^k$ represents the unit outward vector normal to the *k-th* plane ($k$=1,...,$Y$ and $Y$=18 in this case). The vector $\Delta\mu$ collects eighteen plastic multipliers $\Delta\mu^k$. A multi-branch piecewise linear law is usually needed to properly represent the hardening/softening rule.

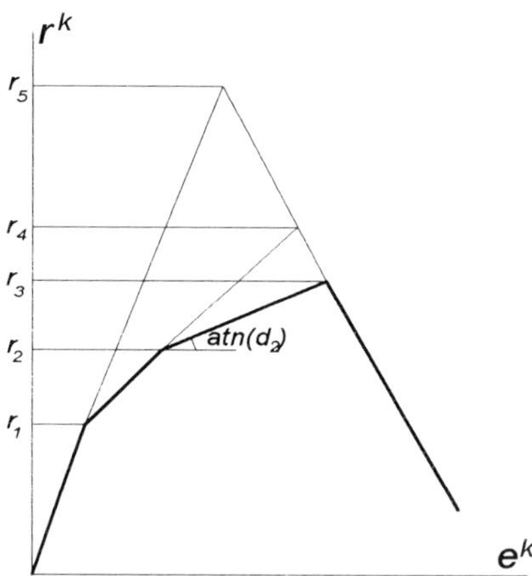


*Figure 3.* Plane hardening law

Thus, several critical stresses should be defined beyond which an inelastic mode becomes active. In this case each multiplier $\Delta\mu^k$ should consist of different contributions. For instance, the piecewise linear approximation of Fig. 3 implies

$$\Delta\mu^k = \Sigma_{j=1,5}\, \Delta\lambda_j^k \tag{2}$$

When the *k-th* plane only is violated, the non-negative contributions $\Delta\lambda_j^k$ must satisfy the following LCP (written in matrix form):

$$\boldsymbol{\phi}^k = [\mathbf{N}^k]^{\mathrm{T}} \{\boldsymbol{\sigma}^{\mathrm{o}} + \mathbf{D}\,\Delta\boldsymbol{\varepsilon}\} - [\mathbf{N}^k]^{\mathrm{T}}\, \mathbf{D}\, \mathbf{N}^k\, \Delta\boldsymbol{\lambda}^k - \mathbf{H}^k\, \Delta\boldsymbol{\lambda}^k - \mathbf{r}^k \leq \underline{0} \tag{3a}$$

$$\Delta\boldsymbol{\lambda}^k \geq \underline{0} \quad ; \quad \{\boldsymbol{\phi}^k\}^{\mathrm{T}} \Delta\boldsymbol{\lambda}^k = 0 \tag{3b,c}$$

with

$$[\mathbf{N}^k]^{\mathrm{T}} = \begin{bmatrix} \{\mathbf{n}^k_1\}^{\mathrm{T}} \\ \{\mathbf{n}^k_2\}^{\mathrm{T}} \\ \{\mathbf{n}^k_3\}^{\mathrm{T}} \\ \{\mathbf{n}^k_4\}^{\mathrm{T}} \\ \{\mathbf{n}^k_5\}^{\mathrm{T}} \end{bmatrix}, \quad \Delta\boldsymbol{\lambda}^k = \begin{Bmatrix} \Delta\lambda_1^k \\ \Delta\lambda_2^k \\ \Delta\lambda_3^k \\ \Delta\lambda_4^k \\ \Delta\lambda_5^k \end{Bmatrix}, \quad \mathbf{r}^k = \begin{Bmatrix} \mathbf{r}_1^k \\ \mathbf{r}_2^k \\ \mathbf{r}_3^k \\ \mathbf{r}_4^k \\ \mathbf{r}_5^k \end{Bmatrix}$$

$$\mathbf{H}^k = \begin{bmatrix} \mathbf{h}_1^k & 0 & 0 & 0 & 0 \\ 0 & \mathbf{h}_2^k & 0 & 0 & 0 \\ 0 & 0 & \mathbf{h}_3^k & \mathbf{h}_4^k & \mathbf{h}_5^k \\ 0 & 0 & \mathbf{h}_4^k & \mathbf{h}_4^k & \mathbf{h}_5^k \\ 0 & 0 & \mathbf{h}_5^k & \mathbf{h}_5^k & \mathbf{h}_5^k \end{bmatrix}$$

The symbol $\sigma^{\mathrm{o}}$ in Equation (3) denotes the stress at the beginning of the current time step. The coefficients $\mathbf{r}_j^k$ represent the distance of the *k-th* plane from the origin of the stress space at which different plastic modes become active, while the hardening/softening parameters $\mathbf{h}_j^k$ are related to the slopes of the linear branches in the $\mathbf{r}^k$-$\mathbf{e}^k$ plot shown in Fig. 3. With simple geometrical considerations, one finds the expression $\mathbf{h}_j = (d_{j-1}d_j)/(d_{j-1}-d_j)$, for $j$=1,2,3 (hardening behaviour). In the case of softening ($j$ = 4,5), the parameters $\mathbf{h}_4$ and $\mathbf{h}_5$ depend on the initial stiffness and one gets $\mathbf{h}_4 = (d_1\, d_3)/(d_1+d_3)$ and $\mathbf{h}_5 = (d\, d_3)/(d+d_3)$.

The above equations have been written in incremental form, as typical for non-linear problems. Therefore, each finite increment of the external loads implies an increment $\Delta\varepsilon$ of the total strain vector at certain locations (that usually coincide with Gauss points), where the constitutive law is to be enforced. This law is satisfied for increments $\Delta\lambda^k$ ($k$=1,...,$Y$), that solve the problem (3). Indeed, the *j-th* multiplier related to the *k-th* plane is $\Delta\lambda_j^k$ ($j$=1,..,5 in the case of Fig. 3) must be zero if $\phi_j^k$<0.

The rank of this LCP is not constant, since the dimension of the vector $\Delta\lambda^k$ depends on the number of branches used to discretise the hardening law. During the analysis, the rank decreases to one when the distance of the *k-th* plane from the origin becomes zero or equals a critical minimum distance (if defined).
Considering all the $Y$ planes of the piecewise linear yield surface, the actual stress increment is obtained from the equation:

$$\Delta\sigma = \mathbf{D}\,\Delta\varepsilon - \mathbf{D}\,\{\Sigma_{k=1,Y}\ \mathbf{N}^k\,\Delta\lambda^k\} = \mathbf{D}\,\Delta\varepsilon - \mathbf{D}\,\mathbf{N}\,\Delta\lambda \tag{4}$$

where $\mathbf{N}=[\,\mathbf{N}^1 \mid \mathbf{N}^2 \mid \ldots \mid \mathbf{N}^Y]$, while $\Delta\lambda$ collects the $Y$ subvectors $\Delta\lambda^k$.
Thus, at each point where the constitutive law must be enforced (often defined *stress point*), the inelastic strains corresponding to a given incremental total strain $\Delta\varepsilon$ can be found by solving the LCP:

$$\phi = \mathbf{N}^{\mathrm{T}}\,\{\sigma^{\mathrm{o}} + \mathbf{D}\,\Delta\varepsilon\} - \mathbf{N}^{\mathrm{T}}\,\mathbf{D}\,\mathbf{N}\,\Delta\lambda - \mathbf{H}\,\Delta\lambda - \mathbf{r} \le \mathbf{0}\quad,\quad \phi^{\mathrm{T}}\,\Delta\lambda = 0\quad,\quad \Delta\lambda \ge 0 \tag{5}$$

Here $\mathbf{H}$ is the block diagonal matrix diag[$\mathbf{H}^k$] and $\mathbf{r}$ collects the subvectors $\mathbf{r}^k$ ($k$=1,...,$Y$).
The iterative solution in terms of $\Delta\lambda$ can be obtained by using Mangasarian's algorithm [16]: convergence is assured if the matrix ($\mathbf{N}^{\mathbf{T}}\ \mathbf{D}\ \mathbf{N} + \mathbf{H}$) is symmetric and positive definite. Considering the hypothesis assumed for the plastic strain determination, this matrix is certainly symmetric, whereas additional conditions would be necessary to have a positive definite matrix.

## 2.1 DAMAGE EVOLUTION LAW

At the end of each time step the vector of the inelastic strains $\Delta\varepsilon^i=\mathbf{N}\,\Delta\lambda$ is determined at each stress point. It can be subdivided in two contributions, $\Delta\varepsilon^p$ and $\Delta\varepsilon^d$, in order to take damage effects into account. The first term represents plastic, non reversible strains, while the second contribution is concerned with strains which are related to damage phenomena and are considered as reversible. Consequently, damage is quantified in this model by updating the material stiffness and compliance matrices. In order to split $\Delta\varepsilon^i$ in two parts, a parameter $\eta$ (with $0 \le \eta \le 1$) is introduced in the following coaxial laws:

$$\Delta\varepsilon^d = \eta\,\Delta\varepsilon^i \quad ; \Delta\varepsilon^p = (1 - \eta)\,\Delta\varepsilon^i \tag{6}$$

At the end of each time interval and at each stress point, the total strain can be expressed in this way:

$$\varepsilon = \mathbf{C}^{\mathrm{o}}\,\sigma + \varepsilon^p + \Delta\varepsilon^d = \varepsilon^p + [\mathbf{C}^{\mathrm{o}} + \Delta C]\,\sigma = \varepsilon^p + \mathbf{C}^{\mathrm{o}}\,\sigma + \Delta C\,\sigma \tag{7}$$

where $\mathbf{C}^{\mathrm{o}}$ (written for an orthotropic material and referred to axes $x_1$-$x_2$ introduced in Fig. 1) is the initial compliance matrix at the beginning of each step, while $\varepsilon$, $\varepsilon^p$ and $\sigma$

are the vectors that collect the total strains, the plastic strains and the stresses at the end of the current step. The orthotropic behaviour, defined from a block partitioned stiffness matrix, suggests that normal and shear components of the strain (stress) tensor should be separately considered. To this aim, the following vectors can be defined: $\sigma^N$ ($\varepsilon^N$), whose non-zero terms are the normal stresses (strains) $\sigma_{11}$ ($\varepsilon_{11}$) and $\sigma_{22}$ ($\varepsilon_{22}$); $\sigma^S$ ($\varepsilon^S$), in which the only non-zero entry is the shear stress (strain) $\sigma_{12}$ ($2\varepsilon_{12}$). Let us also introduce the unit vectors $\mathbf{n}^N$, $\mathbf{n}^S$, $\mathbf{m}^N$, $\mathbf{m}^S$ that give the directions of $\Delta\varepsilon^N$, $\Delta\varepsilon^S$, $\sigma^N$, $\sigma^S$ in the space $\varepsilon_{11}$-$\varepsilon_{22}$-$\gamma_{12}$ and $\sigma_{11}$-$\sigma_{22}$-$\sigma_2$, respectively. With $\rho^N$ and $\rho^S$ defined as the ratios $|\Delta\varepsilon^N|/|\sigma^N|$ and $|\Delta\varepsilon^S|/|\sigma^S|$, (the symbol $|\mathbf{x}|$ denotes the Euclidean norm of $\mathbf{x}$), the damage strain increment can be written as

$$\Delta\varepsilon^d = |\Delta\varepsilon^N|\,\mathbf{n}^N + |\Delta\varepsilon^S|\,\mathbf{n}^S = \rho^N\,|\sigma^N|\,\mathbf{n}^N + \rho^S\,|\sigma^S|\,\mathbf{n}^S \tag{8}$$

In view of eq. (7), the same increment is equal to:

$$\Delta\varepsilon^d = \Delta\mathbf{C}_N\,|\sigma^N|\,\mathbf{m}^N + \Delta\mathbf{C}_S\,|\sigma^S|\,\mathbf{m}^S \tag{9}$$

where $\Delta\mathbf{C}_N$ and $\Delta\mathbf{C}_S$ are the contributions to the compliance matrix $\Delta\mathbf{C}$ characterised by the form [1]:

$$\Delta\mathbf{C}_N = \begin{bmatrix} \Delta\mathbf{C}_{11} & \Delta\mathbf{C}_{12} & 0 \\ \Delta\mathbf{C}_{12} & \Delta\mathbf{C}_{22} & 0 \\ 0 & 0 & 0 \end{bmatrix} \qquad \Delta\mathbf{C}_S = \begin{bmatrix} 0 & 0 & 0 \\ 0 & 0 & 0 \\ 0 & 0 & \Delta\mathbf{C}_{33} \end{bmatrix}$$

Thus, eqns. (9) and (10) imply:

$$\rho^N\,|\sigma^N|\,\mathbf{n}^N + \rho^S\,|\sigma^S|\,\mathbf{n}^S = \Delta\mathbf{C}_N\,|\sigma^N|\,\mathbf{m}^N + \Delta\mathbf{C}_S\,|\sigma^S|\,\mathbf{m}^S \tag{10}$$

Considering the structure of the vectors and tensors of eq. (10), the increment of the compliance matrix related to normal stresses becomes

$$\Delta\mathbf{C}_N = \frac{\rho^N}{\left\{\mathbf{n}^N\right\}^{\mathrm{T}}\mathbf{m}^N}\,[\mathbf{n}^N\,\{\mathbf{n}^N\}^{\mathrm{T}}] \tag{11}$$

whereas the only non-zero component of the $\Delta\underline{C}_S$ matrix is

$$\Delta\mathbf{C}_{33} = \rho\,\frac{\mathbf{n}_3^S\;\mathbf{n}_3^S}{\mathbf{m}_3^S\;\mathbf{m}_3^S} \tag{12}$$

In numerical applications damage levels will be shown by using appropriate contour lines. Such lines will not be drawn by referring to compliance changes, but by introducing a fictitious internal variable whose values range between 1 (elastic behaviour) and 6 (complete failure).

## 3. Numerical tests

In order to check the reliability of the model, first a few simple problems were solved by using a single finite element. By considering the load history of Fig. 4, the model gives the response depicted in Fig. 5: one notes that the different behaviour under compression and tension (typical of experimental tests) has been reproduced. Indeed, initial positive increments of the total strain lead to the tensile limit stress shown by point A in Fig. 5. Next, negative increments give the piecewise linear curve A-B, characterised by changes of slope at p2, p3, p4, where new yield modes are activated. The straight line B-C is related to a loading/unloading path in the elastic range owing to relatively small positive/negative strain increments. Further negative strain increments lead to point D, while subsequent positive increments give the curve D-E. Note that the peak stress associated to the point p5 corresponds to the positive yield limit previously determined (point A). During this phase of the loading process, further softening occurs and the upper yield limit becomes zero (point E). Finally, the negative strain increment E-F (*cf.* Fig. 4) implies an elastic response up to p6 (*cf.* Fig. 5), that denotes the current negative yield limit established at D. After point p6, softening takes place once more, until the lower yield limit is also reduced to zero.

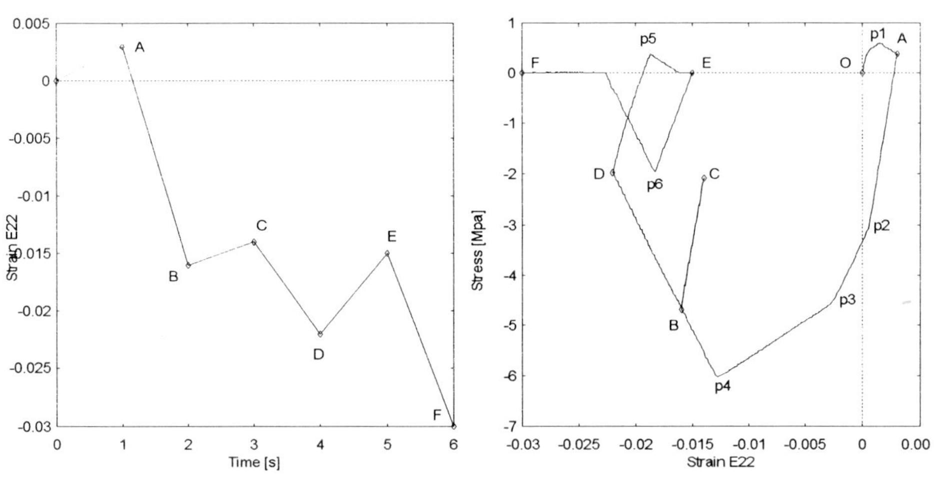


*Figure 4.* Time history

*Figure 5.* Stress-strain curve

In addition, the realistic response of Fig. 6a is obtained by imposing an increasing level of shear alternate strain. On the other hand, if the plane $\alpha$ (see Fig. 2) is not considered and the limit shear strength is only governed by the plane $\beta$, the maximum shear stress continues to increase, toward high non-consistent values (Fig. 6b).

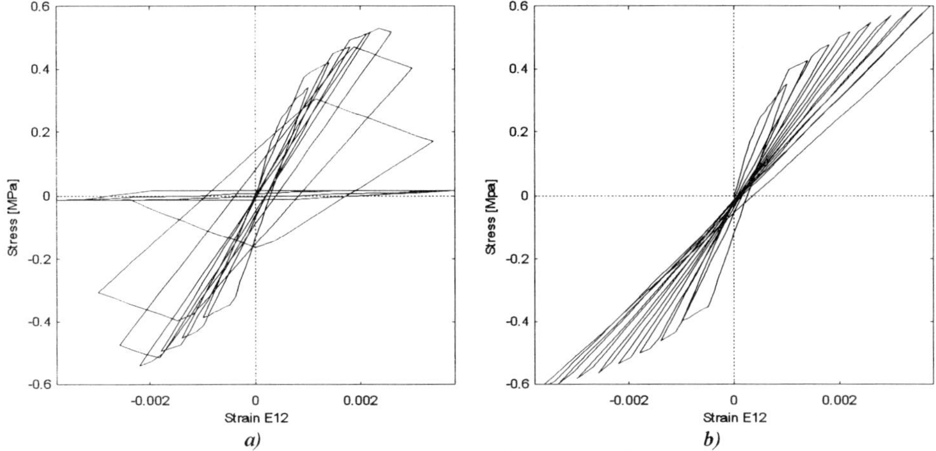

*Figure 6.* Shear stress-strain curve with (a) and without (b) plane 9

After these qualitatively encouraging results, the model was used to simulate shear tests carried out on a slender panel and a squat panel. The two masonry walls, which were equal in width (1 m), were 2 m and 1.35 m high, respectively.

A constant vertical downward load was applied and a cyclic displacement at the top was imposed. The experimental responses are obviously different for the two panels [12]. For the squat panel, the horizontal force *vs.* horizontal displacement plot (Fig. 7a) shows a degradation of stiffness and strength coupled with a significant dissipation. Indeed, its collapse depends on the growth of shear damage in the central zone. On the contrary, the curve for the slender panel (Fig. 7b) is characterised by a smaller dissipation due to a smaller reduction of stiffness and a lack of softening behaviour.

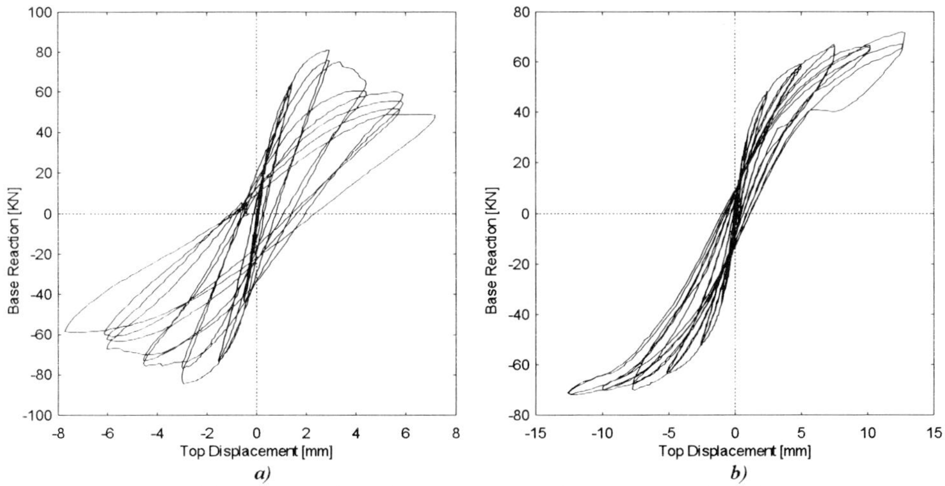

*Figure 7.* Experimental response for squat (a) and slender (b) panel

Also the failure mechanism, which is due to rocking (and, hence, to bending stresses), is different. Figs. 8a,b show the numerical responses for the two panels: it is possible to note a good agreement both in terms of failure loads and of dissipation.
Also the distribution of shear damage for the squat panel (Fig. 9a) and of tensile damage for the slender panel (Fig. 9b) is consistent with the different failure mechanisms.

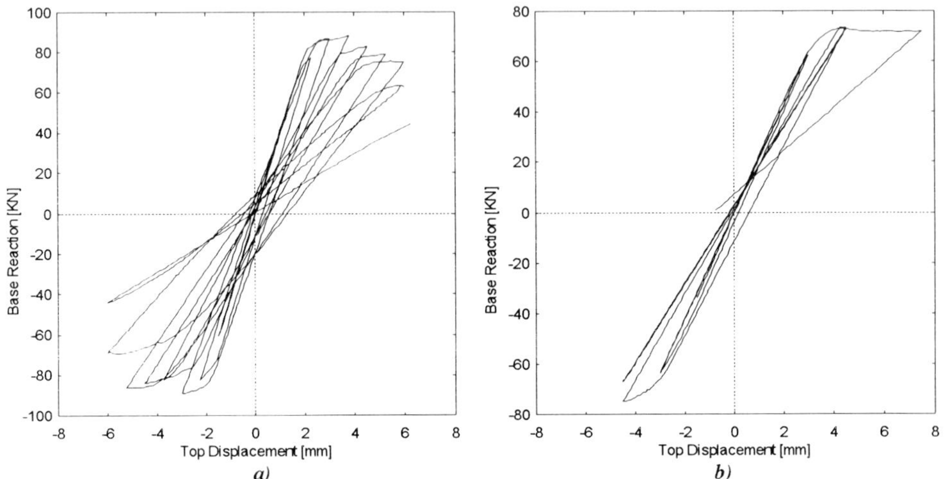

*Figure 8.* Numerical response for squat (a) and slender (b) panel.

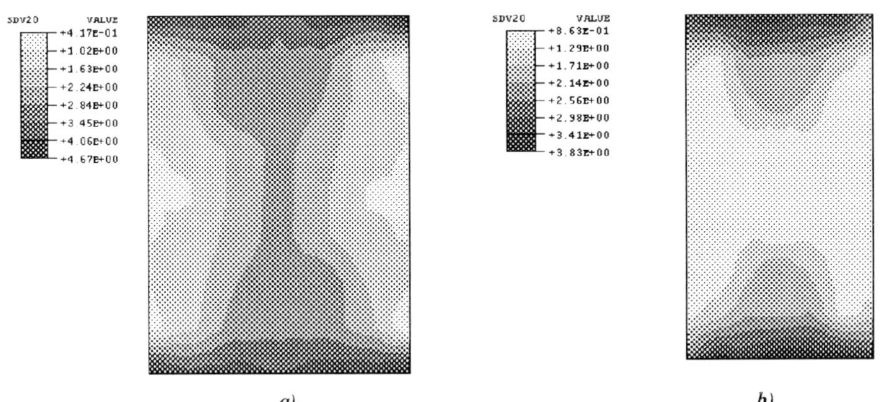

*Figure 9.* Distribution of shear damage for squat panel (a) and of tensile damage for slender panel (b)

This model was also used to simulate the behaviour of a real wall of a two-story masonry building tested at the University of Pavia [3]. A dead vertical load was considered and two equal horizontal forces (which vary in a cyclic way and change in sense) were applied at each floor. The total horizontal loads *vs.* horizontal displacement at the left corner of the second floor curves are reported in Fig. 10. It is possible to note that the experimental (Fig. 10a) and numerical (Fig. 10b) responses are in good agreement.

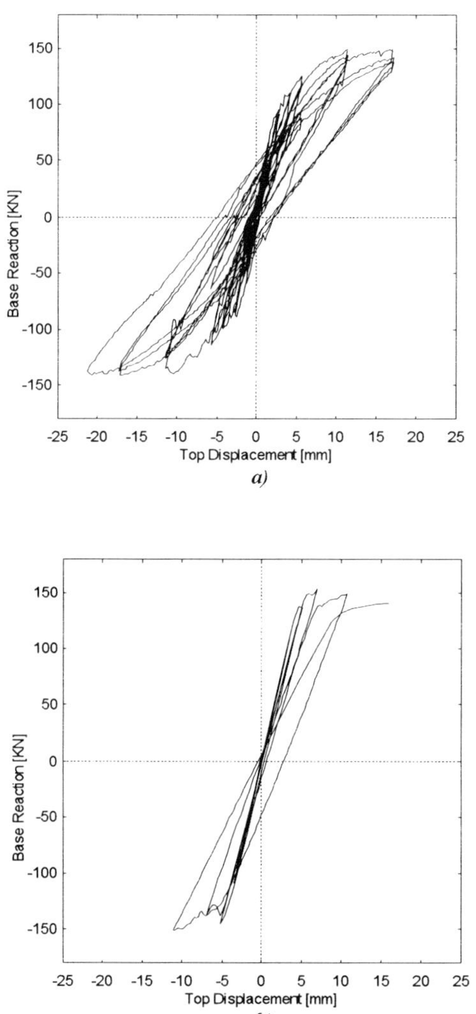


*Figure 10.* Experimental (a) and numerical (b) response of real masonry wall

The numerical damage patterns at failure, depicted in Figs. 11, show a significant shear damage in the lower central zone (Fig. 11a) and a high tensile damage at the bottom (Fig. 11b). These distributions correspond to the experimentally observed failure mechanism. Indeed, after the development of tensile cracks at the bottom of the wall, shear cracks start to appear in the lower central zone and tend to propagate until collapse.

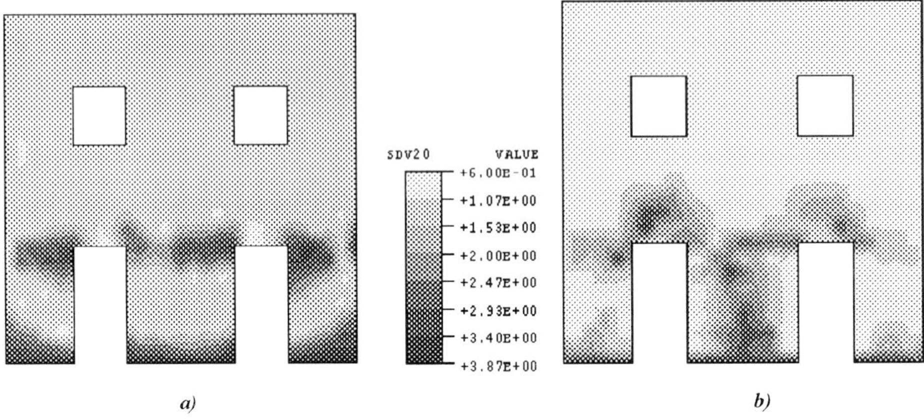


*Figure 11.* Distribution of tensile (a) and shear (b) damage

## 4. Conclusions

A numerical approach has been discussed, which is suitable for the analysis of masonry structures subject to plane stress states and cyclic loads. The approach is based on concepts found in the framework of plasticity and damage mechanics. One piecewise linear yield surface is defined at each stress point and an associated flow rule is assumed for inelastic strains. These strains are split in two portions: a non-reversible part (plastic deformation) and a reversible part, that takes into account damage effects, quantified by updating the stiffness matrix. A hardening/softening rule is also defined. It is governed by piecewise linear curves and implies some shift of the yield planes as inelastic strains tend to develop. In view of a double piecewise linear approximation (both for yield surfaces and hardening/softening behaviour), the constitutive law at each stress point is enforced by solving a Linear Complementarity Problem. The parameters that govern the material response are found in a relatively simple way: experimental data for simple tension, compression and shear are needed. Consequently, the model is suitable for practical applications and even historical building can be considered at the cost of reasonable approximations.

It is also worth noting that the general features of the model appear consistent with the mechanical behaviour of several rock-like materials subject to quasi-brittle fracture. Therefore, its field of potential applications is quite large and, in any case, well beyond the case of brick masonry. The paper, however, is centred on this kind of structural systems and the numerical tests reported in the previous Section are concerned with different patterns of brick masonry walls subject to cyclic load conditions: one squat panel, one slender panel and one large wall with openings. Numerical results are compared with experimental data and show an excellent agreement, so that the model seems to be as a proper tool for the numerical analysis of masonry in the presence of cyclic loads.

**Acknowledgements**

The financial support of the Italian Ministry of University and Scientific Research is acknowledged.

## 5. References

1. Callerio, A. (1998) Comportamento sismico delle murature negli edifici monumentali: un metodo di analisi basato sulla meccanica del danneggiamento, PhD. Thesis, Dep. of Structural Engineering, Politecnico of Milan, Italy, (in Italian)
2. Callerio, A. and Papa, E., (1997) An elastic-plastic model with damage for cyclic analysis of masonry panels, in *Proceedings of 4th Int. Symposium on Computer Methods in Structural Masonry,* Florence, Italy, (in press)
3. Calvi, G.M. and Magenes, G. (1994) Experimental research on response of URM building system, in D.P. Abrams et al. (eds.*), Guidelines for Seismic Evaluation and Rehabilitation of Unreinforced Masonry Buildings*, State University of New York at Buffalo, pp. 3/41-57
4. Chen, W.F. (1982) *Plasticity in reinforced concrete*, McGraw-Hill
5. Dhanasekar, M., Page, A.W. and Kleeman, P.W. (1985) The failure of brick masonry under biaxial stresses, *Proc. Inst. Civ. Engrs.* **79**, pp. 295-313
6. Gambarotta, L. and Lagomarsino, S. (1997) Damage models for the seismic response of brick masonry shear walls. Part I: the mortar joint model and its applications, *Earth. Eng. and Struct. Dynamics* **26**, 423-439
7. Gambarotta, L. and Lagomarsino, S. (1997) Damage models for the seismic response of brick masonry shear walls. Part II: the continuum model and its applications, *Earth. Eng. and Struct. Dynamics* **26**, pp. 441-462
8. Ganju, T.N. (1977) Non-linear finite element analysis of clay brick masonry, in *Proceedings of 6th Australian Conf. on the Mechanics of Structures and Materials*, Univ. of Canterbury, pp. 59-65
9. Lotfi, H.R. and Shing, P.B. (1994) Interface model applied to fracture of masonry structures, *J. Struct. Engrg. ASCE* **120**, 63-80
10. Lourenço, P.B. (1996) Computational strategies for masonry structures, PhD. Thesis, Delft Univ. of Technology
11. Lourenço, P.B., Rots, J.G. and Blaauwendraad, J. (1994) Implementation of an interface cap model for the analysis of masonry structures, in H. Mang, N. Bicanic and R. de Borst (eds.), *Computational Modelling of Concrete Structures*, Pineridge Press, Swansea, pp. 123-134
12. Magenes, G. (1992) Comportamento sismico di murature di mattoni: resistenza e meccanismi di rottura di maschi murari, Ph.D. Thesis, Dept. of Structural Mechanics, Univ. of Pavia, (in Italian)
13. Maier, G. (1970) A matrix structural theory of piecewise linear elastoplasticity with interacting yield planes, *Meccanica* **5**, 54-66
14. Maier, G. and Nappi, A. (1990) A theory of perfectly no-tension discretized structural system, *Eng. Struct.* **12**, 229-233
15. Maier, G., Papa, E. and Nappi, A. (1991) On damage and failure of brick masonry, in *Experimental and Numerical Methods in Earthquake Engineering*, Kluwer, Dordrecht, pp. 223-245
16. Mangasarian, O.L. (1977) Solutions of symmetric linear complementarity problems by iterative methods, *J. of Optimisation Theory & Appl.* **22**, 465-485
17. Middleton, J., Pande, G.N., Liang, J.X. and Kralj, B. (1991) Some recent advances in computer methods in structural masonry, in J. Middleton and G.N. Pande (eds.), *Computer Methods in Structural Masonry*, Books and Journals International, Swansea, U.K., pp. 1-21
18. Nappi, A., Facchin, G. and Marcuzzi, C. (to appear) Structural dynamics: convergence properties in the presence of damage and applications to masonry structure, *Struct. Eng. & Mech.*

19. Ortiz, M. (1985) A constitutive theory for the inelastic behaviour of concrete, *J. Mech. Mat.* **4**, 67-93
20. Page, A.W., Kleeman, P.W. and Dhanasekar, M. (1985) An in-plane finite element model for brick masonry, in S.C. Anand (ed.), *New Analysis Techniques for Structural Masonry*, ASCE Press, New York, pp. 1-18
21. Pande, G.N., Liang, J.X. and Middleton, J. (1989) Equivalent elastic moduli for brick masonry, *Computer & Geotechnics* **8**, 243-265
22. Pietruszczak, S. and Niu, X. (1992) A mathematical description of macroscopic behaviour of brick masonry, *Int. J. of Solids & Struct.* **29**, 531-546
23. Samarashinghe, W., Page, A.W. and Hendry, A.W. (1982) A finite element model for the in-plane behaviour of brickwork, *Proceedings of Inst. of Civ. Engrs* **72**, 171-178
24. Shing, P.B., Klamerus, E., Spaeh, H. and Noland, J.L. (1988) Seismic performance of reinforced masonry shear walls, in *Proceedings of 9th World Conf. on Earthquake Engineering*, Japan, pp. 103-108

# CYCLONE - SYSTEM FOR STRUCTURAL ADAPTATION AND LIMIT ANALYSIS

A. SIEMASZKO*, G. BIELAWSKI+ AND J. ZWOLINSKI+
* *Institute of Fundamental Technological Research, Swiętokrzyska 21, Warsaw, Poland*
\+ *Warsaw University of Technology, Plac Politechniki 1, Warsaw, Poland*

**Abstract**

This paper proposes an extended min-max procedure which evaluates safety factors against inadaptation and ultimate states for structures working under variable repeated loading. The algorithm applied is shown to be highly efficient. The procedure is incorporated into CYCLONE - computational system for structural adaptation and limit analysis. CYCLONE is integrated with commercial finite element codes.

## 1. Introduction

Nuclear and conventional power plants, chemical and petrochemical plants, pipelines, off-shore installations, and ultra high pressure chambers require for their design the application of the adaptation and limit analysis methods. This is recognised by leading design codes for pressure vessels such as BS 5500 and ASME VIII. Inelastic structures under extremely heavy variable repeated loading may work in four different regimes: elastic, adaptation (shakedown), inadaptation (non-shakedown), and ultimate (limit) state. Since for the elastic regime there are no plastic effects at all, whereas for the adaptation regime plastic effects are restricted to the initial loading cycles and then followed by purely elastic behaviour, both regimes are considered as safe working regimes. Hence, the elastic and adaptation regimes constitute a foundation for the structural design.

Design codes of pressure vessels, e.g. BS 5500 and ASME VIII, divide stresses into several categories (e.g. primary, secondary stresses) and provide safety factors according to limits of elastic and adaptation regimes. The classical design-by-rules route is now partially substituted by the design-by-analysis based in general upon a finite element analysis. The elastic limit is obtained by a linear elastic and the ultimate limit by a non-linear incremental analysis. The adaptation limit is provided by a formula, which holds only for simple stress states:



*D. Weichert and G. Maier (eds.), Inelastic Analysis of Structures under Variable Loads,* 135–146.

$$p_{SD} = min(p_U, 2p_E) \tag{1}$$

where $p_{SD}$, $p_U$, $p_E$ are the adaptation, ultimate, and elastic limits, respectively. In general, for combined loading the above formula overestimates the structural safety.

There is growing evidence that the evaluation of parameters for the actual working regime of existing structures is crucial for the assessment of their structural safety and reliability. For a given load variation domain the safety factors (multipliers) against inadaptation and ultimate state should be computed. For given loading histories the non-linear incremental procedures offered by modern finite element codes may be used. However, since the variable loading path must be incrementally simulated, the computational effort may become prohibitively high. Moreover, the exact loading history is often not available.

The adaptation and limit analysis offer a powerful alternative or at least a complementary approach. They provide the accurate adaptation and ultimate limits without specification of exact loading history. It is enough to know only the limits of load variation. Practically, as yet, there are no commercial codes computing the adaptation limit in a general case. To compute the adaptation and ultimate multipliers, this paper proposes an extended min-max procedure, which minimises the maximum values of the elastic stresses by optimal selection of the residual stresses. The procedure is implemented for plane stress/strain and axisymmetric problems and is actually extended to 3D problems. A computational system CYCLONE is developed which is fully integrated with commercial finite element codes. The realistic design problem of a ultra high pressure chamber is discussed in detail. The algorithm applied is shown to be highly efficient.

## 2. Elastic, adaptation, inadaptation, and ultimate states

The adaptation is considered as the most rational safety criterion for elastic-plastic structures subjected to cyclic loading. For the adaptation, plastic strains produced within the first critical load cycle will result in a development of residual stresses as well as material and geometrical hardening. These factors may drastically change properties of critical elements removing a threat of further plastic effects. As a result the structure will work for all further load cycles exclusively within the elastic regime.

If the adaptation limit is exceeded, nonvanishing increments of plastic strain may be observed. In this case, due to accumulation of plastic strains, the inadaptation phenomena will develop: alternating plasticity and/or plastic ratchetting (progressive plastic deformation). The alternating plasticity occurs for alternating sign plastic strain increments and results in a low-cycle fatigue. The ratchetting occurs for monotonic plastic strain increments and results in a ductile or brittle fracture. Of course, the inadaptation effects have to be avoided in the real structural life. However, for some special cases, also the inadaptation can be admitted provided that the number of critical loading cycles is controlled. Examples of such systems are structures subjected to very scarce extreme exploitation loads, structures with very low ratchetting effects, or ultra high pressure chambers. It is obvious that structures working within the inadaptation regime have a limited life time. A limit case for the inadaptation is called the ultimate state (limit load state). It means that applied cyclic loads are so high that the immediate collapse occurs within the first critical load cycle.

There are three important limits separating the states mentioned above: elastic limit, adaptation limit and ultimate limit. Considering actual (given) cyclic or variably repeated loading three safety factors (multipliers) against respective limits can be computed: elastic, adaptation and ultimate multipliers. It always holds that the adaptation multiplier is between elastic and ultimate multipliers (or coincides with one of them).

The insight into post-critical (inelastic) behaviour can be crucial for the accurate evaluation of structural safety and reliability. A mutual relation of elastic, adaptation, and limit carrying capacities can be an important indicative factor during structural design. Performing the design according to the elastic theory may lead to conservative, uneconomical designs since adaptation reserves of strength are not exploited. However, the opposite situation is possible that the elastic approach leads to unsafe designs with a very narrow reserve of strength margins against the inadaptation or the limit state.

To discuss these problems an example of bar with a circular notch is considered, Fig. 1. The radius of the notch is denoted by $r$, the radius of the bar is equal to $3a$, where $a$ denotes the radius of the neck. The bar is subjected to variable tensile forces $p$. The elastic - perfectly plastic material model and the von Mises yield condition are applied. The elastic, adaptation, and ultimate problems are solved.

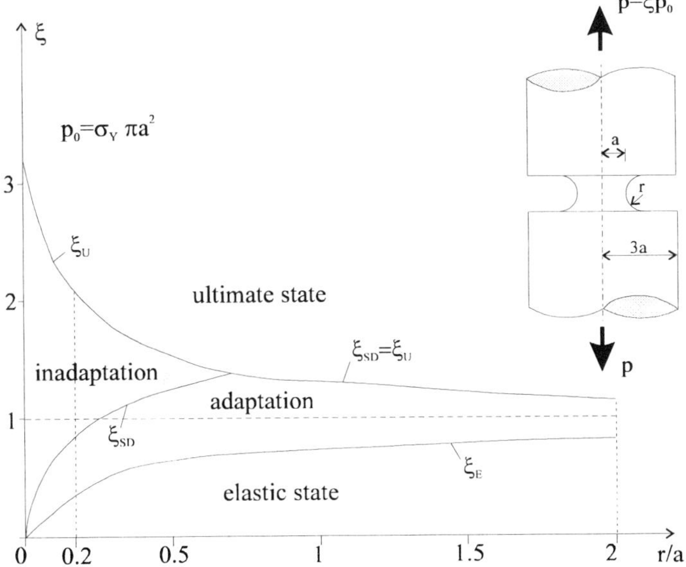


*Figure 1.* Elastic $\xi_E$, adaptation $\xi_{SD}$, and ultimate $\xi_U$ multipliers for the rod with circular notch.

For $r/a = 2$, there is a very narrow adaptation domain and, moreover, there is no inadaptation domain. The structure designed according to the elastic theory will have a low reliability. Some imperfections or unexpected overloading may result in structural failure. Increasing loading by *38%* the ultimate state will be attained. For $r/a = 0.2$ the structure designed according to the elastic theory will be highly reliable, but completely uneconomical. To reach the ultimate state the loading must be increased *4.34* times.

On the other hand, neglecting the variability of loads, i.e. assuming that they are applied once, the limit analysis or non-linear incremental analysis by a FE code admits load levels exceeding considerably the adaptation limit, e.g. *2.2* times in the case $r/a = 0.2$, Fig.1. Even application of high safety factors, decreasing the admissible loads, will cause the inadaptation. The structure will fail after several load cycles. Therefore, the limit analysis and incremental FE analysis (with a single application of loading), providing ultimate multipliers, may be inadequate for the evaluation of structural safety. This example shows that for structural assessment and design it would be desirable to develop a triad of analysis methods allowing to estimate the elastic (first yielding), adaptation, and ultimate multipliers.

## 3. Computational method

There are two basic strategies for identification of adaptation, inadaptation and ultimate states as well as computation of respective limits (safety multipliers). In the first one the non-linear incremental FE procedures are employed. In the second one, called a global plastic analysis the loading history is by-passed and a final asymptotic state is analysed.

The cyclic non-linear incremental analysis requires that the load history must be explicitly specified and simulated. For complicated cyclic loads this approach is currently very computer time consuming [2]. The final steady cycle behaviour evaluated gives only the answer in which regime (elastic, adaptation, inadaptation, ultimate) the structure works. It says nothing about limits separating these regimes. To evaluate safety multipliers against any of these regimes, a trial and error procedure must be used and the incremental procedure must be applied many times.

As a powerful alternative the global plastic analysis methods: the adaptation (shakedown) analysis and limit analysis can be used for which only limits of load variation must be known and the specification of exact loading history is not necessary [1,3,4]. Respective safety multipliers can be easily computed. Majority of the adaptation and limit analysis methods is based on mathematical programming procedures that for real-sized problems are not so efficient [3, 4]. To compute the adaptation and limit multipliers, this paper proposes an extended min-max procedure which minimises the maximum values of the reduced (von Mises) stresses by optimal selection of the residual stresses. The procedure is based upon the statical approach to adaptation and limit load problems.

The adaptation and limit load problem can be formulated as a min-max problem

$$\xi^{-1} = \min_{(q)} \max_{(x)} \frac{f\{\sigma_E(x) + \sigma_R[q(x)]\}}{\sigma_Y[q(x)]} \tag{2}$$

where $\xi$ is the adaptation or ultimate multiplier, $f$ is the yield condition, $\sigma_E$ is the elastic stress corresponding to loads applied (for evaluation of the limit load) or the envelope of elastic stresses (for the adaptation), $\sigma_R$ denotes residual stresses, $\sigma_Y$ is the yield stress which may be dependent on plastic strains $q$ for hardening materials and $x$ are coordinates.

The procedure tries to select such residual stresses $\sigma_R$ which reduce in a most efficient way all peaks (maxima) of elastic stresses $\sigma_E$ within the whole structure. A highly efficient minimisation algorithm [6,7], specially tailored for residual stresses problems is implemented. The algorithm requires the elastic solution $\sigma_E$ from the commercial FE code. Assuming that for the step $t\text{-}1$ of the computation process the plastic strain field $q_{t-1}$ is known the residual and total stresses can be written as

$$\begin{aligned} \sigma_R^{t-1} &= Aq^{t-1} \\ \sigma^{t-1} &= \sigma_E + \sigma_R^{t-1} \end{aligned} \tag{3}$$

Plastic strain increments $\Delta q^t = \lambda^t \, \bar{q}$ are assumed to be proportional to the predicted field $\bar{q}$ which may, for instance, follow the gradient $\partial f / \partial \sigma$, and where $\lambda$ is to be determined. Hence, plastic strains and total stresses at the step $t$ are

$$\begin{aligned} q^t &= q^{t-1} + \Delta q^t \\ \sigma^t = \sigma^{t-1} + \Delta\sigma_R^t &= \sigma^{t-1} + \lambda^t A \bar{q}^t \end{aligned} \tag{4}$$

Substituting the above relations into the von Mises yield functions, quadratic equations $f(\lambda^t)$ are obtained. For all elements the procedure identifies maximal parabola at $\lambda^t = 0$ and moves along it to the minimum or to the intersection point with another parabola. This provides the value of $\lambda^t$ to be used in scaling plastic strain increments. When $\lambda^t$ is determined the actual plastic strains and stresses are computed and the next iteration is initiated.

Since the method imposes residual stresses $\sigma_R$ reducing all peaks (maxima) of elastic stresses $\sigma_E$, when the procedure converges to the optimum, there are more and more elements reaching the same value of yield function $f$. There are some special rules applied to ensure for these elements the equal descent of function $f$. Different scaling procedures are applied at each iteration depending on if the analysis aims at computing the ultimate multiplier $\xi_U$ or the adaptation multiplier $\xi_{SD}$. The procedure is stopped when some convergence criteria are fulfilled.

The method admits that effects of kinematic material hardening resulting from plastic strains can be accounted for as well as the influence of temperature on material properties.

## 4. CYCLONE

The min-max method has been implemented into a computational system called CYCLONE. The computational code is written using the object oriented programming provided by C++ language. The system is installed on PC computers as well as on the

SUN workstation. Actually, standard FE codes as ABAQUS or ANSYS are used as background codes for finite element analysis. The min-max procedure works for plane strains and stresses as well as axisymmetrical problems. Linear finite elements with the constant stress and strain assumption are implemented. Development of 3D capabilities is in progress. It is worth to notice that the min-max algorithm is a general one allowing to solve large realistic problems. The size of problems is limited practically only by computer capabilities. Moreover, the procedure may work for any type of finite elements supported by the background FE code. The necessary matrix of dependence *A,* eq. (3), is generated by CYCLONE. The computational system was tested on problems with several thousands of elements. For the largest solved example the number of elements in yielding, which is critical for the computational time, has reached 600. In this case several hours were required on the PC Pentium computer with 32 MB RAM.

To support plastic analysis and design of structures, an integrated computational environment has been developed. It relies on commercial and in-house pre- and post-processors, finite element analysis, adaptation and limit analysis capabilities. It supports the capability to launch software codes, define geometry by a pre-processor, perform finite element analysis, adaptation and limit analysis, and visualise results by a post-processor. A menu-driven-system in a multi-window graphics environment is provided abiding to currently accepted standards. Open modular architecture of the computational system is maintained enabling each module to be replaced with ease.

The adaptation and limit analysis module must be integrated with a FE code from which they retrieve the following data: initial solution with the elastic stress $\sigma^E$ and in an iterative way the residual stresses $\sigma^R$ corresponding to the imposed plastic strains $q$. The elastic multiplier (safety factor) $\xi_E$ is computed as

$$\xi_E = \sigma_Y / \sigma_E \tag{5}$$

where $\sigma_Y$ is the yield stress and $\sigma_E$ is the maximum von Mises stress in the structure.

At each iteration a new increment of plastic strains and resulting residual stresses are computed and imposed on the actual stress field. When the minimisation procedure converges to optimum the maximum adaptation von Mises stress $\sigma_{SD}$ or the maximum ulti-mate von Mises stress $\sigma_U$ is obtained (more accurately the minimum-maximum von Mises stress). The resulting adaptation multiplier $\xi_{SD}$ or the ultimate multiplier $\xi_U$ are defined as

$$\begin{aligned} \xi_{SD} &= \sigma_Y / \sigma_{SD} \\ \xi_U &= \sigma_Y / \sigma_U \end{aligned} \tag{6}$$

The results, i.e. elastic, residual and total stress as well as plastic strain field can be visualised by a commercial post-processor.

## 5. Example

The min-max procedure presented in this paper has been used to assess the safety of a cylindrical pressure chamber. It is subjected to variable repeated pressure $p_0 = 420\ MPa$. The geometry of the pressure chamber is shown in Fig. 2. Three cases were considered:

- case 1: chamber with the internal pressure applied on the full length $l_1 = 225\ mm$;
- case 2: chamber with the internal pressure applied on the length $l_2 = 100\ mm$;
- case 3: chamber with the spherical end subjected to the internal pressure.

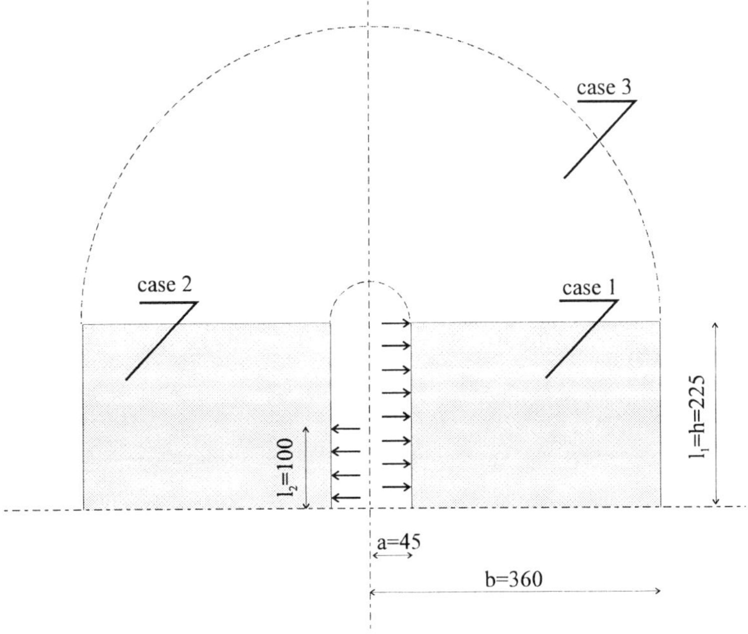


*Figure 2.* High pressure chamber

The non-dimensional pressure parameters are introduced

$$\bar{p} = p / \sigma_Y \tag{7}$$

where $\bar{p}$ is the pressure parameter, p is the actual pressure and $\sigma_Y$ is the yield limit.

The elastic, adaptation, and limit pressure parameters $\bar{p}_E$, $\bar{p}_{SD}$ and $\bar{p}_U$, respectively are computed as follows

$$\bar{p}_E = p_0 / \sigma_E \qquad \bar{p}_{SD} = p_0 / \sigma_{SD} \qquad \bar{p}_U = p_0 / \sigma_U \tag{8}$$

where $p_0$ is the reference (computational) pressure and $\sigma_E, \sigma_{SD}, \sigma_U$ are the corresponding maximum von Mises stresses for elastic, adaptation and ultimate states, respectively.

For the case 1, Fig. 2, there are analytical solutions of the long pipe (with closed ends) subjected to internal pressure (a, b are the internal and external pipe radii)

$$\bar{p}_E = (1 - a^2 / b^2) / \sqrt{3}$$

$$\bar{p}_U = \frac{2}{\sqrt{3}} ln\left(\frac{b}{a}\right) \tag{9}$$

Assuming (1) the following formulae are obtained for the adaptation

$$\bar{p}_{SD} = \frac{2}{\sqrt{3}} ln\left(\frac{b}{a}\right) \quad for\ 1 < b / a < 2.2$$

$$\bar{p}_{SD} = \frac{2}{\sqrt{3}}\left(1 - \frac{a^2}{b^2}\right) \quad for\ b / a > 2.2 \tag{10}$$

The theoretical results are shown in Fig.3.

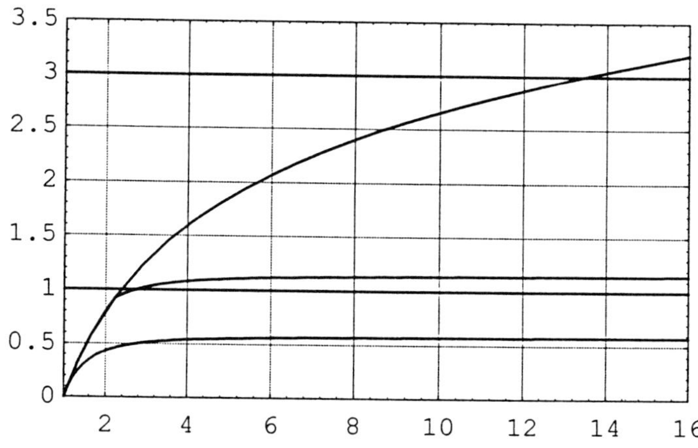


*Figure 3.* Elastic, adaptation and ultimate curves for the long pipe

The computation with CYCLONE gives the following pressure parameters *(b/a = 8.0)*

$$\bar{p}_E = 0.60 \qquad \bar{p}_{SD} = 1.08 \qquad \bar{p}_U = 1.21 \tag{11}$$

Load step = 1 Iteration = 1
Neutral file = K11L.RSE

Variable = VONM
Max. = 347.841
Min. = 20.6229
9 = 329.25
8 = 292.95
7 = 256.65
6 = 220.35
5 = 184.05
4 = 147.75
3 = 111.45
2 = 75.15
1 = 38.85

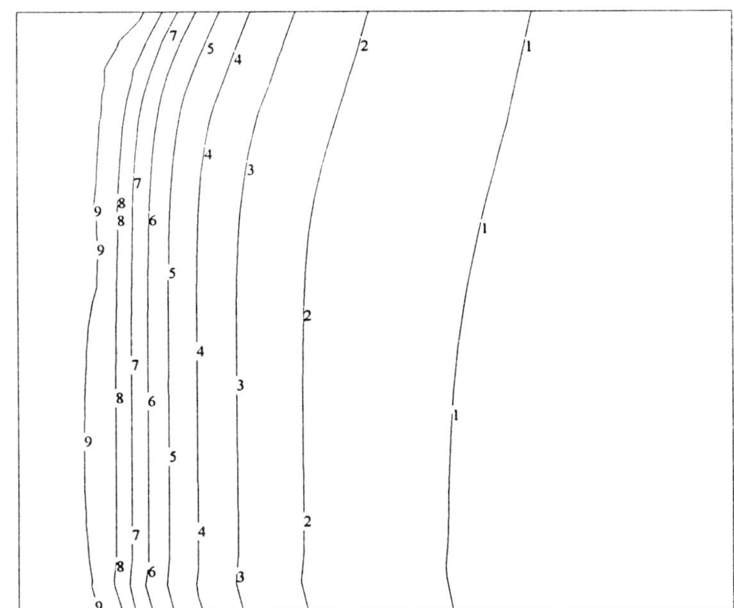

Load step = 2 Iteration = 1
Neutral file = K11L.RSE

Variable = VONM
Max. = 0.38843E-02
Min. = 0.0E+00
9 = 0.3672E-02
8 = 0.324E-02
7 = 0.2808E-02
6 = 0.2376E-02
5 = 0.1944E-02
4 = 0.1512E-02
3 = 0.108E-02
2 = 0.648E-03
1 = 0.216E-03

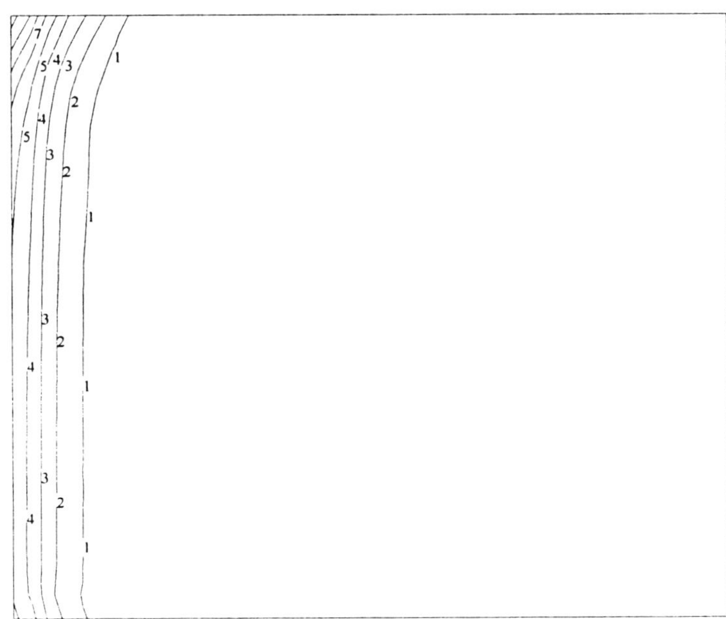

*Figure 4.* Case1. a) Von Mises stresses for the ultimate case; b) the corresponding plastic strains.

Distribution of the von Mises stresses for the ultimate case is shown in Fig.4a and the corresponding plastic strains in Fig. 4b.

The elastic and adaptation pressure parameters (11) are in a good agreement with the theoretical results, cf. Fig.3, however, the adaptation pressure is lower than expected. The ultimate state of the pipe results from the unconstrained flow of the unsupported pipe ends, as it is shown in Fig. 4b. Therefore, the theoretical model of long pipe with closed ends (9) cannot be used here.

To analyse the effect of closure of pipe ends the case 2 was considered. The chamber pressure carrying capacities were raised

$$\bar{p}_E = 0.63 \qquad \bar{p}_{SD} = 1.33 \qquad \bar{p}_U = 3.04 \tag{12}$$

The ultimate pressure parameter $\bar{p}_U$ is still not in a good agreement with the theoretical result for the long pipe with closed ends. It fits more to the Prandtl-Hill solution of the concentrated force for the plane strain state

$$\bar{p}_U = \left(\frac{\pi}{2} + 1\right)\frac{2}{\sqrt{3}} \cong 3 \tag{13}$$

Fig.5a shows the von Mises stresses for the adaptation state. Results similar to that of the case 2 were obtained for the spherical ended chamber, the case 3, however, the ultimate pressure was lower than in the case 2

$$\bar{p}_E = 0.62 \qquad \bar{p}_{SD} = 1.32 \qquad \bar{p}_U = 2.15 \tag{14}$$

Fig.5b shows the Von Mises stress for the ultimate state.

In general, the thick walled chambers *(b/a>2.2)* with free ends (case 1) have the pressure carrying capacity lower than that of supported or closed ends (cases 2 and 3). For the thin walled chambers *(b/a<2.2)*, type of the end support has no big influence on the admissible pressure since $\bar{p}_{SD} = \bar{p}_U$. Comparison of elastic, adaptation and ultimate pressure parameters obtained for different cases are shown in the Table 1. The chamber geometry of the case 2 allows to obtain the highest admissible pressures.

TABLE 1. Elastic adaptation and ultimate pressure parameters for different chamber geometry

| | Case 1 | Case 2 | Case 3 |
|---|---|---|---|
| $\bar{p}_E$ | *0.60* | *0.63* | *0.62* |
| $\bar{p}_{SD}$ | *1.08* | *1.33* | *1.32* |
| $\bar{p}_U$ | *1.21* | *3.04* | *2.15* |

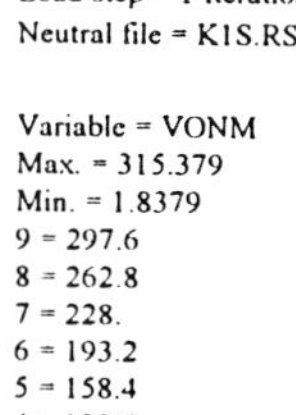


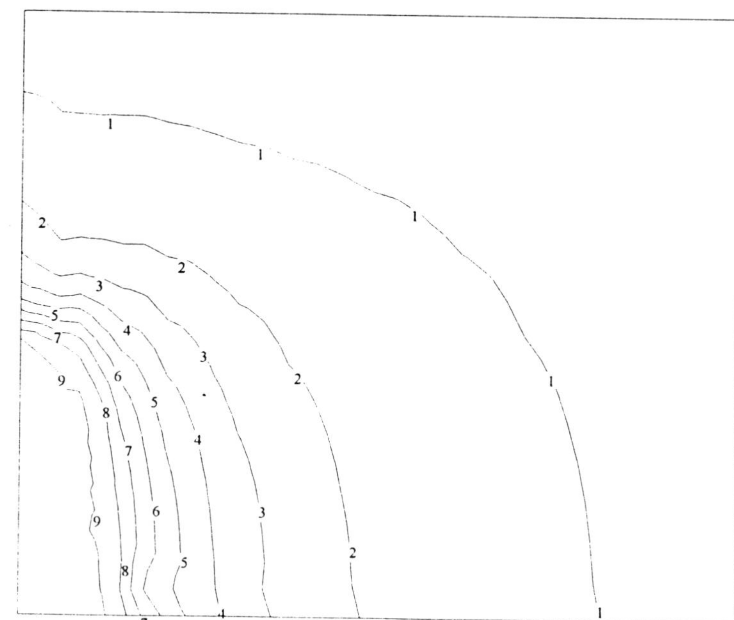

Load step = 1 Iteration = 1
Neutral file = K2L.RSE

Variable = VONM
Max. = 195.474
Min. = 0.553218
9 = 184.2
8 = 162.6
7 = 141.
6 = 119.4
5 = 97.8
4 = 76.2
3 = 54.6
2 = 33.
1 = 11.4

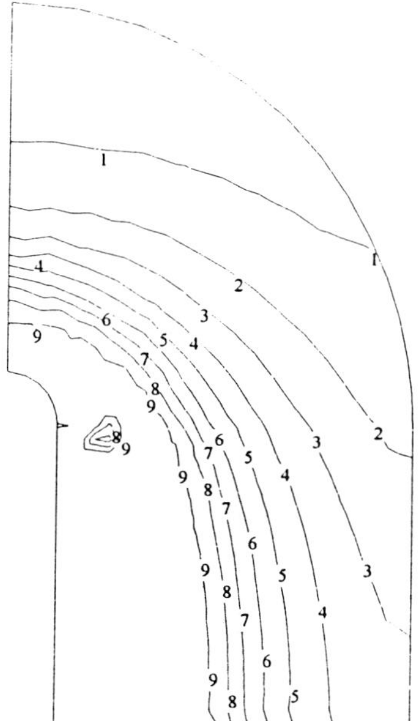

*Figure 5,* a) Case 2. Von Mises stresses for the adaptation;
b) Case 3. Von Mises stresses for the ultimate state.

## 6. Conclusions

This paper presented a computational system CYCLONE which evaluates safety factors for adaptation, inadaptation, and ultimate states for structures working under heavy variable repeated loading. CYCLONE is integrated with commercial finite element codes. The algorithm applied is shown to be highly efficient.

The example of cylindrical pressure chamber with the free, supported and closed ends has been analysed in detail. Results are in good agreement with a theoretical prediction. Practical hints for design of ultra high pressure chambers are derived.

**Acknowledgement**
The partial support of the FNP by the Phare SCI-TECH II PL9611-03.02 project is gratefully acknowledged.

## References

1. Siemaszko A. 1995. Limit, shakedown, post-yield, and inadaptation analyses of discrete plastic structures. In [5]: 279-291.
2. König J.A., Kleiber M. 1978. On a new method of shakedown analysis. *Bull.Ac.Pol.Sci.Ser.Sci.Techn.* 26: 165-171.
3. Maier G. 1970. A matrix structural theory of piecewise-linear plasticity with interacting yield planes., *Meccanica* 5: 55-66.
4. König J.A. 1987., *Shakedown of elastic-plastic structures*. PWN-Elsevier: Warsaw-Amsterdam.
5. Mroz Z., Weichert D., Dorosz S. (eds.) 1995. *Inelastic behaviour of structures under variable loads*. Kluwer: Dordrecht.
6. Zwolinski J., Bielawski G. 1987. An optimal selection of residual stress for shakedown and limit load analysis. *Proc.Conf. Comp. Meth. Struct. Mech.*: 459-462. Jadwisin.
7. Zwolinski J. 1994. Min-max approach to shakedown and limit load analysis for elastic perfectly plastic and kinematic hardening materials. In [5]: 363-380.

# VARIATIONAL PRINCIPLES FOR SHAKEDOWN ANALYSIS

NESTOR ZOUAIN AND JOSÉ LUÍS SILVEIRA
*Mechanical Engineering Department, COPPE, EE / UFRJ*
*Federal University of Rio de Janeiro*
*Caixa Postal 68503, CEP 21945.970, Rio de Janeiro, Brazil.*

**Abstract.** The safety factor for elastic shakedown of structures under variable loads is considered. Variational principles for shakedown analysis are reviewed in a unified presentation, suitable to finite element discretizations, and considering nonlinear yield functions. Extremum principles for bounds to shakedown loads are also presented. A tube under thermo-mechanical loading is used to show analytical solutions, and numerical results obtained with a mixed finite element formulation.

## 1. Introduction

Shakedown analysis is fundamental among direct methods for safety assessment of engineering structures submitted to variable loading. Theoretical foundations of this analysis received considerable attention in essential papers or books by Symonds [1], Koiter [2], Maier [3], König [11], König [5], Weichert [9], Ponter & Karadeniz [10], Polizzotto [14], Pycko & Maier [15], Nayroles [17], Stein & Huang [18], Telega [19], Kamenjarzh [21], Gross-Weege [12], De Saxcé [16], among others.

This paper is devoted to the discussion of the general variational principles of shakedown analysis as basic formulations of the numerical procedures for continuum structures. The computation of the safety factor for elastic shakedown of bodies under variable loads is the main objective considered here. Extremum principles for this purpose are reviewed in a unified presentation, suitable to finite element discretizations, and considering nonlinear yield functions. Variational formulations for the computation of bounds for the elastic shakedown factor are also discussed.



*D. Weichert and G. Maier (eds.), Inelastic Analysis of Structures under Variable Loads,* 147–165.

Failure mechanisms for perfectly plastic bodies under variable loads are: alternating plasticity (plastic shakedown), or incremental collapse (ratcheting), or else instantaneous collapse (plastic collapse). Elastic shakedown analysis (or simply shakedown analysis) prevents all these failure modes.

When alternating plasticity is the only failure phenomenon that can threaten one structure (see Polizzotto [14]), the analysis turns to be considerably simpler. For this case, there exist specific extremum principles allowing the computation of the safety factor against alternating plasticity. Moreover, in the general shakedown analysis, the safety factor against alternating plasticity may be used as an upper bound for the elastic shakedown load factor. These features are discussed in the present paper.

Likewise, the exclusive prevention of incremental collapse may also be considered of interest *per se*, and it also leads to extremum principles for the computation of upper bounds to the elastic shakedown safety factor. We show that an upper bound is obtained when the only incremental collapse mechanisms considered are some simple deformation patterns, denoted SMIC, that we define in this paper. This issue has been investigated by the authors in previous works and we give here a short review of this topic with a new mathematical proof of the basic bounding inequality.

A tube under thermo-mechanical loading is used in the last section to show analytical solutions and also numerical results obtained with a mixed finite element formulation.

## 2. Basic notation

### 2.1. KINEMATICS AND EQUILIBRIUM

The continuum model of a body is defined in an open bounded region $\mathcal{B}$ with regular boundary $\Gamma$. The space $V$ is the set of all admissible velocity fields $v$ complying with homogeneous boundary conditions prescribed on a part $\Gamma_u$ of $\Gamma$. The strain rate tensor fields $D$ are elements of the space $W$, and the tangent deformation operator $\mathcal{D}$ maps $V$ into $W$. Let $W'$ be the space of stress fields $T$, and $V'$ the space of load systems $F$. The equilibrium operator $\mathcal{D}'$, dual of $\mathcal{D}$, maps $W'$ into $V'$. Accordingly, the kinematical and equilibrium relations are written as

$$D = \mathcal{D}v \qquad F = \mathcal{D}'T \tag{1}$$

The set of all self-equilibrated (residual) stress fields is denoted by $S^r$.

To simplify the notation a hat is adopted to denote the local value of any field; for instance: $\hat{v} \equiv v(x)$, $\hat{D} \equiv D(x)$ and $\hat{T} \equiv T(x)$. Then, the internal power for any pair $T \in W'$ and $D \in W$ is given by the duality product

$$< T, D > = \int_{\mathcal{B}} \hat{T} \cdot \hat{D} \; d\mathcal{B} \tag{2}$$

## 2.2. ELASTIC–IDEALLY PLASTIC MATERIALS

The stress $\hat{T}$ at any point $x$ of an elastic-ideally plastic body $\mathcal{B}$ is constrained to fulfill the plastic admissibility condition, i.e. it must belong to the set

$$\hat{P} = \{\hat{T} \quad | \quad f(\hat{T}) \leq 0\} \tag{3}$$

where $f$ is a $\hat{m}-$vector valued function describing the yield criterion. The inequality above is then understood as imposing that each component $f_\ell$ of $f$, which is a regular convex function of $\hat{T}$, is nonpositive.

Likewise, the closed convex set $P$ of plastically admissible stress fields is

$$T \in P \qquad \Longleftrightarrow \qquad \hat{T} \in \hat{P} \quad \forall x \in \mathcal{B} \tag{4}$$

The stress–free state of the body is assumed admissible, i.e. $T = 0 \in P$.

Let us define the plastic dissipation function as

$$\hat{\chi}(\hat{D}^p) = \sup_{\hat{T}^* \in \hat{P}} \; \hat{T}^* \cdot \hat{D}^p \tag{5}$$

and the indicator function $\mathcal{I}_{\hat{P}}(\hat{T})$ of $\hat{P}$, that equals zero for any $\hat{T} \in \hat{P}$ and $+\infty$ otherwise. Then, the constitutive relation between plastic strain rates $\hat{D}^p$ and stresses $\hat{T}$ is written (Maugin [13], Kamenjarzh [21]), for the case of associative plastic flow, as

$$\hat{T} \in \partial\hat{\chi}(\hat{D}^p) \qquad \Longleftrightarrow \qquad \hat{D}^p \in \partial\mathcal{I}_{\hat{P}}(\hat{T}) \tag{6}$$

where the subdifferential $\partial\hat{\chi}(\hat{D}^p)$ is the set of all stress tensors such that

$$\hat{\chi}(\hat{D}^{p*}) - \hat{\chi}(\hat{D}^p) \quad \geq \quad \hat{T} \cdot \left(\hat{D}^{p*} - \hat{D}^p\right) \qquad\qquad \forall \; \hat{D}^{p*} \tag{7}$$

The dissipation $\hat{\chi}$ can be identified as the support function of $\hat{P}$, hence it is sublinear, i.e. convex and positively homogeneous of first degree. It also satisfies $\hat{\chi}(0) \geq 0$ because $\hat{T} = 0 \in \hat{P}$.

The material relations (6) are equivalent to the following classical form. The plastic strain rate is related to the stress, at any point of $\mathcal{B}$, by the normality rule $\hat{D}^p = \nabla f(\hat{T}) \; \dot{\lambda}$ . Here $\nabla f(\hat{T})$ denotes the gradient of $f$, and $\dot{\lambda}$ is the $\hat{m}-$vector field of plastic multipliers. At any point of $\mathcal{B}$, the components of $\dot{\lambda}$ are related to each plastic mode in $f$ by the complementarity condition: $\dot{\lambda} \geq 0$ , $f \leq 0$ , and $f \cdot \dot{\lambda} = 0$ (these inequalities hold componentwise).

There are global relations, in terms of the fields $D$ and $T$, completely analogous to the local relations (5), (6) and (7). For instance, by substituting $\hat{P}$ by $P$ in (5) we obtain $\chi(D) = \sup_{T \in P} < T, D >= \int_{\mathcal{B}} \hat{\chi}(\hat{D})\, d\mathcal{B}$.

The total strain is the sum of elastic and plastic terms, as usual under small deformation assumptions.

## 3. Shakedown analysis

The data of shakedown analysis is a prescribed range of variation $\Delta^0$ which contains any feasible history of external loads, cyclic or not. However, we prefer to represent any external action, either a mechanical or a thermal load, by the stress field which is the unique solution of the corresponding purely (or unlimited) elastic problem. Then, the data for shakedown analysis will be given in terms of a set $\Delta^e$ of (elastic) stress fields representing the domain of variation of mechanical and thermal loads. In this paper $\Delta^e$ is assumed convex and bounded.

Moreover, all shakedown problems can be stated using the pointwise envelope $\Delta$ of the domain of elastic stresses $\Delta^e$, which is defined in the sequel. Consider the set of all the local values of elastic stresses associated to any feasible loading, i.e.

$$\forall x \in \mathcal{B} \qquad \hat{\Delta} = \{T^e(x) \mid \forall T^e \in \Delta^e\} \tag{8}$$

Define now the pointwise envelope of the set $\Delta^e$

$$\Delta = \{T \in W' \mid \hat{T} \in \hat{\Delta} \ \ \forall x \in \mathcal{B}\} \tag{9}$$

As a mechanical interpretation, any (virtual) stress field $T$ in the set $\Delta$ may be sought as collecting local values of elastic stresses produced, at different instants, along a certain admissible load program (cyclic or not).

Any elastic field corresponding to a single feasible load also belongs to the envelope $\Delta$ (i.e. $\Delta^e \subset \Delta$). However, this envelope contains other kind of fields which, for instance, may violate the regularity conditions inherent to elastic solutions.

For instance, consider a beam section under fixed axial load combined with variable bending moments. Stress fields in the elastic domain $\Delta^e$ are represented (plotting direct stress versus transversal coordinate $z$) by a bundle of straight lines passing through a fixed point, corresponding to $z = 0$ and to the mean stress. Therefore, stress fields belonging to the elastic envelope $\Delta$ are represented by any nonlinear curve bounded by the two linear stress profiles which define the limits of the elastic domain.

We use in shakedown formulations the support function of the set $\hat{\Delta}$:

$$\hat{\varphi}(\hat{D}^p) = \sup_{\hat{T}^* \in \hat{\Delta}} \hat{T}^* \cdot \hat{D}^p \tag{10}$$

which is sublinear. In particular, it is eventually negative but always has some positive value because $\hat{\Delta}$ is convex and bounded. The total counterpart $\varphi(D^p) = \int \hat{\varphi}(\hat{D}^p)\, d\mathcal{B}$ also satisfies a relation analogous to the above definition (without hats).

This support function represents the maximum external power, over the assumed rate of deformation, obtained from any feasible loading history.

## 4. Elastic Shakedown

This section is directed towards the computation of the limit factor $\mu$ which amplifies the load domain ensuring elastic shakedown. Consequently, this safety factor $\mu$ prevents against the three modes of failure described in the classical theory of shakedown (Koiter [2]), namely: alternating plasticity, incremental collapse, and plastic collapse.

### 4.1. EQUILIBRIUM PRINCIPLES FOR ELASTIC SHAKEDOWN

The fundamental theorem due to Bleich and Melan states that any load factor $\mu^*$ is safe if there exists a fixed self-equilibrated stress field $T^r$ such that its superposition with any stress belonging to the amplified load domain $\mu^*\Delta$ is plastically admissible. The limit load factor $\mu$ for elastic shakedown is the supremum of all safe factors.

*Equilibrium formulation for elastic shakedown*

$$\mu = \sup_{\substack{\mu^* \in R \\ T^r \in W'}} \left\{ \mu^* \;\middle|\; \begin{array}{l} \mu^* \, \Delta + T^r \subset P \\ T^r \in S^r \end{array} \right\} \tag{11}$$

The notation used in the plastic admissibility constraint above means

$$\omega^* \, \Delta + T^r \subset P \quad \Longleftrightarrow \quad \omega^* \, T + T^r \in P \quad \forall \, T \in \Delta \tag{12}$$

In geometrical terms, solving the equilibrium formulation for elastic shakedown consists in finding the translation $T^r$ in the affine manifold $S^r$, which moves the prescribed set $\Delta$ and allows for the maximal amplification, of this translated load domain, complying with the constraint that the resulting set is entirely contained in the plastically admissible set $P$.

### 4.2. MIXED PRINCIPLES FOR ELASTIC SHAKEDOWN

The variational principle (11), derived from Bleich and Melan's criterion, can be transformed by introducing the self–equilibrium constraint as an exact penalty in the objective functional. This leads to a mixed principle.

*Mixed formulation for elastic shakedown (primal)*

$$\mu \quad = \quad \sup_{\substack{\mu^* \in R \\ T^o \in W'}} \; \inf_{v \in V} \; \{ \, \mu^* + < T^o \, , \, \mathcal{D}v > \quad | \quad \mu^* \, \Delta + T^o \subset P \, \} \tag{13}$$

*Proof* Exact penalization of the equilibrium constraint in Melan's formulation is obtained by using the following indicator function

$$\mathcal{I}_{S^r}(T^o) = - \inf_{v \in V} \; < T^o \, , \, \mathcal{D}v > \tag{14}$$

which introduced in the objective function completes the proof. □

A dual principle is now achieved by formally interchanging the sup and inf operations. Recall that this procedure does not lead automatically to an equivalent problem; i.e. proper conditions should be verified to guarantee the primal–dual equivalence. This is the main subject of the so called min–max theory in convex analysis (Rockafellar [4]).

*Mixed formulation for elastic shakedown (dual)*

$$\mu \quad = \quad \inf_{v \in V} \; \sup_{\substack{\mu^* \in R \\ T^o \in W'}} \; \{ \, \mu^* + < T^o \, , \, \mathcal{D}v > \quad | \quad \mu^* \, \Delta + T^o \subset P \, \} \tag{15}$$

Concerning the proof of the duality equivalence, we observe that limit analysis is the particular case of elastic shakedown when $\Delta$ is a singleton; therefore, we shall find here, at least, the same difficulties encountered in dualizing limit analysis formulations (Christiansen [6], Fremond & Friaa [7]). Namely, that the classic min–max theorems do not apply because it cannot be guaranteed either bounded domains or coercitivity. Likewise in limit analysis, Christiansen's theorem (in Christiansen [6]) is specially suited to this purpose.

### 4.3. KINEMATICAL PRINCIPLE FOR ELASTIC SHAKEDOWN

A new insight on the dual mixed formulation (15) leads to a kinematical principle.

*Kinematical formulation for elastic shakedown*

$$\mu \quad = \quad \inf_{v \in V} \; \widetilde{\mu}(\mathcal{D}v) \tag{16}$$

where

$$\widetilde{\mu}(D) = \sup_{\substack{\mu^* \in R \\ T^o \in W'}} \{ \mu^* + < T^o , D > \quad | \quad \mu^* \Delta + T^o \subset P \} \tag{17}$$

As usual, the kinematical formulation becomes simpler when a particular form of the plastic admissibility is introduced. Furthermore, we show in the next subsection a general form of this principle for the case of discrete models of the structure.

## 4.4. DISCRETE MODELS FOR ELASTIC SHAKEDOWN

We consider in this subsection the case when the loading domain is given, or approximated, by a finite number $n_\ell$ of basic loads. Furthermore, a finite element discretization of the continuum is adopted so as to produce a finite dimensional model.

Since the load domain is assumed polyhedral in this subsection, then the local domain of variable loading $\hat{\Delta}$ (for any point $x$ in the body) is a convex polyhedron. The number of vertices of this local domain may in fact depend on the point of the body considered whenever we eliminate from the $n_\ell$ local elastic stresses associated to basic loads all those stress tensors which are interior (or non-extreme points) to $\hat{\Delta}$.

The additional step in the way to reach a finite number of admissibility constraints in Bleich-Melan's formulation is to select a discrete set $\{x^j;\ j = 1, \ldots, p\}$ of critical points in the body. This is accomplished in accordance with the finite element discretization. For instance, in a mesh of $n_{el}$ mixed (or equilibrium) triangles, where stresses are linearly interpolated, the index $j = 1, \ldots, p$ enumerates all vertices of all triangles and $p = 3\, n_{el}$.

For the sake of simplicity we maintain the same symbols used for fields in the continuum model to denote now finite dimensional global vectors in the finite element model. For instance, $T \in I\!R^q$ denotes in this subsection the global vector of interpolation parameters for stress fields in an statical or mixed formulation. Accordingly, $T^r$ is the finite dimensional vector of global residual stresses. If we use $n_{el}$ mixed triangles where $\hat{q}$ stress components are linearly interpolated in terms of nodal parameters, then each $T$ or $T^r$ has $q = 3\, n_{el}\, \hat{q}$ components.

The local domain of variable stresses $\Delta^j := \hat{\Delta}(x^j)$, which is a convex polyhedron, is described as the convex hull of its vertices, i.e.

$$\Delta^j = \text{co} \{ T^{jk} ; \quad {\scriptstyle k=1,\ldots,m_j} \} \tag{18}$$

where $T^{jk}$ denotes the global stress vector of interpolation parameters representing a vertex of $\Delta^j$. This global vector is composed of the elastic stress produced by a single load case in one vertex of some element and completed with zeros for all other components (this is only to simplify the theoretical presentation).

Accordingly, the plastic admissibility condition $\mu\,\Delta + T^r \subset P$ is equivalent to the following constraints

$$\mu T^{jk} + T^r \in P^j \qquad k=1,\dots,m_j \quad j=1,\dots,p \tag{19}$$

where $P^j$ represents the elastic range of a particular point in the continuum, written in terms of the global vector of stress parameters, i.e. previously selecting the pertinent components.

Finally, the discrete form of compatibility and equilibrium created by the finite element discretization reads

$$D = Bv \qquad B^T T^r = 0 \tag{20}$$

Consequently, the Bleich–Melan's formulation for this case can be simplified as follows.

*Discrete equilibrium formulation for elastic shakedown*

$$\mu = \sup_{\substack{\mu^* \in R \\ T^r \in R^q}} \left\{ \mu^* \;\middle|\; \begin{array}{l} \mu^* T^{jk} + T^r \in P^j \qquad k=1,\dots,m_j \quad j=1,\dots,p \\ B^T T^r = 0 \end{array} \right\} \tag{21}$$

By applying the above notations to the general mixed principle for elastic shakedown (13) we obtain the following finite-dimensional counterpart.

*Discrete mixed formulation for elastic shakedown*

$$\mu = \sup_{\substack{\mu^* \in R \\ T^o \in R^q}} \inf_{v \in \mathbf{R}^n} \left\{ \mu^* + T^o \cdot Bv \;\middle|\; \begin{array}{l} \mu^* T^{jk} + T^o \in P^j \\ k=1,\dots,m_j \quad j=1,\dots,p \end{array} \right\} \tag{22}$$

We state and prove in the sequel the finite-dimensional version of the kinematical principle for elastic shakedown (16). The simplified notation $\Sigma_{jk} := \sum_{j=1}^{p} \sum_{k=1}^{m_j}$ is used from now on.

*Discrete kinematical formulation for elastic shakedown*

$$\mu = \inf_{\substack{v \in R^n \\ D^{jk} \in R^q}} \left\{ \Sigma_{jk}\ \chi^j(D^{jk}) \;\middle|\; \begin{array}{l} \Sigma_{jk}\, D^{jk} = Bv \\ \Sigma_{jk}\ T^{jk} \cdot D^{jk} = 1 \end{array} \right\} \tag{23}$$

where

$$\chi^j(D) = \sup_{T^* \in P^j} T^* \cdot D \tag{24}$$

is the usual dissipation function written in terms of the global strain rate vector $D$, i.e. previously selecting the components related to the point $x_j$.

*Proof* - The discrete version of (17) is

$$\tilde{\mu}(D) = \sup_{\substack{\mu^* \in R \\ T^o \in R^q}} \left\{ \mu^* + T^o \cdot D \;\middle|\; \mu^* T^{jk} + T^o \in P^j \quad {}_{k=1,\dots,m_j \quad j=1,\dots,p} \right\} \tag{25}$$

The constraints above can be imposed by exact penalty as follows

$$\tilde{\mu}(D) = \sup_{\substack{\mu^* \in R \\ T^o \in R^q}} \left\{ \mu^* + T^o \cdot D - \Sigma_{jk}\, \mathcal{I}^j(\mu^* T^{jk} + T^o) \right\} \tag{26}$$

where $\mathcal{I}^j$ is the indicator function of $P^j$.

Using now that

$$\mathcal{I}^j(T^*) = \sup_{D^* \in \boldsymbol{R}^q} \left\{ T^* \cdot D^* - \chi^j(D^*) \right\} \tag{27}$$

we transform (26) in

$$\tilde{\mu}(D) = \tag{28}$$

$$\sup_{\substack{\mu^* \in R \\ T^o \in R^q}} \inf_{D^{jk} \in \boldsymbol{R}^q} \left\{ \mu^* + T^o \cdot D - \Sigma_{jk} \left[ (\mu^* T^{jk} + T^o) \cdot D^{jk} - \chi^j(D^{jk}) \right] \right\}$$

Consequently

$$\tilde{\mu}(D) = \inf_{D^{jk} \in \boldsymbol{R}^q} \left\{ \Sigma_{jk}\ \chi^j(D^{jk}) \;\middle|\; \begin{array}{l} \Sigma_{jk}\, D^{jk} = D \\ \Sigma_{jk}\ T^{jk} \cdot D^{jk} = 1 \end{array} \right\} \tag{29}$$

Substitution of this expression for $\tilde{\mu}(D)$ in (16) leads to (23), so the proof is complete. □

Finally, we state the set of relations characterizing the solution of the variational problems (21), (22), and (23).

*Discrete optimality conditions for elastic shakedown*

$$\Sigma_{jk}\ T^{jk} \cdot D^{jk} = 1 \tag{30}$$

$$\Sigma_{jk}\ D^{jk} = Bv \tag{31}$$

$$B^T T^r = 0 \tag{32}$$

$$\mu^* T^{jk} + T^r \in \partial\chi^j(D^{jk}) \quad {}_{k=1,\dots,m_j \quad j=1,\dots,p} \tag{33}$$

Additionally, the following relations are also fulfilled at the solution.

$$\mu = \Sigma_{jk}\, \chi^j(D^{jk}) \tag{34}$$

$$T^{jk} \in \partial\varphi(D^{jk}) \qquad {\scriptstyle k=1,\dots,m_j \quad j=1,\dots,p} \tag{35}$$

The discretization of the continuum may be based either on the equilibrium, mixed or kinematical formulations, i.e. (11), (13) or (16), thus leading directly to a primal discrete problem of the corresponding type: (21), (22) or (23). This choice strongly determines the matrix of discrete compatibility $B$, among other consequences. Nevertheless, the primal discrete formulation obtained can also be formally transformed into the remaining types of principles or optimality conditions, which result mutually equivalent.

Accordingly, anyone of the discrete forms (21), (22), (23) or (30–33) can be considered in order to motivate the generation of an iterative algorithm for solving the elastic shakedown problem. For the numerical applications in the present work, we developed a mathematical programming algorithm based on the above set of optimality conditions.

## 5. Alternating Plasticity

The subject in this section is the computation of the safety factor which prevents exclusively against plastic shakedown (i.e. alternating plasticity). This failure mechanism produces some plastic deformation after any arbitrary large time, although the net strain increments vanish in some infinite sequence of instants.

For a comprehensive study on this kind of analysis we refer to C. Polizzotto [14]. Following Polizzotto, a load factor $\omega^*$ is safe with respect to plastic shakedown if there exists a fixed stress distribution $T^o$ (not necessarily self–equilibrated) such that, when superposed to any stress belonging to the amplified domain of variation $\omega^*\Delta$, nowhere violates the yield criterion.

The supremum of these safe factors is the limit load factor $\omega$ to prevent plastic shakedown.

*Equilibrium formulation for plastic shakedown*

$$\omega \quad = \quad \sup_{\substack{\omega^* \in R \\ T^o \in W'}} \{\, \omega^* \quad | \quad \omega^*\,\Delta + T^o \subset P \,\} \tag{36}$$

The constraints in this variational principle restrains the values of the stress tensor at each point in the body independently of any other point. This is due to the definition of the set $P$ of plastically admissible stresses and also to the concept of feasible variable stress fields adopted in the definition of the envelope $\Delta$. Consequently, the formulation above admits the following unabridged statical form.

*Equilibrium formulation for plastic shakedown (unabridged form)*

$$\omega \quad = \quad \inf_{x \in \mathcal{B}} \widetilde{\omega}(x) \tag{37}$$

with

$$\widetilde{\omega}(x) \quad = \quad \sup_{\omega^*, \hat{T}^o} \{ \, \omega^* \quad | \quad \omega^* \, \hat{\Delta} + \hat{T}^o \subset \hat{P} \, \} \tag{38}$$

That is, to compute the limit load factor $\omega$ against plastic shakedown consists in solving uncoupled nonlinear problems for each point of the body and retaining the minimum value of the amplifying factors thus obtained.

The statical formulation for elastic shakedown (11) may be obtained from the statical formulation for plastic shakedown (36) by just adding the self-equilibrium constraint, hence

$$\mu \, \leq \, \omega \tag{39}$$

i.e. ensuring elastic shakedown prevents plastic shakedown to occur.

## 6. Simple Mechanisms of Incremental Collapse

Another upper bound for the elastic shakedown safety factor $\mu$ is presented in this section. This bound is the load factor $\rho$ obtained in a simplified safety analysis where the only failure mode to be prevented is a particular kind of incremental collapse, which is denoted simple mechanism of incremental collapse, or SMIC. A mechanism of this class is identified among incremental collapse phenomena, where plastic deformation accumulates in every cycle, by the fact that nowhere in the body a single point undergoes plastic deformation more than once per cycle.

The variational principles defining the safety factor for SMIC are presented now. Afterwards, we prove that they in fact give an upper bound for elastic shakedown. Finally, we justify the mechanical interpretation as related to simple mechanisms of incremental collapse.

*Equilibrium formulation for SMIC*

$$\rho \quad = \quad \inf_{T \in W'} \{ \, \widetilde{\rho}(T) \quad | \quad T \in \Delta \, \} \tag{40}$$

where

$$\widetilde{\rho}(T) \quad = \quad \sup_{\substack{\rho^* \in R \\ T^r \in W'}} \left\{ \rho^* \;\middle|\; \begin{array}{l} \rho^* \, T + T^r \in P \\ T^r \in S^r \end{array} \right\} \tag{41}$$

The value of $\tilde{\rho}(T)$ for a given stress field $T$ is the limit load factor which amplifies the unique load equilibrated with $T$ to produce instantaneous plastic collapse. Thus, computing $\tilde{\rho}(T)$ consists in solving a limit analysis problem (Christiansen [6], Christiansen & Andersen [24], Borges *et al.* [20]) where the stress field $T$ determines the associated reference load, i.e. the data for limit analysis.

In view of (40) and (41), *the safety factor $\rho$ preventing simple mechanisms of incremental collapse is the infimum of the limit load factors corresponding to the loads equilibrated with all stress fields $T$ pertaining to the envelope $\Delta$ of the elastic stresses produced by admissible variations of loads.*

**Proposition 1** *The load factor $\rho$ for simple mechanisms of incremental collapse, given by (40–41), is an upper bound for the safety factor $\mu$ in elastic shakedown analysis, i.e.*

$$\mu \leq \rho \tag{42}$$

*Proof-* Consider a solution $(\mu, T^r)$ of the statical principle (11) for $\mu$. In particular, for any given $T$, the pair $(\mu, T^r)$ satisfies the constraints of problem (41). Hence

$$\mu \leq \tilde{\rho}(T) \qquad \forall T \tag{43}$$

This implies the thesis in view of (40). □

The following mixed and kinematical forms, and the optimality conditions below, were derived in Silveira & Zouain [23].

*Mixed formulation for SMIC*

$$\rho \quad = \quad \inf_{\substack{v \in V \\ T \in W'}} \left\{ \chi(\mathcal{D}v) \quad \middle| \quad \begin{array}{l} < T \, , \, \mathcal{D}v >= 1 \\ T \in \Delta \end{array} \right\} \tag{44}$$

*Kinematical formulation for SMIC*

$$\rho \quad = \quad \inf_{v \in V} \; \{ \; \chi(\mathcal{D}v) \quad | \quad \varphi(\mathcal{D}v) = 1 \; \} \tag{45}$$

*Optimality conditions for SMIC*

$$< T, \mathcal{D}v > \; = \; 1 \tag{46}$$

$$T^r \; \in \; S^r \tag{47}$$

$$\rho \, T + T^r \; \in \; \partial\chi(\mathcal{D}v) \tag{48}$$

$$T \; \in \; \partial\varphi(\mathcal{D}v) \tag{49}$$

Moreover, the following useful relations trivially hold at the solution.

$$\rho = \chi(\mathcal{D}v) \tag{50}$$

$$\varphi(\mathcal{D}v) = 1 \tag{51}$$

We can justify now the name SMIC adopted to identify the particular failure mode that is being exclusively prevented in the bounding principles (40), (44) or (45). Indeed, the optimality condition (48) implies that the critical strain rate field $\mathcal{D}v$, which is kinematically compatible, is related, via the plastic flow law, to a single critical stress field, $T^c = \rho\, T + T^r$. Hence, this strain rate field can be properly called a purely plastic collapse mechanism. Moreover, it is a cumulative mechanism (not necessarily a synchronous one) because the critical stresses at each point do not exist simultaneously but sequentially along the critical cycle of loading. This type of strain rate fields are here called *simple mechanisms of incremental collapse* with respect to more complex mechanisms, also found among the general incremental collapse phenomena. Complex mechanisms of incremental collapse are also kinematically compatible but not purely plastic, as a sum of plastic deformation rates at some points. Examples of simple and complex mechanisms are given in Zouain & Silveira [25].

## 7. Bounding the Elastic Shakedown Factor

The elastic shakedown factor $\mu$ is the main objective of shakedown analysis. Besides, we have found in (39) and (42) that the following upper bounding relationship applies

$$\mu \le \mu_{ub} := \min(\omega, \rho) \tag{52}$$

where $\omega$ and $\rho$ are amplifying factors obtained in separate analyses for alternate plasticity and SMIC, respectively.

The safety factor $\omega$, for alternate plasticity, is much easier to compute than the elastic shakedown factor $\mu$, because it only demands solving a sequence of uncoupled local problems.

The safety factor $\rho$, for SMIC, may be computed by performing a sequence of limit analyses, as presented by Silveira & Zouain [23].

Accordingly, computing the upper bound $\mu_{ub}$ may be convenient. However, the upper bound may also be worthless if we cannot estimate the difference $\mu_{ub} - \mu$. This issue has been the aim in a previous paper (see Zouain & Silveira [25]), and we give in the sequel a very brief description of the results presented in that reference.

A simple plane stress situation, with a constrained deformation and thermo-mechanical loading, i.e. a block described in Section 8, presents an effective gap between $\rho$ and $\mu$. This counterexample proves mathematically that no completely general theorem of strong duality exists for the formally dual problems representing static formulations of elastic shakedown and SMIC.

An elementary bending problem under combined axial force and variable bending moment, that is also considered in Zouain & Silveira [25], presents no gap between $\rho$ and $\mu$. This demonstrates that the favourable situation $\mu = \rho$ is attainable in some regular conditions.

The following sufficient condition is very useful, either in analytical or numerical procedures, to identify, *a posteriori*, whether the solution $\rho$ obtained in the SMIC problem is indeed a final result for the elastic shakedown analysis. This proposition is proven in Zouain & Silveira [25].

*A sufficient condition to ensure that there exists some simple mechanism of incremental collapse which is also critical for elastic shakedown analysis, i.e. that it holds*

$$\mu = \rho \tag{53}$$

*is that the solutions $\rho$ and $T^r$ of the SMIC problem satisfy the following admissibility condition*

$$\rho\Delta + T^r \subset P \tag{54}$$

## 8. A Tube Under Thermo-mechanical loading

We present in this section some numerical and analytical solutions for a tube, made up of a von Mises material, axially restrained, and submitted to independent variations of internal pressure and temperature (uniform across the wall thickness).

The main interest of this example is that the considered tube only presents complex mechanisms of incremental collapse, or else alternate plasticity, in the appropriate ranges of the data.

Numerical results for a finite element model of the tube are presented and compared with analytical expressions in the sequel, although we have not presented in this paper the iterative algorithm that we use to solve the discrete problem of elastic shakedown treated in subsection (4.4) (the numerical procedure is being published elsewhere). This general algorithm for discrete elastic shakedown consists of a Newton-like iteration based on the optimality equations (30), (31), (32), and the flow equation implied by (33). Each iteration is followed by stress scaling in order to ensure plastic admissibility along the iterative process.

Consider a long tube with fixed ends. The internal and external radii are $R_{int}$ and $R_{ext}$, respectively. The internal pressure $p_{int}$ varies between 0 and $\overline{p}_{int}$. Independently, the wall temperature increment (over room temperature) $\Theta$ varies between 0 and $\overline{\Theta}$, but remains uniform across the thickness wall.

The shakedown behavior of this tube is dependent on the Young's modulus $\mathsf{E}$, the Poisson's coefficient $\nu$, the thermal expansion coefficient $c$, and the yield stress $\sigma_Y$. The plastic collapse pressure of the tube is

$$p_c = \frac{2}{\sqrt{3}} \sigma_Y \log \frac{R_{ext}}{R_{int}} \tag{55}$$

The nondimensional parameters adopted for mechanical and thermal loadings are

$$p := \frac{p_{int}}{p_c} \qquad q := \frac{\mathsf{E}\, c\, \Theta}{\sigma_Y} \tag{56}$$

Consequently, the prescribed limits are $\overline{p} := \overline{p}_{int}/p_c$ and $\overline{q} := \mathsf{E}\, c\, \overline{\Theta}/\sigma_Y$.

A finite element discretization of the one-dimensional domain $(R_{int}, R_{ext})$ is adopted here. A mixed interpolation is chosen in each element, and conveniently balanced to avoid the locking phenomenon. Radial velocity $v_r$ is quadratic and continuous, while stresses are discontinuous. Deviatoric stress components $S_r$, $S_\theta$, and $S_z$ are linear, while the mean stress $T_m$ is constant, in each element.

Numerical results for two different tubes are shown in the Bree-type diagram of Figure 1.

In order to compare and validate the numerical results we consider next a simple model for the cylindrical wall that was solved in closed form in Zouain & Silveira [25].

The simplified model of the tube wall is a small block, with edges parallel to $x$, $y$, and $z$ directions. The block is restrained to deform in the $y$ direction, which corresponds to the axial direction in the tube. A variable mechanical load $p_b$, in the range $(0, \overline{p}_b)$, acts in the $x$ direction and represents the hoop stress of the tube. Accordingly, direction $z$ of the block represents the radial direction of the tube and hence is assumed free. The thermal loading is exactly the same as in the tube.

Plastic collapse of the block occurs at the load $p_{bc} = 2\sigma_Y/\sqrt{3}$. So, we define the load parameter for the block as

$$p := \frac{p_b}{p_{bc}} = \frac{\sqrt{3}}{2} \frac{p_b}{\sigma_Y} \tag{57}$$

which is the parameter to be compared with the one previously defined for the tube, in (56). Likewise, the loading domain is determined by $\overline{p} := \overline{p}_b/p_{bc}$, besides $\overline{q} := \mathsf{E}\, c\, \overline{\Theta}/\sigma_Y$.

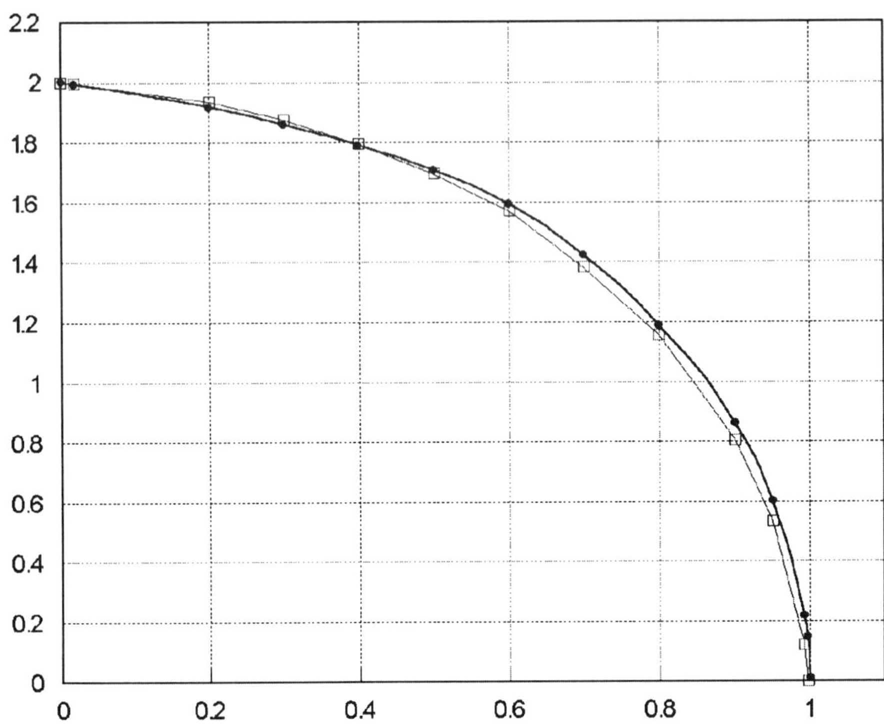


*Figure 1.* Bree-type diagram for: (i) a thin tube (dots, numerical) with $R_{ext}/R_{int} = 1.1$, (ii) a thick tube (squares, numerical) with $R_{ext}/R_{int} = 2$, and (iii) a block (solid line, analytical). Poisson coefficient is $\nu = 0.2$.

The exact solution for the elastic shakedown analysis of the restrained block, given in Zouain & Silveira [25], is

$$\mu = \begin{cases} \dfrac{2\sqrt{3}\,(1-2\nu)\overline{p} + 6\overline{q}}{4(1-\nu+\nu^2)\overline{p}^2 + 3\overline{q}^2 + 2\sqrt{3}\,(1-2\nu)\overline{p}\,\overline{q}} & \text{for } \overline{p} \leq g(\nu)\,\overline{q} \\[2ex] \dfrac{2}{\sqrt{4\overline{p}^2 + \overline{q}^2}} & \text{for } \overline{p} \geq g(\nu)\,\overline{q} \end{cases} \tag{58}$$

where $g(\nu)$ is the ratio of values for $\overline{p}$ and $\overline{q}$ at which both expressions above coincide (more precisely, the smallest positive solution).

These expressions were obtained by finding the exact solution of the optimization problem (11).

The Bree-type diagram for the block, i.e. the curve $(\mu\overline{p}, \mu\overline{q})$ given by (58), is plotted in Figure 1, together with the ones obtained for two different tubes. In these examples Poisson's ratio is $\nu = 0.2$, which gives $g(0.2) = 0.392$. An important condition, in order to obtain a good fitting between the simplified model and the tubes, is that the mechanical load parameters for the tubes and the block are previously divided by their respective plastic collapse loads.

The numerical results for the moderately thin tube, with $R_{ext}/R_{int} = 1.1$, coincide, within 0.5% error, to the block diagram. Only 10 finite elements are enough to reach converged solutions to within 0.2%.

The thick tube with $R_{ext}/R_{int} = 2$ requires 30 finite elements to obtain convergence to 0.2% error. The Bree-type diagram for this thick tube is also very close (roughly 1.5% different) to the simplified model given by the block, as shown in Figure 1.

As predicted by the analytical solution of the shakedown analysis of the block, the mechanisms obtained in the numerical solutions are always complex mechanisms of incremental collapse, except for the case $\overline{p} = 0$ which gives alternating plasticity, and the case $\overline{q} = 0$ when plastic collapse is observed.

## 9. Conclusions

Kinematical, equilibrium, and mixed principles for shakedown analysis are reviewed in this paper, including those defining upper bounds to the safety factor for elastic shakedown.

Discrete counterparts for the general principles allowing the computation of the shakedown factor are also written. In particular, the kinematical formulation for elastic shakedown is discretized, reaching the explicit formulation (23). Moreover, discrete optimality conditions for elastic shakedown are also stated.

Section 6 contains a summary of results concerning simple incremental collapse mechanisms and a new proof of the fundamental bounding inequality (42).

Approximate formulations, mainly aimed to substitute the classical kinematical approach derived from Koiter's second shakedown theorem, are in use for numerical computations (see, for instance, Dang Hung [8], Pham [22]). The present discussion on bounding principles, specially in sections 6 and 7, is intended to contribute to this issue.

The analysis of a constrained tube under thermo-mechanical loading is chosen to show analytical and numerical applications of the general formulations. The numerical solutions presented in this paper are obtained by using the finite element method to discretize the continuum mixed principle for elastic shakedown, and then applying an iterative algorithm motivated by the equivalent discrete optimality conditions. Besides, expression (58) is the exact analytical solution for a small block, representing a simplified model of the tube wall; it was derived previously in Zouain & Silveira [25].

The comparison of this expression with numerical results presented here shows that the analytical Bree-type diagram given by (58) can be taken as a good approximation for a wide range of tubes under the considered loading conditions.

## REFERENCES


1. Symonds, P.S., 1951, "Shakedown in continuous media", J. Applied Mechanics, 18, pp. 85.
2. Koiter, W. T., 1960, "General theorems for elastic-plastic solids", In: Progress in Solids Mechanics (Eds.: I. N. Sneddon and R. Hill), pp.165-221, North-Holland, Amsterdam.
3. Maier, G., 1972, "A shakedown matrix theory allowing for workhardening and second-order geometric effects", In: Foundations of plasticity (Ed.: A. Sawczuk), pp. 417-433, Noordhoff, Leyden.
4. Rockafellar, T., 1973, "Conjugate duality and optimization", Regional Conference Series in Applied Mathematics, 16, published by SIAM.
5. König, J.A., 1979, "On upper bounds to shakedown loads", ZAMM, 59, pp. 349-354.
6. Christiansen, E., 1980, "Limit analysis in plasticity as a mathematical programming problem", Calcolo, 17, pp. 41-65.
7. Fremond M., Friaa A., 1982, "Les méthodes statique et cinématique en calcul à la rupture et en analyse limite", J. de Mécanique théorique et appliquée, 1, No. 5, pp. 881-905.
8. Dang Hung N., 1984, "Sur la plasticité et le calcul des etats limites par elements finis", Doctoral thesis, University of Liege.
9. Weichert, D., 1984, "Shakedown at finite displacements; a note on Melan's theorem", Mech.Res. Comm., 11, pp. 121.
10. Ponter, A.R.S., Karadeniz, S., 1985, "An extended shakedown theory for structures that suffer cyclic thermal loadings: Part 1 - Theory", Journal of Applied Mechanics, ASME, 52, pp. 877-882.
11. König, J.A., 1987, "Shakedown of elastic-plastic structures", Elsevier, Amsterdam.
12. Gross-Weege, J., 1990, "A unified formulation of statical shakedown criteria for geometrically nonlinear problems", Int. J. of Plasticity, 6, pp. 433.
13. Maugin, G.A., 1992, "The thermomechanics of plasticity and fracture", Cambridge University Press.
14. Polizzotto, C., 1993, "On the conditions to prevent plastic shakedown of structures: Part II - The plastic shakedown limit load", Journal of Applied Mechanics, ASME, 60, pp.20.
15. Pycko, S., Maier, G., 1995, "Shakedown theorems for some classes of nonassociative hardening elastic-plastic material models", Int. J. of Plasticity, 11 (4), pp. 367-395.
16. De Saxcé, G., 1995, "A variational deduction of the upper and lower bound shakedown theorems by Markov and Hill's principles over a cycle", In: Inelastic Behaviour of Structures under Variable Loads (Eds.: Z. Mróz, D. Weichert and S. Dorosz), pp. 153-167, Kluwer, London.

17. Nayroles, B., 1995, "Some basic elements of the shakedown theory", In: Inelastic Behaviour of Structures under Variable Loads (Eds.: Z. Mróz, D. Weichert and S. Dorosz), pp. 183-201, Kluwer, London.
18. Stein, E., Huang, Y.J., 1995, "Shakedown for systems of kinematic hardening materials", In: Inelastic Behaviour of Structures under Variable Loads (Eds.: Z. Mróz, D. Weichert and S. Dorosz), pp. 33-50, Kluwer, London.
19. Telega, J.J., 1995, "On shakedown theorems in the presence of Signorini conditions and friction", In: Inelastic Behaviour of Structures under Variable Loads (Eds.: Z. Mróz, D. Weichert and S. Dorosz), pp. 183-201, Kluwer, London.
20. Borges, L.A., Zouain, N., Huespe, A.E., 1996, "A nonlinear optimization procedure for limit analysis", European Journal of Mechanics /A Solids, 15, pp. 487-512.
21. Kamenjarzh, J., 1996, "Limit analysis of solids and structures", CRC Press.
22. Pham D.C., 1997, "Evaluation of shakedown loads for plates", Int. Journal of Mechanical Science, 39 (12), pp. 1415-1422.
23. Silveira, J.L., Zouain, N., 1997, "On extremum principles and algorithms for shakedown analysis", European Journal of Mechanics /A Solids, 16 (5), pp. 757-778.
24. Christiansen, E., Andersen, K.D., 1998, "Computation of collapse states with von Mises type yield condition", Institut for Matematik og Datalogi, Preprint 18, pp. 1-25.
25. Zouain, N., Silveira, J.L., 1999, "Extremum principles for bounds to shakedown loads", European Journal of Mechanics /A Solids, 18 (5), pp. 879-901.

# SHAKEDOWN OF ELASTIC-PLASTIC STRUCTURES WITH NON LINEAR KINEMATICAL HARDENING BY THE BIPOTENTIAL APPROACH

G. DE SAXCE, J.-B. TRITSCH
*Laboratoire de Mécanique de Lille, Lille-I University,*
*F-59655 Villeneuve d'Ascq cedex, France.*
M. HJIAJ
*Mécanique des matériaux et des structures, Polytechnic Faculty of Mons,*
*B-7000 Mons, Belgium.*

## Abstract

In this paper, we study the shakedown behaviour of elastic plastic material with non-linear kinematical hardening rule. The behaviour law taken into account is shown to be a non-associated one. For this purpose, the implicit non-standard material model is introduced through the bipotential approach. After some remarks on the theoretical aspects, an analytical example of a thin walled tube under constant traction and alternating torsion is given as an application and compared with previous results. The obtained solution is proved to be exact.

## 1. Notations

$\dot{\kappa}$ generalised velocity vector.
$\pi$ vector of generalised forces associated to $\dot{\kappa}$.
$\dot{\kappa}'$ velocity vector of internal variables (hardening, damage, phase transition..).
$\pi'$ vector of associated internal variables.
$\sigma$ Cauchy's stress.
$\dot{\varepsilon}^{p}$ plastic strain rate.
$V$ velocity space.
$F$ stress space.
$b$ bipotential.
$\beta$ bifunctional.
$D$ superpotential of dissipation.
$\chi$ superpotential of complementary dissipation.
$\Omega$ domain.
$\tau$ shear stress.
$t$ time variable.



D. Weichert and G. Maier (eds.), *Inelastic Analysis of Structures under Variable Loads,* 167–182.

$\lambda^a$ exact shakedown load factor.
$\lambda^s$ statical load factor.
$\lambda^k$ kinematical load factor.
$K$ elastic domain.
$\sigma^E$ stress field in the ficticious elastic body.
$\sigma^{Eo}$ reference stress field in a ficticious elastic body.
$\bar{\rho}$ time independent self-equilibrated stress field.
$\alpha$ kinematical hardening rule tensor.
$p$ isotropic hardening rule parameter.
$R$ plastic threshold.
$X$ back-stress.
$\sigma_{eq}$ equivalent stress.
$\varepsilon_{eq}$ equivalent strain.
: double contracted tensorial product.
• scalar product.
$\dot{\xi}$ plastic multiplier.
$f$ yield function.

## 2. The mathematical frame

This work is based on the tools of the convex analysis. One of the starting points of this approach is the Legendre-Fenchel inequality (1949) [4][8], which generalises the concept of normality rule. Moreau (1968) [15] introduced the superpotential concept going well together with the concept of normal dissipation. Halphen and Nguyen Quoc (1975) [9] applied this concept to account for linear kinematical hardening in elastic-plastic materials. The result is known as the generalised standard material model and is an associated law type model.

## 3. The bipotential concept

Using bipotentials, we show that many non standard dissipative materials are in fact governed by a normality rule, but in an implicit sense.
Let the generalized velocities be $\dot{\kappa} = (\dot{\varepsilon}^p, \dot{\kappa}') \in V$, including the velocities $\dot{\kappa}'$of additional internal variables (hardening...), and the corresponding associated variables $\pi = (\sigma, \pi') \in F$. The spaces $V$ and $F$ are equipped with locally convex topologies compatible with the duality expressed by a bilinear form $(\dot{\kappa}, \pi) \mapsto \dot{\kappa}\pi$.
A bipotential is a function $b$ from $V \times F$ into [-∞, +∞], separately convex, satisfying the fundamental inequation generalising Fenchel's :

$$\forall(\dot{\kappa}^*, \pi^*) \in V \times F, \quad b(\dot{\kappa}^*, \pi^*) \geq \dot{\kappa}^* . \pi^* \tag{1}$$

The couples $(\dot{\kappa},\pi)$, for which the variables are related by the dissipative law, are qualified as *extremal* in the sense that the equality is reached in the previous relation :

$$b(\dot{\kappa},\pi)=\dot{\kappa}.\pi \tag{2}$$

From (1) and (2), we deduce the following inequalities to be satisfied by the extremal couples :

$$\forall\pi^* \in F,\quad b(\dot{\kappa},\pi^*)-b(\dot{\kappa},\pi)\geq \dot{\kappa}.(\pi^*-\pi) \tag{3}$$

$$\forall\dot{\kappa}^* \in V,\quad b(\dot{\kappa}^*,\pi)-b(\dot{\kappa},\pi)\geq (\dot{\kappa}^*-\dot{\kappa}).\pi \tag{4}$$

Briefly, they are characterised by the following differential inclusions :

$$\dot{\kappa}\in \partial_\pi b(\dot{\kappa},\pi),\quad \pi\in \partial_{\dot{\kappa}} b(\dot{\kappa},\pi), \tag{5}$$

where $\partial_\pi$ ($\partial_{\dot{\kappa}}$ respectively) denotes the subdifferential when partial derivating with respect to $\pi$ (respectively $\dot{\kappa}$). For elastic-plastic behaviour laws, the set of extremal couples is equivalent to that of the material states satisfying the plastic flow law. Physically, the bipotential stands for the plastic dissipation power, and thus, is supposed to be positive.

The bipotential concept sheds a new light on known non-associated laws : Coulomb's dry friction, non associated Drucker-Prager law, the modified Cam-clay model for soil materials and Lemaitre plastic damage model [2][4][6]. On this ground, an extension of usual bound theorems of the limit analysis was proposed by de Saxcé & Bousshine [7].

In the particular case of the generalised standard material model, the bipotential is separable into two parts : a super-potential of dissipation, and its polar function, such that :

$$\forall(\dot{\varepsilon}^p,\sigma)\in V\times F\quad b(\dot{\varepsilon}^p,\sigma)=D(\dot{\varepsilon}^p)+\chi(\sigma) \tag{6}$$

For plastic materials, taking $\chi(\sigma)$ equal to zero if $\sigma$ belongs to the elastic domain $K$ and to infinity otherwise, we may write the flow rule may be given in an explicit way :

$$\dot{\varepsilon}^p\in\partial\chi(\sigma) \tag{7}$$

For $\sigma$ on the yielding surface, (3) gives Hill's inequality :

$$\forall\sigma^*\in K,\quad (\sigma^*-\sigma):\dot{\varepsilon}^P\leq 0 \tag{8}$$

The corresponding superpotential of dissipation $D$ is homogeneous of order one. More generally, the bipotentals representing plastic behaviours are assumed to be homogeneous of order one with respect to $\dot{\kappa}$.

## 4. For materials admitting a bipotential, shakedown bound theorems can be extended

Let $\Omega$ be a solid body with elastic-plastic materials admitting bipotentials, subjected to variable periodic external actions varying between given limits controlled by a load factor $\lambda$. The following question arises : under what conditions does the body shake down ? By numerical step-by-step analysis on simple examples, with a non associated Drucker-Prager material, Chaaba et al [3] observed the existence of time-independent residual stress fields under a critical value $\lambda^a$ of the load factor. Unfortunately, no generalisation of Melan's theorem [14] to material admitting a bipotential has been rigorously proved up to now.

In spite of this, we admit the existence of admissible stress fields $(\overline{\rho},\overline{\pi}')$ in the sense that :

- $\overline{\rho}$ is a residual stress field
- $\overline{\rho}$ and $\overline{\pi}'$ are time-independent and plastically admissible when adding to $\overline{\rho}$ the stress response $\sigma^E = \lambda\sigma^{Eo}$ in the fictitious elastic body :

$$\forall x\in \Omega,\ \forall t,\quad \left(\sigma^E(x,t)+\overline{\rho}(x),\overline{\pi}'(x)\right)=\left(\lambda\sigma^{Eo}(x,t)+\overline{\rho}(x),\overline{\pi}'(x)\right)\in K$$

where $K$ is the elastic domain. On the other hand, we define admissible generalised velocity fields $(\dot{\varepsilon}^P,\dot{\kappa}')$ in the sense that :

-the increment of the plastic strain rate on the load cycle $\Delta\varepsilon^P = \oint \dot{\varepsilon}^P dt$ is kinematically admissible with zero values of the corresponding displacement increment on the supports.

- $\dot{\varepsilon}^P$ is plastically admissible : $\int_\Omega \oint \sigma^E : \dot{\varepsilon}^P dt\, d\Omega > 0$.

As usual, the admissible velocity fields are normalized :

$$\int_\Omega \oint \sigma^{Eo} : \dot{\varepsilon}^P dt\, d\Omega = 1 \tag{9}$$

A possible variational formulation of shakedown problems arises from introducing the so-called bifunctional :

$$\beta_S\left(\dot{\varepsilon}^P,\dot{\kappa}',\overline{\rho},\overline{\pi}',\lambda\right)=\int_\Omega \oint \left\{b\left[\left(\dot{\varepsilon}^P,\dot{\kappa}'\right),\left(\overline{\rho}+\lambda\sigma^{Eo},\overline{\pi}'\right)\right]-\lambda\sigma^{Eo}:\dot{\varepsilon}^P-\overline{\pi}'.\dot{\kappa}'\right\} dt\, d\Omega \tag{10}$$

By virtue of the principle of virtual work, one has, for admissible fields :

$$\int_\Omega \oint \overline{\rho}:\dot{\varepsilon}^P dt\, d\Omega = \int_\Omega \overline{\rho}:\Delta\varepsilon^P d\Omega = 0 \tag{11}$$

A straightforward consequence of (1), (10) and (11) is that for any admissible fields :

$$\beta_S\left(\dot{\varepsilon}^{P*},\dot{\kappa}'^*,\overline{\rho}^*,\overline{\pi}'^*,\lambda\right)\geq 0 \tag{12}$$

In particular, for the exact solution, the constitutive law is exactly satisfied anywhere in $\Omega$ and at any time. As consequence of (2), we have :

$$\beta_S\left(\dot{\varepsilon}^P, \dot{\kappa}', \overline{\rho}, \overline{\pi}', \lambda^a\right) = 0 \tag{13}$$

As we shall see, the previous observation is crucial in the sequel. Special cases are usual standard materials with separable bipotentials (6) and no hardening variable $\kappa'$. The bifunctional splits into two terms :

$$\beta_S\left(\dot{\varepsilon}^P, \overline{\rho}, \lambda\right) = \phi_S\left(\dot{\varepsilon}^P, \lambda\right) + \Pi_S\left(\overline{\rho}, \lambda\right) \tag{14}$$

where

$$\phi_S\left(\dot{\varepsilon}^P, \lambda\right) = \int_\Omega \oint \left[D\left(\dot{\varepsilon}^P\right) - \lambda \sigma^{Eo} : \dot{\varepsilon}^P\right] dt\, d\Omega \tag{15}$$

is the functional of Markov's principle over a cycle and

$$\Pi_S\left(\overline{\rho}, \lambda\right) = \int_\Omega \oint \chi\left(\overline{\rho} + \lambda \sigma^{Eo}\right) dt\, d\Omega \tag{16}$$

is the one of Hill's principle over a cycle, as previously introduced by de Saxcé [5]. Now our goal is to extend the method proposed by de Saxcé [5] for usual materials, to materials admitting bipotentials, and so to establish bound theorems similar to Koiter's [10]. For the exact solution $\left(\left(\dot{\varepsilon}^P, \dot{\kappa}'\right), \left(\overline{\rho}, \overline{\pi}'\right)\right)$, condition (13) combined with the normalisation condition (9) allows one to calculate the value of the shakedown load factor :

$$\lambda^a = \int_\Omega \oint \left\{b\left[\left(\dot{\varepsilon}^P, \dot{\kappa}'\right), \left(\lambda^a \sigma^{Eo} + \overline{\rho}, \overline{\pi}'\right)\right] - \dot{\kappa}' . \overline{\pi}'\right\} dt\, d\Omega \tag{17}$$

## 5. Kinematical bounding theorem

By analogy with (17), for any admissible velocity field $\left(\dot{\varepsilon}^{P^*}, \dot{\kappa}'^*\right)$, the corresponding kinematical load factor is defined by the expression :

$$\lambda^k = \int_\Omega \oint \left\{b\left[\left(\dot{\varepsilon}^{P*}, \dot{\kappa}'^*\right), \left(\lambda^a \sigma^{Eo} + \overline{\rho}, \overline{\pi}'\right)\right] - \dot{\kappa}'^* . \overline{\pi}'\right\} dt\, d\Omega \tag{18}$$

As $\left(\dot{\varepsilon}^{P*}, \dot{\kappa}'^*\right)$ is an admissible velocity field and $\left(\overline{\rho}, \overline{\pi}'\right)$ is the admissible stress field corresponding to the shakedown load $\lambda^a$, one has :

$$\beta_S\left(\dot{\varepsilon}^{P^*}, \dot{\kappa}'^*, \overline{\rho}, \overline{\pi}', \lambda^a\right) \geq 0 \tag{19}$$

Taking into account that $\dot{\varepsilon}^P$ is normalised by (9) and definition (18), we have :

$$\lambda^k \geq \lambda^a \tag{20}$$

That can be considered as the extension of the usual kinematical bounding theorem to materials admitting a bipotential.

## 6. Statical bounding theorem

Let $(\bar{\rho}^*, \bar{\pi}'^*)$ be any admissible stress field corresponding to the statical load factor $\lambda^s$ :

$$\forall x \in \Omega, \forall t \quad (\lambda^s \sigma^{Eo}(x,t) + \bar{\rho}^*(x), \bar{\pi}'^*(x)) \in K \tag{21}$$

let $(\dot{\varepsilon}^P, \dot{\kappa}')$ be the exact admissible velocity field. Then, (12) gives :

$$\beta_S(\dot{\varepsilon}^P, \dot{\kappa}', \bar{\rho}^*, \bar{\pi}'^*, \lambda^s) \geq 0 \tag{22}$$

More explicitly, one has :

$$\lambda^s \leq \int_\Omega \oint \{b[(\dot{\varepsilon}^P, \dot{\kappa}'), (\lambda^s \sigma^{Eo} + \bar{\rho}^*, \bar{\pi}'^*)] - \dot{\kappa}'.\bar{\pi}'^*\} \, dt \, d\Omega \tag{23}$$

From (17) and (23), we deduce the inequality :

$$\begin{aligned} \lambda^a - \int_\Omega \oint \{b[(\dot{\varepsilon}^P, \dot{\kappa}'), (\lambda^a \sigma^{Eo} + \bar{\rho}, \bar{\pi}')] - \dot{\kappa}'.\bar{\pi}'\} \, dt \, d\Omega \geq \\ \lambda^s - \int_\Omega \oint \{b[(\dot{\varepsilon}^P, \dot{\kappa}'), (\lambda^s \sigma^{Eo} + \bar{\rho}^*, \bar{\pi}'^*)] - \dot{\kappa}'.\bar{\pi}'^*\} \, dt \, d\Omega \end{aligned} \tag{24}$$

For the special event of usual standard materials with separable bipotential (6) and no hardening variables, the previous relation degenerates, taking account of (14), (15) and (16) :

$$\lambda^a - \int_\Omega \oint \chi(\lambda^a \sigma^{Eo} + \bar{\rho}) \, dt \, d\Omega \geq \lambda^s - \int_\Omega \oint \chi(\lambda^s \sigma^{Eo} + \bar{\rho}^*) \, dt \, d\Omega \tag{25}$$

For admissible stress fields, the values of the complementary dissipation superpotential $\chi$ vanish :

$$\lambda^a \geq \lambda^s \tag{26}$$

Then, the inequality (24) represents an extension of the usual statical bounds property for standard materials to material admitting bipotentials.

## 7. The plastic flow rule with non-linear kinematical hardening rule admits a bipotential

For this constitutive law, the additional internal variable velocities are $\dot{\kappa}' = (-\dot{\alpha}, -\dot{p})$, where $\dot{\alpha}$ and $\dot{p}$ are respectively the kinematical and isotropic hardening variable rates. The corresponding associated variables are denoted $\pi' = (X, R)$.

Let the stress and the elastic domain be defined by :

$$K = \{\pi = (\sigma, X, R) \;\; \text{such that} \;\; \sigma_{eq}(\sigma - X) - R \leq 0\} \tag{27}$$

where $$\sigma_{eq}(\sigma - X) = \sqrt{\frac{3}{2}(\sigma - X)':(\sigma - X)'} \tag{28}$$

and (…)" means the deviatoric part
The isotropic hardening rule entails that :

$$\varepsilon_{eq}(\dot{\varepsilon}^p) - \dot{p} \leq 0 \tag{29}$$

where $$\varepsilon_{eq}(\dot{\varepsilon}^p) = \sqrt{\frac{2}{3}\dot{\varepsilon}^p : \dot{\varepsilon}^p} \tag{30}$$

The non associated kinematical hardening rule introduced by Armstrong & Frederick [1] and more extensively developped by Chaboche & Lemaitre [11] and Marquis [12] :

$$\dot{\alpha} = \dot{\varepsilon}^p - \frac{3}{2}\frac{X}{X_\infty}\dot{p} \tag{31}$$

gives a more realistic representation of the cyclic plasticity of metals than Prager's rule, and the improved one with a saturation limit surface, in the sense that it better describes the smooth hysteresis shape observed in alternating plastic cycles (see figure 1). Its main drawback is its non-associated nature. Nevertheless, it admits a bipotential equal to :

$$b(\dot{\kappa}, \pi) = \frac{(\sigma_{eq}(X))^2}{X_\infty}\dot{p} \tag{32}$$

when (27), (29), (31) are satisfied and equal to $+\infty$ otherwise.

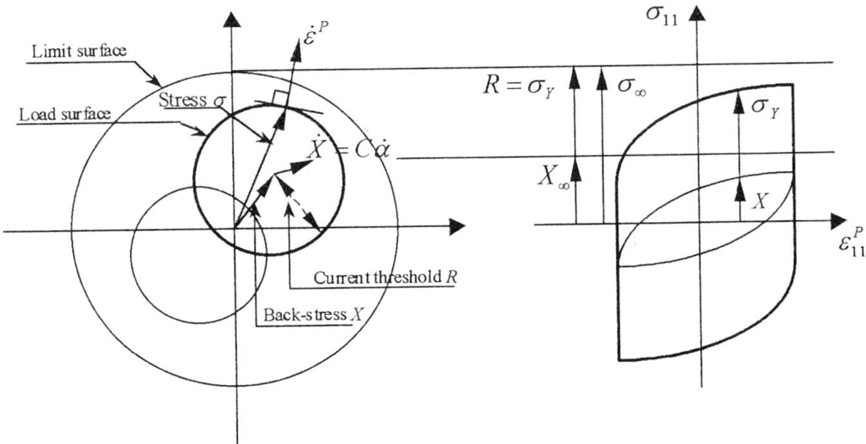


Figure 1

This function allows us to describe the generalized flow rule through the implicit relation:

$$\dot{\kappa} \in \partial_\pi b(\dot{\kappa}, \pi) \tag{33}$$

Now, let us prove that the previously introduced function is a bipotential. For this, we must show that for any set of generalized velocities and stresses fulfilling (23), (25), (27), we have :

$$\frac{(\sigma_{eq}(X))^2}{X_\infty}\dot{p} \geq \sigma : \dot{\varepsilon}^p - X : \dot{\alpha} - R\dot{p} \tag{34}$$

According to the flow rule (29), (31) and to the definition of the elastic domain (27), and remembering the Cauchy-Schwartz inequality, we can write :

$$R\dot{p} \geq R\varepsilon_{eq}(\dot{\varepsilon}^p) \geq \sigma_{eq}(\sigma - X)\varepsilon_{eq}(\dot{\varepsilon}^p) \geq (\sigma - X) : \dot{\varepsilon}^p \tag{35}$$

but
$$(\sigma - X) : \dot{\varepsilon}^p = \sigma : \dot{\varepsilon}^p - X : \left(\dot{\alpha} + \frac{3}{2}\frac{X}{X_\infty}\dot{p}\right)$$

by accounting for the flow rule again, we obtain :

$$\left(\frac{3}{2}\frac{X : X}{X_\infty}\dot{p}\right) \geq \sigma : \dot{\varepsilon}^p - X : \dot{\alpha} - R\dot{p} \tag{36}$$

This last inequality proves that the function proposed above is a bipotential.
In the same spirit, it is possible to show that any extremal couple for the bipotential fonction fulfills the flow rule and, conversely, that any couple satisfying the flow rule is an extremal one for the bipotential function [4].

## 8. Thin walled tube under constant tension and alternating cyclic torsion

The analytical example concerns a thin walled tube subjected to constant tension $\sigma_{11}$ and alternating torsion generating a shear stress state $\sigma_{12}$ (figure 2).
Only considering the stabilized cycle, the plastic threshold $R$ is supposed to be equal to the constant value $\sigma_Y$ :

$$R = \sigma_Y \tag{37}$$

On the other hand, the back stress is linearly dependent on the kinematical variables through :

$$X = \frac{2}{3}C\alpha \tag{38}$$

where $C$ is a constant kinematical hardening modulus. Because of the plane stress state, the shifted stress tensor is as follows :

$$\sigma - X = \begin{bmatrix} \sigma_{11} - \frac{2}{3}X_{11} & \sigma_{12} - X_{12} & 0 \\ \sigma_{12} - X_{12} & \frac{1}{3}X_{11} & 0 \\ 0 & 0 & \frac{1}{3}X_{11} \end{bmatrix} \tag{39}$$

So, the deviatoric part is :

$$(\sigma - X)" = \begin{bmatrix} \frac{2}{3}(\sigma_{11} - X_{11}) & \sigma_{12} - X_{12} & 0 \\ \sigma_{12} - X_{12} & -\frac{1}{3}(\sigma_{11} - X_{11}) & 0 \\ 0 & 0 & -\frac{1}{3}(\sigma_{11} - X_{11}) \end{bmatrix} \tag{40}$$

Accounting for the Von-Mises criterion, the yield function is of the form :

$$f(\sigma_{11},\sigma_{12},X_{11},X_{12}) = (\sigma_{eq}(\sigma - X))^2 - \sigma_Y{}^2 = (\sigma_{11} - X_{11})^2 + 3(\sigma_{12} - X_{12})^2 - \sigma_Y{}^2 \tag{41}$$

such that the yield criterium gives :

$$(\sigma_{11} - X_{11})^2 + 3(\sigma_{12} - X_{12})^2 \le \sigma_Y{}^2 \tag{42}$$

Let us take the following transformed variables :

$$\sigma_{11} = \sigma \ , \ X_{11} = X \ , \ \sqrt{3}\sigma_{12} = \tau \ et \ \sqrt{3}X_{12} = Y \tag{43}$$

The yield function then becomes :

$$f(\sigma,\tau,X,Y) = (\sigma - X)^2 + (\tau - Y)^2 - \sigma_Y{}^2 \tag{44}$$

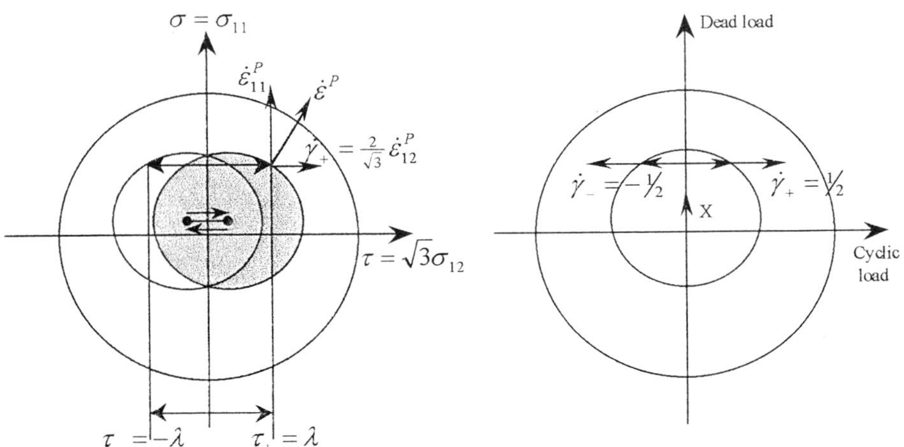


Figure 2

Assuming incompressibility of plastic strains, the tensor of the plastic strain rates is :

$$\dot{\varepsilon}^P = \begin{bmatrix} \dot{\varepsilon}^P_{11} & \dot{\varepsilon}^P_{12} & 0 \\ \dot{\varepsilon}^P_{12} & -\frac{1}{2}\dot{\varepsilon}^P_{11} & 0 \\ 0 & 0 & -\frac{1}{2}\dot{\varepsilon}^P_{11} \end{bmatrix} \tag{45}$$

The tensor of the kinematical internal variable rates has the same form :

$$\dot{\alpha} = \begin{bmatrix} \dot{\alpha}_{11} & \dot{\alpha}_{12} & 0 \\ \dot{\alpha}_{12} & -\frac{1}{2}\dot{\alpha}_{11} & 0 \\ 0 & 0 & -\frac{1}{2}\dot{\alpha}_{11} \end{bmatrix} \tag{46}$$

In the same spirit as previously for stresses, we now take :

$$\dot{\varepsilon}^P_{11} = \dot{\varepsilon} \quad and \quad \frac{2}{\sqrt{3}}\dot{\varepsilon}^P_{12} = \dot{\gamma} \tag{47}$$

so, the cumulated plastic deformation becomes :

$$\dot{p} = \varepsilon_{eq}\left(\dot{\varepsilon}^P\right) = \sqrt{\frac{2}{3}\left(\frac{3}{2}\left(\dot{\varepsilon}^P_{11}\right)^2 + 2\left(\dot{\varepsilon}^P_{12}\right)^2\right)} = \sqrt{\dot{\varepsilon}^2 + \dot{\gamma}^2} \tag{48}$$

For the non linear kinematical hardening rule, we can write :

$$\dot{\alpha}_{11} = \dot{\varepsilon}^P_{11} - \frac{X_{11}}{X_\infty}\dot{p}\,,\ \dot{\alpha}_{12} = \dot{\varepsilon}^P_{12} - \frac{X_{12}}{X_\infty}\dot{p} \tag{49}$$

Putting

$$\dot{\alpha} = \alpha_{11}\,,\ \dot{\beta} = \frac{2}{\sqrt{3}}\dot{\alpha}_{12}\,, \tag{50}$$

and using (43), (47), one gets the following condensed form :

$$\dot{\alpha} = \dot{\varepsilon} - \frac{X}{X_\infty}\dot{p}\,,\ \dot{\beta} = \dot{\gamma} - \frac{Y}{X_\infty}\dot{p} \tag{51}$$

For the sets of dual variables $(\sigma, X, R) \in K$, $(\dot{\varepsilon}^P, -\dot{\alpha}, -\dot{p}) \in K^*$, verifying the non-linear hardening rule, the bipotential function reduces to :

$$b = \frac{\left(\sigma_{eq}(X)\right)^2}{X_\infty}\dot{p} = \frac{X^2 + Y^2}{X_\infty}\sqrt{\dot{\varepsilon}^2 + \dot{\gamma}^2} \tag{52}$$

The unit value is taken as reference shear stress; this allows us to identify the load factor as the maximum shear stress. Therefore, considering cyclic loading, the state will alternate between two shear stress extrema such that, for the maximum of the cycle

$$\sigma = \sigma ,\quad \tau = \lambda \ \left(\tau^0 = 1\right),\quad \dot{\varepsilon} = \dot{\varepsilon}_+ ,\quad \dot{\gamma} = \dot{\gamma}_+ ,\quad \dot{\alpha} = \dot{\alpha}_+ ,\quad \dot{\beta} = \dot{\beta}_+ \tag{53}$$

for the minimum of the cycle, we will get

$$\sigma = \sigma \ ,\ \tau = -\lambda \ \left(\tau^0 = -1\right) ,\ \dot{\varepsilon} = \dot{\varepsilon}_- \ ,\ \dot{\gamma} = \dot{\gamma}_- \ ,\ \dot{\alpha} = \dot{\alpha}_- \ ,\ \dot{\beta} = \dot{\beta}_- \tag{54}$$

For sake of simplicity, a unit volume sample $\Omega$ is now considered, in order to avoid the volume integrals.

## 9. Calculation of the shakedown factor

*Step 1* : It is assumed that the collapse occurs by ratcheting only in traction :

$$\oint \dot{\gamma}\, dt = 0 \tag{55}$$

We note that non vanishing contributions to the time integral are related to the extrema of the collapse cycle. At each extremum, we consider that the velocities are constant during a unit time interval, that leads to :

$$\dot{\gamma}_+ + \dot{\gamma}_- = 0 \tag{56}$$

On the other hand, because of the normalization condition and (54), one has :

$$\oint \tau\dot{\gamma}\, dt = \lambda \oint \tau^o \dot{\gamma}\, dt = \lambda(\dot{\gamma}_+ - \dot{\gamma}_-) = \lambda \tag{57}$$

Consequently, we get

$$\dot{\gamma}_+ = -\dot{\gamma}_- = \frac{1}{2} \text{ , and } \dot{p} = \sqrt{\dot{\varepsilon}^2 + \frac{1}{4}} \tag{58}$$

*Step 2* : we suppose the maxima of the cycle are located on the load surface

$$(\sigma - X)^2 + (\lambda - Y)^2 = \sigma_Y^2 \tag{59.a}$$

$$(\sigma - X)^2 + (-\lambda - Y)^2 = \sigma_Y^2 \tag{59.b}$$

The difference between the two equations gives

$$(\lambda - Y)^2 - (-\lambda - Y)^2 = -4\lambda Y = 0 \tag{60}$$

Because $\lambda$ is non negative, $Y=0$ and the yield criteria becomes :

$$(\sigma - X)^2 + \lambda^2 = \sigma_Y^2 \tag{61}$$

with the following positive solution :

$$\lambda = \sigma_Y \sqrt{1 - \left( \frac{\sigma - X}{\sigma_Y} \right)^2} \tag{62}$$

*Step 3* : Our goal now is to determine the value of $X$ at collapse accounting for the plastic flow and hardening rules. Therefore, we calculate $\dot{\varepsilon}^P$ and $\dot{p}$, in order to get an explicit expression of $X$ through the hardening rule. The plastic yielding rule gives :

$$\dot{\varepsilon} = \dot{\xi} \frac{\partial f}{\partial \sigma} = 2\dot{\xi}(\sigma - X) \tag{63}$$

$$\dot{\gamma} = \dot{\xi} \frac{\partial f}{\partial \tau} = 2\dot{\xi}(\tau - Y) = 2\dot{\xi}\tau \tag{64}$$

In particular, at the extrema of the cycle, we have :

$$\dot{\gamma}_{\pm} = \pm 2\dot{\xi}_{\pm}\lambda \tag{65}$$

Combining with relation (58) of step 1, we find the plastic multiplier is equal to :

$$\dot{\xi}_{\pm} = \frac{1}{4\lambda} \tag{66}$$

then, taking account of the yield criterion (61), one has :

$$\dot{\varepsilon}_{\pm} = \frac{(\sigma - X)}{2\lambda} \text{ , and } \dot{p}_{\pm} = \sqrt{\dot{\varepsilon}_{\pm}^{\,2} + \frac{1}{4}} = \sqrt{\frac{1}{4\lambda^2}\left((\sigma - X)^2 + \lambda^2\right)} = \frac{\sigma_Y}{2\lambda} \tag{67}$$

On the other hand, the non linear hardening rule allows one to write :

$$\dot{\alpha}_{\pm} = \frac{1}{2\lambda}(\sigma - X) - \frac{X}{X_\infty}\frac{\sigma_Y}{2\lambda} = \frac{1}{2\lambda}\left( \sigma - \frac{\sigma_\infty}{X_\infty} X \right) \tag{68}$$

where
$$\sigma_\infty = X_\infty + \sigma_Y \tag{69}$$

A straightforward consequence of the previous developments is :

$$\dot{\alpha}_+ = \dot{\alpha}_- \tag{70}$$

Step 4 : As shown by Martin [13], the actual collapse by rachetting occurs only for a load factor greater than the shakedown one. After a transient phase, the back-stress field tends to a time periodic solution. In other words, the back-stress increment over the collapse cycle vanishes :

$$\Delta X = \oint \dot{X}\, dt = C \oint \dot{\alpha}\, dt = C(\dot{\alpha}_+ + \dot{\alpha}_-) = 0 \tag{71}$$

therefore
$$\dot{\alpha}_+ + \dot{\alpha}_- = 0 \tag{72}$$

Combining with (70) gives :

$$\dot{\alpha}_+ = \dot{\alpha}_- = 0 \tag{73}$$

Consequently, from the expression (63) of $\alpha_\pm$, we deduce the value of the back-stress :

$$X = X_\infty \frac{\sigma}{\sigma_\infty} \tag{74}$$

Then, putting it into (62) we find :

$$\lambda = \sigma_Y \sqrt{1 - \left( \frac{\sigma(\sigma_\infty - X_\infty)}{\sigma_\infty \sigma_Y} \right)^2} \tag{75}$$

which leads to the following expression :

$$\lambda = \sigma_Y \sqrt{1 - \left( \frac{\sigma}{\sigma_\infty} \right)^2} \tag{76}$$

This solution is the same as that given by Chaboche & Lemaitre [11]. In the present work, the solution is deduced from shakedown theory.

## 10. The previous solution is the exact one

*Step 1* : The key idea is to consider the corresponding bifunctional :

$$\beta_S = \oint \left\{ b\left[ (\sigma, X, R), (\dot{\varepsilon}^P, -\dot{\alpha}, -\dot{p}) \right] - \sigma\,\dot{\varepsilon} - \tau\,\dot{\gamma} + X\,\dot{\alpha} + Y\,\dot{\beta} + R\,\dot{p} \right\} dt \tag{77}$$

and to prove its value is zero. Accounting for the expression (52) of the bipotential, and the normalization condition (57), we simplify :

$$\beta_S = \oint \left[ \frac{X^2 + Y^2}{X_\infty} \dot{p} - \sigma\,\dot{\varepsilon} + X\,\dot{\alpha} + Y\,\dot{\beta} + R\dot{p} \right] dt - \lambda \tag{78}$$

*Step 2* : Moreover, for a stabilized cycle, $R = \sigma_Y$, and $Y = 0$ as demonstrated in the previous calculations (60). Then :

$$\beta_S = \oint \left[ \left( \frac{X^2}{X_\infty} + \sigma_Y \right) \dot{p} - \sigma\,\dot{\varepsilon} + X\,\dot{\alpha} \right] dt - \lambda \tag{79}$$

Because the tension stress $\sigma$ acts as a dead load and the back stress $X$ is time independent

$$\beta_S = \oint \left( \frac{X^2}{X_\infty} + \sigma_Y \right) \dot{p}\, dt - \sigma \oint \dot{\varepsilon}\, dt + X \oint \dot{\alpha}\, dt - \lambda \tag{80}$$

Taking into account the remark of section 9, step 1 concerning the time integrals, one has :

$$\beta_S = \left( \frac{X^2}{X_\infty} + \sigma_Y \right) (\dot{p}_+ + \dot{p}_-) - \sigma (\dot{\varepsilon}_+ + \dot{\varepsilon}_-) + X (\dot{\alpha}_+ + \dot{\alpha}_-) - \lambda \tag{81}$$

*Step 3* : Accounting for the explicit expressions (67) and (68) of $\dot{\varepsilon}$ , $\dot{p}_+$, $\dot{\alpha}_+$ previously found, we reduce the bifunctional to :

$$\beta_S = 2 \left[ \left( \frac{X^2}{X_\infty} + \sigma_Y \right) \frac{\sigma_Y}{2\lambda} - \sigma \frac{(\sigma - X)}{2\lambda} + \frac{X}{2\lambda} \left( \sigma - \frac{\sigma_\infty}{X_\infty} X \right) \right] - \lambda$$

$$\beta_S = \frac{1}{\lambda} \left[ -(\sigma - X)^2 - \lambda^2 + \sigma_Y^2 \right] \tag{82}$$

Finally, taking into account that the yield criterion (61) at the extrema of the collapse cycle is satisfied, we prove that the bifunctional vanishes :

$$\beta_S = 0 \tag{83}$$

The theoretical considerations show that the previous analytical solution is the exact one. For the same problem, statical and kinematical approximations of the shakedown factor were previously given by Pycko & Maier [16].

## Conclusions and perspectives

As we have seen, the normality law can be extended in a weakened sense as an **implicit relation** between dual variables by the concept of bipotential.

If we assume the possible existence of time independent residual stress fields and admissible plastic strain fields, we can prove **dual bound theorems** for the plastic materials admitting a **bipotential : the implicit standard materials**.

For the problem of the thin walled tube under constant tension and alternating cyclic torsion, a complete analytical solution was provided and proved to be the exact one according to the previous bound theorems using the bipotential approach.

In the future, extensions and improvements of the previous results are expected on the following topics :

- rigorous proof of the existence of admissible residual stress fields for elastoplastic materials admitting a bipotential;
- construction of numerical algorithms based on the bipotential and the mathematical programing to compute the shakedown load;
- exact or approximated analytical solutions for shakedown problems with the non linear hardening rule such as :

- A thin walled tube under constant traction and alternating cyclic torsion in plane strain conditions.
- Structures with the redundancy degree equal to one, e.g. two parallel bars specimen.
- A thick walled tube under alternating internal pressure.

## Acknowledgements

The authors adress their thanks to Mister Brahim Bouafs, student, who accepted to work on the present developments [17].

## References

1. Armstrong & Frederick, A mathematical representation of the multiaxial Baushinger effect, *C. E. G. B. Report n°RB/B/N 731*, (1966).
2. Bodovillé G., Sur l'endommagement et les matériaux standards implicites, (with abriged english version), to appear in *Comptes rendus de l'Académie des Sciences*, Paris, 1999.
3. Chaaba A., Bousshine L., El Harif A., De Saxcé G., A new approach of shakedown analysis for non-standard elasto-plastic materials by the bipotential theory, *Euromech 385*, Aachen 1998.
4. De Saxcé G., Une généralisation de l'inégalité de Fenchel et ses applications aux lois constitutives, (with abriged english version), *Comptes rendus de l'Académie des Sciences, Paris, 314, série II,* (1992), 125-129.
5. De Saxcé G., A variational deduction of upper and lower bound shakedown theorems by Markov's and Hill's principle over a cycle, *in Inelastic behaviour of structures under variable loads, S.M.I.A. 36*, Z. Mroz, D. Weichert & S. Dorosz (Eds), Kluwer Academic Publishers, Dordrecht, 1995.
6. De Saxcé G., The bipotential method, a new variational and numerical treatment of the dissipative laws of materials, Proc. $10^{th}$ . conf. on mathematical and computer modeling and scientific computing, Boston (USA), 5-8 july 1995.
7. De Saxcé G, L. Bousshine, The limit analysis theorems for the implicit standard materials: Application to the unilateral contact with dry friction and the non associated flow rules in soils and rocks, *Int. J. Mech. Sci.,*40, N°4, (1998), 387-398.
8. Fenchel W., On conjugate convex functions, *Canad. J Math.*, 1, 1949, 73-77.
9. Halphen B. & Nguyen Quoc S., Sur les matériaux standards généralisés, *J. de Mécanique*, **14**, 1975, 39-63.
10. Koiter W.T., General theorems for elastic-plastic solids, *Progress in Solid Mech. , Vol. 1*, I.N. Sneddon, R. Hill Eds., North Holland, Amsterdam, 1960.
11. Lemaitre J., Chaboche J.-L., Mechanics of solid materials, *Cambridge University Press,* Cambridge, 1990.
12. Marquis D., Modélisation et identification de l'écrouïssage anisotrope des métaux, *thèse de 3ième cycle de l'université Pierre et Marie Curie*, Paris VI, 1979.
13. Martin J.B., Plasticity : fundamentals and general results, *Cambridge (Massachusetts) and London, M.I.T. Press*, 1975.

14. Melan E., Theorie statisch unbestimmer Systeme aus ideal plastischem Baustoff, *Sitz Ber. Akad. Wiss. Wien IIa, 145*, (1936), 195-218.

15. Moreau J.J., *Comptes rendus de l'Académie des Sciences*, 267, série A, Paris, (1968), p. 954.

16. Pycko S., Maier G., Shakedown theorems for some classes of nonassociative hardening elastic-plastic material models, *Int. J. Plasticity,* 11, N°4, pp 367-395, (1995).

17. Bouhafs Brahim, Adaptation plastique des matériaux à restauration dynamique, mémoire de D.E.A. de Mécanique, Université de Lille I, 1998.

# SHAKEDOWN AND FATIGUE DAMAGE IN METAL MATRIX COMPOSITES

GEORGE J. DVORAK, DIMITRIS C. LAGOUDAS* AND CHIEN-MING HUANG**
*Center for Composite Materials and Structures and*
*Department of Mechanical and Aeronautical Engineering and Mechanics*
*Rensselaer Polytechnic Institute, Troy, N.Y. 12180*
**Center for Mechanics of Composites, Aerospace Engineering Department*
*Texas A&M University, College Station, TX 77843*
*** 3D/Eey Inc., 700 Galleria Parkway*
*Atlanta, GA 30339*

## Abstract

Fatigue failure of metal matrix composite laminates is often preceded by a substantial loss of stiffness associated with cyclic plastic straining and subsequent low-cycle fatigue crack growth in the matrix. Experimental observations indicate that two damage patterns evolve under cyclic loading beyond the elastic range, one formed by cracks extending along the fibers in off-axis plies, and another consisting of cracks bypassing the fibers at an angle in axially loaded plies. Damage saturation is observed under constant load amplitudes. Guided by these experiments, the damage evolution process analyzed herein is regarded as a shakedown mechanism, and damage saturation as a shakedown state. For a given program of variable cyclic loading, evaluation of crack densities needed for shakedown is formulated as a nonlinear constraint optimization problem, where the total damage in a laminate is evaluated from the minimization of a cost function that corresponds to a measure of total damage. The associated nonlinear constraints are derived from the ply yield criterion, hardening rule, and physically motivated bounds on the damage parameters. Effective elastic stiffness reduction and local stress redistribution predicted by the optimization procedure are compared with experimental measurements on B/Al laminates.

## 1. Introduction

Experimental evaluation of fatigue endurance limits of composite materials has been conducted on many different systems, mostly under uniaxial tension loading



*D. Weichert and G. Maier (eds.), Inelastic Analysis of Structures under Variable Loads,* 183–196.

(Reifsnider and Talug, [12]). However, apart from certain empirical insights, the mechanism of the fatigue damage process has not been quantitatively explored. Therefore, it is difficult to utilize the experimental data in applications involving multiaxial stress states and variable amplitude loading programs. In the present study, we address the behavior of laminates with elastic-plastic matrices and elastic fibers, under loading conditions that cause cyclic plastic straining and low cycle fatigue crack growth in the matrix. Laminates reinforced by boron, silicon carbide, or alumina fibers in aluminum or titanium matrices are examples of such systems.

The fatigue experiments and analytical results of Dvorak and Johnson [1] and Johnson [9] on B/Al laminates had shown absence of significant damage under cyclic loading in the elastic range, evolution of such damage up to saturation under constant amplitude loads exceeding this range, and also damage termination after amplitude reduction to the elastic range. This suggests low cycle fatigue crack growth in the plastically deforming matrix, and crack arrest after restoration of elastic response. The implication is that elastic loading or unloading and shakedown can prevent damage or cause saturation, respectively. Accordingly, the fatigue damage process is regarded as a shakedown mechanism (Maier [11], Symonds [13]), which under a prescribed program of loading, provides for load transfer from the more compliant damaged plies to the stiffer plies (Tarn et al. [14]).

As in standard shakedown analysis, the goal is to find a surface in the overall stress space of the laminate that encloses its elastic response region after some previous history of inelastic deformation. In the initial elastic state, such surface coincides with the initial yield surface of the laminate, which is the internal envelope of the yield surfaces of the individual plies, projected into the overall stress space. During plastic straining, each of the ply yield surfaces undergoes a certain translation such that the current loading point is contained within their envelope. If both kinematic and isotropic hardening were admitted, the ply yield surfaces could also expand, but experiments strongly suggest the aluminum matrix and the plies both harden kinematically (Dvorak, et al., [5]). However, expansion of the ply yield surfaces in the overall space becomes possible if the matrix share of the total ply stress is reduced so that a higher overall stress magnitude is required to cause ply yielding. Such stress redistribution within plies is caused by matrix cracking which reduces the effective matrix stiffness. Of course, ply stiffness and, therefore, the ply stresses are also reduced. Both these effects contribute to stress redistribution within the laminate and thus promote expansion of the projected ply yield surface in the overall stress space. If one or more ply yield surfaces translate and expand in this manner, to contain the overall loading path or program of loading, the laminate will shake down.

Evaluation of the various stress averages used in local and overall yield surface definitions is performed here with reasonably simple micromechanical modeling tools that capture the essential features of plastic deformation in the undamaged and damaged laminates, but disregard the many details that have been observed on micrographs of damaged laminates. In particular, only stress and strain field averages, in the fiber and matrix phases, and in the plies, are used to evaluate local stresses and yield surfaces in both damaged and undamaged laminates. Moreover, we consider only symmetric, balanced laminated plates under in-plane, uniformly distributed tractions that retrace a fixed loading path in the overall load space, at constant temperature. The following material comprises a short description of the modeling efforts related to shakedown analysis of fatigue damage in metal matrix composites, while a more detailed exposition can be found in Dvorak et al. [7].

## 2. Analysis of damaged laminates

In a well-designed composite, matrix cracks should mostly bypass the fibers so that the damage process is confined to the matrix and fiber integrity is not significantly compromised by matrix damage. Thus, with reference to the bimodal plasticity theory (Dvorak and Bahei-El-Din [3], Dvorak, et al., [5]), we assume that under sustained cyclic loading, each mode is associated with a pattern of cyclic plastic straining that promotes growth of a certain type of low-cycle fatigue damage in the matrix. In particular, if the ply loading path repeatedly intersects the Matrix Dominated Mode (MDM) yield surface segment, then the damage mode of this ply is expected to be dominated by matrix cracking on planes parallel to the fiber. On the other hand, if the ply loading path intersects only the Fiber Dominated Mode (FDM) segments of the ply yield surface, cracks will grow in the matrix volume on planes inclined to the fiber direction. Since the fiber mode will be seen to cause relatively small stiffness reductions, analysis of this mode will be simplified by assuming that all such cracks are perpendicular to the fiber axis.

In this sense, one may identify the *Matrix-Dominated Damage Mode* (MDDM) and the *Fiber Dominated Damage Mode* (FDDM). For variable path cyclic loading that may intersect both branches of the ply yield surface, preference should be given to the dominant mode, while keeping in mind that the MDDM may form and evolve more easily, and cause much larger ply stiffness reductions than the FDDM.

As an example of damage mode identification, consider the path $\Delta\overline{\sigma} = \overline{\sigma}^2 - \overline{\sigma}^1$, connecting any two stress states $s$ = 1, 2, prescribed in the overall plane stress space of the undamaged composite laminate. This path may intersect the yield surfaces of one or more plies, and if so, it is accommodated by appropriate translations of the yield surfaces due to kinematic hardening. The corresponding path in the local ply stress space may be found by accounting for the stress concentration factors defined by a selected micromechanical model.

Once a steady-state cyclic motion of the ply surfaces is achieved, the bimodal yield surfaces at either end of the path in the ply stress space $\sigma_i$ may be represented schematically as in Fig. 1. Let the stress vector $\mathbf{r}^s_{i_{fdm}}$ connect the center of the i-th ply yield surface with the intercept of the path with the FDM branch, and let the $\mathbf{r}^s_{i_{mdm}}$ denote the stress connecting the center with the MDM branch intercept.

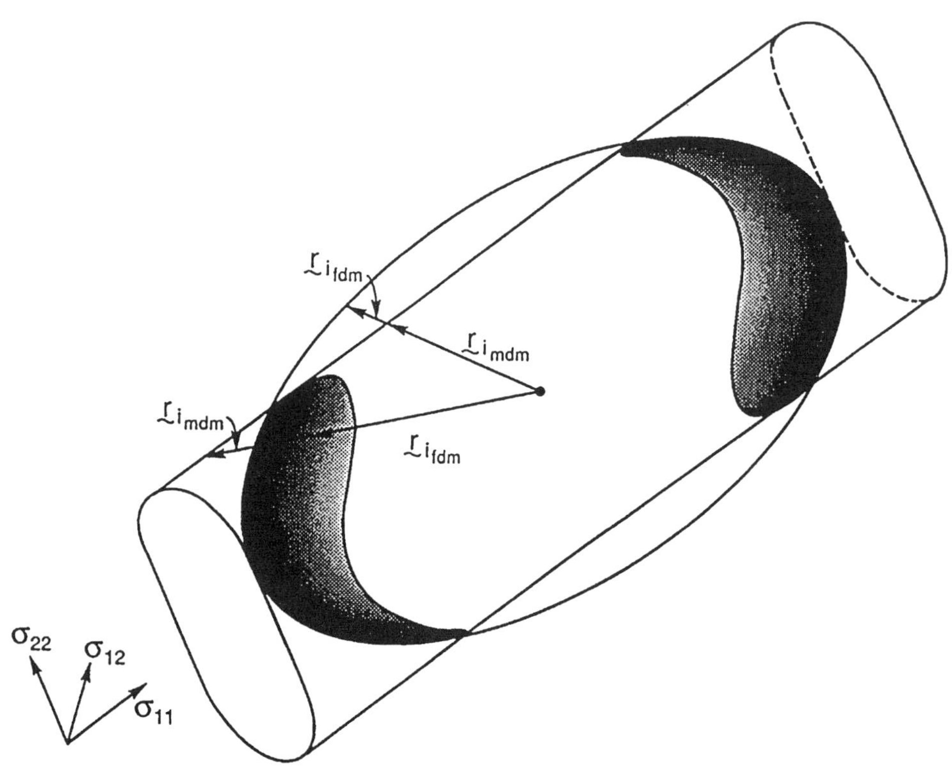


Fig. 1 Schematic of a bimodal yield surface with the cylindrical MDM branch and the ellipsoidal FDM branch (Dvorak et al. [7]).

The condition that $\mathbf{r}_{i_{fdm}}^{s}$ intersects the MDM yield surface is given by the inner product inequality

$$\mathbf{r}_{i_{mdm}}^{s} : \mathbf{r}_{i_{mdm}}^{s} < \mathbf{r}_{i_{fdm}}^{s} : \mathbf{r}_{i_{fdm}}^{s}, \qquad s = 1, 2. \tag{1}$$

Since $\mathbf{r}_{i_{mdm}}^{s}$ is proportional to $\mathbf{r}_{i_{fdm}}^{s}$, we can write

$$\mathbf{r}_{i_{mdm}}^{s} = \eta_i^s \mathbf{r}_{i_{fdm}}^{s} \qquad \mathbf{r}_{i_{mdm}}^{s} : \mathbf{r}_{i_{mdm}}^{s} = \left(\eta_i^s\right)^2 \mathbf{r}_{i_{fdm}}^{s} : \mathbf{r}_{i_{fdm}}^{s} \tag{2}$$

Condition (1) is equivalent to the requirement that $\eta_i^s < 1$, and the damage mode identification is reduced to the evaluation of $\eta_i^s$. Since $\mathbf{r}_{i_{mdm}}^{s}$ ends on the MDM yield surface the evaluation of $\eta_i^s$ relies on the definition described by Dvorak and Bahei-El-Din [3], with the result,

$$\eta_i^s = \frac{2\tau_i^m}{r_{i_{fdm_{22}}}^{s}\left(1 + q_i^{s^2}\right)} \qquad for\left|q_i^s\right| = \left|\frac{r_{i_{fdm_{12}}}^{s}}{r_{i_{fdm_{22}}}^{s}}\right| \leq 1,$$

$$\eta_i^s = \frac{\tau_i^m}{r_{i_{fdm_{12}}}^{s}} \qquad for\left|q_i^s\right| \geq 1, \qquad s = 1, 2, \tag{3}$$

The inequalities specify one of the two branches of the MDM surface where the stress vector $r_{i_{mdm}}^{s}$ is located. In the above formulae $\tau_i^m$ is the in-situ matrix yield stress in simple shear, and the numerical subscripts indicate components of stress.

The damage mode selection criteria outlined above thus imply the following ply damage formation rule: If $\eta_i^s < 1$, $s = 1$ **or** $s = 2$, *longitudinal matrix cracking of density* $\left(\beta_i^l\right)$ *is assumed to occur on planes aligned with fiber direction.* If $\eta^s \geq 1$, for $s =$ 1 **and** $s = 2$, *predominantly transverse matrix cracking of density* $\left(\beta_i^t\right)$ *will develop on planes perpendicular to fiber.* In both instances, the crack density is defined as the ratio of ply thickness to average distance between the cracks; the typical range is $0 < \left(\beta_i^t\right) <$ 1, but higher values are sometimes observed, especially in the transverse mode.

## 3. Saturation state of damage

The composite laminate will shake down if matrix cracks are introduced in individual layers such that the expanded and translated ply yield surfaces contain the prescribed loading path in the overall stress space. This suggests that the laminate has developed a saturation damage state that remains essentially unchanged in the absence of further cyclic plastic straining. Since the shakedown state can be analyzed only in terms of the damage mode selection criteria, there is no need to examine the plastic deformation history, unless the total overall strains are of interest. Moreover, the expanded ply yield surfaces reflect only the stress averages in the matrix, not the stress concentrations at crack tips and interfaces. These effects are neglected here as they appear to have no significant influence on overall response in the saturation damage state. Since any number of shakedown and associated damage states can be created in the laminate for any given loading path or program, it is desirable to introduce certain constraints that provide for a unique shakedown state in the composite structure. In the absence of guiding physical principles, we resort to certain heuristic choices.

To arrive at a realistic criterion, we recall the numerical experiments on motion of laminate yield surfaces under cyclic loading (Dvorak and Wung [4], Dvorak [6]), where a rapid adjustment to steady-state translation was observed within the first few cycles of a fixed cyclic loading program. In contrast, it is well known that many cycles are usually needed for a significant change in crack length or density. Given this disparity in the two response rates, matrix cracking should expand each ply surface only to the minimum size that is necessary for containment of the prescribed loading program. This suggests the hypothesis (Dvorak and Wung [4]) that *the actual crack densities in the plies will reach only the minimum values required for shakedown under the prescribed loading program.*

Therefore, the shakedown analysis for a given program of loading is reduced to finding minimum values of crack densities in the plies, and formulated as a nonlinear optimization problem with a linear objective function and nonlinear constraint inequalities (Künzi et al. [10], Fletcher [8]). Since there is no relative translation of the FDM and MDM segments, and both surfaces are reasonably well approximated by the FDM surface in both the undamaged and damaged plies, the optimization scheme will employ only the FDM surfaces, but both segments will be used to specify the ply damage mode.

Our specific choice is to evaluate the damage parameters $\beta_i^l, \beta_i^t$, and the centers $\overline{\alpha}_i$ of the yield surfaces of the plies i = 1, 2, … N, by requiring that the total damage $\beta$, defined by the objective function

$$\beta\left(\beta_i^t, \beta_i^l, \overline{\alpha}_i\right) = \sum_{i=1}^{N} \beta_i^l + \sum_{i=1}^{N} \beta_i^t, \tag{4}$$

achieves a minimum subject to certain nonlinear constraints. The total damage defined above is the direct sum of the damage densities, or the total damage accumulation. Only one damage mode is admitted in a ply, if both modes were indicated for a more complex path, the matrix mode should be preferred as it develops more easily and has a more significant effect on stiffness.

We now formulate the constraints that are specific to steady state cyclic loading along a fixed straight path in the overall stress space. There are N equality constraints enforcing the requirement that the centers of the overall yield surfaces of the N plies lie on the applied loading path, i.e.,

$$f_i(\overline{\alpha}_i)=\sqrt{\left[\left(\overline{\sigma}^1-\overline{\alpha}_i\right)^T\left(\overline{\sigma}^1-\overline{\alpha}_i\right)\right]}+\sqrt{\left[\left(\overline{\sigma}^2-\overline{\alpha}_i\right)^T\left(\overline{\sigma}^2-\overline{\alpha}_i\right)\right]}-\sqrt{\left[\left(\overline{\sigma}^2-\overline{\alpha}^1\right)^T\left(\overline{\sigma}^2-\overline{\alpha}^1\right)\right]}=0,$$

$$i=1,\,2,\,\dots\,N, \tag{5}$$

when $\overline{\sigma}^1$, $\overline{\sigma}^2$ correspond to the prescribed extreme points of the applied loading path.

Moreover, 2N nonlinear constraint inequalities follow from the requirement that the yield condition is satisfied for every ply.

Finally, we impose 10N linear constraints or bounds on the damage variables and centers of the yield surfaces,

$$\begin{aligned} &\beta_j^{t^l} \le \beta_j^t \le \beta_j^{t^u}, && j=1,\dots N,\\ &\beta_j^{l^l} \le \beta_j^l \le \beta_j^{l^u}, && j=1,\dots N,\\ &\overline{\alpha}_i^l \le \overline{\alpha}_i \le \overline{\alpha}_i^u, && j=1,\dots N. \end{aligned} \tag{6}$$

To be physically acceptable, the lower bounds on the damage variables may not be negative. Moreover the crack densities may not exceed certain experimentally observed magnitudes, which were taken here as $\beta_j^{l^u} \le 1$, $\beta_j^{t^u} \le 3$. The constraints can also incorporate the findings of the damage mode selection construction by assigning zero to both the lower and the upper bounds of the inactive damage mode. The lower and upper bounds of the centers of the yield surfaces have been selected to coincide with the two ends of the loading path.

## 4. Comparison with experimental results

The composite systems that can be analyzed with the present approach are metal matrix composite laminates at room temperature, with aluminum or titanium matrices reinforced by boron or silicon carbide fibers. Other systems may also qualify, providing the matrix fatigue is insignificant in the elastic loading range, but becomes dominant under cyclic plastic straining. However, modifications of the analysis would be needed in applications to systems where extensive fiber debonding is observed.

As an example, we analyze here the experiments of Dvorak and Johnson [1] and Johnson [9] on B/6061 Al laminates under constant amplitude tension-tension cyclic loading at ambient temperature. The principal observation in that work was that no significant fatigue crack growth or other damage process took place when the load amplitude was confined within a shakedown envelope. Matrix fatigue cracks and losses in overall moduli were observed at load amplitudes causing sustained plastic straining, well below the endurance limit. Control experiments at different values of the stress ratio $R = \sigma_{min} / \sigma_{max}$ showed the residual stiffness to depend on the amplitude of the cycle rather than the mean stress.

Typical changes of the measured axial tension moduli under different maximum load levels are shown for a $(0/\pm45/90/0//\pm45/\frac{1}{2}90)_{2s}$ layup in Fig. 2, while similar changes were observed in $0_8$, $(0/90)_s$ , and $(0/\pm45/90)_s$ layups for B/6061-Al laminates. The rapid initial moduli reductions eventually terminate as a saturation damage state is reached. An increase in the load amplitude would lead to further damage and a new saturation state. Fatigue failure of the laminates was typically observed at load amplitudes causing extensive matrix damage in the off-axis plies.

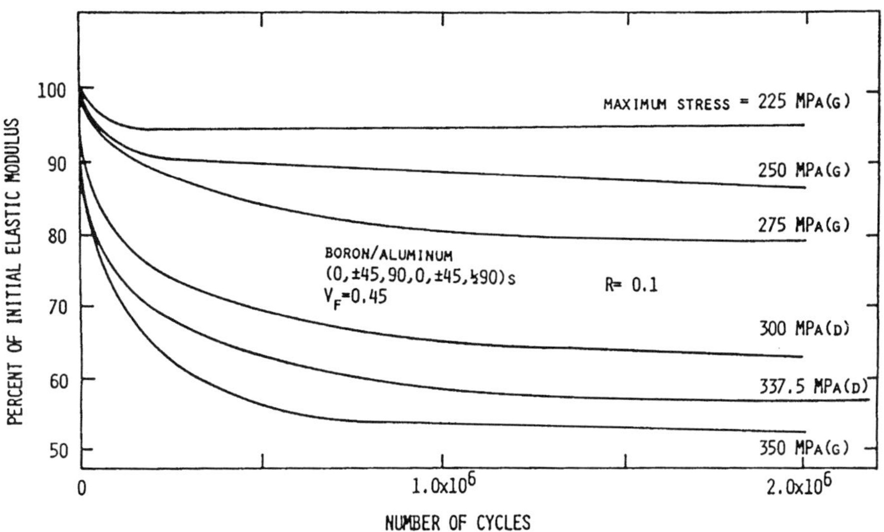


Fig. 2 Loss of stiffness with the number of cycles for various stress amplitudes in a B/Al $(0/\pm45/90/\pm45/\frac{1}{2}90)_{2s}$ laminate with damage. (G) designates specimens without C-scan detectable initial damage. (D) designates specimens with some initial damage (Dvorak and Johnson [1]).

Figures 3 and 4 illustrate the results of damage optimization for a $(0/\pm45/90/0//\pm45/\frac{1}{2}90)_{2s}$, B/6061-A1 laminate under cyclic loading of max $(\overline{\sigma}_{11}) = 250$ MPa and 300 MPa, respectively, with a constant R = $\min(\overline{\sigma}_{11})/\max(\overline{\sigma}_{11}) = 0.1$. The damage bounds were selected as

$$\begin{array}{llllll}\beta_1^{t^l}=0 & \beta_1^{t^u}=3 & \beta_1^{l^l}=0 & \beta_1^{l^u}=0 & \overline{\alpha}_1^l=\overline{\sigma}^1 & \overline{\alpha}_1^u=\overline{\sigma}^2\\ \beta_2^{t^l}=0 & \beta_2^{t^u}=0 & \beta_2^{l^l}=0 & \beta_2^{l^u}=1 & \overline{\alpha}_2^l=\overline{\sigma}^1 & \overline{\alpha}_2^u=\overline{\sigma}^2\\ \beta_3^{t^l}=0 & \beta_3^{t^u}=0 & \beta_3^{l^l}=0 & \beta_3^{l^u}=1 & \overline{\alpha}_3^l=\overline{\sigma}^1 & \overline{\alpha}_3^u=\overline{\sigma}^2,\end{array}$$

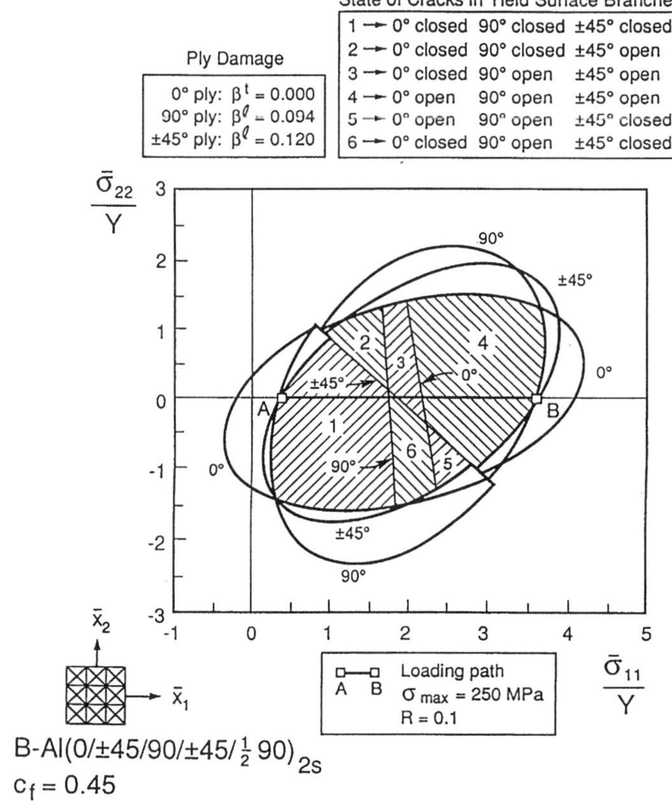


Fig. 3 Shakedown envelopes of a B/Al $(0/\pm45/90/0/\pm45\frac{1}{2}90)_s$ laminate (Dvorak et al. [7]).

where the subscripts 1, 2 and 3 denote the 0°, ±45° and 90° plies, respectively. Note that both the ±45° and 90° plies have longitudinal cracks, while the 0° ply grows transverse cracks.

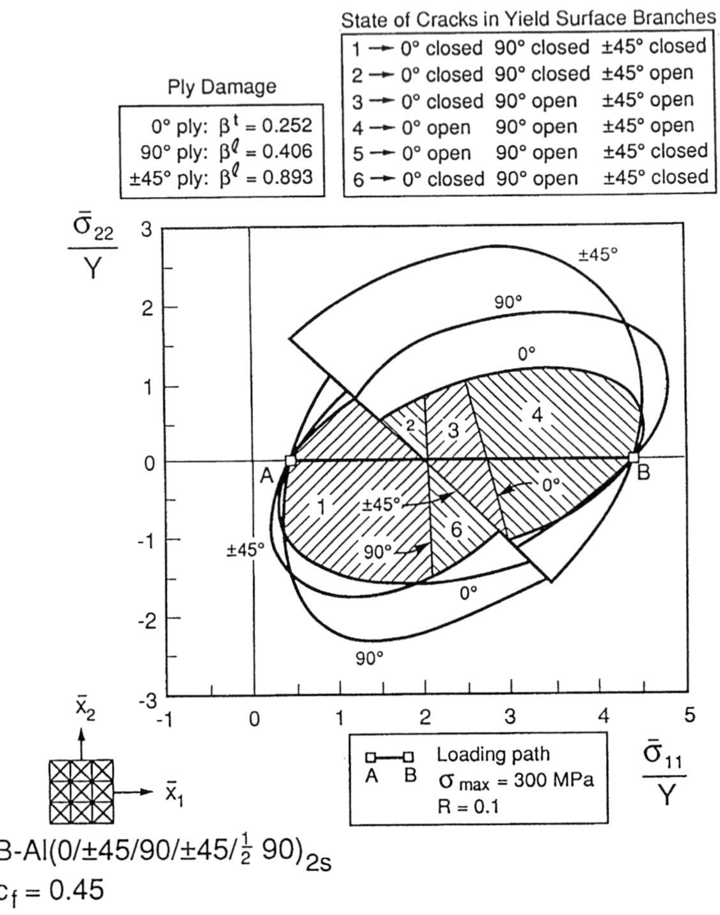


Fig. 4 Shakedown envelopes of B/Al $(0/\pm45/90/0/\pm45\frac{1}{2}90)_s$ laminate (Dvorak et al. [7]).

Development of damage in the laminates results in reduction of overall elastic stiffness. A comparison of the predicted stiffness losses at saturation, as indicated by the optimization procedure, with Dvorak and Johnson [1] experimental measurements appears in Fig. 5 for the $(0/\pm45/90/0/\pm45/\frac{1}{2}90)_{2s}$ laminate. There is a good correlation for the entire range of stress amplitudes applied in the optimization scheme. We note that the analysis may not be applicable when the maximum stress approaches about 90% of the endurance limit, since the experimentally observed frequency of fiber breaks in the 0° plies appears to have an effect on stiffness reduction.

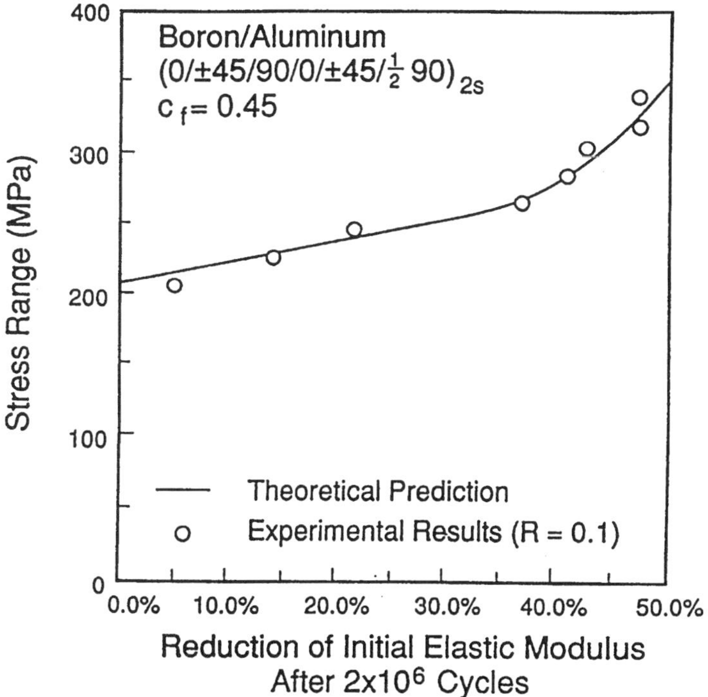


Fig. 5 Reduction in elastic stiffness of a $(0/\pm45/90/0/\pm45\frac{1}{2}90)_s$ B/Al laminate as a function of the applied stress amplitude. Experimental data by Dvorak and Johnson [1] compared with predictions by the damage optimization procedure (Dvorak et al. [7]).

The initiation of fiber breaks in the 0° plies at levels of applied loading close to the endurance limit suggests a mechanism of failure by overloading of the fibers in the 0° plies. This is not unexpected, since damage accumulation and loss of stiffness in the off-axis plies promote transfer of local stresses to the stiffer plies. An illustration is provided in Fig. 6, where the fiber stress in the saturation state is plotted as a function of the maximum applied stress for the $(0/90)_{2s}$, $(0/\pm45/90/0/\pm45/\frac{1}{2}\,90)_{2s}$ and $0_8$ laminates.

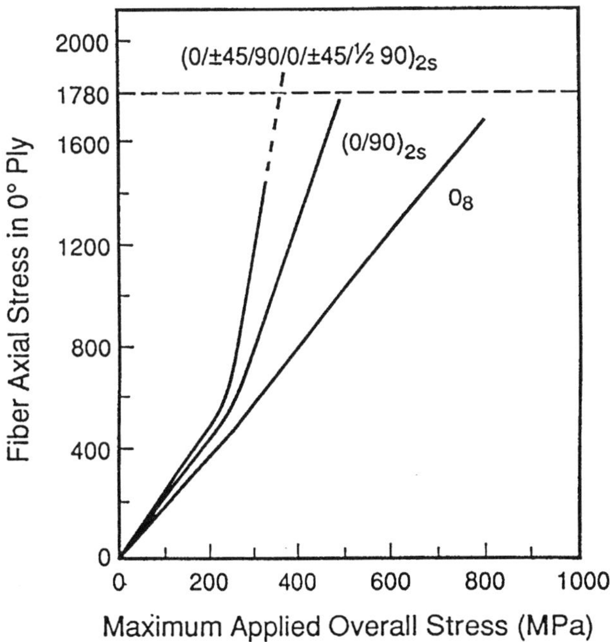


Fig. 6 Axial fiber stress in the 0° plies B/Al laminates of different layup as a function of the maximum overall applied stress, up to the observed endurance limit (Dvorak et al. [7]).

The initial slopes of the three curves correspond to the elastic fiber stress concentration factors with no damage. As the amplitude of the applied load increases, matrix damage is introduced and the slopes continuously increase until they reach certain asymptotic values that correspond to limit magnitudes of the crack densities (Dvorak et al. [2]. The three curves are plotted up to the overall stress levels at the respective laminate endurance limits reported by Dvorak and Johnson [1]; i.e., 800 MPa for the $0_8$, 500 MPa for the $(0/90)_{2s}$ and 375 MPa for the $(0/\pm45/90/0\pm45/\frac{1}{2}90)_{2s}$ layups. The dotted line for the latter laminate indicates extrapolation from 325 MPa, which is the last point computed by the optimization procedure. Considering that the three laminates were reinforced by different batches of boron fibers, and the scatter in experimental endurance limit measurements, it may be deduced from Fig. 6 that the axial fiber stress, reached in the 0-degree fibers at the endurance limit of the laminate, appears to be independent of the layup. The respective magnitudes are 1692, 1775, and 1873 MPa for the axial, cross-ply, and angle-ply laminates. The average of these maximum fiber stresses is 1780 MPa, which may be considered as the approximate *in situ* endurance limit of the fiber.

## 5. Concluding remarks

The procedure described in this work evaluates ply crack densities and the resulting stiffness changes that guarantee elastic shakedown in a laminate under prescribed cyclic stress. Of course, while shakedown is guaranteed, the underlying theorems (Symonds [13]) allow for the actual mechanism to be different from that predicted by the model. A similar procedure could be constructed for a prescribed laminate strain, but since the in-plane ply strains would then be equal to the prescribed strain in each ply, the ply stresses and crack densities would be independent, and no optimization would be required. In principle, the model can be extended to more complex in-plane loading programs applied to elastic-plastic laminated plates, provided that a new set of constraints and damage parameter bounds, specific to the prescribed load, is developed and implemented in the optimization procedure.

## Acknowledgement

This work was supported by grants from the Air Force Office of Scientific Research. Dr. George Haritos served as program monitor.

## References

1. G.J. DVORAK and W.S. JOHNSON, Fatigue of Metal Matrix Composites, *Int. J. Fracture 16*, 585, 1980.

2. G.J. DVORAK, N. LAWS, and M. HEJAZI, Analysis of Progressive Matrix Cracking in Composite Laminates I. Thermoelastic Properties of a Ply with Cracks. *J. Composite Materials, 19*, 216, 1985.
3. G.J. DVORAK and Y.A. BAHEI-EL-DIN, A Bimodal Plasticity Theory of Fibrous Composite Materials, *Acta Mechanica 69*, 219, 1987.
4. G.J. DVORAK and E.C.J. WUNG, Fatigue Damage Mechanics of Metal Matrix Composite Laminates, in *Strain Localization and Size Effect Due to Cracking and Damage* (edited by J. Mazars and Z.P. Bazant), Elsevier, London, 1988.
5. G.J. DVORAK, Y.A. BAHEI-EL-DIN, Y. MACHERET and C.H. LIU, An Experimental Study of Elastic-Plastic Behavior of a Fibrous Boron-Aluminum Composite, *J. Mech. Phys. Solids 36*, 655, 1988.
6. G.J. DVORAK, Plasticity Theories for Fibrous Composite Materials, *Metal Matrix Composites, Vol. 2*, ed by R.K. Everett and R.J. Arsenault, Academic Press, Boston, 1990.
7. G.J. DVORAK, D.C. LAGOUDAS and C.-M. HUANG, Fatigue Damage and Shakedown in Metal Matrix Composite Laminates, *Mechanics of Composite Materials and Structures 1*, 171, 1994.
8. R. FLETCHER, Methods of Nonlinear Constraints, in *Nonlinear Optimization, 1981*, H.J.D. Powell, Ed., Academic Press, pp. 185-212, 1982.
9. W.S. JOHNSON, Mechanism of Fatigue Damage in Boron-Aluminum Composites, in *Damage in Composite Materials, ASTM STP775*, K.L. Reifsnider, editor, American Society for Testing and Materials, Philadelphia, 1982.
10. H.P. KÜNZI, H.G. TZSCHACH AND C.A. ZEHNDER, *Numerical Methods of Mathematical Optimization*, Academic Press, New York, 1968.
11. G. MAIER, Shakedown Theory in Perfect Elastoplasticity with Associated and Nonassociated Flow-Laws: A Finite Element Programming Approach, *Meccanica 4*, 250, 1969.
12. K. L. REIFSNIDER and A. TALUG, Analysis of Fatigue Damage in Composite Laminates, *Int. J. Fatigue*, p. 3, January 1980.
13. P.S. SYMONDS, Shakedown in Continuous Media, *J. Appl. Mech. 18*, 85, 1951.
14. J.Q. TARN, G.J. DVORAK AND M.S.M. RAO, Shakedown of Unidirectional Composites, *Int. J. Solids Structures 11*, 751, 1975.

# ON SHAKEDOWN OF ELASTIC PLASTIC BODIES WITH BRITTLE DAMAGE

B. DRUYANOV AND I. ROMAN
*Graduate School of Applied Science,*
*The Hebrew University of Jerusalem,*
*Givat Ram Campus 91904 Jerusalem, Israel*

## Abstract

Conditions for shakedown of elastic plastic bodies with brittle damage are investigated. In the case of isotropic damage the evolution equation for the damage tensor can be integrated. As a result a one-to-one relation between the damage parameter and maximal value of the damage energy release rate is obtained The consideration of features of the stress path in the stress space leads to necessary shakedown conditions and the notion of core of the limit (stationary) yield condition. The existence of core is a necessary shakedown condition for arbitrary hardening laws. It is sufficient in the case of isotropic hardening and hardening laws similar to it. The notion of core provides a possibility of formulating necessary shakedown conditions for damaged bodies. With a purpose of developing approximate methods able to direct investigating the asymptotic behavior (failure, or non-failure) of damaged bodies, a simplified material model of perfect brittle damage was introduced which ignores the effect of plastic deformation on developing of damage. The mechanical behavior of bodies experienced the development of anisotropy due to microcracks opening and closing (anisotropic damage) is compared with that of the bodies with isotropic damage. It was shown that, in the frames of linear approach, a body with anisotropic damage will shake down, if the isotropically damaged body of the same shape and sizes, and subjected to the same loading program is shaken down.

**Key words:** elastic-plastic bodies, brittle damage, cyclic loading, shakedown

## 1. Introduction

During the last decade the efforts of scientists have been concentrated on the extension of the shakedown theory to realistic material models accounting for the phenomena of strain hardening (Stein et al. 1992, Pycko & Maier 1995, Polizzotto 1994, 1995; Druyanov & Roman 1996,1997, et al.), damage (Hachemi & Weichert



*D. Weichert and G. Maier (eds.), Inelastic Analysis of Structures under Variable Loads,* 197–212.

1992, 1997, 1998; Siemaszko 1993, Weichert & Hachemi 1999, Feng & Liu 1997, Druyanov & Roman 1998a,b 1999, et al.), and viscosity (Klebanov & Boyle 1998, Ponter et al. 1990, Ponter & Leckie 1998 et al.).

The classical Melan theorem (Melan, 1938a, b; Koiter, 1960) is known to provide a necessary and sufficient condition for shakedown of elastic *perfectly* plastic bodies, that is, for their adaptation to cyclic loading. This theorem can be extended to damaged bodies (Hachemi & Weichert, 1992;1997; Druyanov & Roman, 1998a,b). It is assumed in this paper that the damage process is coupled with the plastic one, i.e. the damage process ceases coincidentally with the process of plastic deformation. Nevertheless, the extended Melan theorems provide only necessary conditions for the adaptation of damaged bodies to cyclic loading. Really, the fulfillment of the conditions of these theorems results in the conclusions: the plastic parts of the strain rate tensor components tend to zero, and the plastic dissipation is bounded. These conditions are usually taken as the definition of elastic shakedown for undamaged bodies. Dealing with damaged bodies, one needs to account for the possibility that the body can fail at the transient stage of deformation in the case, if the damage exceeds its critical level before the stationary stage is reached. Thus, the fulfillment of the Melan conditions is not sufficient for assertion that the body under consideration will shakedown.

Accounting for the above reasons, the following definition of shakedown seems to be appropriate for damaged elastic plastic bodies: *A damaged elastic plastic body will shake down in the elastic region (adapt itself to a cyclic loading program), if the plastic dissipation is bounded ($D_p<\infty$ ), and the damage variable does not exceed its critical level.* (See also Hachemi & Weichert 1992, 1997).

In this paper only the conditions for fulfillment of the first part of this definition ($D_p<\infty$) are considered. The term elastic shakedown is kept for this case, as opposed to the case of the general shakedown where the both above requirements are satisfied. Obviously the fulfillment of the conditions of the elastic shakedown is necessary with regard to the general shakedown.

The damage variable $D_n$ is defined as the ratio of damaged and nominal areas of a representative volume element intersection. Damage is called "isotropic", if $D_n$ is a scalar, i.e., if its value does not depend on orientation of the area. This is the ordinary physical definition of the isotropic damage. (See, for example, Lemaitre 1992). In this case $\Delta$ is written instead for $D_n$. If, on the contrary, $D_n$ depends on the orientation, the damage is named "anisotropic".

The approach by Ju (1989) is taken as a starting point for constructing the constitutive material models. These models should be treated as the models with brittle damage because, according to assumption, the conditions of damage evolution and failure depend only on the damage energy release rate.

In the beginning of the paper (Section 2), materials with initially anisotropic damage are considered. Obviously the anisotropy in damage leads in an anisotropy of properties which could be caused by different reasons, and not only by previous damage. The phenomenon of microcracks closing and opening is not taken into account in this Section. Due to this reason, the type of the initial damage anisotropy does not change during deformation. (For example, the orthotropic anisotropy remains orthotropic). The damage variable **D** was found to vary in proportion to a single scalar variable, so that this kind of damage is like isotropic one, and is also termed the isotropic damage. Thus, the concept of isotropic damage proves to be more general than its physical definition.

It is known that, if the phenomenon of microcracks opening and closing is taken into account (Section 3), the initial type of damage anisotropy can change during deformation. This sort of anisotropy is named the anisotropic damage. (Ju 1989, Krajcinovic 1996).

The asymptotic mechanical behavior (failure, or non-failure) of damaged bodies is a functional of the whole deformation process. As a rule, its investigation can be accomplished only by an incremental analysis. However for the assumed material model, the evolution equation for isotropic damage parameter is holonomic, and can be integrated. As a result, a relation between the damage parameter and maximum value of the damage energy release rate can be obtained. This relation facilitates significantly the process of computing of the related boundary-value problems, and the investigation of asymptotic mechanical behavior of bodies with isotropic damage.

The necessary shakedown conditions of the bodies with isotropic damage based on the notion of core are couched in Sections 4, 5, 6. See Druyanov & Roman (1998a,b; 1999).

Engineers need to have in their disposal simplified methods able to fast estimating of asimptotic values of some variables responding for of the body under imvestigation (the asymptotic value of damage threshold, for example).

One of the ways to construction of the estimating methods is the employing of simplified constitutive material models which reflect the basic features of the material behavior, on the one hand, and admit constructing of the estimating methods, on the other hand. This way is widely employed in scientific and engineering practice. For example, the model of perfect plasticity is a very simplified model of plastic body. Nevertheless it is often employed for fast estimating of bearing capacity of structures and necessary loads in forming processes.

In Damage Mechanics of brittle materials the place of perfect plasticity could occupy the model of perfectly brittle isotropic damage. The perfect brittleness means that the effect of plastic deformation on damage process is not taken into account. Notice that the plastic part of free energy is small in comparison with the elastic one, and can be neglected in models designed for estimating objectives. Such models were employed earlier by some authors. See, for example, Cordebois & Sidoroff (1982), Lemaitre (1985).

The comparison between isotropic and anisotropic damage (Section 7) shows that, in the frames of linear approach, a body with anisotropic damage will not fail, if the same body subjected to the same loading program, however with isotropic damage, does not fail. This theorem provides us with a sufficient shakedown condition for the bodies with anisotropic damage, and enables us to avoid the laborious and complex incremental analysis of the asymptotic mechanical behavior of the bodies with anisotropic damage by means of the investigation of mechanical behavior of bodies with isotropic damage

It is supposed that the loading program is prescribed and known in advance. In the more general event, when only the bounds of loading are given, the derived conditions are necessary for shakedown.

## 2. Isotropic Damage

The deformations are assumed to be small, consequently the total strain tensor can be represented as the sum of its elastic and plastic parts: $\varepsilon=\varepsilon^e+\varepsilon^p$.

Let the colon (:) denote the contraction with respect to two indices: $(\mathbf{A:B})_{ijkl}=$ $=a_{ijmn}b_{mnkl}$, or $\mathbf{a:b}= a_{ij}b_{ij}$. The complete contraction of two fourth rank tensors with respect to all indexes is denoted $\mathbf{AB}=a_{ijkl}b_{ijkl}$, whereas $\mathbf{a\cdot b}=a_ib_i$.

The effective stress tensor is defined as $\bar{\sigma}=\mathbf{M}^{-1}:\sigma$ where $\mathbf{M}$ is the transformation tensor of the fourth rank, and $\sigma$ is the tensor of nominal stresses. ($\mathbf{M:M}^{-1}=\mathbf{I}$, $\mathbf{I}$ is the fourth rank unit tensor with the components $I_{ijkl}=0.5(\delta_{ik}\delta_{jl}+\delta_{il}\delta_{jk})$ where $\delta_{ik}$ is the Kronecker symbol).

Let $\mathbf{C}$ denote the current damaged secant (unloading) elastic stiffness tensor, and $\mathbf{C}_0$ be its undamaged value.

The local current free energy is postulated as:

$$\Psi(\varepsilon^e,\chi,\mathbf{C})\equiv\Psi^e(\varepsilon^e,\mathbf{C})+\Psi^p(\chi,\mathbf{C}), \tag{1}$$

where $\Psi^e(\varepsilon^e,\mathbf{C})=0.5\varepsilon^e:\mathbf{C}:\varepsilon^e$ and $\Psi^p(\chi,\mathbf{C})$ are its current elastic and plastic parts correspondingly; and $\chi$ is the vector of internal variables.

The Clausius-Duhem inequality provides us with the equation $\sigma=\partial\Psi/\partial\varepsilon^e=\mathbf{C}:\varepsilon^e$, which, in turn, leads to $\bar{\sigma}=(\mathbf{M}^{-1}:\mathbf{C}):\varepsilon^e$, and with the damage and plastic dissipative inequalities:

$$D^d\equiv-0.5\varepsilon^e:\dot{\mathbf{C}}:\varepsilon^e-\frac{\partial\Psi^p}{\partial\mathbf{C}}\geq 0,\quad D^p\equiv\sigma:\dot{\varepsilon}^p-\dot{\chi}\cdot\frac{\partial\Psi^p}{\partial\chi}\geq 0. \tag{2}$$

According to the principle of strain equivalence (Lemaitre 1992), $\mathbf{M}^{-1}:\mathbf{C}=\mathbf{C}_0$. Thus, $\bar{\sigma}=\mathbf{C}_0:\varepsilon^e$ which results in Hooke's law

$$\varepsilon^e=\mathbf{C}_0^{-1}:\bar{\sigma}. \tag{3}$$

Consequently, $\mathbf{C=M:C}_0$, and $\mathbf{C}:\mathbf{C}_0^{-1}=\mathbf{M}$.

The damage tensor is defined as $\mathbf{D}=\mathbf{I}-\mathbf{M}=\mathbf{I}-\mathbf{C}:\mathbf{C}_o^{-1}$. It is taken as the damage variable.

The existence of a virgin (undamaged) state of the material is a postulate. In this state the damage tensor is **D**=0, and transformation tensor is **M**=**I**.

The yield condition is taken in the form

$$\zeta=\Phi(\overline{\sigma},\chi)=0 \tag{4}$$

where the yield function $\Phi$ is convex in $\overline{\sigma}$. The yield function $\Phi$ is chosen in such a way that the inequality $\Phi<0$ would correspond to the interior of the surface $\Phi=0$.

The plastic part of the strain rate tensor is defined by the associated flow rule

$$\dot{\varepsilon}^p=\dot{\lambda}\frac{\partial\Phi}{\partial\overline{\sigma}},\quad \dot{\lambda}\Phi(\overline{\sigma},\chi)=0,\ \dot{\lambda}\geq 0,\quad \Phi(\overline{\sigma},\chi)\leq 0 \tag{5}$$

where $\dot{\lambda}$ is the plastic consistency parameter.

The evolution of $\chi$ is governed by the equation

$$\dot{\chi}=\dot{\lambda}\,\mathbf{h}(\overline{\sigma},\chi) \tag{6}$$

where $\mathbf{h}(\overline{\sigma},\chi)$ is a given vector-function.

The plastic part of the free energy is assumed to be linear in **C**: $\Psi_p(\chi,\mathbf{C})=\mathbf{C}:\mathbf{C}_o^{-1}\mathbf{I}\Psi_o^p(\chi)$ where $\Psi_o^p(\chi)$ is the plastic free energy of the undamaged material.

The thermodynamic force dual to **C** is

$$-\mathbf{Y}\equiv\partial\Psi/\partial\mathbf{C}=0.5\varepsilon^e\otimes\varepsilon^e+\frac{\partial\Psi^p}{\partial\mathbf{C}}:\mathbf{C}=0.5\varepsilon^e\otimes\varepsilon^e+\Psi_o^p(\chi)\,\mathbf{C}_o^{-1} \tag{7}$$

where $\otimes$ denotes the external product of two tensors.

The damage criterion is formulated depending on the damage energy release rate $\xi$:

$$g\equiv G(\xi)-R\leq 0 \tag{8}$$

where R is the current damage threshold, G is a known function of $\xi$, $dG/d\xi>0$, and

$$\xi=-\mathbf{Y}\mathbf{C}_0=0.5\varepsilon^e:\mathbf{C}_0:\varepsilon^e+\Psi_o^p(\chi) \tag{9}$$

It is assumed that $\dot{R}=\dot{\mu}\partial G/\partial\xi$, $\dot{\mu}g=0$, $\dot{\mu}\geq 0$, $g\leq 0$ where $\mu$ is the damage consistency parameter.

During active damage process one has $\dot{\mu}>0$, and g=G(ξ)-R=0. The differentiation of this equation yields the equality $\dot{\mu}=\dot{\xi}>0$ which holds while damage is in progress. The integration of this equality under the initial condition μ=ξ=0 at t=0 gives the equality μ=ξ while g=0. Hence, μ is the maximum value of ξ which it has assumed by the instant under consideration: $\mu=\max\xi$. On the other hand μ=const as long as g=G(ξ)-R<0, i.e. during unloading and subsequent reloading, until again g=0. Consequently G(ξ)=G(μ) during active damage process.

Notice that equation (8) determines $\mu_o$ - the minimal value of μ under which the damage process starts: $G(\mu_o)=R_o$ where $R_o$ is the initial value of the damage threshold.

It is assumed that the evolution of the damage tensor is governed by the equation

$$\dot{\mathbf{D}}=\dot{\mu}\frac{dG}{d\mu}H(\dot{\lambda})\mathbf{I}=\dot{G}\ H(\dot{\lambda})\mathbf{I}=\dot{R}\ H(\dot{\lambda})\mathbf{I},\ \text{subject to } g=G(\xi)-R=0. \tag{10}$$

where $H(\dot{\lambda})$ is the Heaviside function: $H(\dot{\lambda})=1$ for $\dot{\lambda}>0$, and $H(\dot{\lambda})=0$ for $\dot{\lambda}\le 0$.

Due to the coefficient $H(\dot{\lambda})$, the damage process is coupled with the process of plastic deformation: they start and cease simultaneously.

Equation (10) is holonomic and can be integrated. The integration under initial condition $\mathbf{D}=\mathbf{D}_o$, $R=R_o$ yields

$$\mathbf{D}=\mathbf{D}_o+\mathbf{I}(R-R_o)=\mathbf{D}_o+\mathbf{I}(G(\mu)-R_o)=\mathbf{D}_o+\mathbf{I}\Delta \tag{11}$$

where $\Delta=R-R_o=G(\mu)-R_o$.

Possibly the material under consideration can be initially damaged anisotropically. Obviously, this could be a result of various reasons, not only deformation. However, it assumed that the initial damage tensor $\mathbf{D}_o$ is symmetric, with the symmetry specific for elastic stiffness tensors. Under this condition all the tensors introduced above are of the necessary symmetries.

Equation (11) shows that the damage tensor changes in proportion to the scalar variable Δ, i.e. the type of initial anisotropy remains unchanged. In this respect, the character of the considered damage is similar to the isotropic one, and could be named also isotropic damage.

Equation (11) specifies a one-to-one connection between Δ and μ which is convenient to write in the form

$$\frac{1}{1-\Delta}=\psi(\mu). \tag{12}$$

where

$$\psi(\mu)=1/(1+R_0-G(\mu))=1/(1-R+R_0) \tag{13}$$

is an increasing function of $\mu$.

The transformation tensor is $\mathbf{M}=\mathbf{I}(1-\Delta)-\mathbf{D}_o$, and the relation $\bar{\sigma}=\mathbf{M}^{-1}:\sigma$ is as $\bar{\sigma}=[(1-\Delta)\mathbf{I}-\mathbf{D}_o]^{-1}:\sigma$.

If the material is undamaged at the initial moment, i.e. $\mathbf{D}_o=0$, then the known relation specific for isotropic damage results from (10): $\bar{\sigma}=\sigma/(1-\Delta)$.

Relation (12) makes possible the direct (without a detailed analysis) prediction of the asymptotic behavior of isotropically damaged elastic plastic bodies subjected to cyclic loading (Sections 7,8).

It is assumed that local failure of the material occurs, if the damage threshold R reaches its critical value $R_c$. Thus, the material resists to the applied load until the inequality holds:

$$R<R_c. \tag{14}$$

It is important that equation (8) defines the critical value of the damage energy release rate $\xi_c=\mu_c$ corresponding to $R_c$: $G(\mu_c)=R_c$.

There are two mechanisms affecting the mechanical properties of the material in opposite directions: strain hardening and damage. It is possible to state, without details, that the rate of damage growth increases along with the accumulation of damage. On the contrary, as hardening grows, its rate decreases because of the hardening saturation.

One distinguishes two stages of the damage process (Taltreja 1989). During the initial stage the damage entities (micro-cracks) do not interact; whereas at the second stage an intensive interaction starts which results eventually in damage localization and failure. In the one-dimensional tensile test the start of the second stage corresponds to the beginning of necking. The corresponding point at the $\sigma$~$\varepsilon$ diagram is considered as the point of material instability. Beyond this point the material becomes unstable, and deformation is going on under decreasing load. At the point of material instability the processes of hardening and damage growth are in balance. After the process of material failure starts (Lemaitre 1992).

Hence the point of instability can be considered as the beginning of the failure process. It is reasonable to assume that this point corresponds to the critical value of the damage threshold $R_c$. Under the such approach, $R_c$ is a functional of the deformation process.

## 3. Anisotropic Damage

To account for the phenomenon of microcracks closing and opening, the damage energy release rate (denoted $\eta$ in the case) should be modified.

Following Schreyer (1995) and Krajcinovic (1996) let us consider $\mathbf{P}^+$ - the fourth rank positive projection tensor with the components

$$P^+_{ijmn} = 0.5(S^+_{im}S^+_{jn} + S^+_{in}S^+_{jm}) \tag{15}$$

where $\mathbf{S}^+ = \sum_{k=1}^{3} H(\varepsilon_k)\mathbf{p}_k \otimes \mathbf{p}_k$ is the positive (tensile) spectral projection tensor, $\mathbf{p}_k$ are the directional unit vectors corresponding to the principal directions of the total strain tensor $\varepsilon$; and $\varepsilon_k$ are its corresponding principal components.

The modified energy release rate is defined as

$$\eta = -\mathbf{Y}\,\mathbf{C}^+_0 = 0.5\varepsilon^{e+}:\mathbf{C}^o:\varepsilon^{e+} + \Psi^p_o(\chi)\,\mathbf{C}^{-1}_o\,\mathbf{C}^+_o \tag{16}$$

where $\mathbf{C}^+_o = \mathbf{P}^+:\mathbf{C}_o:\mathbf{P}^+$, and $\varepsilon^{e+}$ is the positive projection of the elastic strain tensor $\varepsilon^e$: $\varepsilon^{e+} = \mathbf{P}^+:\varepsilon^e$. Its principal components are $\varepsilon^{e+}_i = 0.5(\varepsilon^e_i + |\varepsilon^e_i|)$ where $\varepsilon^e_i$ are the principal elastic strain components.

The effect of $\mathbf{P}^+$ is to remove from $\varepsilon^e$ its negative eigenvalue components.

If $\varepsilon_i>0$ for all possible values of the index i, local microcracks are opened in all the principal directions of the strain tensor. In this case $\mathbf{P}^+ = \mathbf{I}$ and $\varepsilon^{e+} = \varepsilon^e$, $\mathbf{C}^+_o = \mathbf{C}_o$. On the other hand, $\mathbf{P}^+=0$, if $\varepsilon_i \le 0$ for all values of i, then $\varepsilon^{e+}=0$, $\mathbf{C}^+_o=0$. This is the case of microcracks closing in all the principal directions. For details see, Krajcinovic (1996), Schreyer (1995), Ju (1989), Ortiz (1985).

The damage criterion is formulated like (8), however the anisotropic damage energy release rate $\eta$ is taken instead of $\xi$:

$$g \equiv G(\eta) - R \le 0 \tag{17}$$

where R is the current damage threshold, G is a known function of $\eta$, and $dG/d\eta>0$.

As previously, it is assumed that $\dot{R} = \dot{\nu}\partial G/\partial\eta$, $\dot{\nu}g = 0$, $\dot{\nu} \ge 0$, $g \le 0$ where $\nu$ is the anisotropic damage consistency parameter.

During active damage process $\dot{\nu} > 0$, and $g=G(\eta)-R=0$. The differentiation of this equation results in the equality $\dot{\nu} = \dot{\eta} > 0$ which holds while damage is in progress. Hence, analogously to the case of isotropic damage, $\nu$ is the maximum value of $\eta$ which it has assumed by the cosidered instant: $\nu = \max\eta$. Otherwise $\nu=\eta$, while $g=0$, and $\dot{\nu} = \dot{\eta} > 0$, e.g. during active damage process. On the other hand $\nu$=const as long as $g=G(\eta)-R<0$, i.e. during unloading and subsequent reloading, until $g=0$. Consequently $G(\eta)=G(\nu)$ during active damage process.

Inequality (17) defines the critical value of the damage energy release rate $\eta_c = \nu_c$ corresponding to $R_c$: $G(\nu_c) = R_c$.

It is assumed that the evolution of the damage tensor is governed by the equation

$$\dot{\mathbf{D}} = \dot{\nu}\frac{dG}{d\nu}\,\mathbf{C}_o^{+} = \dot{G}\,\mathbf{C}_o^{+} = \dot{R}\,\mathbf{C}_o^{+},\ \text{subject to } g=G(\nu)-R=0. \tag{18}$$

This equation is nonholonomic because $\mathbf{C}_o^{+}$ depends on the deformation path. Consequently

$$\mathbf{D}=\mathbf{D}_o+\int_{\nu_o}^{\nu} \mathbf{C}_o^{+}\frac{dG}{d\nu}d\nu \tag{19}$$

The equations of the material model with isotropic damage (Section 2) can be obtained from the equations of this Section, if $\mathbf{P}^{+}=\mathbf{I}$ is taken in the equations.

## 4. Features of the Stress Path and Necessary Conditions for Shakedown

Earlier the authors developed a method for examining of the asymptotic mechanical behavior of elastic plastic bodies with isotropic damage for the material model where damage and plastic deformation processes are uncoupled (Druyanov & Roman 1998, 1999). Below the arguments developed by authors are applied to the assumed material model with isotropic damage (Section 2) for which the coupling between damage and plastic deformation processes is characteristic.

Time independent values of the residual stress tensor $\sigma_s^r$, the damage variable $\Delta_s$, the maximum value of the damage energy release rate $\mu_s$, and the vector of internal variables $\chi_s$ are characteristic for the stationary (post-adaptation) stage of the deformation process $t>t_s$, if it exists, because the body experiences only elastic deformation at this stage. These values and the corresponding yield surface can be named the limit ones. The representative stress point in the effective stress space reaches the limit yield surface repeatedly, but the stress does not causes plastic deformation and damage, and the limit yield surface does not change. This is possible, if the stress path in the effective stress space is either in the interior of the limit yield surface, or is tangent to it, or touches it at some isolated instants. In particular, this is valid at the time points t* corresponding to the instants at which the representative stress point leaves the limit yield surface. These time points will be named the departure instants.

Consequently at these instants

$$\Phi(\bar{\sigma}(t^*),\chi))=0,\ \Phi,_{\bar{\sigma}}(\bar{\sigma}(t^*),\chi)):\dot{\bar{\sigma}}=0. \tag{20}$$

Repeating the arguments developed in (Druyanov and Roman 1997, 1999), it is possible to show that the departure instants are the points of local maximum of the yield function $\zeta(t,\chi)=\Phi(\bar{\sigma}(t),\chi)$ for a fixed value of $\chi$.

This assertion is valid for the departure instants during the whole deformation process. At the post adaptation stage, the quantities $\sigma^r$, $\chi$, $\Delta$, and $\mu$ do not change. To account for this property, it is necessary to return to the nominal stress tensor, and to remember that $\bar{\sigma}=\sigma\,\psi(\mu_s)=(\sigma^E(t)+\sigma_s^r)\,\psi(\mu_s)$ where $\sigma^E(t)$ represents the pure elastic response of the body under consideration to the applied loads without accounting for damage; and $\psi(\mu_s)$ is determined by (13).

Replacing $\bar{\sigma}$ in (9) with the above expression yields in the equation

$$\xi=0.5(\psi(\mu_s))^2(\sigma^E(t)+\sigma_s^r):\mathbf{L}:(\sigma^E(t)+\sigma_s^r)+\Psi_p^0(\chi_s)\,\mathbf{C}_0\,\mathbf{C}_v^{-1} \tag{21}$$

which is valid at the post-adaptation stage.

The quantity $\mu_s$ is determined by (21) as a function of $t^*$, $\sigma_s^r$ and $\chi_s$ :

$$\mu_s=0.5(\psi(\mu_s))^2(\sigma^E(t^*)+\sigma_s^r):\mathbf{L}:(\sigma^E(t^*)+\sigma_s^r)+\Psi_p^0(\chi_s)\mathbf{C}_0\,\mathbf{C}_v^{-1} \tag{22}$$

The limit yield function can be written as follows (4):

$$\zeta_s=\Phi(\psi(\mu_s)(\sigma^E(t)+\sigma_s^r),\chi_s),\ \ t\geq t_s. \tag{23}$$

The function $\sigma^E(t)$ is determined by the solution of the elastic boundary-value problem for the body under consideration without accounting for plastic deformation and damage. Consequently, $\zeta_s$ is a known function of t, $\sigma_s^r$, $\chi_s$.

It is supposed as usual that $\zeta_s$ is a growing function of $\bar{\sigma}$. However due to the cyclic nature of loading, it has points of local extremum with respect to t which are specified by the stress path.

For fixed arbitrary values of $\sigma_s^r$ and $\chi_s$ the function $\zeta_s$ reaches its local maximum values at the points of departure $t=t^*$.

## 5. The Core of the Limit Yield Condition

The quantities $t^*, \sigma^r_s, \chi_s, \mu_s$ depend on the deformation path, and cannot be determined without a detailed computation of the deformation process. However it is possible to derive their a-priori estimates. To that end, let us consider possible limit yield surfaces. If there exists a yield surface which could be a limit one, then $t^o$ are the time points of departure corresponding to this surface, and $\sigma^{ro}$, $\chi^o$, $\mu^o$ are related values of $\sigma^r$, $\chi$ and $\mu$.

The quantity $\mu$ is a function of t, $\sigma^r$ and $\chi$ due to the equation similar to (22)

$$\mu=0.5(\psi(\mu))^2(\sigma^E(t)+\sigma^r):\mathbf{L}:(\sigma^E(t)+\sigma^r)+\Psi^p_o(\chi)\mathbf{C}_o\mathbf{C}_o^{-1}. \quad (24)$$

At a point of the body, the function $\zeta=\Phi(\psi(\mu)(\sigma^E(t)+\sigma^r),\chi)$ can have several points of local maximum with respect to t corresponding to the points of departure for fixed $\sigma^r$ and $\chi$. The related values of $\zeta$ are denoted $\zeta_{max}$.

Depending on $\sigma^r$, the quanitity $\zeta_{max}$ can reach its absolute maximum value (denoted max$\zeta$) under different values of t. With a view to specify a lower estimate for the limit yield surface, let us choose $\sigma^r$ in such a way that the quantity max$\zeta$ would be minimum for a fixed arbitrary value of $\chi$. The related values of $\sigma^r$ and t are $\sigma^{ro}$ and $t^o$, and determined by the solution of the min-max problem of mathematical programming:

$$\zeta^o=\min_{\sigma^r}\max_{t}\zeta \quad (25)$$

under requirements $\nabla\cdot\sigma^r=0$ into the body volume, and $\sigma^r\cdot\nu=0$ at the part of the body surface $S_p$ where tractions are prescribed; $\chi$ is a fixed parameter, $\nu$ is the unit vector of the external normal to $S_p$, and $\nabla$ is the vector with the components $\partial/\partial x_i$.

The solution to (25) provides us with the values of $\zeta^o$, $\sigma^{ro}$ and $t^o$ at every point of the body under consideration. It exists if the function $\zeta$ is convex in $\overline{\sigma}$ for any admissible value of $\chi$.

The departure points are to be placed on the yield surface. This condition imposes a restriction on the values of the internal variables. If there is only one hardening parameter, as in the case of isotropic strain hardening, then this condition specifies its value. Choose $\chi$ so that

$$\zeta^o=0. \quad (26)$$

The related value of $\chi$ is denoted $\chi^o$. The last requirement closes the system of the equations determining the parameters $\mu^o$, $\sigma^{ro}$, $t^o$ and $\chi^o$. These equations are (24), (25) and (26).

Owing to this choice

$$\Phi(\psi(\mu^o)(\sigma^E(t)+\sigma^{ro}),\chi^o)<0 \tag{27}$$

for any t except of $t=t^o$ for which $\Phi=0$.

Inequality (27) leads to the conclusion that the stress path $\bar{\sigma}(t)=\psi(\mu^o)(\sigma^E(t)+$ $+\sigma^{ro})$ is either in the interior of the yield surface $\Phi(\bar{\sigma}, \chi^o)=0$, or coincides with it partly, or touches it.

If $\sigma^{r1}$ differs from $\sigma^{ro}$, then, according to (27) and the condition $\zeta^o=0$, the inequality $\zeta=\Phi(\psi(\mu^1)(\sigma^E(t^1)+\sigma^{r1}),\chi^o)\geq 0$ is valid, where $t^1$ and $\mu^1$ correspond to $\sigma^{r1}$ and $\chi^o$. This inequality implies that at least some of the departure points corresponding to $\sigma^{r1}$ is placed outside of the surface $\Phi(\bar{\sigma},\chi^o)=0$. The yield surface corresponding to $\sigma^{r1}$ is $\Phi(\bar{\sigma},\chi^1)=0$ where $\chi^1$ is specified by the equation $\Phi(\psi(\mu^1)(\sigma^E(t^1)+\sigma^{r1}),\chi^1)=0$. This surface contains the surface $\Phi(\bar{\sigma},\chi^o)=0$ in its interior. Thus, the surface $\Phi(\bar{\sigma},\chi^o)=0$ is inside of any other surface for which shakedown is possible, or coincides with it partly. The surface $\Phi(\bar{\sigma},\chi^o)=0$ may be named *the core* of the limit yield surface. The core exists, if the solution of equations (24), (25) and (26) exist.

The notion of the core is similar to the notions of "sanctuary" (Nayroles and Weichert 1993) and "reduced elastic domain" (Maier 1969).

The actual limit yield condition coincides with the core, or contains it partly or wholly inside.

If the limit yield condition exists, the core and stress path $\bar{\sigma}(t)=$ $=\psi(\mu^o)(\sigma^E(t)+\sigma^{ro})$ exist also, i.e. there is a solution of problem (25) satisfying the requirement $\zeta^o=0$. Really, if the stationary stage of deformation process exists, then there are actual time-independent value of the residual stress tensor $\sigma^r_s$, the internal variables $\chi_S$, maximum value of the damage energy release rate $\mu_S$, and the time points of departure $t^*\geq t_S$. Because the corresponding departure points are on the limit yield surface, the equality is valid $\zeta_{max}=\Phi(\psi(\mu_S)(\sigma^E(t^*_s)+\sigma^r_s),\chi_S)=0$. This value $\zeta_{max}$ is the local maximal value of $\zeta$ with respect to t. If $\sigma^r_s$ provides the minimal value to $\max\zeta$ (the absolute maximum of $\zeta_{max}$), i.e. if $t^*$ and $\sigma^r_s$ are the solution of the problem (26), then the core coincides with the limit yield condition. Otherwise, the residual stresses $\sigma^r$ can be chosen so that min max$\zeta$ would be negative. Then it is possible to choose a value of $\chi$ so that min max$\zeta$=0. The corresponding yield surface is partly or in whole inside the actual limit yield surface because the corresponding departure points are inside it. Anyway, there exists a yield surface which is inside of the actual limit yield surface, or coincides with it. This is the core. Hence the existence of the core is a necessary condition for elastic adaptation.

## 6. Adaptation Conditions

The value of the damage parameter $\Delta^o$ corresponding to the core is specified by (12): $\Delta^o=1-1/\psi(\mu^o)$.

The core could coincide with the actual limit yield surface under some programs of loading. But such coincidence does not hold always. In such case, the actual stress path goes out of the core, and the plastic deformation and damage processes continue outside it.

The damage parameter is a non-decreasing functions of t. Therefore, the value $\Delta^o$ corresponding to the core provides the lower estimate for its limit value:

$$\Delta^o \leq \Delta_S \tag{28}$$

The limit value of the damage parameter is not to exceed its critical value: $\Delta_S \leq \Delta_C$. Consequently, adaptation is possible, if

$$\Delta^o \leq \Delta_C. \tag{29}$$

The internal parameters $\chi$ are non-decreasing functions of the time parameter t. Therefore it is possible to arrange the sequence of $\chi$ corresponding to different values of t: $\chi_1 \leq \chi_2 \leq \chi_3 ...$ for $t_1<t_2<t_3...$ where the inequalities are taken in termwise way.

According to the assumed material model the damage process can occur only along with the process of plastic deformation. If the plastic deformation stops, the damage process ceases as well, and the body experiences only elastic deformation starting from this time on. This case is named the elastic adaptation, or elastic shakedown. However, it is possible to say that the body adapted itself to the applied loading program, only if the limit value of the damage parameter is less than the critical one: $\Delta_S<\Delta_C$.

The known static shakedown (Melan) theorem can be extended to the material models for which the yield surfaces corresponding to subsequent time points encompass each other (Druyanov & Roman 1996b). The material model with isotropic strain hardening is an example. Let us restrict ourselves by this class of material models. In that event the existence of a core is a necessary and sufficient condition for elastic adaptation. Actually, if the core exists, then there exists a stress field which is wholly in the core interior except of some points or segments situated at the yield surface. However, the stress does not induce plastic deformation there. Hence this stress field satisfies the conditions of the extended Melan theorem, and a limit yield surface exists. The limit yield surface encompasses the core, or coincides with it.

If the value of damage parameter $\Delta_s$ corresponding to the core is less than its critical value $\Delta_c$, than not only elastic adaptation will occur, but also the general adaptation is possible.

Thus, in the case under consideration the existence of the core and the fulfillment of the inequality $\Delta_s<\Delta_c$ form the necessary and sufficient condition for general adaptation of the damaged elastic plastic bodies.

## 7. Comparison Between Isotropic and Anisotropic Damage

In this Section a model of perfectly brittle isotropic damage is employed with a purpose of constructing a simplified method able to direct estimating of asymptotic values of variables responding for the asymptotical behavior of the body under imvestigation. To that end the effect of plastic deformation on the development of damage is neglected, i.e. the plastic part of the damage energy release rate is omitted. (See (9) and (16)). Besides a linear approach is accepted, which is to say that the effect of damage on the material properties is neglected.

Let us consider the regular spectral projection tensor $\mathbf{S}=\sum_{k=1}^{3}\mathbf{p}_k\otimes\mathbf{p}_k$. Its components $S_{mn}=\sum_{k=1}^{3}p_{km}p_{kn}$ where $p_{km}=\cos\alpha_{km}$, and $p_{kn}=\cos\alpha_{kn}$, and $\alpha_{km}$ and $\alpha_{kn}$ are the angles between $\mathbf{p}_k$ and the directional unit vectors $\mathbf{e}_m$ and $\mathbf{e}_n$ of a fixed coordinate system. Consequently, $S_{mn}=\cos\alpha_{mn}$ where $\alpha_{mn}$ is the angle between $\mathbf{e}_m$ and $\mathbf{e}_n$. Thus, $S_{mn}=\delta_{mn}$ where $\delta_{mn}$ is the Kroneker symbol, i.e. $S_{mn}=1$, if m=n, and $S_{mn}=0$, if m≠n.

Unlike $S_{mn}$, the tensor $\mathbf{S}^+$ with components $S^+_{mn}$ can assume any value in the interval [0; 1]: $0\le S^+_{mn}\le 1$. Really $S^+_{mn}=0$, if m≠n, and if m=n and $\varepsilon_k\le 0$ for all k. If m=n, then $S^+_{mm}=\sum_{k=1}^{3}H(\varepsilon_k)(\cos\alpha_{km})^2$. If $\varepsilon_k>0$ for all values of k, then $S^+_{mm}=1$. However, if not all $\varepsilon_k$ are positve, then $S^+_{mn}$ can assume any value from the interval [0,1]: $0<S^+_{mn}<1$. At last, $S^+_{mm}=0$ in the case where all the $\varepsilon_k<0$.

Consequently, according to (15),

$$0\le P^+_{ijmn}\le 1. \tag{30}$$

If $\varepsilon_k>0$ for all k, then $\mathbf{P}^+=\mathbf{I}$. Owing to (30) $\boldsymbol{\varepsilon}^{e+}\leq\boldsymbol{\varepsilon}^{e}$, and $\mathbf{C}_0^+\leq\mathbf{C}_0$.

Let us consider two bodies: the first with anisotropic damage, and the second of the same shape and material properties and subjected to the same loading program, however with isotropic damage, and compare their asymptotic mechanical behavior. To that end compare the isotropic damage energy release rate $\xi$, (9), with the anisotropic one $\eta$, (16). Because all the terms in right sides of (9) and (16) are not negative, and the terms with negative $\varepsilon_i$ are absent in (16), it is possible to conclude that the anisotropic energy release rate $\eta$ is not greater than the quasi-anisotropic one $\xi$: $\eta\leq\xi$, and $\max\eta\leq\max\xi$. Because $\mu\equiv\max\xi$, and $\nu\equiv\max\eta$, the inequality holds

$$\nu\leq.\,\mu \tag{31}$$

Consequently $\max\nu\leq\max\mu$. According to the assumption, G is a growing function of its argument. Therefore $G(\max\nu)\leq G(\max\mu)$.

The maximal value of the current damage threshold which it reaches during loading is determined by the *absolute* maximal value of the damage energy release rate $\mathrm{MAX}\xi$ which $\xi$ assumes during loading: $R_{max}=G(\mathrm{MAX}\xi)$. Because $\mathrm{MAX}\xi=\max\mu$ then $R_{max}=G(\max\mu)$.

If the body with isotropic damage does not fail, then $R_{max}=G(\max\mu)<R_c$. Consequently $G(\max\nu)<R_c$.

Due to relation (9) the examination of the asymptotic mechanical behavior of the bodies with isotropic damage is much easier than that of the bodies with anisotropic one. Therefore the above conclusion gives us a chance to facilitate the study of the asymptotic behavior of bodies with damage induced anisotropy.

## References


1. Cordebois, J.P. and Sidoroff, F. (1982) Endommagement anisotrope en élasticité et plasticité, *J. Méc.Théor. Appl.* No. special, 45-59.
2. Druyanov, B. and Roman, I. (1997a) Concept of the limit yield condition in shakedown theory, *Int. J. Sol. Struct.* **34**, 1547-1556.
3. Druyanov, B. and Roman, I. (1997b) Features of the stress path at the post adaptation stage and related shakedown conditions, *Int. J. Sol. Struct.* **34**, 3773-3780.
4. Druyanov, B. and Roman, I. (1998a) On adaptation (shakedown) of a class of damaged elastic plastic bodies to cyclic loading, *Eur. J. Mech., A/ Solids* **17**, 71-78.
5. Druyanov, B. and Roman, I. (1998b) Conditions for adaptation of damaged elastic-plastic bodies to cyclic loading, *Fat. and Fract. Engng. Mater. Struct.* **21**, 631-640.
6. Druyanov, B. and Roman, I. (1999) Conditions for shakedown of damaged elastic-plastic bodies,.*Eur. J. Mech., A/ Solids* **18**, 641-651.
7. Feng, X.Q. and Liu, X.Sh. (1997) On shakedown of three-dimensional elastoplastic strain-hardening structures, *Int. J. Plast.* **12**, 1241-1256.
8. Hachemi, A. and Weichert, D. (1992) An extension of the shakedown theorem to a certain class of inelastic materials with damage, *Arch. Mech.* **44**, 491-498.
9. Hachemi, A. and Weichert, D. 1997, Application of shakedown theory to damaging inelastic material under mechanical and thermal loads, *Int. J. Mech. Sci.* **39**, 1067-1076.

10. Hachemi, A. and Weichert, D. (1998) Numerical shakedown analysis of damaged structures, *Comput. Methods Appl. Mech. Engrg.* **160**, 57-70.
11. Klebanov, J.M. and Boyle, J.T. (1998) Shakedown of creeping structures, *Int.J. Solids Structures* **35**, 3121-3133.
12. Ju, J. W. (1989) On energy-based coupled elastoplastic damage theories: constitutive modeling and computational aspects, *Int. J. Solids Struct.* **25**, 803-833.
13. Koiter, W. T. (1960) General theorems for elastic-plastic bodies, in: I. N. Sneddon and R. Hill (eds.), *Progress in Solid Mechanics*, **1**, North-Holland, Amsterdam.
14. König, J. A. (1987) *Shakedown of Elastic-Plastic Structures,* Elsevier, Amsterdam.
15. Krajcinovic, D. (1996) *Damage Mechanics*, North-Holland, Amsterdam.
16. Lemaitre, J. (1985) A continuous damage mechanics model for ductile fructure. *J.Engng. Mater. Technol.* **107**, 83-89.
17. Lemaitre, J. (1992) *A Course on Damage Mechanics*, Springer, Berlin.
18. Maier, G. (1969) Shakedown theory in elastoplasticity with associated and nonassociated flow laws: a finite element, linear programming approach, *Meccanica* **4**, 1-11.
19. Melan, E. (1938a) Theorie statisch unbestimmter Tragwerke aus idealplastischem Baustoff., *Sitzungsbericht der Akademie der Wissenschaften (Wien) Abt. IIa* **145**, 195-218.
20. Melan, E. (1938b) Zur Plastizität des räumlichen Kontinuums, *Ing.-Arch.* **8**, 116-126.
21. Nayroles, B. and Weichert, D. (1993) La notion de sanctuaire d'elasticite et d'adaptation des structures. *C. R. Acad. Sci. Paris, Serie II (Mec. des solides)* **316**, 1493-1498.
22. Ortiz, M. (1985) A constitutive theory for the inelastic behavior of concrete. Mech. Materials **4**, 67-93.
23. Polizzotto, C. (1995) Elastic-visco-plastic solids subjected to thermal and loading cycles. in: Z. Mroz, D. Weichert, S. Dorosz. (eds.), *Inelastic response of structures under variable loads,* Kluwer Academic Publishers, Dordrecht, pp.95-129.
24. Polizzotto, C. (1994) Steady states and sensitivity analysis in elastic-plastic structures subjected to cyclic loading, *Int. J. Solids. Struct.* **31**, 953-970.
25. Ponter, A.R.S., Karadeniz, S. and Carter, K.F. (1990) Extended upper bound shakedown theory and finite element method for axisymmetric thin shells. In: M. Kleiber, and König J..A. (eds.), *Inalastic analysis of Solids and Structures*,. Pineridge Press, Swansea, UK, 433-449.
26. Ponter, A.R.S. and Leckie, F.A. (1998) Bounding properties of metal matrix composites subjected to cyclic thermal loading., *J. Mech. Phys. Solids* **46**, 697-717.
27. Pycko, S. and Maier, G. (1995) Shakedown theorems for some classes of nonassociated hardening elastic-plastic material models, *Int. J. Plasticity* **11**, 367-395.
28. Schreyer, H.L. (1995) Continuum damage based on elastic projection operators, *Int. J. Damage Mechanics*, **4**, 171-195.
29. Siemaszko, A., (1993) Inadaptation analysis with hardening and damage, *Eur. J. Mech., A/Solids*, **12**, 237-248.
30. Stein, E., Zhang G. and König, J..A. (1992), Shakedown with nonlinear strain-hardening including structural computation using finite element method, *Int. J. Plasticity* **8**, 1-31.
31. Talreja, R. (1989) Damage development in composites: mechanisms and modelling, *J. Strain Anal.* **24**, 215-222.
32. Weichert, D. and Hachemi, A. (1998) Influence of geometrical nonlinearities on the shakedown of damaged structures, *Int. J. Plast.* **14**, 891-907.

# SIMPLIFIED METHODS FOR THE STEADY STATE INELASTIC ANALYSIS OF CYCLICALLY LOADED STRUCTURES

K. V. SPILIOPOULOS
*Institute of Structural Analysis and Seismic Research,*
*Department of Civil Engineering*
*National Technical University of Athens*
*Zografou Campus GR 157-73, Athens, Greece*

## 1. Introduction

Structures, nowadays, in order to increase efficiency, are being pushed to operate in higher and higher levels of loads and temperature. In the design of such structures like nuclear reactors, aircraft gas turbine propulsion engines, etc, a prediction of the inevitable accumulation of creep and plastic strains throughout their life is necessary.

In order to assess this long - term inelastic response of a given structure under a specified loading, detailed time and load stepping calculations must be performed. These analyses turn out to be extremely costly and are often numerically unstable.

In the case of a loading, however, which has a high degree of regularity –being either steady or cyclic- after an initial transient stage, the stresses and strains very often tend towards a steady or cyclic pattern. If this pattern develops early enough in the planned life of the structure then it may well suffice for the assessment of its complete behaviour.

The methods that seek to find the steady state of stress if the load is constant, or the cyclic steady state of stress if the load is cyclic without going through a time stepping analysis are referred in the literature as simplified methods. These methods turn out to be much faster and numerically more stable than the conventional time or load stepping ones and provide much better insight into the inelastic response of the structure. Well known examples of such methods, are the limit and the elastic shakedown analyses of structures. In the case of limit analysis, assuming a purely rigid plastic behaviour right from the start of the calculations, a state of collapse is sought which provides the limiting parameters of the loading. In the case of the elastic shakedown, assuming the material behaviour as elasto-plastic, at the end of the calculations a residual stress distribution constant in time has evolved which guarantees that, despite the initial plastic straining, the long-term behaviour of the structure becomes purely elastic ([1]).

For elevated temperature conditions the effects of creep have to be taken into account. Leckie and Ponter ([2], [3]) have verified theoretically, as well as experimentally, that when the level of loading is below $n/(n+1)$ of the limit load if the



*D. Weichert and G. Maier (eds.), Inelastic Analysis of Structures under Variable Loads,* 213–232.

structure is subjected to constant loading, or $n/(n+1)$ of the shakedown load, if the structure is subjected to cyclic loading (with n being the creep index in the power creep law), then creep effects are the dominant effects in the structure and plasticity has little influence and may be neglected. This boundary is called in the literature the modified shakedown boundary. Within this boundary, a constant in time residual stress distribution added to the elastic cyclic stress in response to the cyclic loading provides an upper bound on the inelastic work (Ponter [4]). For loads above the shakedown limit Ainsworth [5] used a cyclic plasticity solution to provide bounds on the inelastic work.

For load levels within the modified shakedown boundary, Ponter [6] has shown that for cycle times that are very short, compared to some characteristic time of the material, the aforementioned constant in time residual stress distribution coincides with the actual residual stress distribution. The reason is that there is no time for any redistribution inside a cycle. Using average strain rates over a cycle, Ponter and Brown [7] established a simplified method to calculate this rapid cycling solution, arguing that this solution is exact when the cycle time tends to zero.

The first part of the present work addresses the numerical implementation of the simplified methods for the rapid cycling creep problem. A procedure (Chan and Spiliopoulos [8]) is presented which removes the necessity of having to solve the elastic shakedown problem first in order to make sure that the load level is below the modified shakedown boundary and then to solve for the rapid cycling solution considering creep effects only. This procedure is based on a simple way to include plasticity effects together with creep effects (Spiliopoulos [9]). Thus the shakedown boundary in the presence of creep may be calculated. It is also proved that an arbitrary time period may be used which may accelerate the rate of convergence towards the final cyclic steady state solution (Spiliopoulos [10]). Examples of application of a thick walled and a thin walled cylinder are also given.

The above class of simplified methods can be applied only to a limited number of cases of cyclically loaded structures, i.e. to loads whose period is short. When an engineer, however, faces a problem which concerns a structure that is loaded cyclically, he does not know, beforehand, whether the load is of "short", or of "medium" or of "large" period. The only way to tackle problems of this sort appears to be very laborious time stepping methods.

However, in the second part of this paper a new simplified method is developed that may evaluate the steady-state response of structures that are subjected to cyclic loads of any period. The main idea behind this new method is the decomposition of the residual stresses in Fourier series with respect to time. A numerical procedure of iterative nature is then set up to evaluate the various Fourier coefficients. An example of application to a creep dominated simple three-bar structure is also presented.

## 2. On the cyclic stationary stress state

Let us assume that a structure is subjected to a cyclic mechanical loading of period T, i.e.

$$P(t+T) = P(t) \tag{1}$$

In response to this loading, the structure develops a stress system $\sigma_{ij}(t)$, which can be decomposed into two parts: the first part is a cyclic elastic stress $\sigma_{ij}^{el}(t)$ which develops in response to the applied loading assuming a completely linear elastic material behaviour, and the second part is a self-equilibrating stress system $\rho_{ij}(t)$ due to the inelasticity that creep and plasticity introduces in our structure. Thus one can write:

$$\sigma_{ij}(t) = \sigma_{ij}^{el}(t) + \rho_{ij}(t) \tag{2}$$

The strain rates, can also be decomposed into two corresponding terms $\dot{e}_{ij}^{el}$ and $\dot{\varepsilon}_{ijr}$:

$$\dot{\varepsilon}_{ij} = \dot{e}_{ij}^{el} + \dot{\varepsilon}_{ijr} = \dot{e}_{ij}^{el} + \dot{\varepsilon}_{ijr}^{el} + \dot{\varepsilon}_{ijr}^{cr} + \dot{\varepsilon}_{ijr}^{pl} \tag{3}$$

In the above equation the residual strain rate term has been itself decomposed into elastic, creep and plastic parts.

The elastic strain rates are related to their corresponding stress rates by:

$$\dot{e}_{ij}^{el} = C_{ijkl}\dot{\sigma}_{kl}^{el} \qquad\qquad \dot{\varepsilon}_{ijr}^{el} = C_{ijkl}\dot{\rho}_{kl} \tag{4a,b}$$

with $C_{ijkl}$ being the tensor of the elastic constants.

For the creep component, Norton's viscous power law is assumed to hold :

$$\dot{\varepsilon}_{ijr}^{cr} = \frac{1}{n+1}\frac{\partial\phi}{\partial\sigma_{ij}} \tag{5}$$

where $\phi$ is a strictly convex creep surface.

For the plastic part, perfect plasticity is assumed:

$$\dot{\varepsilon}_{ijr}^{pl} = \begin{cases} \lambda\dfrac{\partial f}{\partial\sigma_{ij}} & \text{if } f(\sigma_{ij}) = 0 \text{ and } \dfrac{\partial f}{\partial\sigma_{ij}}\dot{\sigma}_{ij} = 0 \\ 0 & \text{if } f(\sigma_{ij}) < 0 \text{ or if } f(\sigma_{ij}) = 0 \text{ and } \dfrac{\partial f}{\partial\sigma_{ij}}\dot{\sigma}_{ij} < 0 \end{cases} \tag{6}$$

where f is a convex yield surface.

For two different states of stress $\sigma_{ij}$ and $\sigma_{ij*}$, the corresponding creep strain rates, as well as plastic strain rates, satisfy the Drucker's postulate of stability [11]:

$$(\sigma_{ij} - \sigma_{ij*})(\dot{\varepsilon}_{ijr}^{cr} - \dot{\varepsilon}_{ijr*}^{cr}) \geq 0 \tag{7}$$

$$(\sigma_{ij} - \sigma_{ij*})(\dot{\varepsilon}_{ijr}^{pl} - \dot{\varepsilon}_{ijr*}^{pl}) \geq 0 \tag{8}$$

Frederick and Armstrong [12] have stated the following theorem concerning the existence of a cyclic stationary state:

*Theorem 1*. For a cyclic loading of a structure made of Drucker's material, the stresses and the strain rates gradually stabilize to remain unaltered on passing to the next cycle.

For the proof one can use the following positive energy functional related to the residual stresses:

$$E(t)=\frac{1}{2}\int_V C_{ijkl}[\rho_{ij*}-\rho_{ij}][\rho_{kl*}-\rho_{kl}]dV \tag{9}$$

where $\rho_{ij*}$ and $\rho_{ij}$ may be identified as the two states of stress $\rho_{ij}(t+T)$ and $\rho_{ij}(t)$ respectively. The rate of the energy functional is:

$$\dot{E}(t)=\int_V C_{ijkl}[\rho_{ij*}-\rho_{ij}][\dot{\rho}_{kl*}-\dot{\rho}_{kl}]dV=\int_V[\rho_{ij*}-\rho_{ij}][\dot{\varepsilon}^{el}_{klr*}-\dot{\varepsilon}^{el}_{klr}]dV=$$
$$=-\int_V[\rho_{ij*}-\rho_{ij}][\dot{\varepsilon}^{cr}_{klr*}-\dot{\varepsilon}^{cr}_{klr}]dV-\int_V[\rho_{ij*}-\rho_{ij}][\dot{\varepsilon}^{pl}_{klr*}-\dot{\varepsilon}^{pl}_{klr}]dV \tag{10}$$

where use of (2) was made from which the residual stress difference is equal to the actual stress difference which is in equilibrium with the same loading at time t and t+T. Use of the decomposition (3) and of (4b) was made, the fact that total strain rates and elastic strain rates $\dot{e}_{ij}$ are compatible and the Principle of Virtual Work.

From inequalities (7) and (8) it is seen that $\dot{E}\le 0$. However, E is bounded from below ($E\ge 0$) and hence $\dot{E}\to 0$ for large times. This means that both terms in (10) must be equal to zero, i.e. $\rho_{ij}(t+T)\to\rho_{ij}(t)$.

In the case that only creep effects are considered, as in the case of creep dominance below the modified shakedown boundary, a cyclic stationary creep state will be reached as the same arguments would apply for equation (10) in which only the first term will be present.

It can be readily shown that the cyclic stationary state of stress is independent of any initial residual stress distribution or plastic strain in the structure. Suppose that the starred and the unstarred quantities correspond to two distinct states at the same time t >0 which have identical histories P(t) for t >0 but different initial states at t=0. Using the same arguments as above, it follows immediately that $\rho_{ij}(t+T)\to\rho_{ij}(t)$ for large times. The cyclic stationary or steady state is therefore uniquely determined by the cycle of loads.

## 3. Cyclic loading of short period cycles

When the cycle time is short, it is reasonable to assume that inside the cycle no redistribution of stresses can take place and the residual stress system of the cyclic stationary state remains constant in time. This stress system is denoted by $\bar{\rho}_{ij}$. Moreover if compatibility of strains with displacements is assumed at the end of every cycle, then compatibility at any time inside a cycle can reasonably be inferred for the same reason. This suggests a simple way of finding this constant residual stress system:

At some time $t_0$ the cycle ($t_0$-T, $t_0$) has been completed and the residual strain that remains can be found if we integrate (3):

$$\Delta\bar{\varepsilon}_{ijr} = \Delta\bar{\varepsilon}^{el}_{ijr} + \Delta\bar{\varepsilon}^{cr}_{ijr} + \Delta\bar{\varepsilon}^{pl}_{ijr} \tag{11}$$

where :

$$\Delta\bar{\varepsilon}^{el}_{ijr} = C_{ijkl}\Delta\bar{\rho}_{kl}, \qquad \Delta\bar{\varepsilon}^{cr}_{ijr} = \int_0^T \dot{\bar{\varepsilon}}^{cr}_{ijr}\,dt, \qquad \Delta\bar{\varepsilon}^{pl}_{ijr} = \int_0^T \dot{\bar{\varepsilon}}^{pl}_{ijr}\,dt \qquad \text{(11a-c)}$$

with the creep and the plastic terms being both functions of the total stress $\bar{\sigma}_{ij}(t) = \sigma^{el}_{ij}(t) + \bar{\rho}_{ij}$.

The residual strain system $\Delta\bar{\varepsilon}_{ijr}$ must be itself a compatible strain system. This constitutes a boundary value problem that may be solved iteratively. More specifically by multiplying both terms of (11) with $\bar{\rho}_{ij}$ and integrating over the volume of the body, the following equation may be obtained, using the Principle of Virtual Work:

$$0 = \int_V \bar{\rho}_{ij}\Delta\bar{\varepsilon}^{el}_{ijr}\,dV + \int_V \bar{\rho}_{ij}\Delta\bar{\varepsilon}^{cr}_{ijr}\,dV + \int_V \bar{\rho}_{ij}\Delta\bar{\varepsilon}^{pl}_{ijr}\,dV \tag{12}$$

The basic unknown is $\bar{\rho}_{ij}$ of the stationary state solution. Assuming some first approximation ${}^{(v)}\bar{\rho}_{ij}$ to the residual stress where $(v)$ denotes an $v$th iteration, a better approximation can be found from an iterative form of (12):

$$0 = \int_V {}^{(v)}\bar{\rho}_{ij}C_{ijkl}\left({}^{(v+1)}\bar{\rho}_{kl} - {}^{(v)}\bar{\rho}_{kl}\right)dV + \int_V {}^{(v)}\bar{\rho}_{ij}\,{}^{(v)}\Delta\bar{\varepsilon}^{cr}_{ijr}\,dV + \int_V {}^{(v)}\bar{\rho}_{ij}\,{}^{(v)}\Delta\bar{\varepsilon}^{pl}_{ijr}\,dV \tag{13}$$

where relations (11 a-c) are used and a better approximation ${}^{(v+1)}\bar{\rho}_{ij} = {}^{(v)}\bar{\rho}_{ij} + {}^{(v)}\Delta\bar{\rho}_{ij}$.

The iterations will stop when ${}^{(v+1)}\bar{\rho}_{ij} \rightarrow {}^{(v)}\bar{\rho}_{ij}$ and a cyclic solution will therefore have been achieved. If the given loading is below the shakedown boundary in the presence of creep, then the last term of (13) will not exist after final convergence. This steady state solution will, of course be identical to the steady state solution having only creep effects taken into account. The reason is that plastic effects will only affect the residual stress distribution at the initial stages of iteration, something, which as was proved in section 2, does not influence the cyclic stationary state of stress. By including the plasticity effects, though, we may be able to see in one analysis whether shakedown has occurred and therefore not a separate elastoplastic analysis to estimate the shakedown loading is needed. It will be shown later that it is a relatively easy task to include the plasticity effects.

When the steady state is reached with the loading being below the modified shakedown boundary, the only strain remaining is the creep strain which for this reason has to be kinematically admissible.

*Theorem 2.* (Ponter [13]). For an arbitrary cycle history P(t), 0<t<T the kinematically admissible strain field $\Delta\bar{\varepsilon}^{cr}_{ijr}$ given by (11b) is uniquely defined.

The proof is given in [13] and is based on the convexity of the creep surface.

Next a theorem is going to be proved, which is very useful later in numerical applications and makes the convergence towards the cyclic steady state solution very fast.

*Theorem 3.* (Spiliopoulos [10]). Two cyclic loads of different cycling time but having the same variation inside the cycle, will develop identical steady state residual stress systems if the assumption that the residual stress remains constant inside the cycle is valid for both.

Two such loads whose periods differ by two, can be seen in Figure 1.

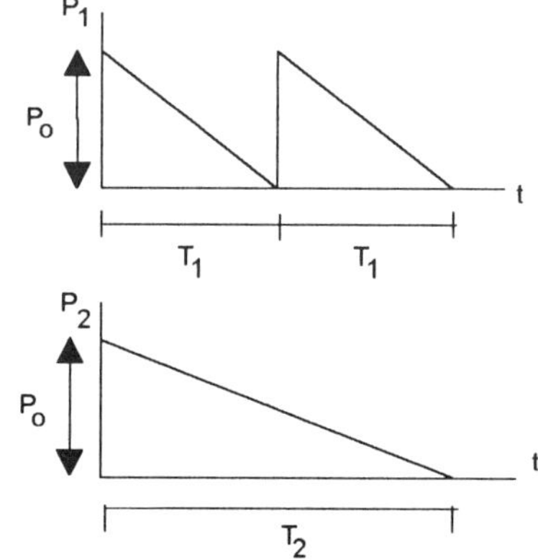


*Figure 1.* Typical cyclic loads with the same variation of different periods.

To prove the above theorem let us consider two loads $P_1(t)$ and $P_2(t)$, whose periods are related by an arbitrary number m so that $T_1 = \frac{T_2}{m} = \frac{T}{m}$. A similar relationship will hold between the two loads and their corresponding elastic stresses, i.e.

$$P_2(t) = P_1\left(\frac{t}{m}\right) \quad , \qquad \sigma^{el}_{ij2}(t) = \sigma^{el}_{ij1}\left(\frac{t}{m}\right) \tag{14a,b}$$

Employing the iterative form (13) of the boundary value problem for both loads we will get if we denote by $(\nu)$ some iteration for load 1 and $(\mu)$ some iteration for loading 2:

$$0 = \int_V {}^{(\nu)}\bar{\rho}_{ij1} C_{ijkl}\left({}^{(\nu+1)}\bar{\rho}_{kl1} - {}^{(\nu)}\bar{\rho}_{kl1}\right)dV + \int_V {}^{(\nu)}\bar{\rho}_{ij1}\,{}^{(\nu)}\Delta\bar{\varepsilon}^{cr}_{ijr1}dV + \int_V {}^{(\nu)}\bar{\rho}_{ij1}\,{}^{(\nu)}\Delta\bar{\varepsilon}^{pl}_{ijr1}dV \tag{15}$$

the second integral may be written using equation (11b):

$$\int_V {}^{(\nu)}\overline{\rho}_{ij1}\,{}^{(\nu)}\Delta\overline{\varepsilon}^{cr}_{ijr1}dV = \int_V {}^{(\nu)}\overline{\rho}_{ij1}\int_0^{T_1}\dot{\varepsilon}^{cr}_{ijr}\left\{\sigma^{el}_{ij1}(t)+{}^{(\nu)}\overline{\rho}_{ij1}\right\}dVdt$$

$$= \int_V \int_0^{T/m} {}^{(\nu)}\overline{\rho}_{ij1}\dot{\varepsilon}^{cr}_{ijr}\left\{\sigma^{el}_{ij1}(t)+{}^{(\nu)}\overline{\rho}_{ij1}\right\}dVdt \tag{16}$$

For loading 2:

$$0 = \int_V {}^{(\mu)}\overline{\rho}_{ij2}C_{ijkl}\left({}^{(\mu+1)}\overline{\rho}_{kl2} - {}^{(\mu)}\overline{\rho}_{kl2}\right)dV + \int_V {}^{(\mu)}\overline{\rho}_{ij2}\,{}^{(\mu)}\Delta\overline{\varepsilon}^{cr}_{ijr2}dV + \int_V {}^{(\mu)}\overline{\rho}_{ij2}\,{}^{(\mu)}\Delta\overline{\varepsilon}^{pl}_{ijr2}dV \tag{17}$$

The second integral may be written, using (11b), (14b) and a change in integration variable $\tau = \frac{t}{m}$:

$$\int_V {}^{(\mu)}\overline{\rho}_{ij2}\,{}^{(\mu)}\Delta\overline{\varepsilon}^{cr}_{ijr2}dV = \int_V {}^{(\mu)}\overline{\rho}_{ij2}\int_0^{T}\dot{\varepsilon}^{cr}_{ijr}\left\{\sigma^{el}_{ij2}(t)+{}^{(\mu)}\overline{\rho}_{ij2}\right\}dVdt$$

$$= m\int_V \int_0^{T/m} {}^{(\mu)}\overline{\rho}_{ij2}\dot{\varepsilon}^{cr}_{ijr}\left\{\sigma^{el}_{ij1}(\tau)+{}^{(\mu)}\overline{\rho}_{ij2}\right\}dVd\tau \tag{18}$$

In the limit $\nu \to \infty$, $\mu \to \infty$ , ${}^{(\nu+1)}\overline{\rho}_{ij1} \to {}^{(\nu)}\overline{\rho}_{ij1} = \overline{\rho}^*_{ij1}$ , ${}^{(\mu+1)}\overline{\rho}_{ij2} \to {}^{(\mu)}\overline{\rho}_{ij2} = \overline{\rho}^*_{ij2}$ . Also since the loading is assumed below the shakedown limit in the presence of creep, the plasticity terms will have disappeared and thus (15) and (17) will become (19) and (20):

$$\int_V \int_0^{T/m} \overline{\rho}^*_{ij1}\dot{\varepsilon}^{cr}_{ijr}\left\{\sigma^{el}_{ij1}(t)+\overline{\rho}^*_{ij1}\right\}dVdt = 0 \tag{19}$$

$$m\int_V \int_0^{T/m} \overline{\rho}^*_{ij2}\dot{\varepsilon}^{cr}_{ijr}\left\{\sigma^{el}_{ij1}(\tau)+\overline{\rho}^*_{ij2}\right\}dVd\tau = 0 \tag{20}$$

Since $\overline{\rho}^*_{ij1}$ and $\overline{\rho}^*_{ij2}$ are constant in time , (19) and (20) become:

$$\int_V \overline{\rho}^*_{ij1} \int_0^{T/m} \dot{\varepsilon}^{cr}_{ijr}\left\{\sigma^{el}_{ij1}(\tau)+\overline{\rho}^*_{ij1}\right\}d\tau dV = 0 \tag{19a}$$

$$\int_V \overline{\rho}^*_{ij2} \int_0^{T/m} \dot{\varepsilon}^{cr}_{ijr}\left\{\sigma^{el}_{ij1}(\tau)+\overline{\rho}^*_{ij2}\right\}d\tau dV = 0 \tag{20a}$$

or

$$\int_V \overline{\rho}_{ij1}^{*} \Delta\overline{\varepsilon}_{ijr1}^{cr*} dV = 0 \tag{19b}$$

$$\int_V \overline{\rho}_{ij2}^{*} \Delta\overline{\varepsilon}_{ijr2}^{cr*} dV = 0 \tag{20b}$$

Since $\overline{\rho}_{ij1}^{*}$ and $\overline{\rho}_{ij2}^{*}$ are self equilibrating, it can be deduced that the creep strains $\Delta\overline{\varepsilon}_{ijr1}^{cr*}$ and $\Delta\overline{\varepsilon}_{ijr2}^{cr*}$ that have resulted from (19a) and (20a) by integrating over the same period T/m are both compatible. But making use of the theorem 2, this system must be unique and therefore $\overline{\rho}_{ij1}^{*} = \overline{\rho}_{ij2}^{*}$ .

## 3.1 NUMERICAL IMPLEMENTATION

The iterative form of (12) and (13) may be converted through the use of the finite element method to a numerical procedure to find the constant in time residual stress system of the cyclic stationary state of a structure. A structure is discretised into finite elements. The elements are assumed to be interconnected at a discrete number of nodal points situated on their boundaries. Bold letters indicate vectors and matrices.

When a cycle of loading is completed, we can express the changes in residual strains $\Delta\boldsymbol{\varepsilon}_r$ in the terms of the changes of the residual nodal displacements $\Delta\mathbf{r}_r$ :

$$\Delta\boldsymbol{\varepsilon}_r = \mathbf{B}\Delta\mathbf{r}_r \tag{21}$$

On the other hand the changes in residual strains may be expressed using equation (11) into the following terms:

$$\Delta\boldsymbol{\varepsilon}_r = \mathbf{D}^{-1}\Delta\boldsymbol{\rho} + \Delta\boldsymbol{\varepsilon}_r^{cr} + \Delta\boldsymbol{\varepsilon}_r^{pl} \tag{22}$$

where **D** is the matrix of elastic constants. The above equation may be solved for $\Delta\boldsymbol{\rho}$ :

$$\Delta\boldsymbol{\rho} = \mathbf{D}(\Delta\boldsymbol{\varepsilon}_r - \Delta\boldsymbol{\varepsilon}_r^{cr} - \Delta\boldsymbol{\varepsilon}_r^{pl}) \tag{23}$$

Since $\Delta\boldsymbol{\rho}$ is in equilibrium with zero loads and the system $\Delta\boldsymbol{\varepsilon}_r$ is compatible, from the Principle of Virtual Work we may write:

$$0 = \int_V \Delta\boldsymbol{\varepsilon}_r^T \Delta\boldsymbol{\rho} dV = \int_V \Delta\mathbf{r}_r^T \mathbf{B}^T \mathbf{D}(\Delta\boldsymbol{\varepsilon}_r - \Delta\boldsymbol{\varepsilon}_r^{cr} - \Delta\boldsymbol{\varepsilon}_r^{pl}) dV$$

$$= \int_V \Delta\mathbf{r}_r^T \mathbf{B}^T \mathbf{D}(\mathbf{B}\Delta\mathbf{r} - \Delta\boldsymbol{\varepsilon}_r^{cr} - \Delta\boldsymbol{\varepsilon}_r^{pl}) dV \tag{24}$$

where use of (21) and (23) was made. Since the vector $\Delta\mathbf{r}$ is arbitrary, we may finally write:

$$(\int_V \mathbf{B}^T \mathbf{D}\mathbf{B}dV)\Delta\mathbf{r}_r = \int_V \mathbf{B}^T \mathbf{D}\Delta\varepsilon_r^{cr} dV + \int_V \mathbf{B}^T \mathbf{D}\Delta\varepsilon_r^{pl} dV \tag{25}$$

The left-hand side of (25) is the stiffness matrix of the structure, whereas the right-hand side are equivalent nodal loads that are due to creep and plasticity effects correspondingly.

The linear elastic problem is solved first in a separate analysis and the elastic stresses are calculated at the Gauss points of the structure. Then the following iterative procedure may be established: Assuming an initial constant in time distribution of residual stresses ${}^{(0)}\boldsymbol{\rho}$, normally zero, we add them to the elastic stresses which are functions of time and the total stresses are calculated at the Gauss points. By performing a numerical time integration over a complete cycle, the creep effects in the right-hand side of (25) are calculated. The plasticity effects may also be estimated, in an approximate way as will be explained below, and the right-hand side of (25) may be established. Equation (25) is then solved for the increments of $\Delta\mathbf{r}_r$, which will give rise to increments of residual stresses that may be calculated using (23). The increments are added to ${}^{(0)}\boldsymbol{\rho}$ and a new residual stress distribution is obtained. The process is repeated until $\Delta\boldsymbol{\rho}$ becomes negligible and a stationary residual stress state is found.

The numerical procedure is very much accelerated, if not the real but a fictitious period is used for the time integration, something that theorem 3 gives us the right to do. Also the stiffness matrix is the normal linear stiffness matrix of the structure and needs be assembled and decomposed only once. This makes the procedure easy to be used within a general finite element program.

The way to check whether convergence has been achieved is done through the error norm of the vector of the residual stresses. More specifically the following quantity, which measures the maximum over all elements relative difference of the Euclidean norm of the residual stresses over two successive iterations, is used:

$$\frac{\max\limits_{\text{all elem.}} \left\| {}^{(\nu+1)}\boldsymbol{\rho} - {}^{(\nu)}\boldsymbol{\rho} \right\|_2}{\left\| {}^{(\nu)}\boldsymbol{\rho} \right\|_2}$$

This quantity is checked against a pre-specified error tolerance.

## 3.2 IMPLEMENTATION OF PLASTIC EFFECTS

As was aforementioned, the inclusion of plasticity effects in an approximate way makes possible to decide whether a given loading is below the shakedown boundary in the presence of creep. In this section a simple and effective way to include these effects is given [9]. A von Mises type of material, with $\sigma^y$ as the uniaxial yield stress has been assumed, although any other yield criterion may be used. Since there is no change of the residual stress inside a cycle, in the course of iterations, we add, after the completion of a cycle, to the current values of the residual stresses the maximum or the minimum values of the elastic stresses at each Gauss point, and calculate the equivalent stress $\bar{\sigma}^{max}$ and $\bar{\sigma}^{min}$. After we have checked for all the Gauss points of the structure, if we find that $\bar{\sigma}^{max} > \bar{\sigma}^{min} > \sigma^{y}$, the maximum values of elastic stresses are considered to the most critical, otherwise if $\bar{\sigma}^{min} > \bar{\sigma}^{max} > \sigma^{y}$ the minimum values of the elastic stresses are considered to be the most critical. We proceed further with the most critical values of the elastic stresses either $\boldsymbol{\sigma}_{*}^{el} = \boldsymbol{\sigma}_{max}^{el}$ or $\boldsymbol{\sigma}_{*}^{el} = \boldsymbol{\sigma}_{min}^{el}$, and perform a radial return operation (Figure 2):

Let us suppose that the initial state of residual stresses is $\boldsymbol{\rho}^{in}$. The final state of stress of the total stress, after a cycle has been completed, had the material responded elastically, is $\boldsymbol{\sigma}^{A} = \boldsymbol{\sigma}_{*}^{el} + \boldsymbol{\rho}^{in}$. A radial return of ratio r is performed so that the total stress is brought back on the yield surface. If we denote by $\boldsymbol{\rho}^{fin}$ the true final, after the correction, residual stresses, the following equation must hold:

$$(\boldsymbol{\sigma}_{*}^{el} + \boldsymbol{\rho}^{in}) - r(\boldsymbol{\sigma}_{*}^{el} + \boldsymbol{\rho}^{in}) = \boldsymbol{\sigma}_{*}^{el} + \boldsymbol{\rho}^{fin} \tag{26}$$

from which we can derive:

$$\boldsymbol{\rho}^{fin} = \boldsymbol{\rho}^{in} - r(\boldsymbol{\sigma}_{*}^{el} + \boldsymbol{\rho}^{in}) \tag{27}$$

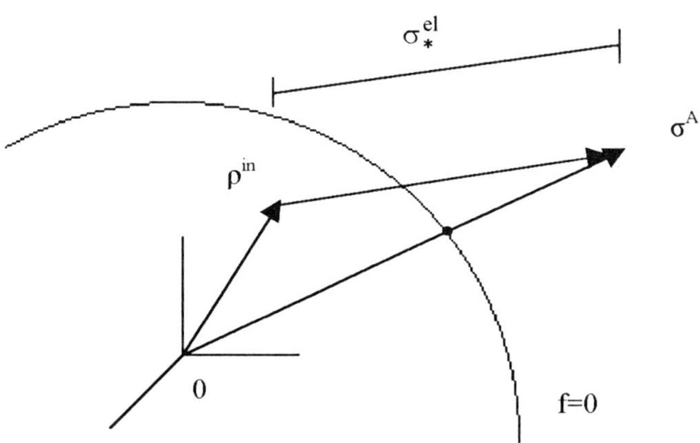


*Figure 2.* Radial return rule for a perfectly plastic material

The ratio r is determined from the condition that the final state of stress lies on the yield surface. We must have therefore:

$$f\{(1-r)(\boldsymbol{\sigma}_{*}^{el} + \boldsymbol{\rho}^{in})\} = 0 \tag{28}$$

The unbalanced force $r(\boldsymbol{\sigma}_*^{el} + \boldsymbol{\rho}^{in})$ is then redistributed into the form of nodal forces giving the plastic term of equation (25).

## 3.3 NUMERICAL APPLICATIONS

The aforementioned numerical procedure has been applied to a thick and a thin cylinder. The multiaxial equivalence to the uniaxial case of creep law $\dot{\varepsilon}^{cr} = K\sigma^n$, was done using the equivalent stress method.

### *3.3.1 Thick Cylinder*

A thick cylinder under internal pressure was considered first. The material constants used were $K = .636\text{x}10^{-10}$ (SI units), with a creep index n=3.0 and Young's modulus E=21000 dN/mm$^2$. A uniaxial yield stress $\sigma^y = 24\text{dN}/\text{mm}^2$ was used. Plane strain conditions were assumed and the material was assumed elastically and plastically incompressible. This was achieved numerically by taking Poisson's ratio $\mu = 0.4999$.

The structure was discretized into 48 eight-noded isoparametric elements (Figure 3), with 2x2 Gauss integration points for each of them. This type of element, with the reduced integration used, showed no "locking" phenomenon due to the incompressibility of the material that was assumed.

The variation of the load with time is also shown in Figure 3.

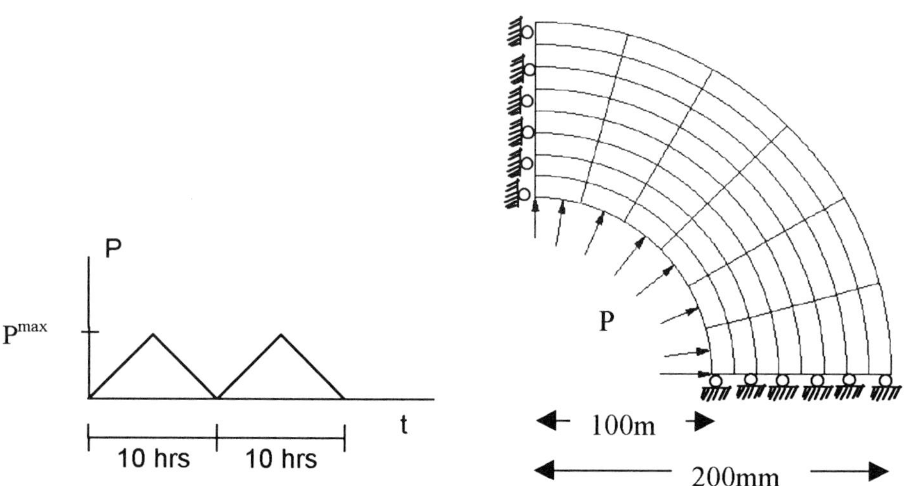


*Figure 3.* Load variation with time and finite element discretisation

By modifying the data so that the load remains constant the computer program that was written was tested against creep constant loading condition for which we have analytical results (Kraus [14]). The right way of implementing the plastic effects was also assured by assuming a constant load of 15 dN/mm$^2$ and pure elastoplastic material.

The analytical results that exist [15] in this case show a very good agreement with the steady state residual hoop stress that was obtained by the program (Figure 4 (a)).

The elastic shakedown load for this cylinder turns out to be (Lubliner [15]]):

$$P^{sh} = \frac{\sigma^y}{\sqrt{3}} \ln \frac{b^2}{a^2} = 19.2 \text{ dN/mm}^2 \tag{29}$$

where a=100mm and b=200 mm are the inner and the outer radii of the cylinder.

The cyclic internal pressure of Figure 3 with $P^{max} = 15 \text{dN}/\text{mm}^2$ was then considered and the cyclic stationary state of the residual hoop stress along the radius of the cylinder can be seen in Figure 4(b). A dramatic drop in the number of iterations takes place if one uses a much bigger, fictitious, cycle time than the actual one something which we are allowed to do because of theorem 3. Identical results were obtained by either considering creep effects only or creep effects together with plasticity. More specifically the plasticity effects were present at the initial applications of cycles and eventually they disappeared. This occurred because the load is obviously below the shakedown load in the presence of creep. For load levels greater than this value, convergence within the specified tolerance started becoming tedious and definitely above 15.5 dN/mm$^2$ there was no convergence to a final steady state and stresses at some Gauss points kept repeating themselves, always exceeding the yield surface, indicating that this is approximately the value of the shakedown loading in the presence of creep. The modified shakedown loading $n/(n+1)P^{sh}$ is equal to 14.4 dN/mm$^2$ and is quite near this value and, although a little conservative, is on the safe side.

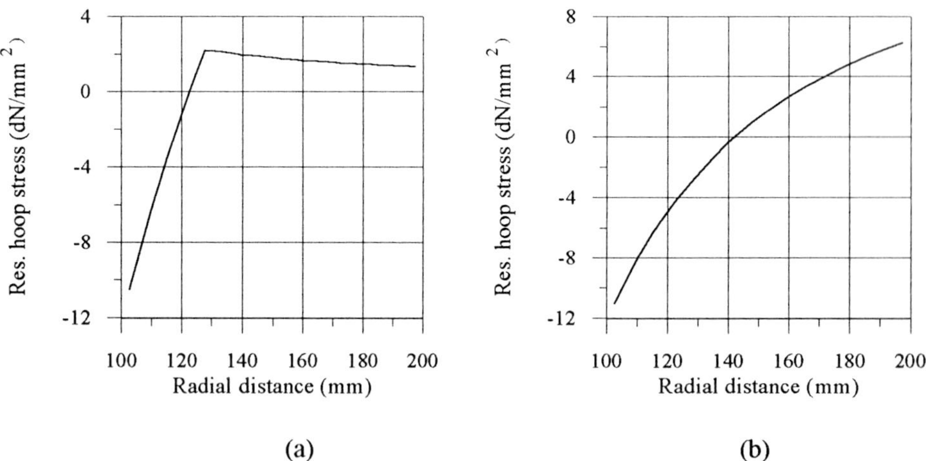


*Figure 4.* Distribution for (a) constant load pure elastoplastic material, (b) cyclic load creep or creep-plastic.

### 3.3.2 Thin Cylinder

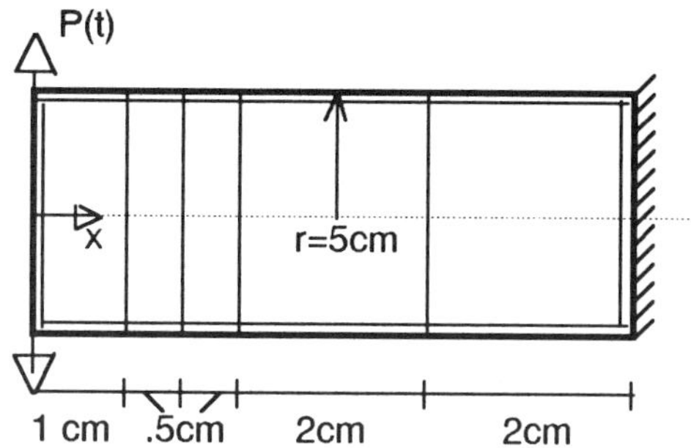


*Figure 5.* Geometry and finite element discretization of a thin cylinder

The next example of application is a thin cylinder (Figure 5), which is loaded by an axisymmetric edge cyclic loading. The structure was discretised using 24 conical frustra elements (Grafton and Strome [16]), with 20 of them from x=0 to x=1cm.

The thickness of the cylinder is h=0.1mm and the material data used: Young's modulus E= $10^7$ N/$cm^2$, Poisson's ratio $\mu = 0.3$, n=3.0, K=.636x$10^{-10}$ (SI units).

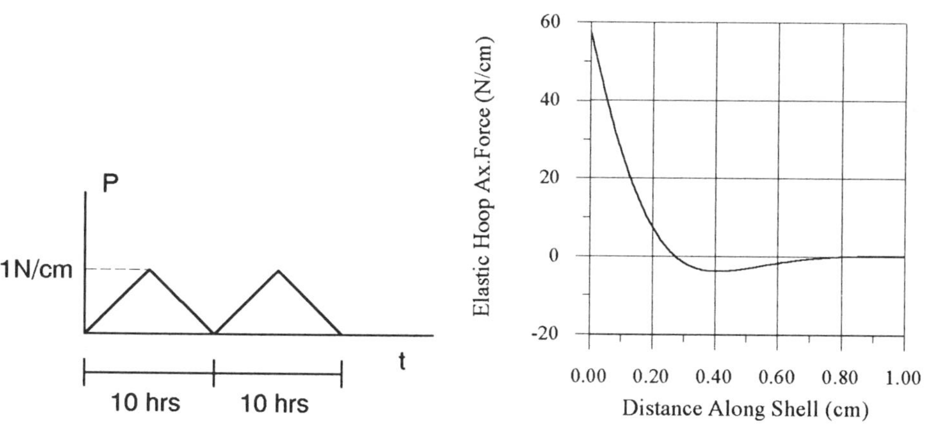


*Figure 6.* Variation of load with time and elastic results for the maximum value of load.

Because of the cyclic symmetry of the structure and the loading, the stress vector consists of two axial forces $N_s$ and $N_\theta$ and two bending moments $M_s$ and $M_\theta$. The elastic distribution of the hoop axial force along the length of the cylinder for the maximum value of the load is shown in Figure 6(b). It can be easily seen the effects of the load are localised and extend in a small region of the cylinder.

The following von Mises type equivalent force equation was used, in the formulation, for either the creep or the yield surface. This equation was used also in Cyras [17]:

$$N_s^2 - N_s N_\theta + N_\theta^2 + \frac{16}{h^2} M_s^2 - \frac{16}{h^2} M_s M_\theta + \frac{16}{h^2} = N_y^2 \tag{30}$$

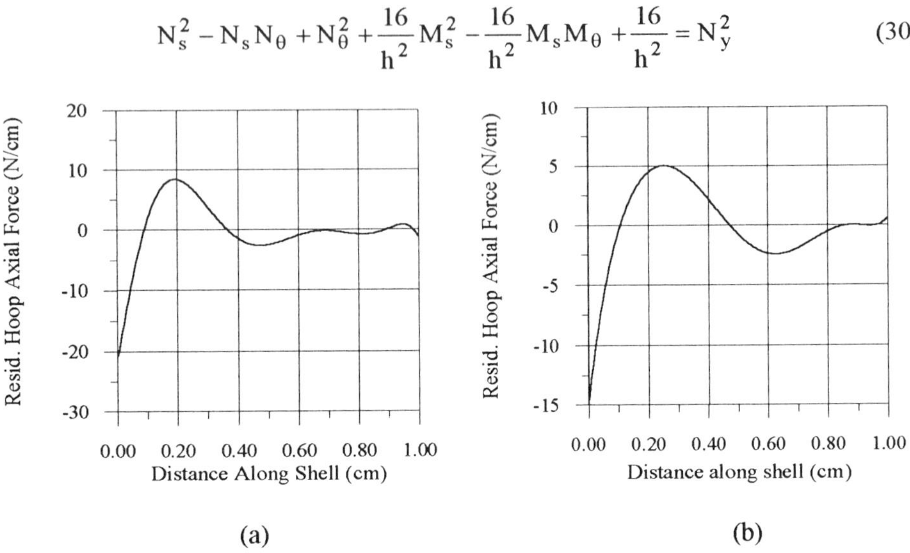


*Figure* 7. Steady state distribution for (a) constant load elastoplastic behaviour, (b) cyclic load creep or creep-plastic.

Similar results as with the previous example were obtained. In the left part of Figure 7 one may see the distribution of the residual hoop stress in the case of constant load equal to 1N/cm assuming pure elastoplastic behaviour with $N_y$= 36N/cm.

In Figure 7(b) the residual cyclic stationary hoop axial force distribution for the case of the cyclic loading of Figure 6 is shown when creep effects are considered. This level of loading combined with the yield force $N_y$=36N/cm, proves to be within the shakedown boundary in the presence of creep, since identical stress distributions are obtained if we include plasticity effects. These effects appear, as in the previous example, during the initial cycles and eventually disappear. This level of loading proves to be an approximate shakedown boundary in the presence of creep, since when the yield force $N_y$ was reduced, there was no convergence towards a final steady state solution within the specified tolerance and once again stresses kept repeating themselves exceeding the yield surface.

## 4. Cyclic loading of any period cycles

When the cycle period is not short one can not assume that the residual stress remains constant inside the cycle. Nevertheless, according to the theorem of Frederick and Armstrong [12] (section 2), the residual stress distribution of a structure subjected to cyclic loading of period T becomes also cyclic with time having the same period T as the applied loading, when the cyclic stationary state has been reached.

The cyclic behaviour of these residual stresses at the cyclic stationary stress state is the basis of a new simplified method that is developed in the remaining part of this work. The method concerns cyclic loads of any period cycles and is applied to a simple structure.

## 4.1 FOURIER ANALYSIS OF RESIDUAL STRESSES

It is known, from mathematics, that any periodic function may be represented by its Fourier series (see for example, Tolstov [18]).

We can write therefore:

$$\rho_{ij}(t) = \frac{a_o}{2} + \sum_{k=1}^{\infty} (a_k \cos\frac{2k\pi t}{T} + b_k \sin\frac{2k\pi t}{T}) \tag{31}$$

where the coefficients $a_o$, $a_k$ and $b_k$, k=1,2,... are called Fourier coefficients of the Fourier series. The main task is to evaluate these coefficients. In a classical Fourier analysis problem these coefficients can be evaluated when the function is known. In our case, though, it is this very function $\rho_{ij}(t)$ we seek to find.

Let us, however, differentiate with respect to time equation (31). Then we get:

$$\dot{\rho}_{ij}(t) = \frac{2\pi}{T}\sum_{k=1}^{\infty}\left\{(-ka_k)\sin\frac{2k\pi t}{T} + kb_k \cos\frac{2k\pi t}{T})\right\} \tag{32}$$

The above equation can be expanded to give:

$$\dot{\rho}_{ij}(t) = \frac{2\pi}{T}\left\{(-a_1)\sin\frac{2\pi t}{T} + (-2a_2)\sin\frac{4\pi t}{T} + (-3a_3)\frac{6\pi t}{T} + \ldots + (-ka_k)\sin\frac{2k\pi t}{T} + \right.$$

$$\left. + b_1\cos\frac{2\pi t}{T} + 2b_2\cos\frac{4\pi t}{T} + \ldots + kb_k\cos\frac{2k\pi t}{T} + \ldots\right\} \tag{32a}$$

a) Multiplying both terms of 32(a) by $\sin\frac{2k\pi t}{T}$ and integrating over the period T:

$$\frac{T}{2\pi}\int_0^T \dot{\rho}_{ij}\sin\frac{2k\pi t}{T}dt = (-a_1)\int_0^T \sin\frac{2\pi t}{T}\sin\frac{2k\pi t}{T}dt + (-2a_2)\int_0^T \sin\frac{4\pi t}{T}\sin\frac{2k\pi t}{T}dt +$$

$$+ \ldots + (-ka_k)\int_0^T \sin^2\frac{2k\pi t}{T}dt + \ldots + b_1\int_0^T \cos\frac{2\pi t}{T}\sin\frac{2k\pi t}{T}dt +$$

$$+ \ldots + (kb_k)\int_0^T \cos\frac{2k\pi t}{T}\sin\frac{2k\pi t}{T}dt + \ldots \tag{33}$$

The above equation may be greatly simplified if one makes use of the orthogonality relations between trigonometric functions:

$$\int_0^T \cos\frac{2k\pi t}{T}\sin\frac{2m\pi t}{T}dt = 0$$

$$\int_0^T \sin\frac{2k\pi t}{T}\sin\frac{2m\pi t}{T}dt = 0$$

$$\int_0^T \cos\frac{2k\pi t}{T}\cos\frac{2m\pi t}{T}dt = 0 \tag{34}$$

$$\int_0^T \sin^2\frac{2k\pi t}{T}dt = \int_0^T \cos^2\frac{2k\pi t}{T}dt = \frac{T}{2}$$

The only remaining term in (33) is the one concerning the $a_k$ coefficient and therefore equation (33) can be solved for $a_k$:

$$a_k = -\frac{1}{k\pi}\int_0^T \dot{\rho}_{ij}(t)\sin\frac{2k\pi t}{T}dt \tag{35}$$

b) Multiplying both terms of equation (32a) with $\cos\frac{2k\pi t}{T}$ and integrating over the period T:

$$\frac{T}{2\pi}\int_0^T \dot{\rho}_{ij}\cos\frac{2k\pi t}{T}dt = (-a_1)\int_0^T \sin\frac{2\pi t}{T}\cos\frac{2k\pi t}{T}dt + (-2a_2)\int_0^T \sin\frac{4\pi t}{T}\cos\frac{2k\pi t}{T}dt +$$

$$+\ldots+(-ka_k)\int_0^T \sin\frac{2k\pi t}{T}\cos\frac{2k\pi t}{T}dt + \ldots + b_1\int_0^T \cos\frac{2\pi t}{T}\cos\frac{2k\pi t}{T}dt +$$

$$+\ldots+(kb_k)\int_0^T \cos^2\frac{2k\pi t}{T}dt \tag{36}$$

Equation (36) may be greatly simplified, if we make use of equations (34), so that we can then solve for the coefficient $b_k$:

$$b_k = \frac{1}{k\pi}\int_0^T \dot{\rho}_{ij} \cos\frac{2k\pi t}{T} dt \tag{37}$$

c) If we integrate $\dot{\rho}_{ij}$ over the period T, we get:

$$\int_0^T \dot{\rho}_{ij} dt = \rho_{ij}(T) - \rho_{ij}(0) = \left(\frac{a_0(T)}{2} + \sum_{k=1}^{\infty} a_k(T)\right) - \left(\frac{a_0(0)}{2} + \sum_{k=1}^{\infty} a_k(0)\right) \tag{38}$$

where the expression (31) at the beginning and at the end of the period was used. The last equation may be used to evaluate the coefficient $a_0$.

By satisfying equilibrium and compatibility at an instant inside the cycle, an expression of $\dot{\rho}_{ij}$ can be established in terms of the residual stress $\rho_{ij}$, which in turn is expressed through its Fourier series. By performing numerical integration over a cycle, equations (35), (37) and (38) may be used in an iterative manner, to find the various Fourier coefficients. The approach will be better explained in the one-dimensional numerical example that follows.

## 4.2 NUMERICAL APPLICATION

An example of application of the method presented above is the three bar structure which is shown in Figure 8. The structure is subjected to a cyclic load P(t), which is applied at node 4. The amplitude of loading is assumed to be below the shakedown boundary in the presence of creep and therefore only creep effects are considered. All the members of the truss have equal cross section A and are made of the same material of modulus of elasticity E. This is a one-dimensional stress problem and the creep law is given by the equation:

$$\dot{\varepsilon}^{cr} = K\sigma^n \tag{39}$$

The stress states, in the two inclined bars, are identical, due to symmetry. If at an instant of time inside the cycle, one enables the one-dimensional equivalent to equations (2) and (3), and at the same time satisfies the conditions of equilibrium and compatibility, the following expression can be obtained for the time derivative of the residual stress in the inclined bars:

$$\dot{\rho}_1 = \frac{EK}{2+\sqrt{2}}\left\{\left(\frac{2}{2+\sqrt{2}}\frac{P(t)}{A} - \sqrt{2}\rho_1(t)\right)^n - 2\left(\frac{1}{2+\sqrt{2}}\frac{P(t)}{A} + \rho_1(t)\right)^n\right\} = C * f(t) \tag{40}$$

whereas from equilibrium considerations the residual stress in the vertical bar 2 is given by:

$$\rho_2 = -\sqrt{2}\rho_1 \tag{41}$$

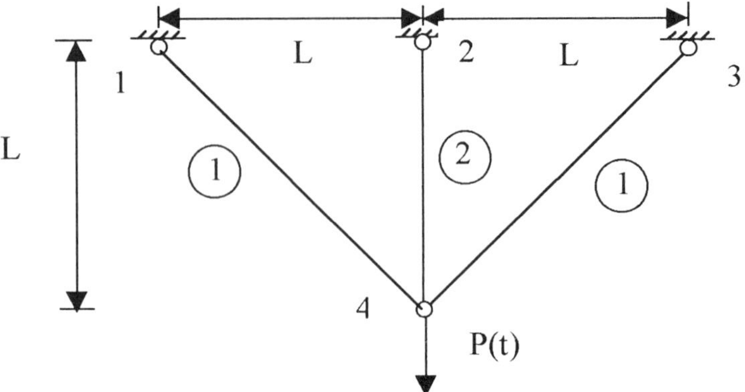


*Figure 8.* Three bar truss example

If we express $\rho_1(t)$ in (40) in terms of its Fourier series (equation (31)) we can get an iterative form for the Fourier coefficients using (35), (37) and (38):

$$a_k^{(\mu+1)} = -\frac{C}{k\pi}\int_0^T [f^{(\mu)}(t)]\sin\frac{2k\pi t}{T}\,dt$$

$$b_k^{(\mu+1)} = \frac{C}{k\pi}\int_0^T [f^{(\mu)}(t)]\cos\frac{2k\pi t}{T}\,dt \tag{42}$$

$$\frac{a_0^{(\mu+1)}}{2} = -\sum_{k=1}^{\infty} a_k^{(\mu+1)} + \frac{a_0^{(\mu)}}{2} + \sum_{k=1}^{\infty} a_k^{(\mu)} + \int_0^T C[f^{(\mu)}(t)]\,dt$$

where $(\mu+1)$ and $(\mu)$ denote the corresponding iterations and $f^{(\mu)}(t)$ is the value of the quantity in the bracket of (40) evaluated at iteration $\mu$.

An application has been made using a load variation of the form: $P(t) = 100\sin^2 \pi t/T$ (in kN), which is plotted in Figure 9 over four periods. The following material data were used: $K = .68*10^{-8}$ (SI units), $E = .2*10^5\,kN/cm^2$, n=3.0, $A = 1\,cm^2$.

Results for two different periods are reported in Figure 10. Complete agreement was observed with the results produced by orthodox time stepping procedures.

It can be seen (Fig. 10(a)) that the cycle period of 1hr is a "short" period, with the residual state of stress being more or less constant with time around the value of 9.50, whereas for the 100 hours period (Fig. 10(b)) the residual stress has quite a big variation with time.

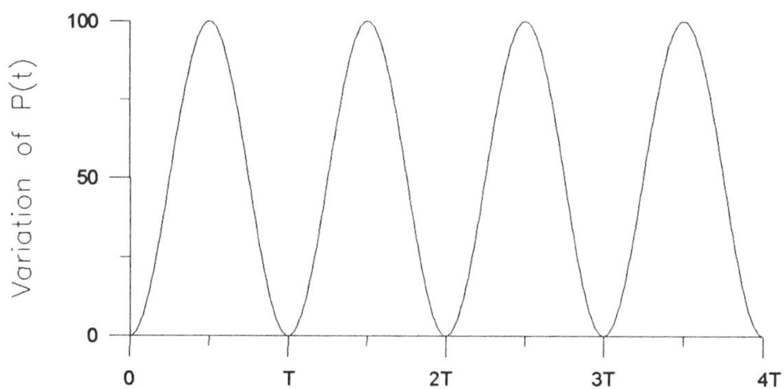


*Figure 9.* Load variation with time over four periods.

In order to calculate the various integrals of (42) numerical integration was used. It was found out that twenty integration points inside a cycle were enough to produce accurate results. In the case of the 100 hrs period not more than the first five terms of the Fourier series needed to be considered. Convergence to the final steady state was achieved within a few iterations.

The simplified method described above was developed in this paper for the general case of multi-axial residual stress. It may, therefore, be applied to any structure. At the time of writing of this paper, however, it was not possible to have results for the case of a general structure. In a forthcoming paper, the method will be applied to general structures and full details of the iterative numerical method used and its convergence characteristics will be given.

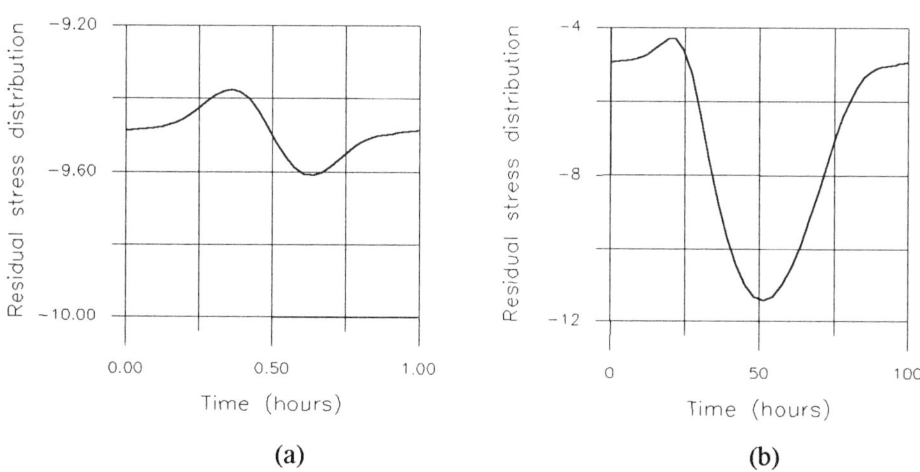


*Figure 10.* Cyclic state residual stress distribution (in $kN/cm^2$) in vertical member 2 for: (a) 1 hr period, (b) 100 hrs period.

## 5. Summary and conclusions

The use of simplified methods of inelastic analysis for structures loaded cyclically makes possible to estimate the steady state of stress without following cumbersome time stepping calculations.

In the first part of this work the problem of rapid cycling creep is considered in which one may assume a constant in time residual stress distribution. It is mathematically proved that under this assumption the iterative numerical procedure towards the final steady state can be greatly accelerated by choosing a fictitious much bigger period than the real one. It is also shown that plasticity effects may be added in the iterations in an easy way and thus establish whether the load will lead to shakedown in the presence of creep. Results of applications to two structures using finite elements are also presented.

In the second part of the present work a new simplified method is developed that may be used for a cyclic loading having a cycle of any period. The method is based on expressing the steady state residual stress distribution through Fourier series whose coefficients are calculated by iterations. It has very good convergence characteristics and although it is currently applied to a simple three-bar creeping structure, it has the potential of application to any structure and results will be reported shortly.

## 6. References

1. Melan, E. (1938) Zur Plastizität des räumlichen Kontinuums, *Ing. Archiv.* **8**, 116-126.
2. Leckie, F. A. and Ponter, A.R.S. (1970) Deformation bounds for bodies which creep in the plastic range, *J. Appl. Mech.* **37**, 426-430.
3. Leckie, F. A. and Ponter, A.R.S. (1972) Theoretical and experimental investigation of the relationship between plastic and creep deformation of structures, *Arch. Mechanics* **24**, 419-437.
4. Ponter, A.R.S. (1972) Deformation, displacement, and work bounds for structures in a state of creep and subject to variable loading, *J. Appl. Mech.* **39**, 953-958.
5. Ainsworth, R.A. (1977) Bounding solutions for creeping structures subjected to load variations above the shakedown limit, *Int. J. Solids Structures* **13**, 971-980.
6. Ponter, A.R.S. (1976) The analysis of cyclically loaded creeping structures for short cycle times, *Int. J. Solids Structures* **12**, 809-825.
7. Ponter, A.R.S. and Brown, P.R. (1978) The finite element solution of rapid cycling creep problems, *Int. J. Num. Meth. Engin.* **12**, 1001-1024.
8. Chan, A.S.L. and Spiliopoulos, K.V. (1987) A simplified method of solution for the short cycle creep-plasticity problem, *Comp. Meth. Appl. Mech. Engng* **60**, 257-274.
9. Spiliopoulos, K.V. (1993) Numerical implementation of smplified methods of inelastic analysis of structures subjected to short period loads, *Proc 12th SmiRT Conf.*, **L11/7**, 273-278.
10. Spiliopoulos, K.V. (1984) Estimation of accumulated creep deformation for structures subjected to cyclic change of loading in the plastic range, *Ph.D. Thesis*, University of London.
11. Drucker, D.C. (1959) A definition of stable inelastic material, *J. Appl. Mech.* **26**, 101-106.
12. Frederick, C. O. and Armstrong P.J. (1966) Convergent internal stresses and steady cyclic states of stress, *Journ. of Strain Analysis* **1**, 154-169.
13. Ponter, A.R.S. (1973) On the stress analysis of creeping structures subject to variable loading, *J. Appl. Mech.* **40**, 589-594.
14. Kraus, H (1980) *Creep analysis*, John Wiley & Sons, USA.
15. Lubliner , J (1990) *Plasticity theory*, Macmllan Publishing Company, USA.
16. Grafton, P.E. and Strome, D.R. (1963) Analysis of axisymmetrical shells by the direct stiffness method, *A.I.A.A Journ.* **1**, 2342-2347.
17. Cyras, A. A. (1983) *Mathematical models for the analysis and optimization of elastoplastic structures*, Ellis Horwood Ltd, England.
18. Tolstov, G.P. (1962) *Fourier series*, Dover Publications, New York.

# DIRECT FINITE ELEMENT KINEMATICAL APPROACHES IN LIMIT AND SHAKEDOWN ANALYSIS OF SHELLS AND ELBOWS

A.M. YAN and H. NGUYEN-DANG
*LTAS-University of Liège*
*Rue Ernest Solvay 21, 4000-Liège, Belgium*

**Abstract**

Theoretical and numerical methods of limit and shakedown analysis are presented for the applications to shell and pipe structures under mechanical and thermal loading. The kinematical theorem and its numerical implementation are developed. A method of simplified shakedown analysis is proposed. The numerical results for shell and pipe structures are compared with theoretical predictions.

## 1. Introduction

Structural integrity and cost of pipelines and pressure vessels are of major concern in the nuclear, oil and other industries. Limit and shakedown analysis constitutes a very useful tool for engineering design. Its fundamental theorems were established long time ago, see e.g. [9,13,14,16,17] etc. However, some numerical difficulties were encountered, especially in shakedown analysis with non-linear yield function. When many finite elements are used to discretize a pipe elbow structure, for example, one may face a large-scale optimisation problem and the computing cost may be high. On the other hand if sophisticated structural elements are used to discretize the plate, shell and pipe structures, improved computing precision and efficiency may be obtained. However, the solid finite elements should be used to study thermal loading through the thickness of structures. In classical kinematical shakedown analysis, linear programming techniques are applied using piece-wise linear (Tresca) or linearized yield criterion, e.g. [8,15,19] etc. The use of von Mises criterion leads to a non-linear mathematical programming problem. Its numerical application to practical engineering with complex structures and loading remains a challenge.

In this work, we present briefly the development of a modified upper bound limit and shakedown formula, which is applicable with any type of kinematically admissible finite elements. Some numerical methods of implementation are demonstrated. The temperature dependence of the yield limit of the material is considered. For the applications, three kinds of finite elements, such as pipe-elbow element, general shell element and usual quadratic solid element, are used in both limit and shakedown analysis. A simplified shakedown method is suggested in the case of pure mechanical loading. Various numerical applications to shell and pipe structures are illustrated to show the efficiency of the present methods.



*D. Weichert and G. Maier (eds.), Inelastic Analysis of Structures under Variable Loads,* 233–254.

## 2. A Modified Kinematical Formulation

### 2.1. LIMIT ANALYSIS

Limit analysis deals with proportional and monotonic loads (**f**: volume force, **t**: surface force). The objective is to determine the plastic limit loads of structures, beyond which the plastic collapse happens. The classical upper bound limit analysis was based on Markov's variational principle, cf. [22]. It states:

*Among all kinematically admissible and incompressible velocity fields* $\dot{\mathbf{u}}$, *the solution* $\dot{\mathbf{u}}'$ *corresponding to the limit state renders the following function an absolute minimum:*

$$\Pi(\dot{\mathbf{u}}) = \int_V D(\dot{\varepsilon}_{ij})\,\mathrm{d}V - \left(\int_V \mathbf{f}^{\mathrm{T}}\dot{\mathbf{u}}\,\mathrm{d}V + \int_{S_t} \mathbf{t}^{\mathrm{T}}\dot{\mathbf{u}}\,\mathrm{d}S\right) \tag{1}$$

For Saint-Venant-Levy-Mises material, we have the plastic dissipation function:

$$D(\dot{\varepsilon}_{ij}) = s_{ij}\dot{e}_{ij} = \sqrt{2}k_v(\dot{\varepsilon}_{ij}\dot{\varepsilon}_{ij})^{1/2} \tag{2}$$

where

$$k_v = \sigma_y/\sqrt{3}, \quad \dot{\varepsilon}_{ij} = \frac{1}{2}(\dot{u}_{i,j} + \dot{u}_{j,i})$$

where $s_{ij}, \dot{e}_{ij}, \dot{\varepsilon}_{ij}$ are stress deviator, strain rate deviator and strain rate tensor respectively; $\sigma_y$ is the yield limit of the material. It was noticed that the hypothesis of incompressibility, although it is reasonable for plastic deformation of metal, introduces some numerical difficulties [18]. In the case of using plane stress, plate and shell-type elements, this hypothesis can be naturally satisfied by using the Kirchhoff-Love hypothesis. However, it is not the case for usual solid finite elements such as plane strain, axisymmetric and 3-D elements. To overcome this difficulty Jiang [7] applied a regularisation method (Norton-Hoff algorithm). The incompressibility condition was imposed in a complementary constraint on strain variables. Liu *et al.* [12] used a penalty function algorithm in the optimisation process. In the present work, we introduce a fictitious volume strain power to provide a modification of Markov's variational principle [26] such that eq. (1) becomes:

$$\Pi(\dot{\mathbf{u}}) = \int_V (D(\dot{e}_{ij}) + \bar{k}\dot{e}_m^2)\,\mathrm{d}V - \left(\int_V \mathbf{f}^T\dot{\mathbf{u}}\,\mathrm{d}V + \int_{S_t} \mathbf{t}^T\dot{\mathbf{u}}\,\mathrm{d}S\right) \tag{3}$$

where

$$\bar{k} = \frac{\bar{K}}{2}, \qquad \bar{K} = \frac{\bar{E}}{3(1-2\nu)}$$

$\bar{E}$ is a fictitious linear viscous modulus; $\bar{K}$ the corresponding bulk modulus; $\nu$ Poisson's ratio, $\dot{e}_m$ volume strain rate. With this formulation the volume strain rate $\dot{e}_m$ no longer must be null. However, it approaches to zero after an optimisation process. Hence, the limit load solution is nearly independent of the fictitious material constant $\bar{k}$. On the other hand, $\bar{k}$ may be replaced by a Lagrange multiplier $\bar{\sigma}$, which may be physically identified as one half of hydrostatic pressure. So the modified Markov functional may be constructed by the following formulation

$$\Pi(\dot{\mathbf{u}},\overline{\sigma}) = \int_V (D^p(\dot{e}_{ij}) + \overline{\sigma}\dot{e}_m)\mathrm{d}V - (\int_V \overline{\mathbf{f}}^{\mathrm{T}}\dot{\mathbf{u}}\mathrm{d}V + \int_{S_t} \overline{\mathbf{t}}^{\mathrm{T}}\dot{\mathbf{u}}\mathrm{d}S) \tag{4}$$

Alternatively, $\bar{k}$ may also be taken as a penalty function coefficient. Consequently, the present method is numerically equivalent to the method by Liu *et al.* [12]. An upper bound formula of limit load multiplier can be deduced directly from the modified Markov variational functional (3):

$$\alpha_l = \min \alpha^+, \qquad \alpha^+ = W_{in} \tag{5a}$$

with

$$W_{in} = \int_V \left[D^p(\dot{e}_{ij}) + \bar{k}\dot{e}_m^2\right]\mathrm{d}V = \int_V \left[\sqrt{2}k_v(\dot{e}_{ij}\dot{e}_{ij})^{1/2} + \bar{k}\dot{e}_m^2\right]\mathrm{d}V \tag{5b}$$

$$W_{ex} = \int_V \overline{\mathbf{f}}_0^T\dot{\mathbf{u}}\mathrm{d}V + \int_{S_t} \overline{\mathbf{t}}_0^T\dot{\mathbf{u}}\mathrm{d}S = 1 \tag{5c}$$

$W_{in}$ represents the internally dissipated deformation energy rate; $W_{ex}$ the external load power. Therefore, the calculation of the limit load multiplier becomes a minimisation problem of a functional.

This general formulation of limit analysis allows us to use any kinematical finite elements with the following principal terms

- Displacement rate vector: $\dot{\mathbf{u}} = \mathbf{N}\dot{\mathbf{q}}_e$ (6)
- Strain rate vector: $\dot{\boldsymbol{\varepsilon}} = \mathbf{B}\dot{\mathbf{q}}_e$ (7)
- Second invariant of strain rate deviator:

$$\begin{aligned} J_2(\dot{e}_{ij}) &= \frac{1}{2}\dot{e}_{ij}\dot{e}_{ij} = \frac{1}{2}\dot{\varepsilon}_{ij}\dot{\varepsilon}_{ij} - \frac{1}{6}\dot{e}_m^2 = \dot{\boldsymbol{\varepsilon}}^T\overline{\mathbf{D}}_p\dot{\boldsymbol{\varepsilon}} - \dot{\boldsymbol{\varepsilon}}^T\overline{\mathbf{D}}_v\dot{\boldsymbol{\varepsilon}} \\ &= (\mathbf{B}\dot{\mathbf{q}}_e)^T\overline{\mathbf{D}}_p\mathbf{B}\dot{\mathbf{q}}_e - (\mathbf{B}\dot{\mathbf{q}}_e)^T\overline{\mathbf{D}}_v\mathbf{B}\dot{\mathbf{q}}_e = \dot{\mathbf{q}}_e^T\mathbf{D}_p\dot{\mathbf{q}}_e - \dot{\mathbf{q}}_e^T\mathbf{D}_v\dot{\mathbf{q}}_e \end{aligned} \tag{8}$$

with $$\mathbf{D}_p = \mathbf{B}^T\overline{\mathbf{D}}_p\mathbf{B}, \quad \mathbf{D}_v = \mathbf{B}^T\overline{\mathbf{D}}_v\mathbf{B} \tag{9}$$

Where **N**, **B**, $\dot{\mathbf{q}}_e$, $\overline{\mathbf{D}}_p$ and $\overline{\mathbf{D}}_v$ are respectively the interpolation matrix, the strain matrix, the elementary nodal velocity vector and the coefficient matrices. Therefore, (5b-5c) can be written in the following matrix form by assembling all elements of the structure:

$$W_{in} = \sum_e \int_{V_e} \left(2k_v\left[\dot{\mathbf{q}}_e^T\mathbf{D}_p\dot{\mathbf{q}}_e - \dot{\mathbf{q}}_e^T\mathbf{D}_v\dot{\mathbf{q}}_e\right]^{1/2} + \bar{k}\dot{\mathbf{q}}_e^T\mathbf{D}_v\dot{\mathbf{q}}_e\right)dV \tag{10}$$

$$W_{ex} = \sum_e\left[\int_{V_e}(\mathbf{N}\dot{\mathbf{q}}_e)^T\overline{\mathbf{f}}_0 dV + \int_{S_{et}}(\overline{\mathbf{N}}\dot{\mathbf{q}}_e)^T\mathbf{t}_0 dS\right] = \mathbf{g}^T\dot{\mathbf{q}} \tag{11}$$

$$\dot{\mathbf{q}}_e = \mathbf{L}_e\dot{\mathbf{q}}, \qquad \mathbf{g} = \sum_e \mathbf{L}_e\mathbf{g}_e \tag{12}$$

where $\dot{\mathbf{q}}$ and $\mathbf{g}$ are respectively the global node-velocity vector and the corresponding load vector; $\mathbf{g}_e$ is the elementary load vector; $\mathbf{L}_e$ is the matrix of localisation.

## 2.2. SHAKEDOWN ANALYSIS

Now we consider a general case where $n$ loads vary independently in certain ranges to form a convex loading domain $\mathcal{D}$. This domain is characterised by $m$ loading vertices ($m$=$2^n$)

$$\overline{P}_k^0 \in I_k^0 = \left[\overline{P}_k^-, \overline{P}_k^+\right] = \left[\mu_k^-, \mu_k^+\right] P_k^0 \qquad k=1, n \tag{13}$$

where $P_k^0$ is $k$-th nominal load, and $\mu_k^-, \mu_k^+$ are the lower and upper bounds of variation.

The objective is to find a shakedown limit $\alpha_s$ multiplying $\overline{P}_k^0$ as $\alpha_s \overline{P}_k^0$, such that the plastic deformation that occurred in the first loading cycles will cease to develop and the structure returns to an elastic behaviour. This shakedown limit can be determined by a direct upper bound analysis basing on a kinematical theorem of Koiter. It may be demonstrated that Koiter's shakedown criterion may be obtained by an integration of Markov's limit criterion over a time cycle $\tau$. So the above modification of Markov's theorem may lead to a modified Koiter's formulation [24]:

For a given loading domain $\mathcal{D}$ leading to the fictitious elastic solution $\sigma_{ij}^*$, the shakedown limit $\alpha_s$ is the minimum value defined by (14b)

$$\alpha_s = \min W_{in} \tag{14a}$$

$$W_{in} = \int_\tau \mathrm{d}t \int_V \left(\sqrt{2} k_v (\dot{e}_{ij} \dot{e}_{ij})^{1/2} + \bar{k} \dot{e}_m^2 \right) \mathrm{d}V \tag{14b}$$

$$W_{ex} = \int_\tau dt \int_V \sigma_{ij}^* \dot{\varepsilon}_{ij}^p \, \mathrm{d}V = 1 \tag{14c}$$

$$\Delta\varepsilon_{ij} = \int_\tau \dot{\varepsilon}_{ij}^p \, \mathrm{d}t \quad \text{is kinematically admissible} \tag{14d}$$

We note that when the loading domain has only one vertex, this formula degenerates into a limit analysis related to the modified Markov's theorem by cancelling the integration on time. On the other hand, as long as the varying temperature field exists, it should be expressed as thermal elastic stresses. If the yield limit of the material is considered as temperature-dependent, we write $k_v = k_v^t$ which changes during the loading cycle. The finite element discretization of (14) has a similar form as limit analysis but with an integral over time. In Section 3, we will present some practical numerical methods of shakedown analysis.

# 3. Numerical Methods of Shakedown Analysis

## 3.1 UNIFIED SHAKEDOWN LIMIT METHOD (USL)

It is to find directly the shakedown limit $\alpha_s$ defined by the minimum of incremental plasticity limit $\alpha_i$ and alternating plasticity limit $\alpha_f$, such as $\alpha_s = \min(\alpha_i, \alpha_f)$. The method is based on the following convex-cycle theorems [10]:

- **Theorem 1**

  *Shakedown will happen for a loading domain $\mathcal{D}$ if and only if it occurs on the convex envelope of $\mathcal{D}$.*

- **Theorem 2**

  *Shakedown will happen for an arbitrarily varying load path within the domain $\mathcal{D}$ if it happens for a loading cycle passing through all vertices $\hat{P}_k$ of the frontier of $\mathcal{D}$.*

According to these two theorems, we consider a cyclic loading instead of an arbitrarily loading history and consider the stress and strain rate fields only at each vertex of the loading domain instead of integration over the time cycle. Fig. 1 shows an example of loading path in the case of three independently varying loads. We note that the path order (to pass the all vertices) has no importance. A continuous loading cycle is hence discretized a set of loading vertices:

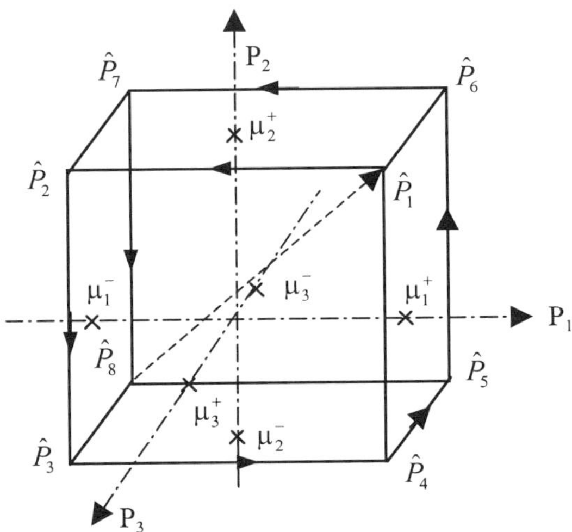


*Fig.1.* Critical cycles of load for shakedown analysis (Three varying loads)

$$\mathbf{P}(\mathbf{x},t) = \sum_{k=1}^{m} \delta(t_k)\hat{\mathbf{P}}_k(\mathbf{x}), \qquad \delta(t_k) = \begin{cases} 1 & \text{if} \quad t = t_k \\ 0 & \text{if} \quad t \neq t_k \end{cases}, \qquad \sum_{k=1}^{m} t_k = \tau \tag{15}$$

where $m=2^n$ is the number of vertices of load domain; $n$ the number of varying loads. Correspondingly, stresses and strains are represented by:

$$\sigma_{ij}^{*}(t) = \sum_{k=1}^{m} \delta(t_k)\sigma_{ij}^{k*}, \qquad \dot{\varepsilon}_{ij}(t) = \sum_{k=1}^{m} \delta(t_k)\dot{\varepsilon}_{ij}^{k} \tag{16}$$

At each instant (load vertex), the kinematical condition may not be satisfied. However, the accumulated strain in a load cycle:

$$\Delta\varepsilon_{ij} = \sum_{k=1}^{m} \dot{\varepsilon}_{ij}^{k} \tag{17}$$

is compatible in the sense of Koiter.

By writing the applied loads (including thermal loads) in their elastic solution $\sigma_{ij}^{*}$ and discretizing into finite elements as (6-9), we obtain an applicable upper bound formulation of shakedown limit:

$$\alpha_s = \min \alpha^{+}, \quad \alpha^{+} = W_{in} \tag{18a}$$

$$\text{s.t.} \quad W_{ex} = 1 \tag{18b}$$

$$\sum_{k=1}^{m} \dot{\mathbf{q}}_e^k = \mathbf{L}_e \Delta\dot{\mathbf{q}} \tag{18c}$$

where

$$W_{in} = \sum_{k}^{m} \sum_{e}^{n_e} \int_{V_e} \left( \frac{2\sigma_p^k}{\sqrt{3}} \left[ (\dot{\mathbf{q}}_e^k)^T \mathbf{D}_p \dot{\mathbf{q}}_e^k - (\dot{\mathbf{q}}_e^k)^T \mathbf{D}_v \dot{\mathbf{q}}_e^k \right]^{1/2} + \bar{k} (\dot{\mathbf{q}}_e^k)^T \mathbf{D}_v \dot{\mathbf{q}}_e^k \right) \mathrm{d}V \tag{18d}$$

$$W_{ex} = \sum_{k}^{m} \sum_{e}^{n_e} (\mathbf{g}_e^k)^T \mathbf{q}_e^k \tag{18e}$$

$$\mathbf{g}_e^k = \int_{V_e} \mathbf{B}^T \sigma^{k*} \, \mathrm{d}V \tag{18f}$$

where $\dot{\mathbf{q}}_e^k$ and $\mathbf{g}_e^k$ are, respectively, the nodal velocity and load vector of element relative to load vertex $k$. $\sigma_y^k$ is the yield limit of material at $k$-th actual load-temperature point. With this formula, the temperature-dependence of yield limit can be examined. It should be pointed out that the direct use of (18) poses a disadvantage of increasing the number of variables in comparison with limit analysis. So an effective numerical method is significant to reduce the problem size, which will be illustrated by a separated paper.

### 3.2. SEPARATED SHAKEDOWN LIMIT METHOD (SSL)

It is well known that incremental plasticity (ratchetting) and alternating plasticity (plastic fatigue) are two different failure modes. The former behaves usually as global plastic flow, while the latter occurs often as a local alternating plastic phenomenon. It is useful in practice to distinct these two different failure modes and the corresponding loading limits from each other. For this, we split any plastic strain history $\dot{\varepsilon}_{ij}^{p}(\mathbf{x},t)$, which leads to a kinematically admissible plastic strain increment after a periodic interval (0,τ), into its two components [11].

$$\dot{\varepsilon}_{ij}^{p}(\mathbf{x},t) = \dot{\bar{\varepsilon}}_{ij}(\mathbf{x},t) + \dot{\hat{\varepsilon}}_{ij}(\mathbf{x},t) \tag{19}$$

- The first term represents *a perfectly incremental collapse process*, in which a kinematically admissible plastic strain increment is accumulated in a monotonic way:

$$\dot{\bar{\varepsilon}}_{ij}(\mathbf{x},t)=\dot{\Lambda}(\mathbf{x},t).\Delta\bar{\varepsilon}_{ij}(\mathbf{x}) \tag{20a}$$

$$\Delta\bar{\varepsilon}_{ij}=\bar{\varepsilon}_{ij}(\mathbf{x},\tau)-\bar{\varepsilon}_{ij}(\mathbf{x},0) \tag{20b}$$

$$\dot{\Lambda}(\mathbf{x},t)\geq 0\,,\qquad \int_0^\tau \dot{\Lambda}(\mathbf{x},t)\,\mathrm{d}t=1 \tag{20c}$$

- The second term represents *an alternating plasticity process,* defined by (19) and

$$\Delta\hat{\varepsilon}_{ij}(\mathbf{x})=\int_0^\tau \dot{\hat{\varepsilon}}(\mathbf{x},t)\,\mathrm{d}t=0 \tag{21}$$

This alternating plasticity process should correspond to an alternating stress process for isotropic material.

Based on the above idea presented in [11], we formulate and implement two separate shakedown limit criteria with the modified Koiter theorem.

3.2.1. *Criterion for incremental collapse*

It appears that the smallest upper bound of incremental limit could be attained when the external power (14c) attains its maximum and the internal dissipation (14b) takes a minimum. To this end, the function $\dot{\Lambda}(\mathbf{x},t)$ is selected in such a way that $\dot{\Lambda}(\mathbf{x},t)\neq 0$ *only when the product* $\sigma^*_{ij}(\mathbf{x},t)\Delta\bar{\varepsilon}_{ij}(\mathbf{x})$ *takes its maximum value for a given load domain* $\mathcal{D}$. So the external power (14c) may be written as below:

$$W_{ex}=\int_\tau \mathrm{d}t\int_V \sigma^*_{ij}(\mathbf{x},t)\dot{\Lambda}(\mathbf{x},t)\Delta\bar{\varepsilon}_{ij}(\mathbf{x})\,\mathrm{d}V=\int_V \bar{\sigma}^*_{ij}(\mathbf{x})\Delta\bar{\varepsilon}_{ij}(\mathbf{x})\mathrm{d}V \tag{22a}$$

where

$$\bar{\sigma}^*_{ij}(\mathbf{x})\Delta\bar{\varepsilon}_{ij}(\mathbf{x})=\max\left(\sigma^*_{ij}(\mathbf{x},t)\Delta\bar{\varepsilon}_{ij}(\mathbf{x})\right) \tag{22b}$$

Owing to the above special choice of $\dot{\Lambda}(\mathbf{x},t)$, the internal energy dissipation (14b) over a time period is written as

$$W_{in}=\int_\tau \mathrm{d}t\int_V\left(\dot{\Lambda}(\mathbf{x},t)D^p(\Delta\bar{\varepsilon}^p_{ij})+\bar{k}\dot{\Lambda}^2(\mathbf{x},t)(\Delta e_m)^2\right)\mathrm{d}V=\int_V\left(D^p(\Delta\bar{\varepsilon}^p_{ij})+\bar{k}(\Delta e_m)^2\right)\mathrm{d}V \tag{23}$$

If the loading domain is prescribed by (13), namely $n$ independently varying loads, (22) can be written in a practical form (assuming that $\sigma^{k*}_{ij}=\sigma^*_{ij}(P^0_k)$ is elastic solution for $k$-th nominal load $P^0_k$):

$$W_{ex}=\int_V\sum_{k=1}^n \bar{\mu}_k\sigma^{k*}_{ij}(\mathbf{x})\Delta\bar{\varepsilon}_{ij}(\mathbf{x})\,\mathrm{d}V \tag{24a}$$

$$\bar{\mu}_k=\begin{cases}\mu^+_k & \text{if}\quad \sigma^{k*}_{ij}(\mathbf{x})\Delta\bar{\varepsilon}_{ij}(\mathbf{x})\geq 0\\ \mu^-_k & \text{if}\quad \sigma^{k*}_{ij}(\mathbf{x})\Delta\bar{\varepsilon}_{ij}(\mathbf{x})<0\end{cases} \tag{24b}$$

We use the following finite element discretization:

$$\Delta\bar{\varepsilon} = \mathbf{B}\mathbf{q}_e, \qquad \mathbf{g}_e^k = \int_{V_e} \bar{\mu}_k \mathbf{B}^T \sigma^{k*} \,\mathrm{d}V \tag{25}$$

The incremental limit will be found by a mathematical programming process:

$$\alpha_i^+ = \min W_{in} \tag{26a}$$

$$\text{s.t.} \qquad W_{ex} = 1 \tag{26b}$$

where

$$W_{in} = \sum_{e=1}^{n_e} \int_{V_e} \left[ 2k_v \left( \mathbf{q}_e^T \mathbf{D}_p \mathbf{q}_e - \mathbf{q}_e^T \mathbf{D}_v \mathbf{q}_e \right)^{1/2} + \bar{k} \mathbf{q}_e^T \mathbf{D}_v \mathbf{q}_e \right] \mathrm{d}V \tag{26c}$$

$$W_{ex} = \sum_{k=1}^{n} \sum_{e=1}^{n_e} (\mathbf{g}_e^k)^T \mathbf{q}_e \tag{26d}$$

$$\mathbf{q}_e = \mathbf{L}_e \mathbf{q} \tag{26e}$$

$\bar{\mu}_k$ in (25) is defined by (24b). Since $\bar{\mu}_k$ may vary during the optimisation process, (26) constitutes a non-linear objective function with a non-stable constraint. $k_v$ should correspond to the actual temperature at each integration point if it is considered temperature dependent. This method shows an advantage of having fewer variables to optimise (independent of the number of varying loads). It is approximately equivalent to limit analysis. The variable constraint generally does not pose numerical difficulty.

### 3.2.2. *Criterion for alternating plasticity*

Starting from Koiter's theorem and the definition of alternating plasticity (21), the structural safety factor against the alternating plasticity may be obtained by the *minimisation* of local $\alpha_f(\mathbf{x})$ defined by the following equations [11]:

$$\frac{1}{\alpha_f(\mathbf{x})} = \max_{\dot{\varepsilon}_{ij}^p} \int_0^\tau \left[ \sigma_{ij}^*(\mathbf{x},t) \dot{\varepsilon}_{ij}^p(\mathbf{x},t) \right] \mathrm{d}t \tag{27a}$$

$$\text{s. t.} \quad \int_0^\tau D^p(\dot{\varepsilon}_{ij}^p) \,\mathrm{d}t = 1 \tag{27b}$$

$$\int_0^\tau \dot{\varepsilon}_{ij}^p(\mathbf{x},t) \,\mathrm{d}t = 0 \tag{27c}$$

Here we consider only the alternating part of the plastic strain rate ($\dot{\hat{\varepsilon}}_{ij}$ is simply written as $\dot{\varepsilon}_{ij}^p$). By using Lagrange multipliers $\varphi$ and $\rho_{ij}$ (residual stress field), (27) is transformed into an unconstrained maximisation problem:

$$\frac{1}{\alpha_f^+} = \max_{\dot{\varepsilon}_{ij}^p, \varphi, \rho_{ij}} \int_0^\tau \left( \sigma_{ij}^* \dot{\varepsilon}_{ij}^p + \varphi[D^p(\dot{\varepsilon}_{ij}^p) - \frac{1}{\tau}] + \rho_{ij} \dot{\varepsilon}_{ij}^p \right) \mathrm{d}t \tag{28}$$

By taking its stationary condition and using the alternating property of the stress fields corresponding to an alternating strain rate, the alternating plasticity limit may be finally represented as follows:

$$\alpha_f = \min_{\mathbf{x}} \frac{1}{F\left(\sum_{k=1}^{n} \mu_k \sigma_{ij}^{k0*}(\mathbf{x})\right)} , \qquad |\mu_k| = \frac{\mu_k^+ - \mu_k^-}{2} \tag{29}$$

where $F = F(\sigma_{eq} / \sigma_y)$ represents Mises equivalent stress function. $\sigma_{ij}^{k0*}$ is the elastic response to $k$-th nominal load $P_k^0$ of (13). The sign of $\mu_k$ should be fixed such that it renders maximum the value of function $F$. So we see that a direct use of shakedown theorem appears unnecessary to find the alternating limit. In fact, (29) means that the plastic fatigue limit is determined by the condition that anywhere in the structure, the maximum varying magnitude of fictitious elastic equivalent stress $\Delta\sigma_{eq}$ can not exceed two times the yield limit of the material. The calculation may be based only on an elastic analysis.

It is easy to see from (29) that constant (or monotonic) loads have no influence on the plastic fatigue limit. Pycko & Mroz [21] gave also a proof for this conclusion through a so-called "min-max" shakedown analysis. If the mechanical loads are proportional to temperature variation and the yield limit of the material is considered as temperature-dependent, (29) may be represented in another simple form

$$\alpha_f = \frac{\sigma_y^{t_1} + \sigma_y^{t_2}}{\max_{\mathbf{x}}(\Delta\sigma_{eq})} \tag{30}$$

where $\sigma_y^{t_1}$ and $\sigma_y^{t_2}$ are the values of the yield point corresponding to the actual temperature at the beginning and at the end of the half-cycle.

### 3.3 SIMPLIFIED SHAKEDOWN ANALYSIS (SSA)

Since the alternating plasticity limit can generally be found on the basis of elastic analysis, we consider here only the incremental plasticity limit. It is generally known that shakedown analysis is more expensive than limit analysis. By a direct use of the USL method for example, the number of variables increases rapidly with varying-load number $n$ by $2^n$. Simplified methods are therefore practically meaningful.

As we know, the instantaneous collapse generally takes place at a somewhat larger load than the incremental limit. However this situation does not always happen. The real incremental plasticity occurs only when the necessary conditions, as stated by Gokhfeld & Cherniavsky [4], is fulfilled to ensure that the *nonisochronism* (nonsimultaneity) does in fact take place, i.e. the plastic yielding does not commence in *different zones* at the same time point of loading cycle. In particular, the incremental collapse is absent if during loading cycle the internal actions computed elastically are of the *same sign* in the all elements of structure. In this situation, the structural plastic collapse happens only in the form of instantaneous plasticity. Although it is difficult in some cases to justify if really incremental collapse occurs, Ponter [20] pointed out that "*no pressure vessel problem has been shown to suffer incremental collapse below the limit load*" under proportional mechanical loading.

However some non-proportional loading problems do exhibit incremental collapse. For pressure vessel applications without thermal loads, such possibilities may be ignored. Currently, we could propose the following assumptions:

- **Assumptions**

*1) If every load (varying independently) causes a uniform stress distribution, the incremental limit coincides with the minimum instantaneous limit corresponding to the most unfavourable loading vertex.*

*2) If some of the loads cause non-uniform stress distribution, the limit analysis corresponding to the most unfavourable loading vertex generates generally an upper bound approximation of the incremental limit.*

Assumption 1 is clearly justified because the defined stress field under the considered loading modes can not satisfy the incremental plasticity condition of Gokhfeld & Cherniavsky. For assumption 2, we suppose that the difference between the instantaneous limit corresponding to the most unfavourable loading vertex and the incremental limit depend on the stress distribution of loads. If no large concentration of elastic stress (or strain) exists, the incremental plasticity may be approximated by a special limit analysis. In this case we use a practical Simplified Shakedown Analysis (SSA) as described as follow.

Consider a structure subjected to $n$ mechanical loads varying independently with $m = 2^n$ loading vertices. Instead of using (18), we carry out some separate limit analyses corresponding to $m$ load vertices to obtain $m$ limit load factors $\alpha_k^+$, ($k$=1, $m$). The minimum of them is an upper bound of or equal to the real incremental limit factor $\alpha_i$. We present this method in the following formulation:

$$\alpha_i \le \min_k \alpha_k^+ \tag{31a}$$

$$\alpha_k^+ = \min W_{in}^k\,, \quad W_{in}^k = \int_V \left[ \sigma_y^k \left(\frac{2}{3}\dot{e}_{ij}^k \dot{e}_{ij}^k\right)^{1/2} + \bar{k}(\dot{e}_m^k)^2 \right] \mathrm{d}V \tag{31b}$$

$$\text{s.t.} \quad W_{ex}^k = 1\,, \quad W_{ex}^k = \int_V \sigma_{ij}^{*k} \dot{\varepsilon}_{ij}^k \,\mathrm{d}V \tag{31c}$$

$$\dot{\varepsilon}_{ij}^k \text{ is kinematically admissible}$$

This method consists of finding a combined loading mode or loading vertex that leads to the smallest limit factor by a usual limit analysis. This solution is equal or close to the incremental limit of the structure according to the above assumptions. In many practical cases, the most dangerous vertex of loading domain is obvious. The computation by this method may be completed by only one limit analysis. In some other cases, we need to do several independent but similar computations. So the shakedown analysis for some complex problems is simplified.

## 4. Numerical Application

The presented kinematical formulation is general and can be implemented with any kinematical finite element. The computing code ELSA has been developed for this purpose. For the present pipe and shell structure applications, we use the following three types of elements: 1) A new pipe elbow element developed on basis of the previous work [25]; 2) A general shell element using generalised variables for an axisymmetric shell structure under symmetric or non symmetric loading; 3) A solid axisymmetric element for shakedown analysis of thermal loading through the thickness. In this section, we present some typical limit and shakedown application to various pipe and shell structures. Numerical results are compared with analytic solutions found in the literature or deduced by ourselves.

## 4.1. PLASTIC LIMIT PRESSURE OF A TORISPHERICAL VESSEL HEAD

A pressure vessel head consists of a part of sphere, a spherical torus and a cylinder. Two types of axisymmetric elements are used: 1) solid elements as Fig. 2; and 2) shell elements with similar discretization as solid ones. The limit analysis solutions are compared in Table 1 with the incremental solution obtained with SAMCEF software.

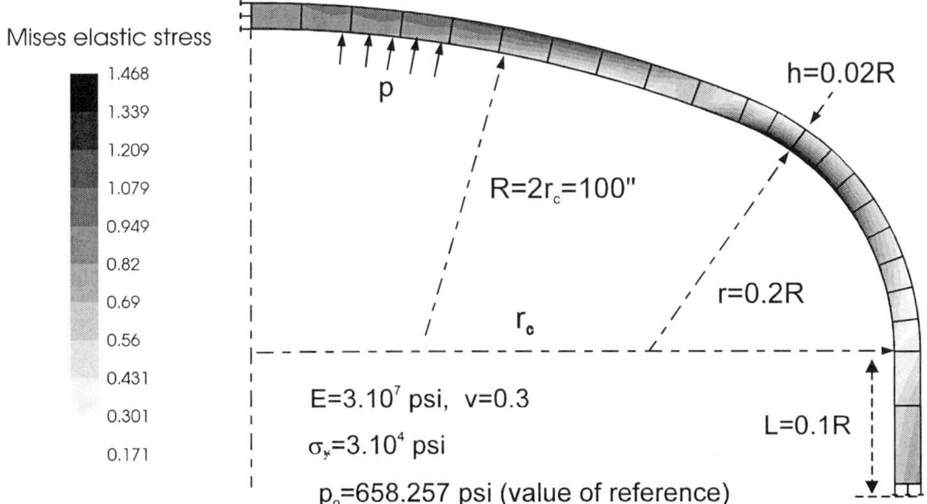


*Fig. 2.* Torispherical vessel head under internal pressure, finite element meshing and Mises elastic stress field (*R* and *r* are the mean radii)

Although the maximum elastic stress is at the joint between the sphere and the torus as shown in Fig. 2, the plastic limit pressure of the structure may be approximated by the limit solution of the sphere part: $p_l = 2\sigma_y \ln(R_o / R_i)$, where $R_i$ and $R_o$ are internal and external radii of the sphere, respectively. The length of the cylindrical part is shown to have no obvious influence on limit load solutions by numerical tests. Using solid elements, the obtained results change a little with Gauss integral-point series. According to experience in calculation, using an under-integration 2×2 gives generally accurate results. The results obtained by shell element with generalised variables approaches to that by solid elements with over- (or just) integration. On the other hand, concerning shakedown analysis when the pressure varies, the incremental plasticity and alternating plasticity will happen in different regions. The latter happens in the joint of the sphere and the torus. So we know that these two types of loading: increasing monotonically or varying arbitrarily, may sometimes lead to different failure modes in the different parts of structures.

Table 1. Plastic limit pressure of a torispherical vessel head, $p_o$=658.257

| $\alpha_l = p_l / p_o$ | Integral order | Number of elements | Method |
|---|---|---|---|
| 1.792 | 2×2 | 272 quadratic elements | "Step by step" by SAMCEF |
| 1.924 | 3×3 | | |
| 1.931 | 4×4 | | |
| 1.790 | 2×2 | 22 quadratic elements | Direct analysis by ELSA |
| 1.805 | 3×3 | | |
| 1.813 | 4×4 | | |
| 1.835 | 7×1 | 22 shell elements | |
| 1.823 | - | - | Sphere's solution |

4.2. LIMIT INTERACTION OF CYLINDRICAL PIPE UNDER COMBINED LOADING

A thin-walled straight pipe is subjected to a combination of a bending moment $M$, internal pressure $p$ and an axial force $F$, Fig. 3.

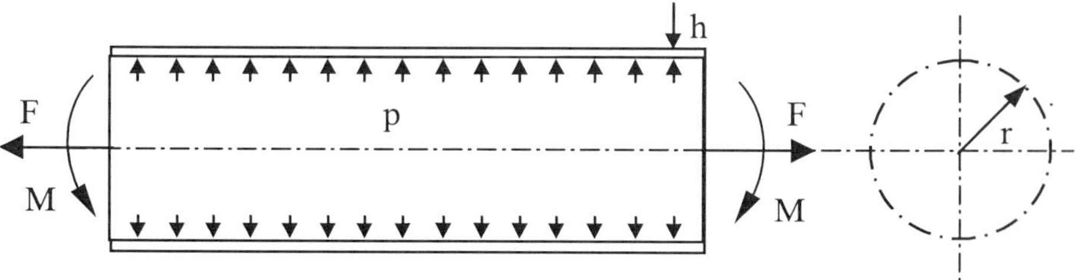


*Fig. 3.* Straight pipe under bending, internal pressure and axial forces

We define the following non-dimensional load parameters:

$$m = M/M_l\,,\ M_l = 4r^2h\sigma_y\,;\quad n_\varphi = p/p_l\,,\ p_l = \sigma_y h/r\,;\quad n_x = F/F_l\,,\ F_l = 2\pi\sigma_y rh \qquad (32)$$

Using a static model, it is easy to find two analytic solutions as (33-34) with, respectively, Tresca and von Mises criteria. We compare solution (33) with the numerical results in Fig. 4, showing a good agreement.

- **Analytic solution with von Mises' criterion:**

$$m = \frac{\sqrt{4-3n_\varphi^2}}{2}\cos\left[\frac{n_\varphi - 2n_x}{\sqrt{4-3n_\varphi^2}}\frac{\pi}{2}\right] \qquad (33)$$

- **Analytic solution with Tresca's criterion:**

$$m = \frac{2 - n_\varphi}{2} \cos\left[\frac{n_\varphi - 2n_x}{2 - n_\varphi} \frac{\pi}{2}\right] \tag{34}$$

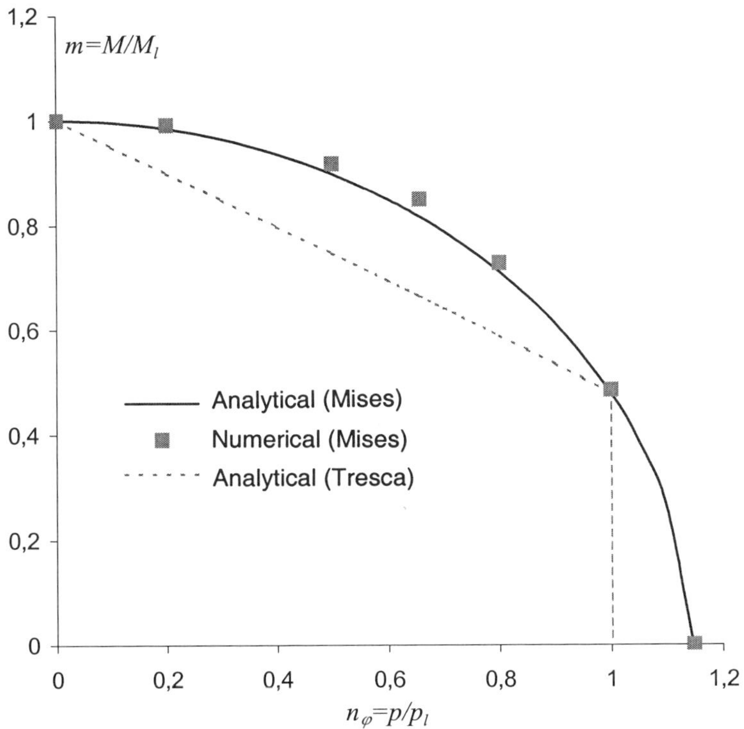


*Fig. 4.* Interaction of bending moment, internal pressure and axial tension
Note: 1) $r_i$=270 mm, $h$=60 mm; 2) $F = \pi r_i^2 p$ (a long end-closed pipe)

## 4.3 SHAKEDOWN LIMIT OF A THIN-WALLED PIPE SUBJECTED TO INTERNAL PRESSURE AND AXIAL LOAD (cf. fig. 3)

This problem was analytically studied by Cocks & Leckie [3] with simplified yield criterion. We will illustrate by this example the application of the Simplified Shakedown Analysis (SSA) proposed in §3.3 for both analytical and numerical analyses. The shakedown domain is defined by a monotonic internal pressure and a cyclic axial force:

$$p = \alpha_s p_0 \quad \text{and} \quad F \in \alpha_s [-1,1] F_0$$

If the axial loading is also monotonic as $p = \alpha_l p_0$ and $F = \alpha_l F_0$, the shakedown domain becomes a plastic collapse limit.

Starting from a general solution (33) of straight pipe and taking $m$=0 because the bending moment is absent, we have

$$\frac{n_\varphi - 2n_x}{\sqrt{4-3n_\varphi^2}} = \pm 1 \tag{35}$$

• **Instantaneous collapse limit domain with Mises' criterion**

$$\frac{p^2}{p_l^2} + \frac{F^2}{F_l^2} - \frac{p}{p_l}\frac{F}{F_l} = 1 \tag{36}$$

where $p_l$, $F_l$ are defined in (32) for a long pipe without the end-constraining effect.

In the case of varying axial force $F$(-1,1) we look for the shakedown limit. According to the assumptions proposed in §3.3, the incremental plastic limit in a generalised meaning is identical to the lowest instantaneous plastic limit corresponding to the most unfavourable load vertex. On the other hand, the alternating plastic limit depends only on the elastic limit of the axial force. Using the proposed SSA method, the shakedown limit may be found by such a combination of loading: the axial force is equal to -$F$ and the internal pressure remains as $p$. This leads to a lowest plastic collapse limit. So taking $n_x$ in (35) as - $n_x$, we obtain

$$\frac{p^2}{p_l^2} + \frac{F^2}{F_l^2} + \frac{p}{p_l}\frac{F}{F_l} = 1 \tag{37}$$

As this limit interaction is lower than the alternating plastic limit, the following solution is a real shakedown limit:

• **Shakedown limit domain with Mises' criterion:**

$$\frac{p}{p_l} = -\frac{F}{2F_l} + \frac{1}{2}\sqrt{4-3\frac{F^2}{F_l^2}} \tag{38}$$

Cocks & Leckie [3] used a method due to Gokhfeld & Cherniavsky [4], which involves the calculation of limit state for a yield limit decreased by the varying range of the elastic stress corresponding to cyclic loading. The linearized criterion of Hodge (a type of normalised Tresca's yield surface) was used. They obtained (39).

• **Shakedown limit domain by Cocks and Leckie with Tresca-type criterion**:

$$\frac{p}{p_l} = 1 - \frac{F}{F_l} \tag{39}$$

We would point out that this solution may be also deduced by using the present method as described above with Tresca's criterion. In fact, taking $m$ = 0 in (34) (no bending moment):

$$\frac{2-n_\varphi}{2}\cos\left[\frac{n_\varphi - 2n_x}{2-n_\varphi}\frac{\pi}{2}\right] = 0 \tag{40}$$

and taking $n_x$ in (40) as $-n_x$ (which corresponds to the most unfavourable combination of loading), we obtain immediately the shakedown limit that is same to Cocks and Leckie's solution:

$$\frac{n_\varphi + 2n_x}{2 - n_\varphi} = 1 \quad \text{or} \quad \frac{p}{p_l} = 1 - \frac{F}{F_l} \tag{41}$$

Now we verify above solutions by direct finite element calculations. Two pipe elements are used to discretize the structure considering the symmetry. Von Mises' criterion is adopted. Three methods are used together with ELSA software: USL (Unified-shakedown limit, eq.(18)), SSL (Separated-shakedown limit, eq.(26)) and SSA (Simplified shakedown analysis, eq.(31)). They give almost same results (difference <0.8%). Only the first two methods are reported to show their efficiency. The excellent agreement between numerical and analytic solutions is shown in Fig. 5, with the maximum error < 1%. Finally, one can observe that the limit load domain is strongly reduced in comparison with the shakedown domain in the case of cyclic loading.

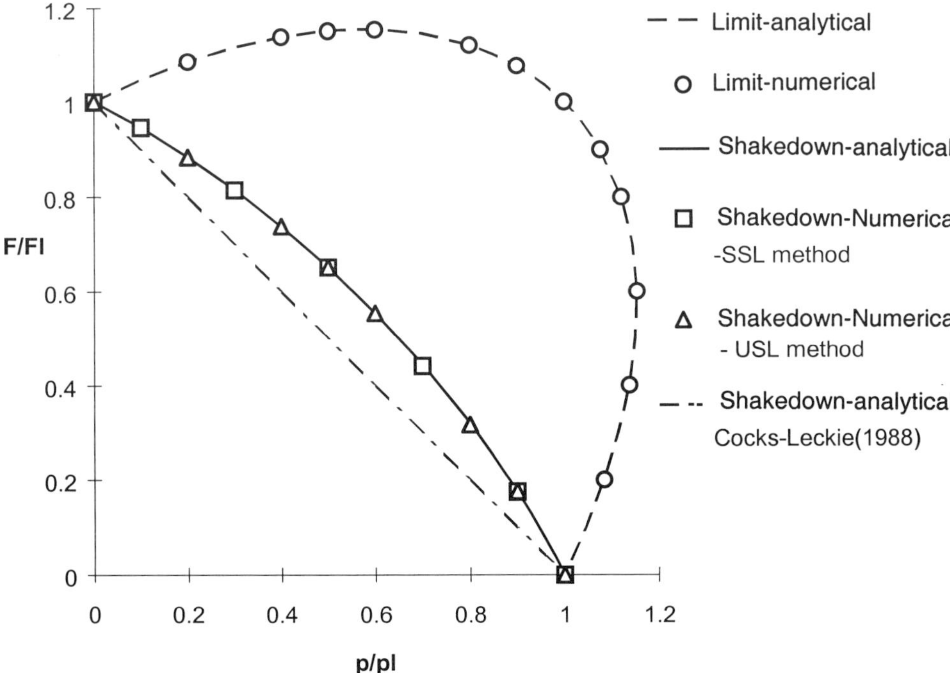


*Fig. 5.* Limit and shakedown domains of cylinder to internal pressure and axial force

Now we examine the USL SSL and SSA methods in another loading condition:

$$p = \alpha_s p_0 \text{ (monotonic) and } F \in \alpha_s [0,1] F_0 \text{ (cyclically varying)}$$

According to the proposed SSA method, cf. §3.3. eq.(31), the shakedown domain of Fig. 6 will be determined by the plastic collapse limits corresponding to the most unfavourable loading combinations. Such a combination of loading is easily found as follows:

- For any value of $F \in F_l$ (0, 1), the maximum internal pressure attains as $p = p_l$, corresponding to the most unfavourable axial force as $F$=0. So we get a bound as line A of Fig. 6.
- For any internal pressure $p \leq p_l$ constant, the maximum of axial force $F_{max}$ (varying between (0, 1)$F_{max}$) is defined by the same limit formula as (36). So we obtain another bound as line B of Fig. 6.

This prediction is checked by numerical calculations. Fig. 6 shows a good agreement between the simplified solutions as above (using SSA method) and the calculations using USL and SSL methods.

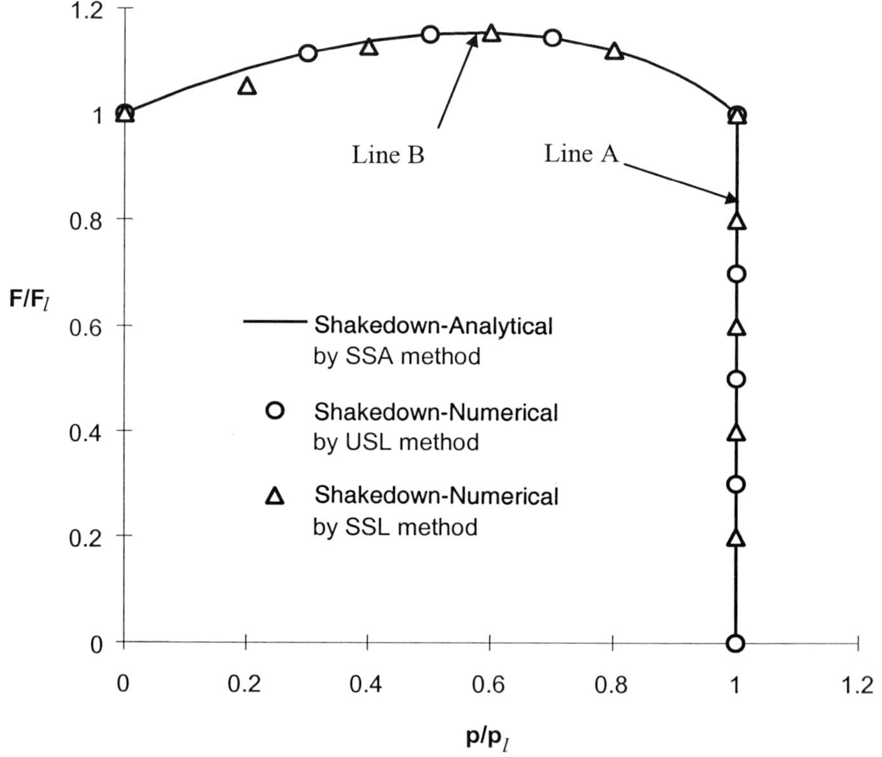


*Fig. 6.* Shakedown domains of cylinder to internal pressure $p$ and axial force $F$(0,1)

## 4.4 LIMIT ANALYSIS OF PIPE ELBOW STRUCTURES UNDER COMPLEX LOAD

### 4.4.1. *Plastic limit of 90° elbow subjected to in-plane bending (cf. Fig. 8)*

A curved pipe (elbow) is subjected to a bending moment $M_I$ in its symmetric plane. The end effect is not included because the analytic solution is known in this case. Calladine [2] proposed a so-called lower bound solution (42a) for a highly curved pipe. This solution was well supported by much numerical and experimental work in the literature. For a slightly curved pipe ($\lambda$>0.7), (42b) was proposed in [27]. Hence, (42) constitutes a complete solution of in-plane elbow with free-end condition (torus).

$$\alpha_0 = 0.9346\lambda^{2/3} \qquad \text{for } \lambda < 0.7 \tag{42a}$$

$$\alpha_0 = \cos(\frac{\pi}{6\lambda}) \qquad \text{for } \lambda \geq 0.7 \tag{42b}$$

where

$$\alpha_0 = M_I / M_l \qquad \lambda = Rh / r^2 \tag{42c}$$

*R, h* and *r* are respectively the curve radius of pipe, thickness and mean radius of the crossed section of pipe. $M_I$ is the in-plane limit moment of the curved pipe; $M_l$ is the limit solution of a straight pipe (32). See Fig. 7, we obtain an excellent agreement between analytic and numeric solutions for all $\lambda$ values.

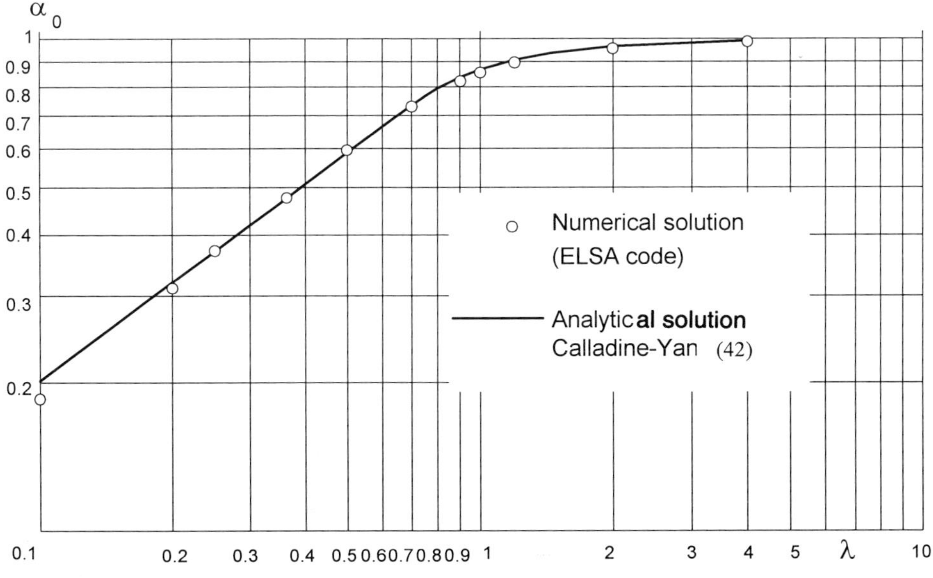


*Fig. 7* Limit factor of in-plane bending moment of elbow; $\alpha_0 = M_I / M_l$

4.4.2 *Pipe elbow structure under complex loading (in-plane and out-of-plane bending, internal pressure and axial force)*

A pipe elbow structure under complex loading is shown in Fig. 8. Different combinations of loading are dealt with. In-plane bending moment $M_I$ is in the direction of closing the elbow. It may be very expensive to solve this problem by a traditional elastic-plastic calculation using three dimensional or shell elements. However as it is shown in Fig. 9, using the present direct programming method with a few developed elbow elements, satisfactory solutions are obtained. Typically only 9 elbow elements are used for whole structure (5 for elbow and 2 for each straight pipe), or 5 elements for one half of the structure if out-of-plane bending moment $M_{II}$ is absent. For a thin-walled pipe, the interaction limit solution can be approximated by a simple formula [27]

$$\frac{M}{M_o} = \frac{1}{2}\sqrt{4 - 3(\frac{p}{p_l})^2} \tag{43}$$

where $p_l$ is the limit pressure of a curved pipe:

$$p_l = \frac{1 - r/R}{1 - r/2R}\frac{\sigma_y h}{r} \tag{44}$$

$M$ and $M_o$ are respectively the plastic limit moment with and without internal pressure $p$ (and end force $F$). In Fig.9 the analytic solution of Goodall [5] is also presented to compare with the present results.

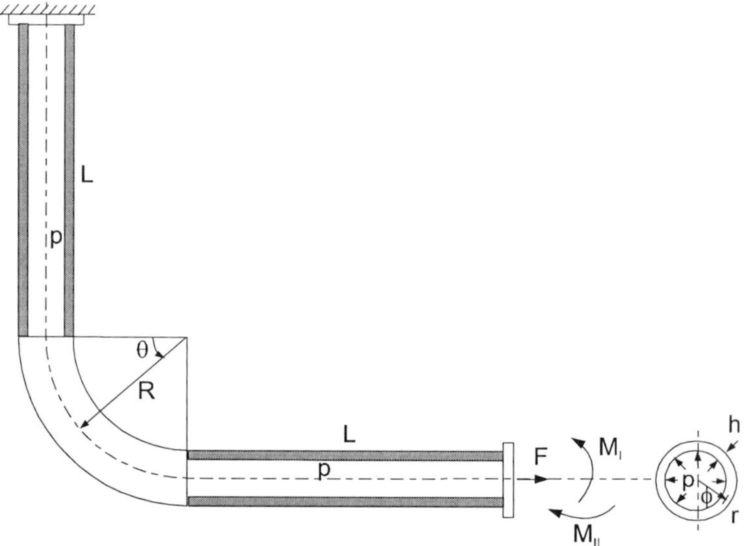


*Fig. 8* Elbow with prolongation under combined loading

## 4.5 SHAKEDOWN OF SPHERE SHELL UNDER RADIAL THERMAL LOADING AND INTERNAL PRESSURE

Thermal loading has important effects on shakedown behavior, see e.g. [4,6,23]. Here we present only an example of a thick-walled sphere ($k$=$b/a$=2.5, $b$ and $a$ are respectively external and internal radii) subjected to radial thermal loading (temperature) $T$ and internal pressure $p$, both varying independently as:

$p$: varying (0,1)$p$ or constant; $T$: varying (0,1)$T$ defined by $T = T_0 \dfrac{b/r - 1}{k - 1}$

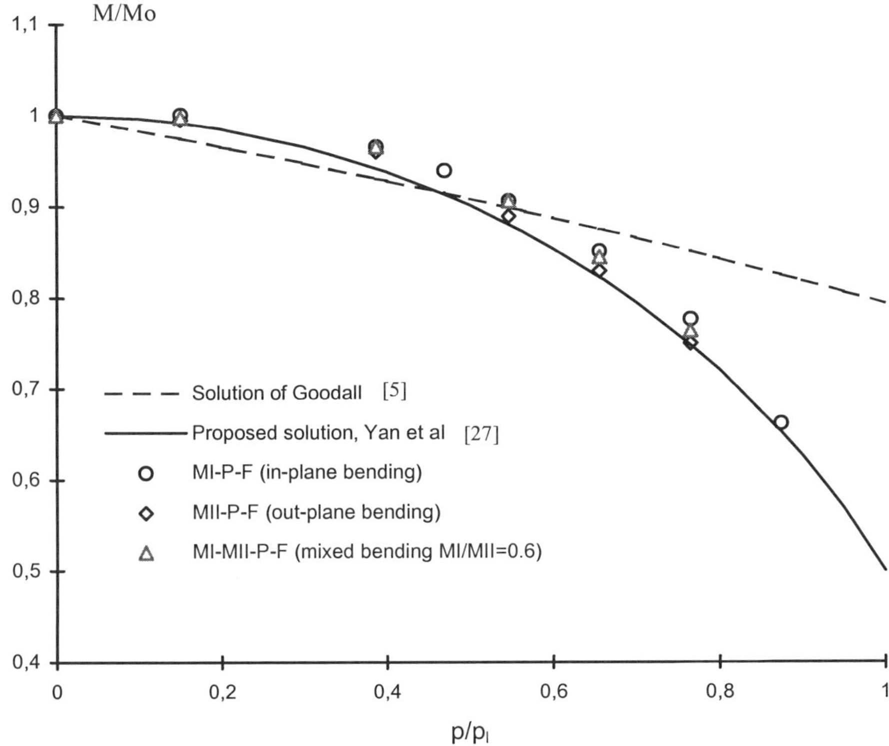


*Fig.9* Interaction solution of general bending and internal pressure of end-closed pipe elbow ($M_I$, $M_{II}$: in plane and out-of-plane bending moment; p: internal pressure; F: axial force due to p)

Due to the central symmetry, only one fourth of the sphere is modeled by axisymmetric quadratic finite elements. As an example, the data of a 316L(N) steel is used. The yield limit of material is described as temperature dependent in an explicit form:

$$\sigma_y^t = -6\times10^{-7}T + 9.6\times10^{-4}T^2 - 0.5599T + 230.65 \text{ (MPa)} \tag{45}$$

The analytic solution of shakedown limit was reported by Gokhfeld & Cherniavsky [4]. The comparison between analytical and numerical results of shakedown is represented in BREE diagram 10-11.

First we consider the yield limit of material as temperature-independent, and take its minimum value corresponding to the highest temperature. The numerical results are in excellent agreement with analytic solutions. BREE diagram of Fig. 10 is subdivided into some sub-regions that correspond to the different modes of deformation:

1) Completely elastic behavior in regions A;

2) Shakedown happens in region B if internal pressure varies arbitrarily or in (B+C) if the pressure is constant. In this region, the structure plastified in initial loading cycles will return to elasticity. So the structure may be considered safe;

3) Alternating plasticity in region D (or D+C if $p$ varies); Possibly the structure will fail by fatigue crack in the internal skin of the sphere after a finite time or finite number of loading cycles;

4) Incremental plasticity in region E. The structure fails due to excessive radial plastic deformation.

5) Beyond these regions, the structure will fail in a possibly mixed mode.

Now we consider the yield limit of the material as temperature-dependent. The results are represented by the solid points and the dashed line in Fig 10. As shown, the shakedown limit now is higher than the previous calculation with a constant yield limit. This means that if we take the lowest yield limit at highest temperature, the obtained shakedown limits are generally on the side of safety.

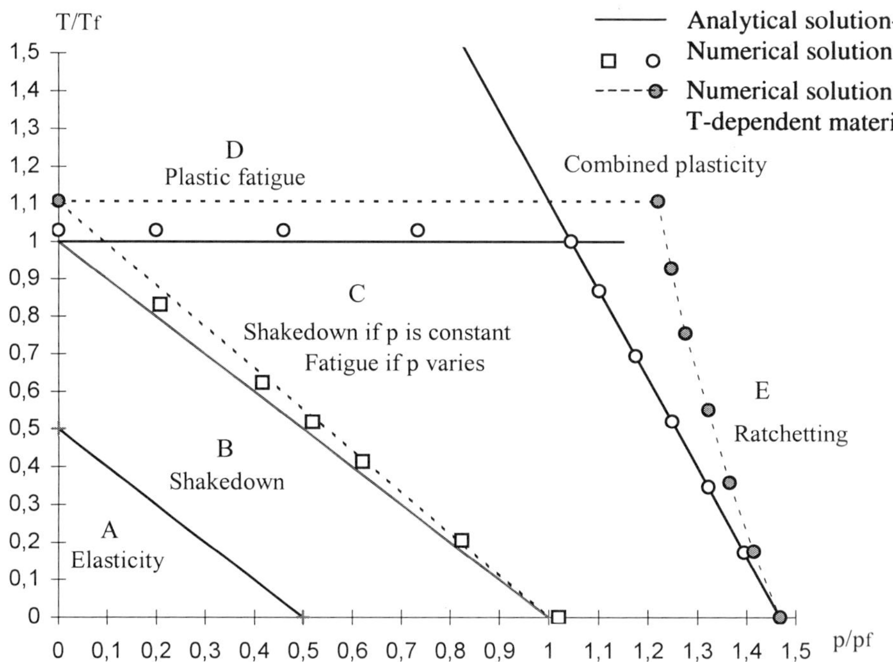


*Fig.10.* BREE diagram of sphere with temperature-independent or -dependent material
Take the yield limit of the material at highest temperature in analytic prediction

In order to give a more accurate prediction, we modify the theoretical solution simply by using the mean yield limit corresponding to a mean temperature. In this way, we see a better agreement between analytic and numeric solution as shown in Fig. 11. It should be pointed out that shakedown analysis with temperature-dependent material properties is a complex and being developed subject [1]. Its practical application to practical problem may need new theoretical and numerical effort. The present paper shows that a simplified analysis based on classical kinematical method is completely possible.

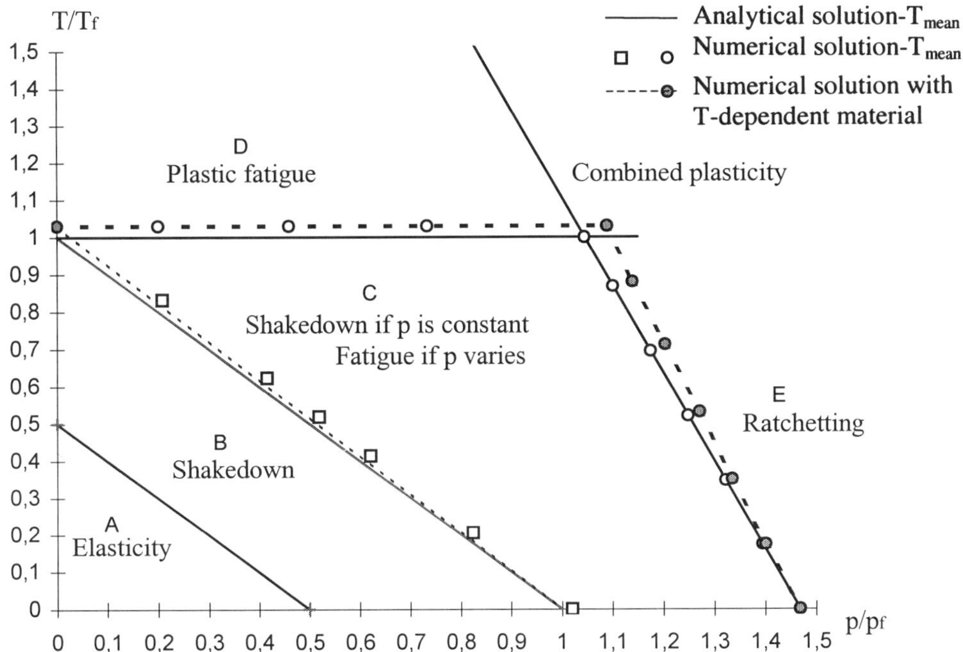


*Fig. 11.* BREE diagram of sphere with temperature-independent or -dependent material
Take the mean yield limit of the material in the prediction

## 5. Conclusions

In this paper, we discussed kinematical limit and shakedown theorems and their numerical implementation. The modified Markov theorem and the corresponding modified Koiter theorem are formulated in a general form of finite element discretization. The incompressibility of the plastic deformation does not pose any difficulty. Concerning shakedown analysis, temperature-dependence of yield limit of the material is included in theoretical and numerical formulation. Several numerical methods are presented. The separated-shakedown limit method (SSL) may specify two different inadaptation modes, and numerically it is approximately equivalent to limit analysis. The unified-shakedown limit method (USL) shows generally high precision but sometimes the difficulty appears in determining a local plastic fatigue. So an alternative use of these two methods is recommended. In some special cases of shell-type structures without thermal loading, we have shown that the proposed simplified shakedown analysis (SSA) is useful as an approximate analysis of incremental plasticity. This method permits us to estimate easily incremental limit in some cases. Numerically it is also less expensive than a usual shakedown analysis. Further studies are expected to specify its applicability domain.

Although the presented formulation is general, special attention is paid to some applications to shell and pipe structures. For such simple or more complex structures under simple or complex loading, limit and shakedown solutions are obtained accurately and easily due to the use of special elements. The calculations with temperature dependent material show the efficiency of the developed method.

## Acknowledgement

Since 1998, part of this work has been supported by European Community through the BRITE-EuRam project LISA BE97-4547, Contract BRPR-CT97-0595

## References

1. Borino G & Polizzotto C: Shakedown theorems for a class of materials with temperature-dependent yield stress, in Owen DRJ, Oñate E & Hinton E (ed.) *Computational Plasticity,* CIMNE, Barcelona, Part 1, 1997, 475-480
2. Calladine CR: Limit analysis of curved tubes, *Journal Mech. Eng. Science* **16** (1974), 85-87
3. Cocks ACF & Leckie FA: Deformation bounds for cyclically loaded shell structures operating under creep condition, *Journal of applied Mechanics* **55** (1988), 509-516
4. Gokhfeld DA & Cherniavsky OF: *Limit analysis of structures at thermal cycling*, Sijthoff & Noordhoff, The Netherlands, 1980
5. Goodall IM: Lower bound limit analysis of curved tubes loaded by combined internal pressure and in-plane bending moment, CEGB-RD/B/N4360, Central electricity generating board, UK, 1978
6. Gross-Weege J & Weichert D: Elastic-plastic shells under variable mechanical and thermal loads, *Int. J. Mech. Sci.* **34**(1992), 863-880
7. Jiang GL: Non linear finite element formulation of kinematic limit analysis, *Int. J. Num. Meth. Engng.* **38** (1995), 2775-2807
8. Karadeniz S & Ponter ARS: A linear programming upper bound approach to the shakedown limit of thin shells subjected to variable Thermal loading, *Journal of strain analysis* **19** (1984), 221-230
9. Köiter WT: General theorems for elastic plastic solids, in Sneddon I. N. & Hill R.(eds.): *Progress in Solid Mechanics,* Nord-Holland, Amsterdam, 1960,165-221
10. König JA & Kleiber M: On a new method of shakedown analysis, *Bull. Acad. Pol. Sci., Sér. Sci. Techn.* **4** (1978), 165-171
11. König JA: *Shakedown of elastic-plastic structure*, Elsevier & PWN-Polish Scientific Publishers, 1987
12. Liu YH, Cen ZZ. & Xu BY: A numerical method for plastic limit analysis of 3-D structures, *Int. J. Solids Structures* **32** (1995), 1645-1658
13. Maier G: A shakedown matrix theory allowing for work hardening and second-order geometric effects, In Sawczuk A. (ed.): Foundations of plasticity, Vol. I, Noordhoff, Leyden, 1972, 417-433
14. Martin JB: *Plasticity,* MIT Press, Cambridge, 1975
15. Morelle P: Numerical shakedown analysis of axisymmetric sandwich shells: an upper bound formulation, *Int. J. Num. Meth. Engng.* **23** (1986), 2071-2088
16. Nguyen-Dang H: Direct limit analysis via rigid-plastic element computer methods, *Comput. Meth. Appl. Mech. Engng.* **8** (1976), 81-116
17. Nguyen-Dang H & König JA: A finite element formulation for shakedown problems using a yield criterion of the mean, *Comput. Meth. Appl. Mech. Engng.,* **8** (1976), 179-192
18. Nguyen-Dang H: Sur la plasticité et calcul des états limites par éléments finis, *Thèse de doctorat spéciale,* Collection des publications N°.98, Université de Liège, Belgique, 1984
19. Nguyen-Dang H & Morelle P: Plastic shakedown analysis, in Smith D.L. (ed.), *Mathematical programming methods in structural plasticity,* Springer-Verlag Wien-New York, 1990,183-205
20. Ponter ARS: Shakedown and ratchetting below the creep range, Nuclear science and technology, Report EUR 8702, 1983
21. Pycko S & Mroz Z: Alternative approach to shakedown as a solution of min-max problem, *Acta Mechanica* (1992), 205-222
22. Washizu K: *Variational methods in elasticity & plasticity*, Third edition, Pergamon Press, 1982
23. Weichert D & Gross-Weege J: The numerical assessment of elastic-plastic sheets under variable mechanical and thermal loads using a simplified two-surface yield condition, *Int. J. Mech. Sci.* **30** (1988), 757-767
24. Yan AM & Nguyen-Dang H: Shakedown of structures by improved Koiter's theorem, *Proceeding of 4th National Congress on Theoretical & Applied Mechanics,* Leuven, 22-23 May, 1997, 449-452
25. Yan AM, Jospin RJ & Nguyen-Dang H, An enhanced pipe elbow element-Application in plastic limit analysis of pipe structures, *Int. J. Num. Meth. Engng.* **46** (1999), 409-431
26. Yan AM & Nguyen-Dang H, Limit analysis of cracked structures by mathematical programming and finite element technique, *Computational Mechanics*, **43** (1999) (to appear soon)
27. Yan AM, Nguyen-Dang H. & Gilles Ph: Practical estimation of the plastic collapse limit of curved pipe subjected to complex loading, *Structural Engineering and Mechanics*, **8**(4) (1999), 421-438

# SHAKEDOWN AND DAMAGE ANALYSIS APPLIED TO ROCKET ENGINES

T. HASSINE *, G. INGLEBERT *, M. PONS **
* *ISMCM-CESTI, groupe Tribologie*
*3 rue Fernand Hainaut, F93407 St-Ouen Cedex*
** *CNES*
*Rond Point de l'Espace, F91023Evry Cedex, France*

## Abstract

Two modular finite element software package have been developed. One, A2DF-6S, deals with high cycles fatigue (HCF) and elastic shakedown response to cyclic loading. The second, A2DI, deals with a step by step thermo-elasto-viscoplastic analysis response under multiaxial loading coupled with damage. The HCF software evaluates the shakedown status, using standard generalised models for the material behaviour law and the MASSI method (Method for Simplified Analysis of Inelastic Structures), and estimates the life of the studied part through multiaxial fatigue criteria. The step by step software evaluates the state of structures under thermal mechanical loading involving wide ranges of temperatures, using the Walker and Freed law, and a Kachanov law for damage accumulation.

## 1. Introduction

Evaluation of the life of cryogenic space engine components is a real challenge. Special tools have been developed from elastic shakedown and high cycle fatigue to take into account strong damage accumulation under severe thermomechanical loading. This work was initiated in the GDR 0916 (Research Group: CNES, CNRS, SEP) and continued in the GRT, Tenue Mécanique, (Research Group on Technology, CNES).
In the shakedown range, high reliability is needed, even if a small number of each component is used. Statistical approaches like Weibull law are too expensive to be done. The main problems occur in contacts, with rolling-sliding (for non lubricated ball bearings of the turbopumps) or fretting fatigue. Structural analysis coupled to the study of the fatigue behaviour of the materials in use has been performed. Special attention has been given to the influence of initial stresses and their evolution.
On the other hand, some parts, as the combustion chamber work under so severe conditions that only a step by step thermo-elasto-viscoplastic analysis with damage accumulation can help to understand, describe and optimise its performance.



*D. Weichert and G. Maier (eds.), Inelastic Analysis of Structures under Variable Loads,* 255–267.

We present in the first part of this paper the shakedown analysis and the associated software, A2DF-6S, the introduction of the initial residual stresses and the damage analysis via some multiaxial criteria (HCF).
The second part of the paper deals with the thermo-inelastic analysis of structures under thermal/mechanical loading involving wide ranges of temperatures and the introduction of the damage accumulated during the loading (A2DI software). Attention is given to the characterisation of the inelastic and damage materials constants.

## 2. Shakedown and damage analysis (HCF) - A2DF-6S

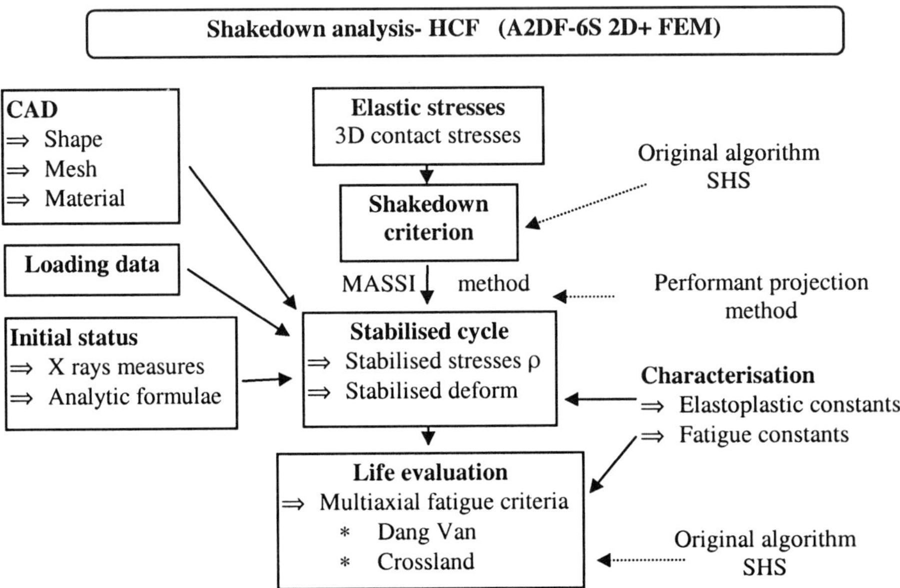


figure 1 : HCF Software Organigram

We developed a modular software composed of : 2D finite elements design and mesh, definition and calculation of the loading path in the elastic domain (combination of classical FEM loading and analytically 3D calculated contact stresses are available) and evaluation of the shakedown status, introduction of the initial residual stresses in the upper layers of the mechanical part (assuming that they are generated by initial plastic strains which vary only with depth), estimation of the stabilised residual stresses and strains in the elastic shakedown, evaluation of the life expectation through multiaxial fatigue criteria.

## 2.1 SHAKEDOWN ANALYSIS

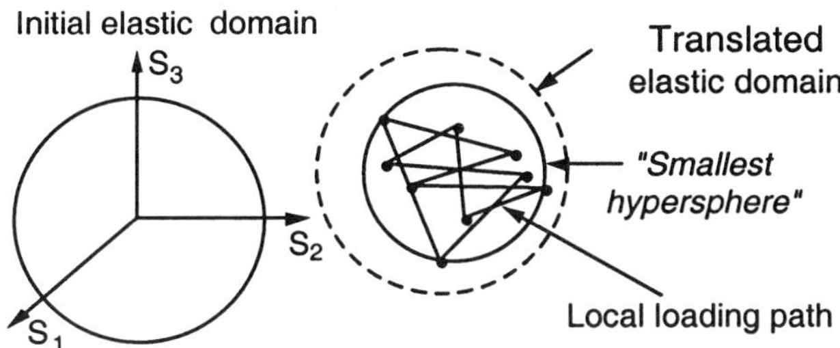


figure 2 : Elastic shakedown

Shakedown analysis has been initiated by Melan in 1938. His theorem provides a powerful tool to investigate the shakedown state of a structure under cyclic loading. According to it and to further works on shakedown, for kinematic hardening materials, the loading path, whatever its shape, must be small enough to be contained in a translated elastic domain (fig. 2). So stabilised constant plastic strain and residual stress fields can be obtained [1].

### *2.1.1 Shakedown Status*

Evaluation of the shakedown status relies upon an original algorithm first presented in [1] to obtain the smallest hypersphere embracing the loading stress path ; its radius should be smaller than the yield strength. This smallest hypersphere is also needed in the evaluation of the multiaxial criteria. A new algorithm [2] has been developed to find out this smallest hypersphere. Based on a projected gradient method, this algorithm is very efficient ; tested on a 486DX2-66 P.C. a few minutes are sufficient to find the exact solution of the problem on a set including 1500 points with 6 components per points.

The stabilised cycle is estimated using standard generalised models for the material behaviour law and the MASSI method (Method for Simplified Analysis of Inelastic Structures) [1], with a performant method to obtain the necessary local projections in the stress space [3].

Particular attention has been given to make easy comparisons with experimental data. General deformation, plastic strains, residual stresses, location of fatigue damage initiation are available.

### *2.1.2 The MASSI method (Method for Simplified Analysis of Inelastic Structures)*

Our analysis applies to Standard Generalised Materials [5] ; each volume element of the studied materials is considered as a structure made of perfectly plastic mechanisms (P.M.) in a linear elastic matrix. Complex hardening law may be achieved for these materials through the coupling between the perfectly plastic mechanisms and the surrounding elastic matrix [4], [5]. To demonstrate the algorithm, we shall use one of the simplest models in this field, linear kinematic hardening. The tensile test gives a bilinear curve, and the material properties are defined by the yield strength $s_y$ and a tensile hardening modulus h.

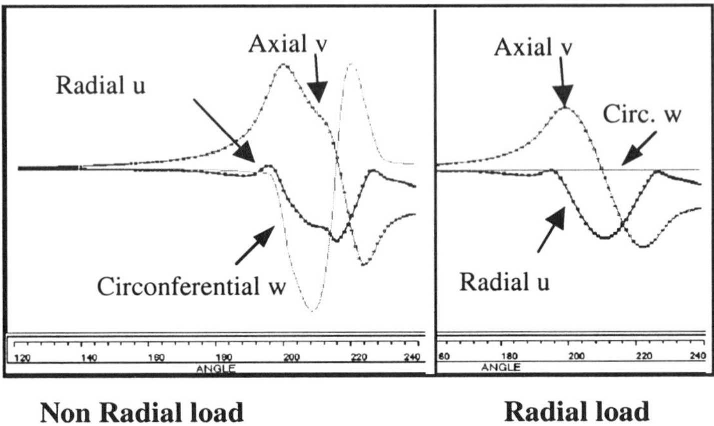


figure 3 : Residual displacements of an internal ball bearing ring

General theorems for these materials ensure that elastic shakedown will occur in a structure, if the radius of the smallest hypersphere embracing the local elastic loading path for each plastic mechanism is inferior to the yield strength (fig. 3). So the stabilised evolution of the structure will be elastic after some plastic evolution.

### *2.1.2.1 Searching for the smallest hypersphere (SHS)*

We start from a given set of points defining the loading path in stress space. Let us denote $E = S_i$ $(1 \leq i \leq n)$, this set of points. For a three dimensional physical problem under small strains hypothesis, we work in a six dimensional space with Von Mises norm (Euclidean norm associated to a scalar product).

Building the smallest hypersphere containing the given set of points (let call it SHS) may be achieved through different ways.

Papadopoulos (1987) remarked that in an n dimensional space, at most n+1 points were needed to build the hypersphere. An exact way of building the smallest hypersphere is thus to test all possible hyperspheres built on two to n+1 points among the N points of the set. The first hypersphere containing the whole set is the right one. It may be shown that the calculation time of this algorithm may be expressed as a polynomial in N whose degree $\lambda$ varies between 2 and n+1, depending on the positions of the points. It is easy to see that, for great numbers of points, this algorithm may be very lengthy.

It is clear that fast and exact calculation of the SHS must rely on some other principle. Our algorithm relies on a well known minimisation algorithm, the projected gradient method which, in our case, may also be easily illustrated.

Let us express the search of SHS as a minimisation problem. Let C and R be the centre and radius of the SHS, then our problem is to find C such that :

$$R = \underset{S_i \in E}{\mathrm{Max}} \ \overrightarrow{S_iC}^2 \quad \text{minimal} \tag{2}$$

At first glance, our problem looks like a highly non linear one for which projected gradient method is far from optimal. However, let us suppose that we know a point $S_m$ of E belonging to the surface of the SHS (let call it a forcing point). The problem (2) may be rewritten as :

$$R = \overrightarrow{S_mC}^2 \ \text{minimal} \tag{3a}$$

under the constraints :

$$\overrightarrow{S_mS_i}^2 - 2.\overrightarrow{S_mS_i}.\overrightarrow{S_mC} \leq 0, \forall S_i \in E \tag{3b}$$

Here the constraints express the condition that the sphere centred on C and of radius R must contain all points of the set E.
Problem (3) is a quadratic minimisation problem under linear constraints for which the projected gradient method is nearly optimal.
Here the projected gradient method consists in displacing the centre C of the candidate sphere in a direction given by the projection of the gradient $-dR/dC = \overrightarrow{CS_m}$ on the subspace defined by the intersection of the planes of the active constraints, where a constraint i is said to be active if :

$$\overrightarrow{S_mS_i}^2 - 2.\overrightarrow{S_mS_i}.\overrightarrow{S_mC} = 0 \text{ and } \overrightarrow{S_mS_i}.\overrightarrow{S_mC} \geq 0 \tag{4}$$

Let us define the subset A of E as the set of points belonging to E and associated with the active constraints :

$$A = \left\{ S \in E, S_i \neq S_m / \left( \overrightarrow{S_mS_i}^2 - 2.\overrightarrow{S_mS_i}.\overrightarrow{S_mC} = 0 \text{ and } \overrightarrow{S_mS_i}.\overrightarrow{S_mC} \geq 0 \right) \right\} \tag{5}$$

The displacement direction of C is then defined by the projection of $\overrightarrow{CS_m}$ on the subspace :

$$\left\{ \vec{U} / \overrightarrow{S_mS_i}.\vec{U} = 0, \forall S_i \forall A \right\} \tag{6}$$

The centre C is displaced in the projected gradient direction until the minimum of R in this direction is reached or another constraint becomes active. The solution is obtained after several such displacements when it becomes impossible to further reduce R without violating the constraints. It may be shown that the exact solution of problem (3) is reached after at most n displacements, n being the dimension of the space.
Now, to solve completely problem (2), we are lead to find a forcing point $S_m$ belonging to the surface of the SHS. As the SHS is unknown, we don't know any point belonging to its surface. Let us remark, however, that there exist at least two of them between the N points of the set E and that we can still try to make a reasonable initial guess that we shall refine afterwards.

Indeed, if we start solving problem (3) with $S_{m0} \in E$ as a first guess for the forcing point and $C^{(0)}$ as a first candidate centre for the SHS, we will converge to a solution $C_0$ of problem (3) provided that the couple ($C^{(0)}$, $S_{m0}$) satisfies the constraints (3b). The couple ($C_0$, $S_{m0}$) defines a new smaller sphere which is not necessarily the SHS but on which surface have appeared new points $S_j$ defining with point $S_{m0}$ a subset $D_0$ of the set E.

We may know choose another forcing point $S_m$ different from $S_{m0}$ from $D_0$ and solve problem (3) again. If the centre of the candidate sphere does not change, we choose another untested point $S_{mj} \in D_0$ until we find one that permits to reduce the radius of the candidate sphere. If this is the case, we have a new subset $D_j$ of E, whose points belong to the surface of the candidate sphere ($C_j$, $S_{mj}$) and from which we can extract other forcing points to test.

It is easy to see that this process finally converges to a candidate sphere ($C_k$, $S_{mk}$) and a subset $D_k$ whose points have all been tested without any change. The sphere ($C_k$, $S_{mk}$) is the smallest hypersphere containing the whole set E because its radius cannot be further reduced without violating the constraints.

In practice, we choose the initial centre $C^{(0)}$ as the centre of the largest segment between two points of the set E and the initial forcing point $S_{m0}$ as the farthest point of E from $C^{(0)}$. This initial guess ensures that the constraints are satisfied (all points of E belong to the initial candidate sphere). We may see the whole process as constituted of two embedded loops, the innermost loop being the projected gradient method for minimising problem (3) and the outermost testing all possible forcing points.

Figures 4 and 5 illustrate the process in a 2 dimensional case. The shaded areas are forbidden by the constraints.

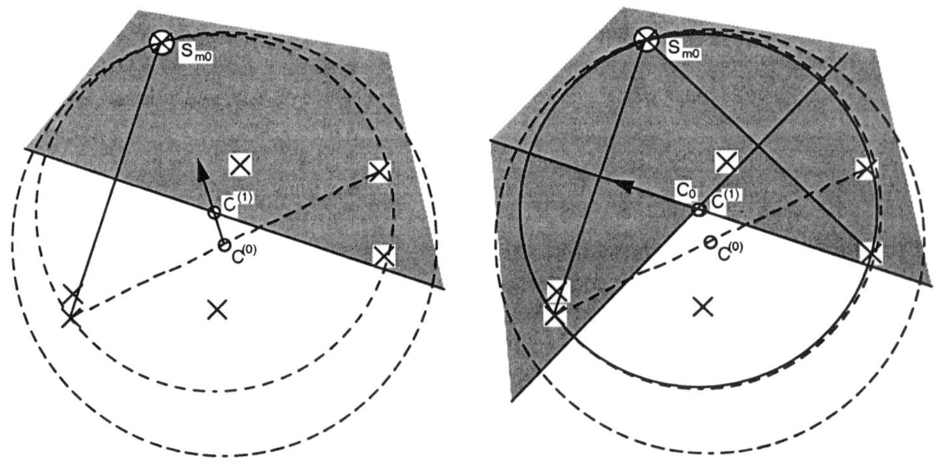


figure 4: First minimisation loop (innermost) steps 1 and 2

It can be seen that in this case the first forcing point belongs effectively to the SHS which is obtained after only one minimisation loop by the projected gradient method.
It may be shown that the number of displacements of the centre C needed by the whole process is less than N×n where N is the number of points in the set and n the dimension of the space. In many practical cases, it is only of the order of the dimension n of the space.

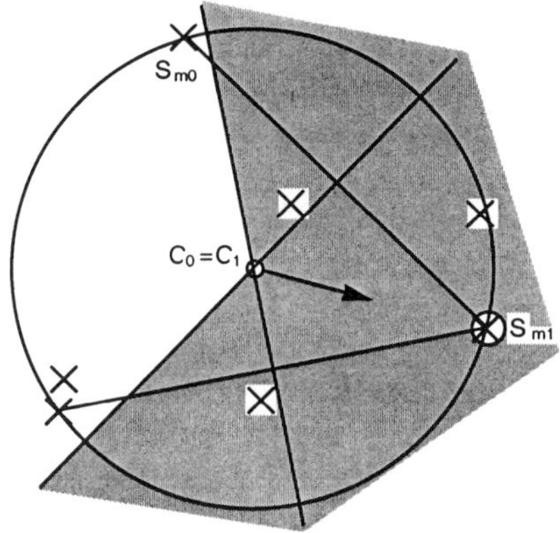


figure 5 : Forcing point loop (outermost)

When compared to the algorithms mentioned in the literature, our method appears to be a very efficient way of finding the exact smallest hypersphere containing a given set of points. Indeed the minimisation process is often faster than the initial guess which is an $N^2$ process involving the calculation of all the distances between pairs of points in the set E.

### *2.1.2.2 Projection : iterative algorithm based on minimisation principles*

For linear kinematic hardening materials, the constant stabilised residual stresses and plastic strains may be determined through a projection of a reasonable initial state on the limit shakedown area (fig 6).
Intermediate transformed variables Y allow to rewrite the plasticity criterion in term of elastically calculated stresses and Y instead of actual stresses and plastic strain :

$$\sqrt{\frac{3}{2}\left(s_{ij} - C\varepsilon^{p}_{ij}\right)\left(s_{ij}\right) - C\varepsilon^{p}_{ij}\right)} = \sqrt{\frac{3}{2}\left(s^{el}_{ij}\right) - Y_{ij}\right)\left(s^{el}_{ij} - Y_{ij}\right)} \qquad (7)$$

The projection problem is written as a minimisation problem. We still use a projected gradient algorithm as we did for the SHS calculation [2].
It is shown that, in an n dimensional space, the whole process converges to the exact solution in a finite number of iterations which is at most a multiple of n.

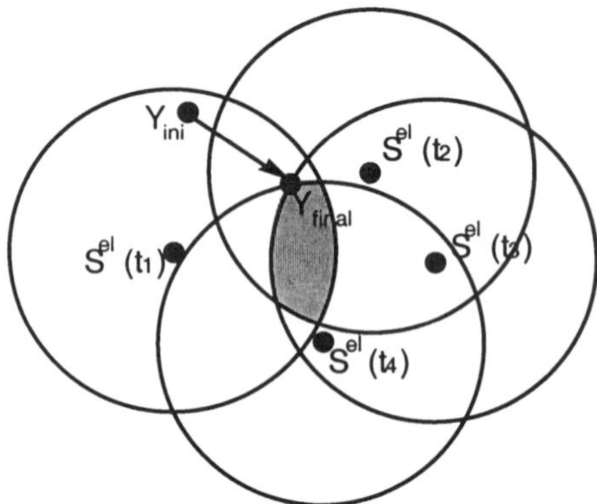


figure 6 : initial and final values of Y variables in elastic shakedown case

## 2.2 INTRODUCTION OF THE INITIAL STATE

The residual stresses evaluation in a plastic-coated structure has to take into account the initial state created during the elaboration or the finishing of surfaces of this structure. If we use a simple representation of the plastic behaviour of the material, such as a linear kinematic hardening law, the plastic domain and plastic stresses can be evaluated as soon as the plastic strains are known. In this case, the plastic strain is the unique variable characterising the hardening. Unfortunately, the only possible measurable quantities are the initial residual stresses. They can be measured with X-Ray diffraction [6] or with extending gauges. Plastic strains are then analytically calculable from measured stresses, in the absence of other initial strains, assuming simple geometry (such as tubes, cylinders or plates), and an elastic core.

For each point of the structure, the distance to the pre-stressed surface is calculated. The associated plastic strains are introduced as an initial state for the element as a function of that distance.

## 2.3 DAMAGE ANALYSIS (HCF)

To evaluate the life of a component, two steps are needed :

- evaluation with respect to a multiaxial fatigue by using the criteria of Crossland, Papadopoulos [7] or Dang-Van [8]. These criteria are defined as linear combinations of a local hydrostatic pressure (strongly dependant of residual stresses), and of an elastic shear-stress range; the critical values of these two fatigue parameters varies for each criterion. The smallest hypersphere containing the local loading path, calculated for the elastic shakedown condition, is a compulsory element for the evaluation of the shear-stress range, whatever the criterion.

- evaluation of the life time : Wöhler curves and Goodman or Haigh diagrams define the life time of test samples submitted to uniaxial load. For a multiaxial load, which is the case of a ball bearing ring, a correction of the evaluated life time is done using the complete local loading path.

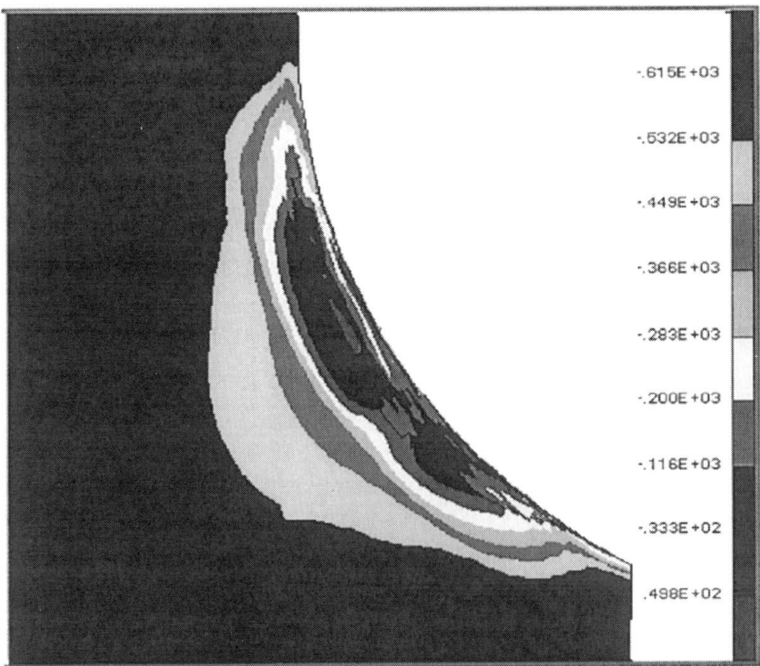


figure 7 : Isovalues of Dang Van's fatigue criterion in a ball bearing ring

## 2.4 OTHER APPLICATIONS

Development of this software and techniques for life time evaluation was performed for bearings. The dedicated software A2DF6S or A2DFSC allows calculations of bearings and associated characterisation experiments [9] (disc on disc and ball on plane). The techniques could apply to a much wider variety of problems; so a 2D finite element code including the initial stresses, elastic shakedown (MASSI) for any combination of classical loading plus "rolling loads" (3D punctual or linear contact in motion along the third direction, normal to screen plane direction), fatigue and life prediction from multiaxial criteria and easy comparison with the residual stresses measurements (all calculated stresses in the volume on which measurements have been performed are plotted).
Application to carbonitrided gears, rails, fretting fatigue studies and various components were already made [10].

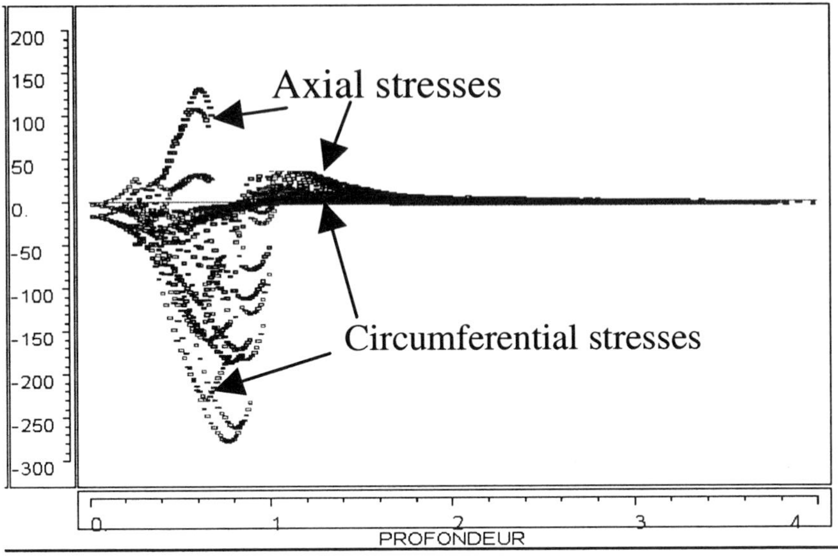


figure 8 : Residual stresses under a contact zone for comparison with Xray measures

## 3. Short life analysis

A 2D finite elements code has been developed on PC in order to perform thermal/inelastic analysis of structures under thermal mechanical loading involving wide ranges of temperatures. Both damage and inelastic strains are introduced. The step by step response in the inelastic range is calculated, using an asymptotic procedure for integration. Stability of this algorithm was verified. Dependence of material properties with temperature is taken into account. During the step by step calculation, a stress-strain response for a selected point of the structure may be followed on the PC screen. Stresses, strains and damage for up to ten steps per calculation may be stored for further examination.

### 3.1 INELASTIC AND DAMAGE STEP BY STEP ANALYSIS

According to the modular scheme chosen for coding, various inelastic behaviour laws can be incorporated. Thus, this code should be a powerful tool for researchers studying new models for the behaviour of materials. The first inelastic behaviour law introduced in the code is the unified strain inelastic behaviour law proposed by Walker and Freed [11]. For damage, a simple Kachanov law (1958) for creep damage is used. More inelastic and damage behaviour laws were selected and will be incorporated later on.
Special attention is devoted to define systematic identification procedures for any behaviour law selected for incorporation in the code. Classical identification tests such

as monotonous or cyclic tensile tests, creep tests, relaxation tests,... are reviewed and specific tools have been created for identification of material constants on the assumption of Walker and Freed, Kachanov, or other models for material behaviour.

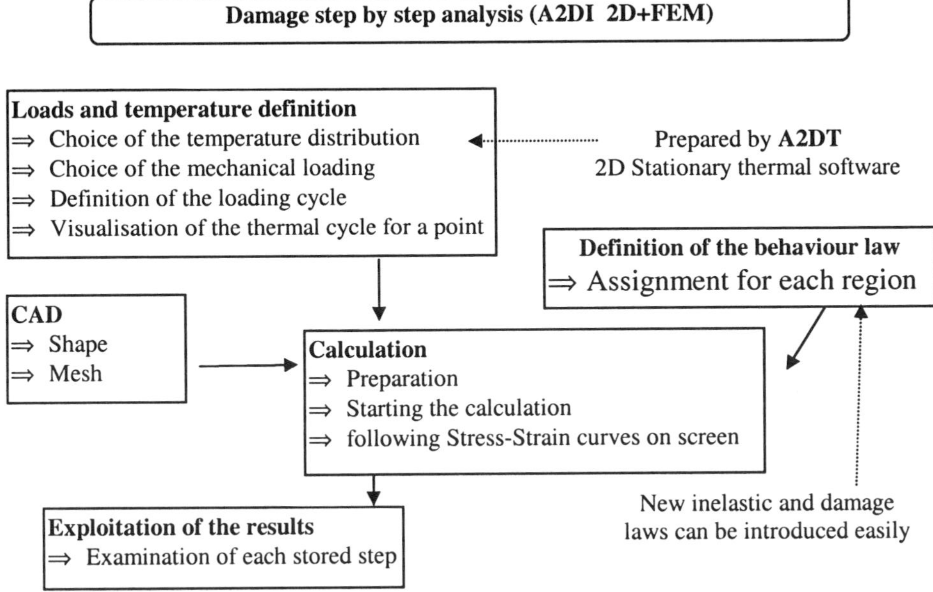


figure 9 : Step by step Software Organigram

## 3.2 APPLICATION TO A ROCKET ENGINE COMBUSTION CHAMBER

To illustrate the efficiency of the code, calculation of the thermal mechanical evolution of a combustion chamber of the rocket engine is performed. First, identification of the material constants of the Walker and Freed and the Kachanov behaviour law on the available tests on the constitutive material of the chamber are performed. Thus the identification procedure can be tested on real experimental data. A good agreement can be observed between experimental and calculated test responses at various strain rates and temperatures. Special attention is devoted to tests involving coupled thermal and mechanical loading such as coupled cooling and tensile tests. Damage material constants are identified by creep experiments.
Then calculation on the combustion chamber is performed. The thermal loading ranges from liquid hydrogen temperature (20K) to some 800K or 1000K coupled with mechanical pressures. This thermal loading generates extremely high temperature gradients and severe inelastic strain and damage. A good agreement is found between calculated and observed strains as well as between predicted and observed location of damage initiation.

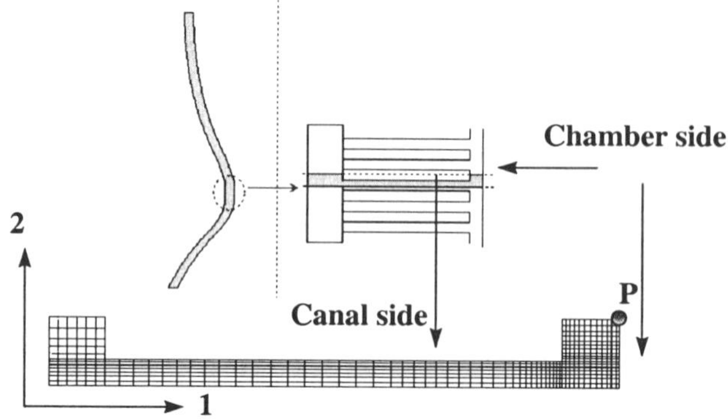


figure 10 : Mesh of a half canal of a combustion chamber

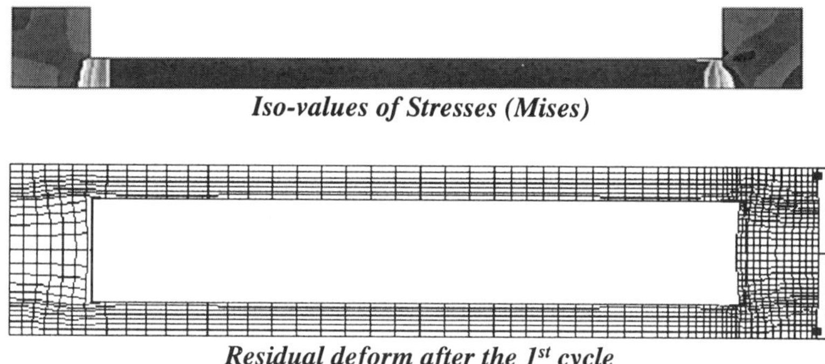


figure 11 : Some results of a combustion chamber

## 4. Acknowledgement

Thanks to CNES (French National Space Agency), SEP (European Society for Propulsion) and CNRS (French National Research Centre)

## 5. References

1. Inglebert, G., Frelat, J. : Quick analysis of inelastic structures using a simplified method, *Nuclear Engineering and Design*, volume 116 (1989), p. 281-291

2. Inglebert, G. & al, : A new algorithm to perform shakedown analysis on a structure under non radial loading, *Mathematical Modelling and Scientific Computing*, Volume 6, Special issue : proceedings of the 10th ICMMSC, 5-8 july 1995, Boston, Massachusetts.

3. Inglebert, G. & al, : Direct estimation of stabilised state of a structure under non radial loading, *Mathematical Modelling and Scientific Computing*, Volume 6, Special issue : proceedings of the 10th ICMMSC, 5-8 july 1995, Boston, Massachusetts.

4. Zarka J., Engel J.J., Inglebert G. (1980). On a simplified inelastic analysis of structures, *Nuclear Engineering and Design* (North Holland), 57., pp. 333-368.

5. Halphen B. and Nguyen Q.S. (1975). Sur les matériaux standards généralisés, *J. de Mécanique*, 14., n°1, pp. 39-63.

6. L. Castex. " Etude par diffractométrie X d'aciers au voisinage de la limite d'endurance et à différents stades de la fatigue ". *Ph.D Thesis*, University of Bordeaux, Dec, 1987.

7. Papadopoulos I; (1987). Fatigue polycyclique des métaux : une nouvelle approche, *E.N.P.C. thesis*, Ecole Nationale des Ponts et Chaussées, Paris.

8. Dang Van K;, Griveau B., Message O. (1989). On a new multiaxial fatigue limit criterion : theory and applications. In : *Biaxial and Multiaxial Fatigue*, EGF 3 (M.W. Brown and K.J. Miller ed.), pp. 479-496, Mechanical Engineering Publications, London.

9. Audrain V., Hassine T., Inglebert G., Gras R., Blouet J. (1993). Characterisation of elastoplastic behaviour applied to the contact, *6th international congress of tribology*, Budapest, August 30th - September 2nd.

10. A.C. Batista, A.M. Dias, P. Virmoux, G. Inglebert, T. Hassine, J.C. Leflour, J.L. Lebrun - Characterisation of mechanical properties and residual stresses in hardened materials - $10^{th}$ Experimental Mechanics, Lisbonne, Juillet 1994, pp 739-744;

11. A. D. Freed and K. P. Walker, Viscoplastic model development with an eye towards characterisation, in *Material Parameter Estimation for Modern Constitutive Equations*. L. A. Bertram & al editions, MD Vol : 43, and AMD Vol : 168, ASME, New York, 1993, pp 57.D

# RELIABILITY ANALYSIS OF ELASTO-PLASTIC STRUCTURES UNDER VARIABLE LOADS

M. HEITZER
*Forschungszentrum Jülich GmbH*
*Institute of Safety Research and Reactor Technology*

AND

M. STAAT
*Fachhochschule Aachen Div. Jülich*
*Labor Biomechanik*

**Abstract.** Structural reliability analysis is based on the concept of a limit state function separating failure from safe states of a structure. The paper discusses some difficulties of different reliability methods for FEM-discretized nonlinear structures. It is proposed that theorems of limit and shakedown analysis are used for a direct definition of the limit state function for failure by plastic collapse or by inadaptation. Shakedown describes an asymptotic and therefore time invariant structural behaviour under time variant loading. The limit state function and its gradient is obtained from a mathematical optimization problem. For application to large FEM models a basis reduction method is used. The method is implemented into a general purpose FEM code. Combined with FORM highly effective, robust and precise analyses could be performed for high-reliabilty problems.

## 1. Introduction

Design and assessment of engineering structures imply decision making under uncertainty of the actual load carrying capacity of a structure. Uncertainty may originate from random fluctuations of significant physical properties, from limited information and from model idealizations of unknown credibility. Structural reliability analysis deals with all these uncertainties in a rational way. Reliability assessment of structures requires on the one hand mechanical models and analysis procedures that are capable of mod-



*D. Weichert and G. Maier (eds.), Inelastic Analysis of Structures under Variable Loads,* 269–288.

eling limit states accurately. On the other hand, full coverage of the present random variables is also necessary for a meaningful reliability assessment. The mechanical and stochastic model depends on the definition of the limit state. For instance, if the limit state of the structure is defined with respect to plastic collapse, then Young's modulus, hardening modulus and secondary stress need not be modelled as random variables, because they all do not influence the limit load. Conversely, elastic buckling is governed by Young's modulus, secondary stress, and geometry imperfections. Damage

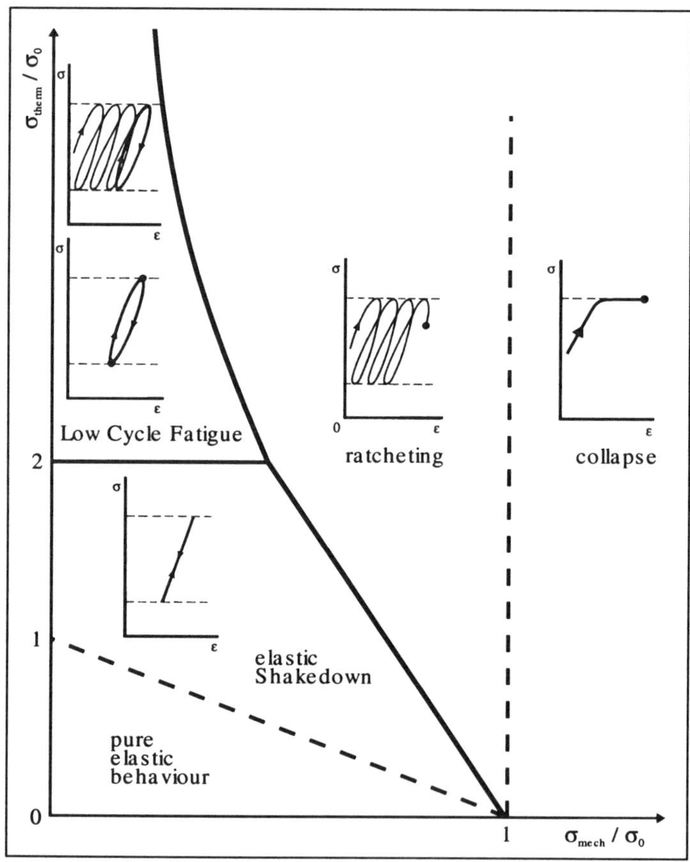


*Figure 1.* Bree-Diagram of pressurized thin wall tube under thermal loading [19]

accumulation in LCF or plastic strain accumulation in ratcheting are evolution problems. Shakedown theorems transform them into much simpler time independent problems. In principle the possible structural responses, which are presented as icons in the Bree-Diagram (see Figure 1 and [4]) may be reproduced in a detailed incremental plastic analysis. However, this assumes that the details of the load history (including any residual stress) and of the constitutive equations are known. Often it is economically not justified or even impossible to obtain all these necessary information. It is an advantage under uncertain conditions, that limit and shakedown analyses need only few key information of material behaviour and of load history. Geometrical nonlinearities and bounds on inelastic deformations pose open problems [12], [20].

Our probabilistic approach is based on the lower bound theorem of limit and shakedown analysis. A limit state function for reliability analysis is introduced for the First Order Reliability Method (FORM).

## 2. Theoretical basis

The behaviour of a structure is influenced by various typically uncertain parameters (loading type, loading magnitude, dimensions, or material data,...). All parameters are described by random variables which are collected in the vector of basic variables $\mathbf{X} = (X_1, X_2, ...)$. We will restrict ourselves to those basic variables $X_j$ for which the joint density $f_X(x_1, \ldots, x_n)$ exists and the joint distribution function $F(\mathbf{x})$ is given by

$$F(\mathbf{x}) = P(X_1 < x_1, \ldots, X_n < x_n) = \int_{-\infty}^{x_1} \cdots \int_{-\infty}^{x_n} f_X(t_1, \ldots, t_n) dt_1 \ldots dt_n. \quad (1)$$

The deterministic safety margin $R-S$ is based on the comparison of a structural resistance (threshold) $R$ and loading $S$ (which is usually an invariant measure of local stress at a hot spot or in a representative cross-section). With $R$, $S$ functions of $\mathbf{X}$ the structure fails for any realization with non-positive limit state function $g(\mathbf{X})$, i.e.

$$g(\mathbf{X}) = R(\mathbf{X}) - S(\mathbf{X}) \begin{cases} < 0 & \text{for failure} \\ = 0 & \text{for limit state} \\ > 0 & \text{for safe structure} \end{cases} \quad (2)$$

Different definitions of limit state functions for various failure modes are suggested in Table 1. The limit state $g(\mathbf{x}) = 0$ defines the limit state hyper surface $\partial V$ which separates the failure region $V = \{\mathbf{x} | g(\mathbf{x}) < 0\}$ from the safe region. Figure 2 shows the densities of two random variables $R, S$, which are generally unknown or difficult to establish. The failure probability $P_f = P(g(\mathbf{X}) \leq 0)$ is the probability that $g(\mathbf{X})$ is non-positive. This means

$$P_f = P(g(\mathbf{X}) \leq 0) = \int_V f_X(\mathbf{x}) d\mathbf{x}. \quad (3)$$

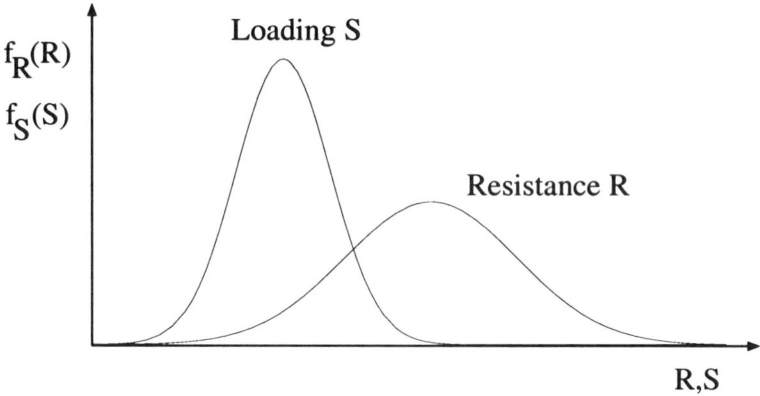


*Figure 2.* Basic $R - S$ problem in $f_R, f_S$ presentation on one axis

In general it is not possible to calculate $P_f$ analytically from the failure integral, because of the complex structure of the failure region $V$. Additionally, it is not necessary that the limit state function is given explicitly but only in algorithmic form. A FE–analysis of structures with one initial data set gives only one value of the limit state function.

TABLE 1. Different limit state functions

| Analysis | Resistance $R$ | Loading $S$ | Limit state function |
|---|---|---|---|
| Elastic strength | yield stress $\sigma_y$ | equivalent stress $\hat{\sigma}$ | $g = \sigma_y - \hat{\sigma}$ |
| Serviceability | displ. threshold $u_0$ | displacement $u$ | $g = u_0 - u$ |
| Fatigue | critical damage $D_{cr}$ | accum. damage $D$ | $g = D_{cr} - D$ |
| Elastic stability | buckling load $F_{cr}$ | compressive load $F$ | $g = F_{cr} - F$ |
| Elastic vibration | eigen frequency $\omega_0$ | harm. exitation $\Omega$ | $g = \omega_0 - \Omega$ |
| Brittle fracture | fract. toughness $K_{IC}$ | SIF $K_I$ | $g = K_{IC} - K_I$ |
| Limit load | limit load | applied load $P$ | |
| | $P_y = \alpha_y P_0$ | $P = \alpha_0 P_0$ | $g = \alpha_y - \alpha_0$ |
| Shakedown | shakedown domain | applied domain | |
| | $\mathcal{L}_{SD} = \alpha_{SD} \mathcal{L}_0$ | $\mathcal{L}_a = \alpha_a \mathcal{L}_0$ | $g = \alpha_{SD} - \alpha_a$ |

## 2.1. MONTE-CARLO-SIMULATION (MCS)

The MCS is a well known method for the evaluation of the failure probability $P_f$. For use together with the simulation methods there are less strict requirements on the analytical properties of the limit state function and functions of the algorithmic type (like "black box") can be used. The straight-forward (or crude) MCS become generally costly for small probabilities. In the mean it thus takes about $1/P_f$ simulations to obtain one outcome of $\mathbf{X}$ in the failure region $V$, such that for typically failure probabilities less than $10^{-4}$ the computational effort of crude MCS increases quickly with reliability [2] but not with the number of basic variables (contrary to FORM/SORM). Importance Sampling or other variance reduction techniques should be used to reduce the computational effort [3]. MSC is an approximate solution of the exact stochastic problem.

## 2.2. FIRST/SECOND ORDER RELIABILITY METHOD (FORM/SORM)

First and Second Order Reliability Methods (FORM/SORM) are analytical probability integration methods. Therefore, the defined problem has to fulfill the necessary analytical requirements (e.g. FORM/SORM apply to problems, where the set of basic variables is continuous). Because of the large computational effort of MCS due to small failure probabilities ($10^{-4}$ to $10^{-8}$), any effective analysis is based on FORM/SORM [8]. The failure probability is computed in three steps.

- Transformation of basic variable $\mathbf{X}$ into the standard normal vector $\mathbf{U}$,
- Approximation $V_a$ of the failure region $V$ in the $\mathbf{U}$–space,
- Computation of the failure probability due to the approximation $V_a$

### 2.2.1. *Transformation*

The basic variables $\mathbf{X}$ are transformed into standard normal variables $\mathbf{U}$ (zero mean, unit standard deviation). Such a transformation is always possible for continuous random variables. If the variables $X_i$ are mutually independent, with distribution functions $F_{X_i}$, each variable can be transformed separately by the Gaussian normal distribution $\Phi$ into $\mathbf{U}_i = \Phi^{-1}[F_{X_i}(\mathbf{x}_i)]$. For dependent random variables analogous transformations can be used [8]. The function $G(\mathbf{u}) = g(\mathbf{x})$ is the corresponding limit state function in $\mathbf{U}$–space. The dimension of the $\mathbf{U}$–space depends on the dependencies of the random variables $\mathbf{X}_i$ and is not necessarily equal to the dimension of the $\mathbf{X}$–space. However, the transformation to $\mathbf{U}$–space is exact and not an approximation [3].

### 2.2.2. *Approximation*

In FORM a linear approximation $V_a$ of the failure region $V$ is generated. The failure region $V$ is approximated at a point $\mathbf{u}_0 \in \partial V$ with the normal $\boldsymbol{a} = \nabla_u G(\mathbf{u}_0)$

$$V_a = \{\mathbf{u} | \nabla_u^T G(\mathbf{u}_0)\mathbf{u} + a_0 \leq 0\} = \{\mathbf{u} | \boldsymbol{a}^T \mathbf{u} + a_0 \leq 0\} \tag{4}$$

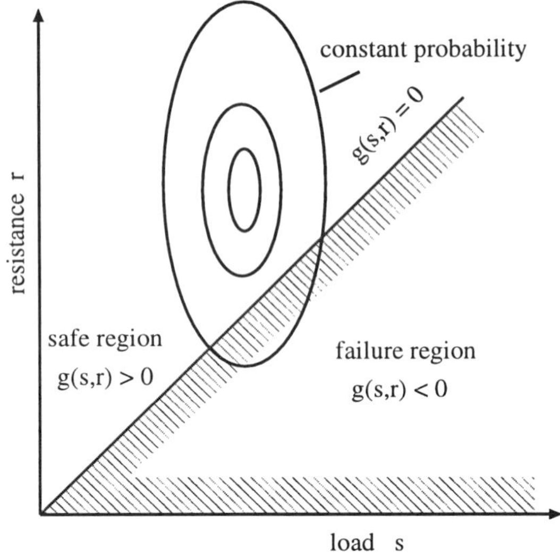


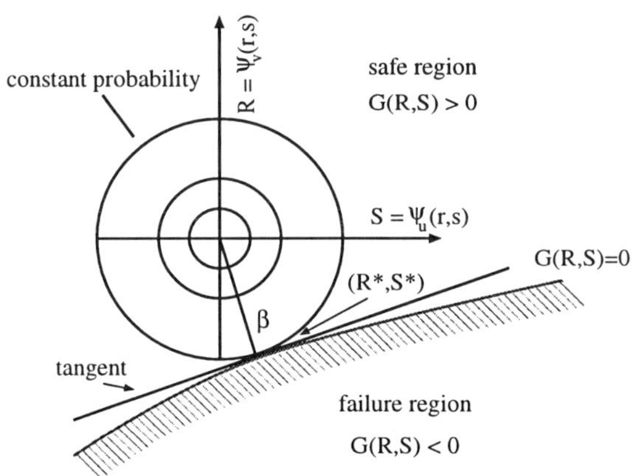


*Figure 3.* Transformation into normally distributed random variables

The limit state hyper surface $\partial V_a$ is represented by the normal form

$$\partial V_a = \left\{ \mathbf{u} \;\middle|\; \boldsymbol{a}^T \mathbf{u} + a_0 = 0 \right\} = \left\{ \mathbf{u} \;\middle|\; \boldsymbol{\alpha}^T \mathbf{u} + \beta = 0 \right\}, \tag{5}$$

with $\boldsymbol{\alpha} = \boldsymbol{a}/|\boldsymbol{a}|$ and $\beta = a_0/|\boldsymbol{a}|$, such that $|\boldsymbol{\alpha}| = 1$. The vector $\boldsymbol{\alpha}$ is proportional to the sensitivities $\nabla_u G(\mathbf{u}_0)$. The failure event $\{\mathbf{u} \in V_a\}$ is equivalent to the event $\{\boldsymbol{\alpha}^T \mathbf{u} \leq -\beta\}$, such that an approximation of the failure probability $P_f$ is

$$\mathrm{P}_f = \mathrm{P}(\boldsymbol{\alpha}^T \mathbf{U} \leq -\beta) = \Phi(-\beta) \tag{6}$$

because the random variable $\boldsymbol{\alpha}^T\mathbf{U}$ is normally distributed. The failure probability depends only on $\beta$, such that it is called safety index. If it is possible to derive $\beta$ analytically from the input data, the probability $P_f$ is calculated directly from the function $\boldsymbol{\Phi}$.

If the limit state function is linear in $\mathbf{U}$-space a quadratic approximation of the failure region $V$ gives closer predictions of $P_f$. These second order methods (SORM) may be either based on a nonlinear optimization algorithm or on correction of a FORM analysis. FORM/SORM give the exact solution to an approximate problem. The numerical effort depends on the number of stochastic variables but not on $P_f$ (in opposition to MCS).

2.2.3. *Computation*

To apply FORM/SORM one or several likely failure points on the limit state surface in $\mathbf{U}$–space must be identified. These points are defined by having a locally minimum distance to the origin. Therefore, a nonlinear constrained optimization problem must be solved [2]

$$\beta = \min \mathbf{u}^T\mathbf{u} \quad \text{such that} \quad \{\mathbf{u} \mid G(\mathbf{u}) \leq 0\}, \tag{7}$$

which usually needs the gradient of $G(\mathbf{u})$.

The design point $\mathbf{u}^* \in \partial V_a$ is the point, which is the solution of problem (7), or which is closest to the origin. The limit state function $G(\mathbf{U})$ is approximated by its linear Taylor series in point $\mathbf{u}_0 \in \partial V$

$$G(\mathbf{u}) = G(\mathbf{u}_0) + \nabla_u G(\mathbf{u}_0)(\mathbf{u} - \mathbf{u}_0) \tag{8}$$

in order to generate the tangent hyperplane in point $\mathbf{u}_0$. Let $\mathbf{u}_k$ be an approximation of the design point $\mathbf{u}^*$. If $\nabla_u G(\mathbf{u}_k) \neq 0$ holds, the following iterative procedure is defined

$$\mathbf{u}_{k+1} = \frac{\nabla_u G(\mathbf{u}_k)}{|\nabla_u G(\mathbf{u}_k)|^2}\left[\mathbf{u}_k^T \nabla_u G(\mathbf{u}_k) - G(\mathbf{u}_k)\right] \tag{9}$$

as a search algorithm for the design point $\mathbf{u}^*$.

The derivates are determined by

$$\nabla_u G(\mathbf{u}) = \nabla_u g(\mathbf{x}) = \nabla_x g(\mathbf{x}) \nabla_u \mathbf{x}. \tag{10}$$

If the structural problem is solved by the Finite Element Method (FEM) this gradient information is obtained from a sensitivity analysis, which consumes much computing time.

Extension of this type of reliability analysis to plastic structural failure faces several problems which are not present in linear elastic analysis: Local stress has no direct relevance to plastic failure and structural behaviour becomes load-path dependent. No straight-forward $g(\mathbf{X})$ is obtained from standard incremental analysis if failure is assumed by plastic collapse, by ratcheting or by plastic shakedown (LCF). It is even more difficult to obtain the gradient of $g(\mathbf{X})$. Therefore, MCS (improved by importance sampling or by some other means of variance reduction) is generally used in connection with incremental nonlinear reliability analyses. However, the computing effort of MCS is extreme.

### 2.3. RESPONSE SURFACE METHODS (RSM)

Repeated FEM analyses are the most time consuming part in both, MCS and FORM/SORM. Therefore the limit state function is replaces by a simple function, which is obtained as the approximation to the function values resulting from only few FEM analyses. Usually a linear or quadratic polynomial of the basic variables is employed. Starting from some values of the limit state function a fit is generated. Adopting the simpler response functions allows more efficient simulation or parameter studies. However, further research is needed to make RSM robust for high-reliable problems of many basic variables [3].

### 2.4. SYSTEMS RELIABILITY

Linear elastic material models do not allow to define a limit state function such that it is able to describe e.g. collapse or buckling, because the yield stress $\sigma_y$ or the ultimate stress $\sigma_u$ is a fictitious parameter in these models. Exceeding $\sigma_y$ or $\sigma_u$ at any location can be associated with collapse only for statically determined structures. Most real structures are statically undetermined. They are thus safer, because redundancy allows some load carrying capacity beyond partial collapse of a structural member (or section). A local threshold concept would have to define a composite limit state function in terms of stress at different locations such as the plastic hinges in a frame structure. If these locations are known a-priori, the definition of the limit state function of a series system would be possible. Failure occurs if a sufficient number of hinges 1 AND 2 AND ... have been detected.

Different sequences of hinge development may lead to different kinematic mechanisms of collapse modes. For example a plane portal frame may fail in beam OR in sway OR in combined mode. This is a parallel system in fault-tree representation. Reliability analysis of such parallel series systems may be based on the failure modes approach (event-tree representation) or on the survival modes approach (failure-graph representation). Analysis is possible with MCS and with FORM/SORM but it causes additional complications. The analyst is required to identify the complete system representation. Algorithms for automatic generation of the significant failure modes work properly for truss structures [15]. However, more complex structures may not be considered as consisting of a finite number of members with lumped parameters (e.g. beams). In a FEM discretization a series of finite elements may be formed which must all fail in order to define a possible collapse mechanism. The definition of such series systems is neither straight-forward nor unique. Moreover, the resulting system may be large and complex.

These difficulties are avoided by the direct limit and shakedown approach, which formulates the limit state function by a proper mathematical problem of plastic failure. It remains to define a parallel system in the typical situation that more than one failure mode is possible. According to the Bree-diagram Fig. 1 the thin tube may fail locally by LCF at low mechanical stress when crossing the shakedown limit. At higher mechanical stress it may fail globally by ratcheting. Using different starting points in a FORM/SORM analysis $n$ different design points $u_i^*$ may be obtained and collected in the vector $\mathbf{u}^*$, leading to different $\beta_i$-factors and failure probabilities $P_{fi} = \Phi(-\beta_i)$ for the respective failure modes. For linearized limit state functions a matrix $\mathbf{R}$, of correlations coefficients for $\mathbf{u}^*$ and $\boldsymbol{\beta}$, may be obtained. Then $P_f$ may be estimated from the $n$-dimensional standard multinormal distribution $\Phi_n(-\boldsymbol{\beta};\mathbf{R})$. With little numerical effort also first-order series bounds for the cases of fully dependent and fully independent failure modes may be used

$$\max_{i=1..n}\{P_{fi}\} \leq P_f \leq 1 - \prod_{i=1}^{n}(1 - P_{fi}). \tag{11}$$

Different methods have been proposed to find all significant failure modes or at least the most dominant one [14].

## 3. Concepts of limit and shakedown analysis

An objective measure of the loss of stability may be based on the loss of stable equilibrium [5]. A system is said to be in a critical state of neutral equilibrium or collapse if the second-order energy dissipation vanishes,

$$\int_{\Omega} \dot{\varepsilon} : \dot{\sigma} d\Omega = 0, \tag{12}$$

for at least one kinematically admissible strain-rate field $\dot{\varepsilon}$.

In a FEM discretization using a varying (symmetric part of the) stiffness matrix $\mathbf{K}$ or an appropriate update of it for nonlinear analysis this occurs, if

$$\dot{\mathbf{u}}\mathbf{K}\dot{\mathbf{u}} = 0 \tag{13}$$

holds, for at least one admissible nodal velocity vector $\dot{\mathbf{u}}$. This is equivalent with the limit state function $g(\mathbf{X})$ =det$\mathbf{K}$ = 0. A sufficient condition is, that the smallest eigenvalue of $\mathbf{K}$ vanishes. Both limit state functions are numerically expensive and suffer from hard numerical problems (round-off and truncation error, non-uniform dependence on basic variables and possibly non-smoothness). A limiting structural stiffness may be used (e.g. half the elastic stiffness matrix $\mathbf{K}_0$) in the sense of the Double Elastic Slope Method (DESM) of ASME Code, Sect. III, NB-3213.25) on the basis of an appropriate matrix norm

$$g(\mathbf{X}) = ||\mathbf{K}(\mathbf{X})|| - 0.5||\mathbf{K}_0(\mathbf{X})||. \tag{14}$$

Such complications do not occur if plastic collapse modes are identified by limit analysis. Moreover the DESM introduces the elastic properties into the plastic collapse problem, which is mechanically questionable. However, the stiffness approach may be employed for failure modes like buckling, which in turn fall outside of limit analysis.

Static limit load theorems are formulated in terms of stress and define safe structural states giving an optimization problem for safe loads. The maximum safe load is the limit load avoiding collapse. Alternatively, kinematic theorems are formulated in terms of kinematic quantities and define unsafe structural states yielding a dual optimization problem for the minimum of limit loads [10]. Any admissible solution to the static or kinematic theorem is a true lower or upper bound to the safe load, respectively. Both can be made as close as desired to the exact solution. If upper and lower bound coincide, the true solution has been found. The limit load factor is defined in (15) by $\alpha = \mathbf{P}_l/\mathbf{P}_0$, where $\mathbf{P}_l = (\mathbf{b}_l, \mathbf{p}_l)$ and $\mathbf{P}_0 = (\mathbf{b}_0, \mathbf{p}_0)$ are the plastic limit load and the chosen reference load, respectively. Here we have supposed that all loads ($\mathbf{b}$ body forces and $\mathbf{p}$ surface loads) are applied in a monotone and proportional way. The theorems are stated below.

### 3.1. STATIC OR LOWER BOUND LIMIT LOAD ANALYSIS

Find the maximum load factor for which the structure is safe. The structure is safe against plastic collapse if there is a stress field $\boldsymbol{\sigma}$ such that the equilibrium equations are satisfied and the yield condition is nowhere violated. We could then formulate a maximum problem:

$$\begin{array}{lrll} \max & \alpha & & \\ \text{s. t.} & \phi(\boldsymbol{\sigma}) \leq & \sigma_y & \text{in } \Omega \\ & \operatorname{div}\boldsymbol{\sigma} = & -\alpha\mathbf{b}_0 & \text{in } \Omega \\ & \boldsymbol{\sigma}\,\mathbf{n} = & \alpha\mathbf{p}_0 & \text{on } \partial\Omega_\sigma \end{array} \tag{15}$$

for the structure $\Omega$, traction boundary $\partial\Omega_\sigma$ (with outer normal $\mathbf{n}$), yield function $\phi$, body forces $\alpha\mathbf{b}_0$ and surface loads $\alpha\mathbf{p}_0$.

The shakedown analysis starts from Melan's lower bound theorem [13]. In the shakedown analysis the equilibrium conditions and the yield criterion of the actual stresses have to be fulfilled at every instant of the load history.

### 3.2. STATIC OR LOWER BOUND SHAKEDOWN ANALYSIS

Find the maximum load factor for which the structure is safe. The structure is safe against LCF or ratcheting if there is a stress field $\boldsymbol{\sigma}(t)$ such that the equilibrium equations are satisfied and the yield condition is nowhere and at no instant $t$ violated. We could then formulate a maximum problem:

$$\begin{array}{ll} \max & \alpha \\ \text{s. t.} & \phi(\boldsymbol{\sigma}(t)) \leq \sigma_y \quad \text{in } \Omega \\ & \operatorname{div}\boldsymbol{\sigma}(t) = -\alpha\mathbf{b}_0(t) \quad \text{in } \Omega \\ & \boldsymbol{\sigma}(t)\ \mathbf{n} = \alpha\mathbf{p}_0(t) \quad \text{on } \partial\Omega_\sigma \end{array} \tag{16}$$

for the structure $\Omega$, traction boundary $\partial\Omega_\sigma$ (with outer normal $\mathbf{n}$), yield function $\phi$, body forces $\alpha\mathbf{b}_0(t)$ and surface loads $\alpha\mathbf{p}_0(t)$ for all $\mathbf{b}_0(t)$,$\mathbf{p}_0(t)$ in a given initial load domain $\mathcal{L}_0$.

The maximum problems (15) and (16) are solved by splitting the stresses $\boldsymbol{\sigma}$ and $\boldsymbol{\sigma}(t)$ into fictitious elastic stresses $\boldsymbol{\sigma}^E$, $\boldsymbol{\sigma}^E(t)$ and time invariant residual stresses $\boldsymbol{\rho}$.

### 3.3. DISCRETIZATION AND OPTIMIZATION

The structure is divided into $N$ finite elements with $NG$ Gaussian points. PERMAS calculates the load dependent elastic stress vectors $\boldsymbol{\sigma}_i^E(t)$ at point $i$ by means of a displacement method. The discretized homogeneous equilibrium conditions of the residual stresses can be noted as (see [21])

$$\sum_{i=1}^{NG} \mathbf{C}_i \boldsymbol{\rho}_i = \mathbf{0}, \tag{17}$$

where the $\mathbf{C}_i$ are system dependent element matrices. This equation represents a discretized formulation of the equilibrium equations. In convex load domains $\mathcal{L}$ in the form of a polyhedron every load $\mathbf{P}(t)$ can be represented as a convex combination of the $NV$ vertices $\mathbf{P}(j)$, i.e.

$$\mathbf{P}(t) = \sum_{j=1}^{NV} \mu_j \mathbf{P}(j) \quad \text{with } 0 \leq \mu_j \quad \text{and} \quad \sum_{j=1}^{NV} \mu_j = 1. \tag{18}$$

From convex optimization theory (see [6]) follows, that the yield condition only has to be satisfied in the vertices of $\mathcal{L}$, such that the original problem is transferred in a time–independent one. So the problem (16) should be transformed with the stresses $\boldsymbol{\sigma}_i^E(j)$ as fictitious elastic response to vertex $\mathbf{P}(j)$ at the Gaussian point $i$ into

$$\begin{array}{ll} \max & \alpha \\ \text{s. t.} & \phi[\alpha\boldsymbol{\sigma}_i^E(j)+\boldsymbol{\rho}_i] \le \sigma_{y,i}^2 \quad i=1,\dots,NG, \quad j=1,\dots,NV. \end{array} \tag{19}$$

The unknowns of the problem are $\alpha$ and the residual stresses $\boldsymbol{\rho}_i$. So this is a large scale optimization problem for a realistic Finite–Element–Model. Collecting the matrices $\mathbf{C}_i$ to the maximum rank rectangular global matrix $\mathbf{C}$ and the residual stresses $\boldsymbol{\rho}_i$ to a global vector $\boldsymbol{\rho}$ condition (17) reads $\mathbf{C}\boldsymbol{\rho}=\mathbf{0}$. The kernel of the linear form $\mathbf{C}\boldsymbol{\rho}$ is a linear subspace $\mathcal{B}_d$ of the space of all residual stresses. With a basis $\{\mathbf{b}_1,\dots,\mathbf{b}_d\}$ of $\mathcal{B}_d$ every residual stress $\boldsymbol{\rho}\in\mathcal{B}_d$ (i.e. every $\boldsymbol{\rho}$ that satisfies (17)) can be represented as

$$\boldsymbol{\rho}=\sum_{j=1}^{d} y_j\mathbf{b}_j. \tag{20}$$

Collecting the $\mathbf{b}_k$ and $y_k$ to $\mathbf{B}_d$ and $\boldsymbol{y}_d$ respectively, the equilibrium conditions are fulfilled. The final optimization problem reads

$$\begin{array}{ll} \max & \alpha \\ \text{s.t.} & \phi[\alpha\boldsymbol{\sigma}_i^E(j)+\mathbf{B}_d\boldsymbol{y}_d] \le \sigma_{y,i}^2 \quad \forall\ i,j. \end{array} \tag{21}$$

It is convex, because the restrictions and the objective function are convex. Thus every local minimum is a global minimum (see [6]). Instead of infinitely many time–dependent restrictions and unknowns in the continuous case after the complete discretization there are only $NG\times NV$ time–independent restrictions and $d+1$ unknowns.

This problem is now solved in a recursive manner. Starting from initial residual stresses $\boldsymbol{\rho}^{(0)}=\mathbf{0}$ and shakedown–factor $\alpha^{(0)}=1$ in the $k$–th step of the algorithm $r$ basis vectors $\mathbf{b}^{r,k}$ from $\mathcal{B}_d$ with $r\ll d$ are selected and the problem

$$\begin{array}{ll} \max & \alpha^{(k)} \\ \text{s.t.} & \phi[\alpha^{(k)}\boldsymbol{\sigma}_i^E(j)+\boldsymbol{\rho}^{(k-1)}+\mathbf{B}_r^{(k)}\boldsymbol{y}^{(k)}] \le \sigma_{y,i}^2 \quad \forall\ i,j \end{array} \tag{22}$$

is solved in the reduced residual stress space $\mathcal{B}_r$ which is spanned by the $r$ basis vectors $\mathbf{b}^{r,k}$. After solving the sub–problem with the solution $(\alpha^{(k)},\boldsymbol{y}^*)$ the $k$–th residual stresses are updated to

$$\boldsymbol{\rho}^{(k)}=\boldsymbol{\rho}^{(k-1)}+\mathbf{B}_r^{(k)}\boldsymbol{y}^*. \tag{23}$$

In every step $\alpha^{(k)}$ is improved. This basis reduction method was developed in [18] and extended in [22], [7].

In every step $k$ the stresses at the Gaussian point $i$ to the load vertex $\mathbf{P}(j)$ at the beginning of the step are

$$\boldsymbol{\sigma}_i^{(k-1)}(j) = \alpha^{(k-1)}\boldsymbol{\sigma}_i^E(j) + \boldsymbol{\rho}_i^{(k-1)}. \tag{24}$$

A load vertex $\mathbf{P}(j)$ of the load domain $\mathcal{L}$ is called *active* if a Gaussian point $i$ exists with $\phi(\boldsymbol{\sigma}_i^{(k-1)}(j)) = 0$. Adding a load increment $\Delta\alpha^{(k)}\mathbf{P}(j)$ for every active load vertex the system will yield further. By an equilibrium iteration in every step $n$ of the iteration the stress field $\boldsymbol{\sigma}^n(j)$ is in equilibrium with the external load $(\alpha^{(k-1)} + \Delta\alpha^{(k)})\mathbf{P}(j)$. Residual stresses are obtained as the difference between the stresses $\boldsymbol{\sigma}^n(j)$ and $\boldsymbol{\sigma}^1(j)$, because the external load in every iteration step is the same. Thus $n$ equilibrium iterations generate $n-1$ residual stresses. Performing only a few equilibrium iterations guarantees linear independence of the residual stresses. Thus a basis of the reduced subspace and therefore a solution of the problem is obtained. In every step of limit analysis 3–6 residual stresses are computed, all belonging to the only load vertex. In shakedown analysis a maximum total number of 6 residual stresses are calculated for all vertices.

In the $k$–th step the problem is solved by a SQP–method (Sequential Quadratic Programming) with augmented Lagrangian type line search function (see [17]). Armijo's step length rule and BFGS matrix update are used. Because of the small numbers of unknowns and the large number of restrictions, the quadratic sub–problems are solved by an active–set–strategy (see [6]). Derivatives are calculated analytically avoiding automatic differentiation methods. All computations were performed using version 4 of PERMAS [16] (implementation in version 7 is in preparation).

For kinematic hardening material the limit load factor is obtained by the same optimization problems with yield stress $\sigma_y$ replaced by ultimate stress $\sigma_u$. For homogeneous material ($\sigma_{y,i} = \sigma_y, \forall i$) the computed limit load is a linear function of the chosen failure stress.

## 4. Plastic failure and reliability analysis

Probabilistic limit and shakedown analyses were pioneered in Italy [1]. Further work seemed to remain restricted to stochastic limit analysis of frames based on linear programming [2], [9], [11], [23]. The present contribution extends plastic reliability analysis towards nonlinear programming, shakedown, and a general purpose large-scale FEM approach using lower bound theorems of limit and shakedown load to define a limit state function $g(\mathbf{X})$ for reliability analysis by FORM. $R$ and $S$ are defined by the limit or shakedown load factor and the applied load factor, respectively. The resulting large-scale optimization problem is transferred to a relatively small one by the basis reduction method described above.

The solution of the limit or shakedown problems (15), (16) is a linear function of the failure stress $\sigma_y$ or $\sigma_u$ if a homogeneous material distribution is assumed. If the structure has a heterogeneous material distribution we obtain in different Gaussian points $i$ eventually different failure stresses $\sigma_{y,i}$ or $\sigma_{u,i}$. Then the limit load is no more a linear function of the failure stresses. In this case the derivates of the limit state function may not be computed directly from the linear function of the failure stresses. The Lagrange multipliers of the optimization problem (15) yield the gradient information of $g(\mathbf{X})$ without any extra computation. This is derived from a variation of $\sigma_{y,i}$ or $\sigma_{u,i}$ as the right hand side of problem (15) (see [6]).

In principle the limit load and shakedown analysis have the following form with $\boldsymbol{G}$ as vector of all inequality restrictions $\phi$, the failure stresses $\boldsymbol{\delta} = (\sigma_{y,1}^2, \ldots, \sigma_{y,NG}^2)$ and the variables $\boldsymbol{\xi} = (\alpha, \boldsymbol{\rho}_1, \ldots, \boldsymbol{\rho}_{NG})$

$$\begin{array}{ll} \max & f(\boldsymbol{\xi}) \\ \text{s. t.} & \boldsymbol{G}(\boldsymbol{\xi}) \leq \boldsymbol{\delta} \end{array} \tag{25}$$

The influence of the failure stresses $\sigma_{y,i}$ on the load factor $\alpha$ is dominated by the derivates $\partial\alpha/\partial\sigma_{y,i}$ or by $\partial f(\boldsymbol{\xi})/\partial\boldsymbol{\delta}$. These derivates could be generated by the following theorem [6]

**Theorem**

Let $\boldsymbol{\xi}^*$ be the solution of problem (25) with the Lagrange multipliers $\boldsymbol{\lambda}^* = (\lambda_1^*, \ldots, \lambda_{NG}^*)$ for an initial right hand side $\boldsymbol{\delta}_0$. Then exists a continuous solution $\boldsymbol{\xi}^*(\boldsymbol{\delta})$ in a neighbourhood of $\boldsymbol{\delta}_0$ and it holds

$$\nabla_{\boldsymbol{\delta}} f(\boldsymbol{\xi}^*(\boldsymbol{\delta}))|_{\boldsymbol{\delta}=\boldsymbol{\delta}_0} = -\boldsymbol{\lambda}^*. \tag{26}$$

with the operator $\nabla_{\boldsymbol{\delta}}(\cdot)$ of all partial derivates of the components of $\boldsymbol{\delta}$.

For the limit and shakedown analysis it follows

$$\frac{\partial\alpha}{\partial\sigma_{y,1}} = -\lambda_1^*, \ldots, \frac{\partial\alpha}{\partial\sigma_{y,NG}} = -\lambda_{NG}^*, \tag{27}$$

such that the influences of the failure stresses on the limit and shakedown load factor $\alpha$ could be obtained by the solution of the problems (15), (16). The flowchart in Figure 4 contains the logical connections of the main analysis steps.

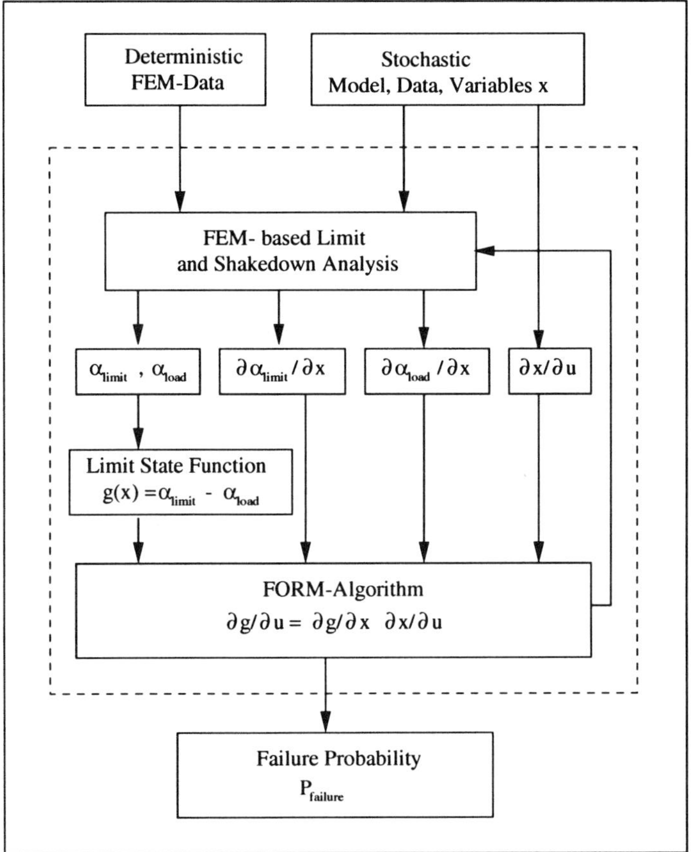


*Figure 4.* Flowchart of the probabilistic limit load analysis

## 5. Examples

The plastic reliability problem can be solved analytically if the limit load is known and $R$ and $S$ are both normally or log-normally distributed. The plate with a hole problem is to test correctness and numerical error.

### 5.1. LIMIT LOAD ANALYSIS

In case of a square plate of length $L$ with a hole of diameter $D$ (see Figure 5) and $D/L = 0.2$ subjected to the uniaxial tension $p$ the exact limit load for plane stress is given by $R = (1 - D/L)\sigma_y$ with the yield stress $\sigma_y$.

Thus the limit load $R$ depends linearly of the realization $\sigma_y$ of the yield stress basic variable X. The load $S$ is a homogeneous uniaxial tension on one side of the plate. The magnitude of the tension is the second basic

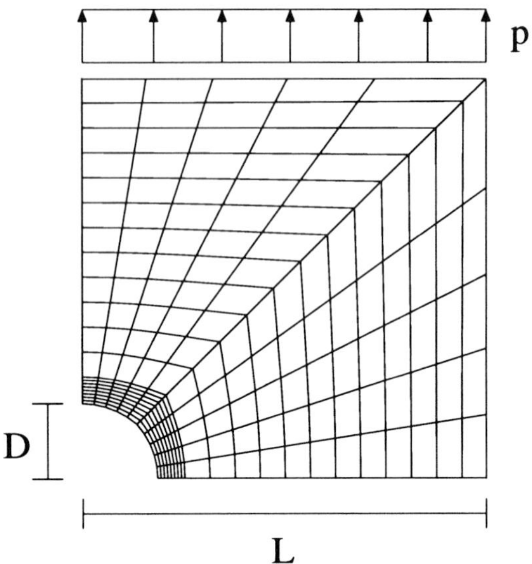


*Figure 5.* Finite element mesh of plate with a hole

variable Y. The limit load $P_{lim}$ of every realization $y$ of Y is

$$P_{lim}(y) = (1 - D/L)\ y. \tag{28}$$

The limit state function is defined by

$$g(x, y) = P_{lim} - P = (1 - D/L)\, y - x. \tag{29}$$

The normally distributed random variables X and Y with means $\mu_r$, $\mu_s$ and deviations $\sigma_r^2$, $\sigma_s^2$ respectively, yield with $x = \sigma_r\ u_r + \mu_r$ and $y = \sigma_s\ u_s + \mu_s$ the transformed limit state function

$$g(u_r, u_s) = ((1 - D/L)\mu_s - \mu_r) + (1 - D/L)\sigma_s u_s - \sigma_r u_r. \tag{30}$$

With realizations $\boldsymbol{u} = (u_r, u_s)^T$ of the new random variable $\mathbf{U}$ it may be written

$$g(\boldsymbol{u}) = \frac{(-\sigma_r, (1 - D/L)\sigma_s)}{\sqrt{\sigma_r^2 + (1 - D/L)^2\sigma_s^2}}\ \boldsymbol{u} + \frac{(1 - D/L)\mu_s - \mu_r}{\sqrt{\sigma_r^2 + (1 - D/L)^2\sigma_s^2}}, \tag{31}$$

such that the safety index $\beta$ (with $D/L = 0.2$) is

$$\beta = \frac{(1 - D/L)\mu_s - \mu_r}{\sqrt{\sigma_r^2 + (1 - D/L)^2\sigma_s^2}} = \frac{0.8\mu_s - \mu_r}{\sqrt{\sigma_r^2 + 0.64\sigma_s^2}} \tag{32}$$

In Figure 6 the failure probabilities $P_f = \Phi(-\beta)$ are shown versus $\mu_r/\mu_s$. The numerical value of $P_f$ of the limit analyses are compared with the analytic values resulting from the exact solution. Both variables are normally distributed with standard deviations $\sigma_r = 0.1\mu_r$ and $\sigma_s = 0.1\mu_s$.

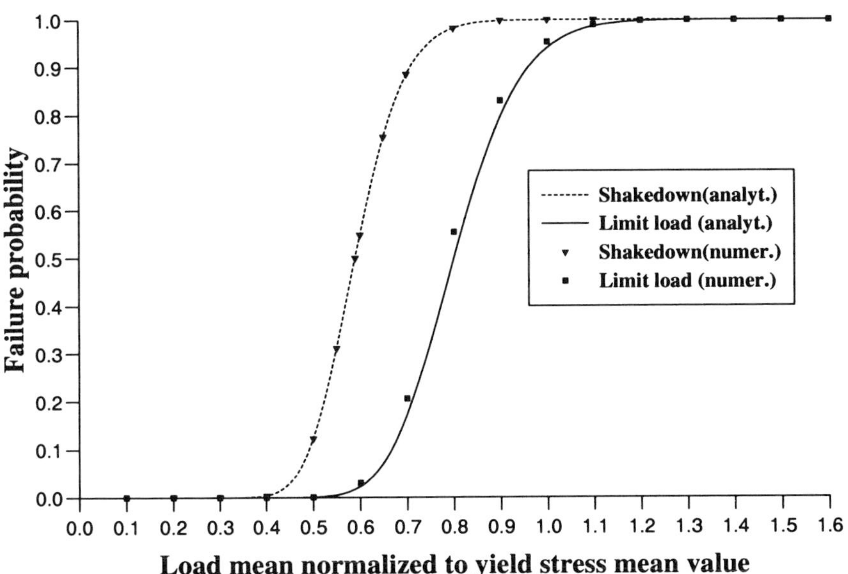


*Figure 6.* Comparison of numerical with analytical results for $\sigma_r = 0.1\mu_r$, $\sigma_s = 0.1\mu_s$

The lower bound theorem generates loads which are safe, but due to the termination error of the iteration they are 1 to 2 % below the analytical limit loads. This error is amplified in the probabilistic analysis. The errors of the FORM calculations and of the numerical limit analyses are included in the results (see Figure 6 and Table 2). The errors are acceptable for highly reliable components, because the wellknown tail sensitivity problem is much more severe. The calculated failure probabilities correspond very well with the analytical probabilities if the analytical limit loads are reduced by 2% to obtain $P_f$ (anal.-2%). This shows that the main part of the observed errors results from the deterministic limit analyses. SORM would give no improved results with a linear limit state function $g(\boldsymbol{u})$. Linearity may be lost, if X or Y are not normally distributed.

Much more severe deviations of the computed failure probabilities have to be expected if other limit state functions were used such as the extension of the plastic zone to a chosen size (e.g. 80% of a ligament) or the DESM (equation (14). Moreover, such limit states give the wrong impression that the stochastic plastic collapse load is sensitive to the basic variable Young's modulus.

### 5.2. SHAKEDOWN ANALYSIS

In the shakedown analysis a convex load domain $\mathcal{L}$ is analyzed. The tension $p$ cycles between zero and a maximal magnitude of $p_0$. Only the amplitudes but not the uncertain full load history enters the solution.

$$0 \leq \ p \leq \ \alpha\lambda p_0, \qquad 0 \leq \lambda \leq 1. \tag{33}$$

In the first simple reliability analysis the maximal magnitude $p_0$ is a random variable, but the minimum magnitude zero is held constant. The results of the FORM calculation are compared with an analytical approximation of the shakedown load in Table 2.

TABLE 2. Comparison of numerical and analytical results for $\sigma_r = 0.1\mu_r$, $\sigma_s = 0.1\mu_s$

| Limit load analysis | | | | Shakedown analysis | | |
|---|---|---|---|---|---|---|
| $\mu_r/\mu_s$ | $P_f$ (num.) | $P_f$ (anal.) | $P_f$ (anal.-2%) | $\mu_r/\mu_s$ | $P_f$ (num.) | $P_f$ (anal.) |
| 0.2 | 2.643E-13 | 1.718E-13 | 2.640E-13 | 0.2 | 1.943E-10 | 1.943E-10 |
| 0.3 | 3.843E-09 | 2.426E-09 | 4.063E-09 | 0.3 | 5.964E-06 | 5.963E-06 |
| 0.4 | 6.112E-06 | 3.872E-06 | 6.416E-06 | 0.4 | 3.877E-03 | 3.877E-03 |
| 0.5 | 1.093E-03 | 7.364E-04 | 1.128E-03 | 0.5 | 1.227E-01 | 1.229E-01 |
| 0.6 | 3.049E-02 | 2.275E-02 | 3.118E-02 | 0.55 | 3.108E-01 | 3.111E-01 |
| 0.7 | 2.067E-01 | 1.734E-01 | 2.112E-01 | 0.59 | 5.000E-01 | 5.000E-01 |
| 0.8 | 5.550E-01 | 5.000E-01 | 5.567E-01 | 0.6 | 5.485E-01 | 5.485E-01 |
| 0.9 | 8.305E-01 | 7.969E-01 | 8.344E-01 | 0.65 | 7.538E-01 | 7.538E-01 |
| 1.0 | 9.544E-01 | 9.408E-01 | 9.554E-01 | 0.7 | 8.858E-01 | 8.858E-01 |
| 1.1 | 9.900E-01 | 9.863E-01 | 9.903E-01 | 0.8 | 9.828E-01 | 9.828E-01 |
| 1.2 | 9.981E-01 | 9.972E-01 | 9.981E-01 | 0.9 | 9.980E-01 | 9.980E-01 |
| 1.3 | 9.996E-01 | 9.995E-01 | 9.996E-01 | 1.0 | 9.997E-01 | 9.997E-01 |
| 1.4 | 9.999E-01 | 9.999E-01 | 9.999E-01 | 1.1 | 9.999E-01 | 9.999E-01 |

Because of the local failure of the plate in the ligament points of the hole, the shakedown factor $\alpha_{SD}$ corresponding to the initial yield load $p_i$ is equal to 2 (see [7], [22]). Therefore, from the yield load $p_i = 0.2949\sigma_y$ resulting from the deterministic FEM-computation follows that the FEM–approximation of the shakedown load is $0.5897\sigma_y$. The implemented shakedown analysis with the basis reduction technique gives very good results for the reliability analysis of the plate (listed in Table 2), because the deterministic shakedown factor 2 is reached in 3 to 5 steps nearly identically.

Additionally, the shakedown reliability analysis needs less computing time than the limit load reliability analysis. The results of the shakedown reliability analysis show a decrease in reliability in comparison with the limit load reliability results. For a load level of $0.4\mu_r$ the reliability decrease by 3 orders of magnitude. This means that the reliability of the structure depends very strongly on the loading conditions, such that the assessment of the load carrying capacity has to be done very carefully.

## 6. CONCLUSIONS

Limit and shakedown theorems of plastic structural failure provide unique definitions of limit state functions. In combination with FEM and with FORM, failure probabilities of passive components are obtained with sufficient precision at very low computational efforts compared to incremental analyses with MCS. In this approach sensitivities need no extra FEM analysis. The remaining numerical error may be estimated or reduced by the additional use of upper bound theorems. It is most important for the analysis under uncertainty that limit and shakedown analyses are based on a minimum of information concerning the constitutive equations and the load history. In fact the shakedown problem is made time invariant. This reduces the costs of the collection of statistical data and the need to introduce stochastic models to compensate the lack of data. Further research is also addressed to more realistic material modelling including non-linear kinematic hardening and continuum damage.

*Acknowledgment: This research has been partly funded by the European Commission through the Brite-EuRam Programme, Project BE 97-4547, Contract BRPR-CT97-0595.*

## References

1. Augusti, G., A. Baratta, and F. Casciati (1984) *Probabilistic Methods in Structural Engineering*. London: Chapman and Hall.
2. Bjerager, P. (1989) Plastic systems reliability by LP and FORM. *Computers & Structures* **31**, 187–196.
3. Bjerager, P. (1991) Methods for structural reliability computations. In: *Course on Reliability Problems: General Principles and Applications of Solids and Structures, International Center for Mechanical Sciences, Udine, Italy, 1990.* CISM lecture notes, Springer, Wien, 1991.
4. Bree, J. (1967) Elastic–plastic behaviour of thin tubes subjected to internal pressure and intermittent high–heat fluxes with applications to fast–nuclear–reactor fuel

elements. *Journal of Strain Analysis* **2**, 226–238.
5. Carmeliet, J. and R. de Borst (1995) Gradient damage and reliability: instability as limit state function. In: F. Wittmann (ed.): *Fracture mechanics of concrete structures. Vol.2.* Freiburg: Aedificatio.
6. Fletcher, R. (1987) *Practical Methods of Optimization.* New York: John Wiley & Sons.
7. Heitzer, M. (1999) Traglast– und Einspielanalyse zur Bewertung der Sicherheit passiver Komponenten. Ph.D. thesis, RWTH Aachen.
8. Hohenbichler, M., S. Gollwitzer, W. Kruse, and R. Rackwitz (1987) New light on first– and second–order reliability methods. *Structural Safety* **4**, 267–284.
9. Klingmüller, O. (1979) Anwendung der Traglastberechnung für die Beurteilung der Sicherheit von Konstruktionen. Ph.D. thesis, Forschungsberichte aus dem Fachbereich Bauwesen **9**, Universität Essen Gesamthochschule.
10. König, J.A. (1987) *Shakedown of Elastic-Plastic Structures.* Amsterdam and Warsaw: Elsevier and PWN.
11. Locci, J. M. (1995) Automatization of limit analysis calculations, application to structural reliability problems. In: M. Lemaire, J.-L. Favre, and A. Mébarki (eds.): *Proceedings of the ICASP7 Conference, Paris, 10–13.7.1995, Vol. 2: Applications of Statistics and Probability.* Rotterdam, Brookfield: A. A. Balkema pp. 1095–1110.
12. Maier, G. et al. (1995) *Bounds and estimates on inelastic deformations: a study of their practical usefulness.* Bruxelles: CEC Report EUR 16555EN.
13. Melan, E. (1936) Theorie statisch unbestimmter Systeme aus ideal–plastischem Baustoff. *Sitzungsbericht der Österreichischen Akademie der Wissenschaften der Mathematisch–Naturwissenschaftlichen Klasse IIa* **145**, 195–218.
14. Melchers, R. (1987) *Structural Reliability - Analysis and Prediction.* Chichester: Ellis Horwood.
15. Murotsu, Y., S. Shao, and S. Quek (1991) Some studies on automatic generation of structural failure modes. In: A. der Kiureghian and P. Thoft-Christensen (eds.): *Reliability and optimization of structural systems '90, Proceedings of the 3rd IFIP WG 7.5 Conference.* Springer, Berlin.
16. PERMAS (1988) *User's Reference Manuals.* Stuttgart: INTES Publications No. 202, 207, 208, 302, UM 404, UM 405.
17. Schittkowski, K. (1981) The nonlinear programming method of Wilson, Han and Powell with augmented Lagrangian type line search function. *Numerische Mathematik* **38**, 83–114.
18. Shen, W. (1986) Traglast– und Anpassungsanalyse von Konstruktionen aus elastisch, ideal plastischem Material. Ph.D. thesis Universität Stuttgart.
19. Staat, M. and M. Heitzer (1997) Limit and shakedown analysis for plastic safety of complex structures. *Transactions of SMiRT* **14**, B02/2.
20. Weichert, D. and Hachemi, A. (1998) Influence of geometrical nonlinearities on the shakedown of damaged structures. *Int. J. Plast.* **14**, 891–907.
21. Stein, E., G. Zhang, and R. Mahnken (1993) Shakedown analysis for perfectly plastic and kinematic hardening materials. In: E. Stein (ed.): *Progress in computational analysis of inelastic structures.* Wien: Springer, pp. 175–244.
22. Zhang, G. (1991) Einspielen und dessen numerische Behandlung von Flächentragwerken aus ideal plastischem bzw. kinematisch verfestigendem Material. Ph.D. thesis Universität Hannover.
23. Zimmermann, J. J. (1991) Analysis of structural system reliability with stochastic programming. Ph.D. thesis John Hopkins University, Baltimore, Maryland.

# UPPER BOUNDS ON POST-SHAKEDOWN QUANTITIES IN POROPLASTICITY[(*)]

G. COCCHETTI and G. MAIER
*Department of Structural Engineering, Technical University (Politecnico)*
*Piazza L. da Vinci 32, 20133 Milan, Italy*

**Abstract**
In this paper various inequalities are established in coupled poroplasticity. These provide upper bounds that can be computed directly for various history-dependent post-shakedown quantities. The main features of the constitutive and computational models considered are as follows: two-phase material; full saturation; piecewise linearization of yield surfaces and hardening; associativity; linear Darcy law; finite element space-discretization in Prager's generalized variables. The results achieved are illustrated by comparative numerical tests.

## 1. Introduction

In previous papers [2] [3], the fundamental shakedown theorems of elasto-plasticity have been extended to poroplasticity on the basis of a piecewise linearization of poroplastic models for fully saturated two-phase materials. On the same constitutive basis, this paper presents theorems which provide upper bounds on history-dependent post-shakedown quantities and illustrates their use.

Motivations for the present study arise in diverse technologies. In particular, in dam engineering, masonry or concrete dams with diffused cracks filled by pressurized water have been successfully analyzed as poroplastic systems [8] [3].

Methods that lead to upper bounds are useful for the following reasons. (*a*) Shakedown (in the sense that the cumulative dissipated energy is bounded in time) represents a necessary but not always sufficient condition for the integrity or "safety" of a structure subjected to a loading domain of variable repeated external actions. In fact,

---

(*) Dedicated to the memory of Professor P. D. Panagiotopoulos.



*D. Weichert and G. Maier (eds.), Inelastic Analysis of Structures under Variable Loads,* 289–314.

even if a structure shakes down it may become unserviceable because of the development of excessive inelastic strains or configuration changes. Such possibility has to be kept in mind especially in the presence of limited material ductility, which is a frequent circumstance in engineering applications of poroplasticity. On the other hand, serviceability can be guaranteed, within the limitations and approximations of the adopted mathematical model, if an upper bound on meaningful deformations is found to be less than a critical threshold. (*b*) In poroplasticity, there is an additional specific motivation for bounding procedures of this kind: the convenient assumption of constant permeability for the linearized (Darcy's) diffusion law can *a posteriori* be corroborated if an upper bound on the first invariant of the inelastic strain tensor turns out to show a limited effect of it on permeability. In fact, the permeability of a two-phase fully-saturated porous medium primarily depends on the dilatancy or compactancy of the solid skeleton [13].

The shakedown (SD) theory and its manifold extensions and computational techniques in elastoplasticity are the subject of an abundant literature condensed in books such as [11] [4] [9] [14] [12]. No survey of previous work on SD in general will be attempted herein. Specific antecedents of the results expounded in this paper can be found in early contributions to SD analysis in engineering plasticity [16] [20] and in recent publications on poroplasticity [17] [2] [3].

## 2 Continuum and space discrete formulations

2.1 The class of "piecewiselinear" poroplastic constitutive models (with linear yield functions and hardening for a multiplicity of yield modes, as an approximation of nonlinear ones, see [15]) can be described by the following relation set:

$$\varphi_\alpha = N^\sigma_{ij\alpha}\,\sigma_{ij} + N^p_\alpha\, p - Y_\alpha \le 0, \qquad Y_\alpha = Y^0_\alpha + H_{\alpha\beta}\,\lambda_\beta \tag{1}$$

$$\dot{\varepsilon}^p_{ij} = N^\sigma_{ij\alpha}\,\dot{\lambda}_\alpha, \qquad \dot{\zeta}^p = N^p_\alpha\,\dot{\lambda}_\alpha \tag{2}$$

$$\dot{\lambda}_\alpha \ge 0, \qquad \varphi_\alpha\dot{\lambda}_\alpha = 0, \qquad \alpha = 1 \ldots n_y \tag{3}$$

where the quantities denoted by $N$, $Y^0$ and $H$ are constants; $\varphi_\alpha$ and $\lambda_\alpha$ represent yield function and plastic multiplier for the $\alpha$-th mode; $Y^0_\alpha$ the relevant (positive) initial "yield limit"; $\sigma_{ij}$, $\varepsilon_{ij}$, $p$ and $\zeta$ denote total stress tensor, strain tensor of the solid skeleton, pressure of the pore liquid and fluid content, respectively; a dot marks time derivatives and superscript $p$ inelastic addends.

Elastic addends of strains and fluid content (marked by superscript $e$) are linearly related to stresses $\sigma_{ij}$ and pressure $p$ through the classical poroelastic (Biot) constitutive law, governed in the isotropic case by four material parameters ($G$, $\nu$, $\alpha$, $M$); namely:

$$2G\varepsilon_{ij}^{e} = \sigma_{ij} - \frac{\nu}{1+\nu}\sigma_{hh}\delta_{ij} + \alpha\frac{1-2\nu}{1+\nu}p\delta_{ij} \tag{4}$$

$$\zeta^{e} = \frac{\alpha}{2G}\frac{1-2\nu}{1+\nu}\sigma_{hh}^{e} + m^{-1}p, \qquad m^{-1} = \left(\frac{1}{M} + \frac{3\alpha^{2}}{2G}\frac{1-2\nu}{1+\nu}\right) \tag{5}$$

The usual additivity hypothesis concerns the inelastic and poroelastic kinematic addends:

$$\varepsilon_{ij} = \varepsilon_{ij}^{e} + \varepsilon_{ij}^{p}, \qquad \zeta = \zeta^{e} + \zeta^{p} \tag{6}$$

Let the above constitutive model be associated with the linear field equations of compatibility (linear kinematics, i.e. infinitesimal strains, is assumed), equilibrium, mass conservation and diffusion. These equations read, respectively ($\tau$ denoting time, $x_i$ Cartesian coordinates and commas derivatives with respect to them):

$$\varepsilon_{ij} = \tfrac{1}{2}\left(u_{i,j} + u_{j,i}\right) \qquad \text{in } \Omega \tag{7}$$

$$\sigma_{ij,i} + \hat{b}_j(x_h,\tau) = 0 \qquad \text{in } \Omega \tag{8}$$

$$\dot{\zeta} = -q_{i,i} \qquad \text{in } \Omega \tag{9}$$

$$q_i = k_{ij}\,\pi_j \qquad \pi_j = -p_{,j} + \hat{f}_j(x_h,\tau) \qquad \text{in } \Omega \tag{10}$$

Boundary and initial conditions can be expressed as follows:

$$u_i = \hat{u}_i(x_h,\tau) \quad \text{on } \Gamma_u, \qquad \sigma_{ij}n_i = \hat{t}_j(x_h,\tau) \quad \text{on } \Gamma_\sigma \tag{11}$$

$$p = \hat{p}(x_h,\tau) \quad \text{on } \Gamma_p \qquad q_i n_i = \hat{q}(x_h,\tau) \quad \text{on } \Gamma_q \tag{12}$$

$$p = p^{0}(x_h) \quad \text{in } \Omega \text{ at } \tau = 0 \tag{13}$$

The above formulated initial-boundary value problem of poroplastic analysis for continua exhibits as a non traditional feature the piecewiselinear specialization (1)-(3) of the inelastic part of poroplastic constitutive relationships. These relationships are discussed in full generality, as for their physical origin and their material parameters, in recent treatises such as [7] [13]. Only the meanings of new symbols in (4)-(13) are specified here: the (open) domain occupied by the system is denoted by $\Omega$, its boundary by $\Gamma$; $\Gamma_u$ and $\Gamma_\sigma$ represent the (complementary, disjoint) portions of $\Gamma$ where displacements $\hat{u}_i$ of the solid skeleton and tractions $\hat{t}_i$, respectively, are assigned

($\Gamma_u \cup \Gamma_\sigma = \Gamma$, $\Gamma_u \cap \Gamma_\sigma = \varnothing$). Similarly, $\Gamma_p$ and $\Gamma_q$ are the parts of $\Gamma$ ($\Gamma_p \cup \Gamma_q = \Gamma$, $\Gamma_p \cap \Gamma_q = \varnothing$) on which fluid pressure $\hat{p}$ and flux $\hat{q} = n_i\, q_i$, respectively, are assigned, $n_i$ being the outward unit normal to the boundary $\Gamma$ (assumed as smooth) and $q_i$ being the flux vector (liquid volume per unit time and unit crossed area orthogonal to axis $x_i$).
Besides the boundary data $\hat{u}_i$, $\hat{t}_i$, $\hat{p}$ and $\hat{q}$, the external actions (marked by caps) include: bulk body forces $\hat{b}_i$; fluid specific weight $\hat{f}_i$ (per unit liquid volume); initial field $p^0$ over $\Omega$. In Darcy law (10a), $\pi_i$ denotes the "filtration force" defined by (10b) in terms of pressure gradient and fluid specific weight. Finally, $k_{ij} = k_{ji}$ will denote the permeability tensor of the porous material, which is assumed as constant in time.

2.2 The space discretization of the above formulated initial-boundary value problem is carried out below by a multifield modelling in generalized variables (not by the traditional displacement approach to finite element analysis). In fact, such modelling (see e.g. [5] [17]) preserves the essential features of the continuum problem and, hence, provides a suitable basis for mechanically meaningful theoretical developments such as the bounding techniques pursued herein.
Denoting interpolation matrices by $\mathbf{\Psi}$ with subscript specifying the field concerned, in matrix notation the above multifield modelling materializes as follows:

$$\boldsymbol{\varepsilon} = \mathbf{\Psi}_\varepsilon\, \overline{\boldsymbol{\varepsilon}}, \quad \boldsymbol{\sigma} = \mathbf{\Psi}_\sigma\, \overline{\boldsymbol{\sigma}}; \qquad \zeta = \mathbf{\Psi}_\zeta\, \overline{\boldsymbol{\zeta}}, \quad p = \mathbf{\Psi}_p\, \overline{\mathbf{p}} \tag{14}$$

$$\boldsymbol{\chi} = \mathbf{\Psi}_\chi\, \overline{\boldsymbol{\chi}}, \quad \boldsymbol{\eta} = \mathbf{\Psi}_\eta\, \overline{\boldsymbol{\eta}}; \qquad \boldsymbol{\pi} = \mathbf{\Psi}_\pi\, \overline{\boldsymbol{\pi}}, \quad \mathbf{q} = \mathbf{\Psi}_q\, \overline{\mathbf{q}} \tag{15}$$

$$\boldsymbol{\lambda} = \mathbf{\Psi}_\lambda\, \overline{\boldsymbol{\lambda}}, \qquad \boldsymbol{\varphi} = \mathbf{\Psi}_\varphi\, \overline{\boldsymbol{\varphi}} \tag{16}$$

The pairs of vectors $\overline{\boldsymbol{\varepsilon}}$ and $\overline{\boldsymbol{\sigma}}$, $\overline{\boldsymbol{\zeta}}$ and $\overline{\mathbf{p}}$, $\overline{\boldsymbol{\chi}}$ and $\overline{\boldsymbol{\eta}}$, $\overline{\boldsymbol{\pi}}$ and $\overline{\mathbf{q}}$, $\overline{\boldsymbol{\lambda}}$ and $\overline{\boldsymbol{\varphi}}$ which govern pairs of conjugate modelled fields according to (14)-(16), acquire the meaning of generalized variables "in Prager's sense" if, in each pair, the latter interpolation is derived from the former through the relationship (where l = latter, f = former):

$$\mathbf{\Psi}_\mathrm{l}(\mathbf{x}) = \mathbf{\Psi}_\mathrm{f}(\mathbf{x}) \left[ \int_\Omega \mathbf{\Psi}_\mathrm{f}^\mathrm{T}(\mathbf{x})\, \mathbf{\Psi}_\mathrm{f}(\mathbf{x})\, d\Omega \right]^{-1} \tag{17}$$

The finite element discretization in space of the continuum problem of Sec. 2.1 through the above modelling can be based either on the variational formulation proposed in [17] or, more in general, on the Galerkin weighted residual procedure adopted in [2] [3]. Referring to these sources for details, only the resulting ordinary differential equations which govern the discretized poroplastic system is given below:

$$\overline{\boldsymbol{\varphi}}=\overline{\mathbf{N}}^{\sigma\mathrm{T}}\overline{\boldsymbol{\sigma}}+\overline{\mathbf{N}}^{p\mathrm{T}}\overline{\mathbf{p}}-\overline{\mathbf{Y}}\leq\mathbf{0},\qquad \overline{\mathbf{Y}}=\overline{\mathbf{Y}}^{0}+\overline{\mathbf{H}}\overline{\boldsymbol{\lambda}} \tag{18}$$

$$\dot{\overline{\boldsymbol{\varepsilon}}}^{p}=\overline{\mathbf{N}}^{\sigma}\dot{\overline{\boldsymbol{\lambda}}},\qquad \dot{\overline{\boldsymbol{\zeta}}}^{p}=\overline{\mathbf{N}}^{p}\dot{\overline{\boldsymbol{\lambda}}} \tag{19}$$

$$\dot{\overline{\boldsymbol{\lambda}}}\geq\mathbf{0},\qquad \overline{\boldsymbol{\varphi}}^{\mathrm{T}}\dot{\overline{\boldsymbol{\lambda}}}=0 \tag{20}$$

$$\overline{\boldsymbol{\varepsilon}}^{e}=\overline{\mathbf{E}}^{-1}\overline{\boldsymbol{\sigma}}+\overline{\mathbf{P}}\,\overline{\mathbf{p}} \tag{21}$$

$$\overline{\boldsymbol{\zeta}}^{e}=\overline{\mathbf{P}}^{\mathrm{T}}\overline{\boldsymbol{\sigma}}+\overline{\mathbf{m}}^{-1}\overline{\mathbf{p}} \tag{22}$$

$$\overline{\boldsymbol{\varepsilon}}=\overline{\boldsymbol{\varepsilon}}^{e}+\overline{\boldsymbol{\varepsilon}}^{p},\qquad \overline{\boldsymbol{\zeta}}=\overline{\boldsymbol{\zeta}}^{e}+\overline{\boldsymbol{\zeta}}^{p} \tag{23}$$

$$\overline{\boldsymbol{\varepsilon}}=\overline{\mathbf{B}}\,\overline{\mathbf{u}} \tag{24}$$

$$\overline{\mathbf{B}}^{\mathrm{T}}\overline{\boldsymbol{\sigma}}=\hat{\overline{\mathbf{b}}}+\overline{\mathbf{t}} \tag{25}$$

$$\dot{\overline{\boldsymbol{\zeta}}}=\overline{\mathbf{G}}^{\mathrm{T}}\overline{\mathbf{q}}-\overline{\mathbf{q}}_{\Gamma} \tag{26}$$

$$\overline{\mathbf{q}}=\overline{\mathbf{k}}\,\overline{\boldsymbol{\pi}},\qquad \overline{\boldsymbol{\pi}}=-\overline{\mathbf{G}}\,\overline{\mathbf{p}}+\hat{\overline{\mathbf{f}}} \tag{27}$$

A one-to-one correspondence relates Eqs. (18)-(27) to Eqs. (1)-(10) and permits to avoid an explicit definition of meanings and symbols in the former equation set. However, it is worth noting that all variable vectors include all the relevant generalized variables in the finite element aggregate, except vectors $\overline{\mathbf{t}}$ and $\overline{\mathbf{q}}_{\Gamma}$: in fact, these contain boundary variables only (tractions and outward fluxes, respectively) and zeroes elsewhere. Such vectors arise from an integration by parts which transforms the equilibrium and divergence operators, so that the discrete equilibrium and mass conservation equations, respectively, acquire matrix operators which are transposed of those in the discretized compatibility and Darcy equations, respectively.
Then, let Eqs. (21)-(27) be rewritten in the following compact form through subsequent substitutions of the variables $\overline{\boldsymbol{\varepsilon}},\overline{\boldsymbol{\varepsilon}}^{e},\overline{\boldsymbol{\zeta}},\overline{\boldsymbol{\zeta}}^{e},\dot{\overline{\boldsymbol{\zeta}}},\overline{\boldsymbol{\pi}}$ and $\overline{\boldsymbol{\sigma}}$:

$$\overline{\mathbf{K}}\,\overline{\mathbf{u}}-\overline{\mathbf{L}}\,\overline{\mathbf{p}}=\hat{\overline{\mathbf{b}}}+\overline{\mathbf{t}}+\overline{\mathbf{B}}^{\mathrm{T}}\,\overline{\mathbf{E}}\,\overline{\boldsymbol{\varepsilon}}^{p} \tag{28}$$

$$\overline{\mathbf{L}}^{\mathrm{T}}\dot{\overline{\mathbf{u}}}+\overline{\mathbf{S}}\dot{\overline{\mathbf{p}}}+\overline{\mathbf{V}}\,\overline{\mathbf{p}}=\overline{\mathbf{P}}^{\mathrm{T}}\overline{\mathbf{E}}\dot{\overline{\boldsymbol{\varepsilon}}}^{p}-\dot{\overline{\boldsymbol{\zeta}}}^{p}+\overline{\mathbf{G}}^{\mathrm{T}}\,\overline{\mathbf{k}}\,\hat{\overline{\mathbf{f}}}-\overline{\mathbf{q}}_{\Gamma} \tag{29}$$

where: $$\overline{\mathbf{K}}=\overline{\mathbf{B}}^{\mathrm{T}}\,\overline{\mathbf{E}\mathbf{B}},\quad \overline{\mathbf{L}}=\overline{\mathbf{B}}^{\mathrm{T}}\,\overline{\mathbf{E}\mathbf{P}},\quad \overline{\mathbf{S}}=\overline{\mathbf{m}}^{-1}-\overline{\mathbf{P}}^{\mathrm{T}}\,\overline{\mathbf{E}\mathbf{P}},\quad \overline{\mathbf{V}}=\overline{\mathbf{G}}^{\mathrm{T}}\,\overline{\mathbf{k}\mathbf{G}} \tag{30}$$

Let us consider now the boundary data (11)-(12), namely the given pressures $\hat{\overline{\mathbf{p}}}$, tractions $\hat{\overline{\mathbf{t}}}$, fluxes $\hat{\overline{\mathbf{q}}}_{\Gamma}$ and displacements $\hat{\overline{\mathbf{u}}}$ at boundary nodes, by assuming for brevity $\hat{\overline{\mathbf{q}}}_{\Gamma}=\mathbf{0}$ and $\hat{\overline{\mathbf{u}}}=\mathbf{0}$. Let these data be introduced into Eqs. (28)-(29), i.e. into vectors $\overline{\mathbf{p}},\overline{\mathbf{t}},\overline{\mathbf{q}}_{\Gamma}$ and $\overline{\mathbf{u}}$, respectively. It is understood that, simultaneously, the individual equations corresponding to boundary unknowns are dropped (for possible later use). With the consequent new interpretation of their meaning, Eqs. (28)-(29) can be rewritten as follows:

$$\overline{\mathbf{K}}\,\overline{\mathbf{u}} - \overline{\mathbf{L}}\,\overline{\mathbf{p}} = \hat{\overline{\mathbf{c}}} + \overline{\mathbf{B}}^{\mathrm{T}}\,\overline{\mathbf{E}}\,\overline{\boldsymbol{\varepsilon}}^{p} \tag{31}$$

$$\overline{\mathbf{L}}^{\mathrm{T}}\dot{\overline{\mathbf{u}}} + \overline{\mathbf{S}}\,\dot{\overline{\mathbf{p}}} + \overline{\mathbf{V}}\,\overline{\mathbf{p}} = \hat{\overline{\mathbf{d}}} + \overline{\mathbf{P}}^{\mathrm{T}}\,\overline{\mathbf{E}}\,\dot{\overline{\boldsymbol{\varepsilon}}}^{p} - \dot{\overline{\boldsymbol{\zeta}}}^{p} + \overline{\mathbf{G}}^{\mathrm{T}}\,\overline{\mathbf{k}}\,\hat{\overline{\mathbf{f}}} \tag{32}$$

having set:

$$\hat{\overline{\mathbf{c}}} = \hat{\overline{\mathbf{b}}} + \hat{\overline{\mathbf{t}}} + \overline{\overline{\mathbf{L}}}\,\hat{\overline{\mathbf{p}}}, \qquad \hat{\overline{\mathbf{d}}} = \overline{\overline{\mathbf{S}}}\,\dot{\hat{\overline{\mathbf{p}}}} + \overline{\overline{\mathbf{V}}}\,\hat{\overline{\mathbf{p}}} \tag{33}$$

The association of Eqs. (31)-(33) with the plastic constitutive relationships (18)-(20) and with the initial conditions (13) imposed in all nodes for pressure, leads to a relation set which governs the discrete model of the poroplastic system in point subjected to a given history of external actions.

## 3. Static shakedown theorems

3.1 In the present poroplastic context, the boundedness in time of the cumulative dissipated energy characterizing shakedown (SD) is expressed by the inequality:

$$D = \lim_{\tau\to\infty}\int_0^{\tau}\int_{\Omega}\dot{\mathcal{D}}(x_i,\tau')\,d\Omega\,d\tau' = \lim_{\tau\to\infty}\int_0^{\tau}\int_{\Omega}\left(\sigma_{ij}\,\dot{\varepsilon}_{ij}^{p} + p\dot{\zeta}^{p} - H_{\alpha\beta}\,\lambda_{\alpha}\,\dot{\lambda}_{\beta}\right)d\Omega\,d\tau' < \infty \tag{34}$$

Here $\dot{\mathcal{D}}$ denotes the energy dissipation rate per unit volume. As in classical hardening plasticity, the energy dissipation rate amounts to the difference between the work rate (which includes in poroplasticity the contribution of the pressure through the permanent fluid content rate) and the rate of the Helmholtz' "free" energy connected with the rearrangements of the material texture at the microscale (see e.g. [7] [12] [14] [15]).
By substituting Eq. (2) into the integrand of (34) and subtracting from it the complementarity equation (3b) account taken of Eq. (1), the shakedown criterion (34) reads:

$$\lim_{\tau\to\infty}\int_{\Omega} Y_{\alpha}^{0}\,\lambda_{\alpha}(\tau)\;d\Omega = Y_{\alpha}^{0}\lim_{\tau\to\infty}\int_{\Omega}\lambda_{\alpha}(\tau)\;d\Omega < \infty \tag{35}$$

In accordance with the multifield finite element discretization in space carried out in Sec. 2.2, the shakedown criterion (35) becomes:

$$\mathbf{Y}^{0\mathrm{T}}\left[\lim_{\tau\to\infty}\boldsymbol{\lambda}(\tau)\right] < \infty \tag{36}$$

3.2 As in many engineering situations, let the external actions be subdivided into "dead loads" (typically self-weights) and "live loads" consisting of basic variable repeated external actions multiplied by a common factor $\mu$. Consequently, the fictitious linear poroelastic response to a given loading history can be conceived as the superposition of two addends such as (with self evident meaning of symbols):

$$\overline{\boldsymbol{\sigma}}^E = \mu \overline{\boldsymbol{\sigma}}^{E'}(\tau) + \overline{\boldsymbol{\sigma}}^{E''}, \quad \overline{\mathbf{p}}^E = \mu \overline{\mathbf{p}}^{E'}(\tau) + \overline{\mathbf{p}}^{E''} \tag{37}$$

All the external actions will be assumed henceforth as periodic. The fields marked by $E$, like those in Eqs. (37), will be interpreted as the steady state periodic poroelastic response to the external actions (after the transient regime due to the initial conditions has died off for practical purposes), namely as the solution to Eqs. (31)-(33) with $\overline{\boldsymbol{\varepsilon}}^p \equiv \mathbf{0}$, $\overline{\boldsymbol{\zeta}}^p \equiv \mathbf{0}$ and with the given boundary conditions.

The main objective of SD analysis is to evaluate the safety factor $s$ with respect to non shakedown (either incremental collapse or alternating plasticity), that is to evaluate the critical value $s$ such that for $\mu \leq s$ shakedown occurs, i.e. Eq. (36) holds, and for $\mu > s$ it does not occur.
The poroelastic stress and pressure asymptotic response (37a, b) to periodic loads can be computed in closed form (see details in Appendix of [3]) and can be used to compute their maxima projections on the normals to all yielding planes ("envelope vector"), namely:

$$\overline{\mathbf{M}} \equiv \max_{\tau \geq 0} \left\{ \overline{\mathbf{N}}^{\sigma \mathrm{T}} \overline{\boldsymbol{\sigma}}^E(\tau) + \overline{\mathbf{N}}^{p \mathrm{T}} \overline{\mathbf{p}}^E(\tau) \right\} = \mu \overline{\mathbf{M}}' + \overline{\mathbf{M}}'' \tag{38}$$

Consider now the solution of Eqs. (31)-(33) with $\hat{\overline{\mathbf{t}}} = \mathbf{0}$ and $\hat{\overline{\mathbf{p}}} = \mathbf{0}$, for plastic strains $\overline{\boldsymbol{\varepsilon}}^p$ constant in time regarded as the only external actions, having set all time derivatives to zero. This solution (marked by superscript $s$) represents a fictitious poroelastic stationary state of the system not influenced by initial conditions; it can be expressed as follows:

$$\overline{\mathbf{u}}^s = \overline{\mathbf{K}}^{-1} \overline{\mathbf{B}}^{\mathrm{T}} \overline{\mathbf{E}} \overline{\boldsymbol{\varepsilon}}^p = \overline{\mathbf{K}}^{-1} \overline{\mathbf{B}}^{\mathrm{T}} \overline{\mathbf{E}} \overline{\mathbf{N}}^{\sigma} \overline{\boldsymbol{\lambda}}; \quad \overline{\mathbf{p}}^s = \mathbf{0}; \quad \overline{\mathbf{q}}^s = \mathbf{0}; \quad \overline{\boldsymbol{\sigma}}^s = \overline{\mathbf{Z}} \overline{\mathbf{N}}^{\sigma} \overline{\boldsymbol{\lambda}} \tag{39}$$

having set:

$$\overline{\mathbf{Z}} = \overline{\mathbf{E}} \overline{\mathbf{B}} \overline{\mathbf{K}}^{-1} \overline{\mathbf{B}}^{\mathrm{T}} \overline{\mathbf{E}} - \overline{\mathbf{E}} \tag{40}$$

Matrices $\overline{\mathbf{E}}$ and $\overline{\mathbf{Z}}$ are, for the whole element aggregate, the drained ($\overline{\mathbf{p}} = \mathbf{0}$) elastic stiffness matrix and the influence matrix of the plastic strains on selfstresses, respectively. Both turn out to be symmetric, the former positive definite, the latter negative semidefinite.

3.3 On the basis of the preceding developments, it was proved in [3] that the safety factor $s$ (i.e. the live-load history amplification above which the cumulative dissipation is unbounded in time) is provided by the (common) optimal value of the following dual linear programming (LP) problems:

$$s = \max_{\mu, \boldsymbol{\lambda}} \{\mu\}, \text{ subject to: } \quad \mu\,\overline{\mathbf{M}}' - \overline{\mathbf{A}}\,\overline{\boldsymbol{\lambda}} \le \overline{\mathbf{Y}}^0 - \overline{\mathbf{M}}'', \quad \mu \ge 0, \quad \overline{\boldsymbol{\lambda}} \ge \mathbf{0} \tag{41}$$

$$s = \min_{\boldsymbol{\lambda}} \left\{ \left( \overline{\mathbf{Y}}^0 - \overline{\mathbf{M}}'' \right)^{\mathrm{T}} \overline{\boldsymbol{\lambda}} \right\}, \text{ subject to: } \quad \overline{\mathbf{M}}'^{\mathrm{T}} \overline{\boldsymbol{\lambda}} = 1, \quad \overline{\mathbf{A}}\,\overline{\boldsymbol{\lambda}} = \mathbf{0}, \quad \overline{\boldsymbol{\lambda}} \ge \mathbf{0} \tag{42}$$

where:

$$\overline{\mathbf{A}} = \overline{\mathbf{H}} + \overline{\mathbf{A}}_0, \quad \overline{\mathbf{A}}_0 = -\overline{\mathbf{N}}^{\sigma\mathrm{T}}\, \overline{\mathbf{Z}}\, \overline{\mathbf{N}}^{\sigma} \tag{43}$$

While omitting here for brevity the proofs (available in [2] [3]), the adopted assumptions are gathered below: (*a*) infinitesimal deformations; (*b*) piecewise linearization of poroplastic material model; (*c*) associative flow rules; (*d*) symmetric and positive semidefinite hardening matrix; (*e*) constant permeability; (*f*) finite element discretization by generalized variables in Prager's sense; (*g*) periodicity in time of external actions.

The following circumstances are worth noting: hypothesis (*d*), not specified in what precedes, is required by the omitted proofs (see e.g. [3]); assumptions (*c*) and (*d*) together are sufficient for the stability in Drucker's sense of the poroplastic model.

As for the mechanical interpretation of the theoretical results embodied in the above LP problems, the former primal problem, Eq. (41), formulates an extension to poroplasticity of Melan's static theorem; the latter (dual), Eq. (42), can be shown to be equivalent to a poroplastic version of the kinematic SD theorem due to Symonds, Neal and Koiter (see [21] [19] [10]).

It is worth noting that the plastic multiplier vector $\overline{\boldsymbol{\lambda}}$ and the load factor $\mu$ together act as optimization variables. In Eq. (41) vector $\overline{\boldsymbol{\lambda}}$ generates selfstresses constant in time and constrained to satisfy the yield inequalities when superposed on the poroelastic response; in Eq. (42) vector $\overline{\boldsymbol{\lambda}}$ (of variables which are dual to $\mu$ and $\overline{\boldsymbol{\lambda}}$ in (41)) governs the incremental collapse or alternate yielding mechanisms. In both static and kinematic approaches, the irreversible changes of fluid content play no role.

## 4. Upper bounds on post shakedown quantities

4.1 Subject to the validity of the assumptions (*a*)-(*g*), Sec. 3.3, underlying the shakedown theorems stated there, under the further hypothesis that shakedown does occur under given live and dead loads, some meaningful residual history-dependent quantities will be shown to be bounded from above by suitable values obtainable via a direct, nonevolutionary procedure. In particular, two kinds of such quantities will be considered below: a residual (permanent, plastic) displacement $u_i^p(\hat{\mathbf{x}})$ of some pre-selected point $\hat{\mathbf{x}}$ in a chosen direction $i$; the plastic volumetric deformation $\varepsilon_V^p(\hat{\hat{\mathbf{x}}})$ of the solid skeleton in some suitably fixed point $\hat{\hat{\mathbf{x}}}$.

An upper bound on the former quantity, with appropriate choice of $\hat{\mathbf{x}}$ and $i$ dictated by engineering judgement, can be useful in practice in order to check serviceability and/or the assumed negligibility of geometric effects. An upper bound on the latter quantity in the suitably chosen location $\hat{\hat{\mathbf{x}}}$ may be used for checks of the validity of two hypotheses: (*i*) compliance with limited ductility of the solid skeleton; (*ii*) constant permeability in Darcy law (which might be adjusted on the basis of the computed $\varepsilon_V^p$ for an iterated shakedown analysis).

In the framework of the space discretization and constitutive piecewise linearization adopted in what precedes, a post shakedown quantity of both the above kinds turns out to be a linear function of the post-shakedown generalized plastic multiplier vector $\overline{\boldsymbol{\lambda}}$, namely:

$$u_i^p(\hat{\mathbf{x}}) = \mathbf{s}^{\mathrm{T}} \boldsymbol{\Psi}_u(\hat{\mathbf{x}}) \overline{\mathbf{K}}^{-1} \overline{\mathbf{B}}^{\mathrm{T}} \overline{\mathbf{E}} \overline{\mathbf{N}}_\sigma \overline{\boldsymbol{\lambda}} \tag{44}$$

$$\varepsilon_V^p(\hat{\hat{\mathbf{x}}}) = \mathbf{r}^{\mathrm{T}} \boldsymbol{\Psi}_\varepsilon(\hat{\hat{\mathbf{x}}}) \overline{\mathbf{N}}_\sigma \overline{\boldsymbol{\lambda}} \tag{45}$$

In Eq. (44), $\mathbf{s}$ is the Boolean vector which selects, by its nonzero entry, the desired component of the modelled displacement field governed by vector $\overline{\mathbf{u}}^p$, which expressed like in Eq. (39a). Similarly, in Eq. (45), $\mathbf{r}$ denotes a Boolean vector which extracts and sums the modelled normal strains in $\hat{\hat{\mathbf{x}}}$ out of the modelled plastic strain field, account taken of Eq. (19a). Clearly, a linear dependence on vector $\overline{\boldsymbol{\lambda}}$ holds also for other quantities, such as residual stress and strain components and permanent fluid content.

Henceforth, reference will be made to any of the above considered quantities, denoted by $Q$ and expressed as a linear combination of plastic multipliers through a coefficient vector $\overline{\mathbf{R}}$:

$$Q=\overline{\mathbf{R}}^{\mathrm{T}}\overline{\boldsymbol{\lambda}} \tag{46}$$

Another post-shakedown quantity of possible technical interest is represented by the value attained, asymptotically in time, by the time-cumulative dissipated energy per unit volume, $\mathcal{D}(\breve{\mathbf{x}})$, in some point $\breve{\mathbf{x}}$ of the considered poroplastic body. In the present context of piecewiselinear poroplastic and finite element models, it can be easily shown that $\mathcal{D}(\breve{\mathbf{x}})$ is a convex quadratic function of vector $\overline{\boldsymbol{\lambda}}$ and that upper bounds on it can be achieved by straightforward generalization of the theoretical developments and conclusions concerning any quantity $Q$ covered by Eq. (46).

4.2 On the basis of what precedes (Section 4.1), the following statement will be proven below (see e.g. as early antecedents in plasticity [20] [16]).

*Theorem 1.*

The post shakedown value $Q$ of a quantity linearly dependent, through the coefficient vector $\overline{\mathbf{R}}$, on the residual plastic multiplier vector $\overline{\boldsymbol{\lambda}}$, is bounded from above by the inequality:

$$Q\leq\tilde{Q}_1=\min_{\overline{\boldsymbol{\lambda}}^*}\ \max_{\overline{\boldsymbol{\lambda}}}\left\{\overline{\mathbf{R}}^{\mathrm{T}}\overline{\boldsymbol{\lambda}}\right\},\quad \text{subject to:} \tag{47}$$

$$\overline{\mathbf{M}}'-\left(\overline{\mathbf{A}}_0+\overline{\mathbf{H}}\right)\overline{\boldsymbol{\lambda}}\leq\overline{\mathbf{Y}}^0-\overline{\mathbf{M}}'',\qquad \overline{\boldsymbol{\lambda}}\geq\mathbf{0} \tag{48}$$

$$s\,\overline{\mathbf{M}}'-s\left(\overline{\mathbf{A}}_0+\overline{\mathbf{H}}\right)\overline{\boldsymbol{\lambda}}^*\leq\overline{\mathbf{Y}}^0-\overline{\mathbf{M}}'',\qquad \overline{\boldsymbol{\lambda}}^*\geq\mathbf{0} \tag{49}$$

$$\frac{s-1}{s}\left[\left(\overline{\mathbf{Y}}^0-\overline{\mathbf{M}}''\right)^{\mathrm{T}}\overline{\boldsymbol{\lambda}}+\frac{1}{2}\overline{\boldsymbol{\lambda}}^{\mathrm{T}}\overline{\mathbf{H}\boldsymbol{\lambda}}\right]-\overline{\boldsymbol{\lambda}}^{*\mathrm{T}}\overline{\mathbf{A}}_0\,\overline{\boldsymbol{\lambda}}+\frac{1}{2}\overline{\boldsymbol{\lambda}}^{\mathrm{T}}\overline{\mathbf{A}}_0\,\overline{\boldsymbol{\lambda}}\leq 0 \tag{50}$$

The following remark provides an interpretation of the above statement. The necessary static condition for SD leads to the constraints (48) for the unknown actual plastic multipliers generated in the modelled system along its evolution under external actions. A maximization under these constraints of the quantity to bound as a function of $\overline{\boldsymbol{\lambda}}$ would lead to a usually large and useless upper bound on it. However, suppose that an energy inequality is established between the actual unknown evolution of the system and the fictitious situation generated by plastic multipliers $\overline{\boldsymbol{\lambda}}^*$ constant in time and such that the sufficient SD condition is fulfilled for the live loads amplified by the safety factor $s$. Vector $\overline{\boldsymbol{\lambda}}^*$, which influences the sought upper bound through the above energy inequality, can be employed as minimization variables in order to improve that bound.

This remark provides guidelines for the rather laborious proof given below to the above statement and centered on the search for the aforementioned energy inequality.

*Proof.* The rather lengthy formal proof of the above theorem will be divided in four points for clarity.

(*a*) Making use of the constitutive relations (18)-(20) in generalized variables, the dissipated energy (integrated in space over the modelled body) can be given by the following expressions:

$$\dot{D}=\overline{\boldsymbol{\sigma}}^{\mathrm{T}}\dot{\overline{\boldsymbol{\varepsilon}}}^{p}+\overline{\mathbf{p}}^{\mathrm{T}}\dot{\overline{\boldsymbol{\zeta}}}^{p}-\overline{\boldsymbol{\lambda}}^{\mathrm{T}}\overline{\mathbf{H}}\dot{\overline{\boldsymbol{\lambda}}}=\left(\overline{\boldsymbol{\sigma}}^{\mathrm{T}}\overline{\mathbf{N}}^{\sigma}+\overline{\mathbf{p}}^{\mathrm{T}}\overline{\mathbf{N}}^{p}-\overline{\boldsymbol{\lambda}}^{\mathrm{T}}\overline{\mathbf{H}}\right)\dot{\overline{\boldsymbol{\lambda}}}= \\ =\left(\overline{\boldsymbol{\varphi}}+\overline{\mathbf{Y}}^{0}\right)^{\mathrm{T}}\dot{\overline{\boldsymbol{\lambda}}}=\overline{\mathbf{Y}}^{0\mathrm{T}}\dot{\overline{\boldsymbol{\lambda}}}\geq 0 \tag{51}$$

whence integration in time and the assumption of poroelastic behaviour at the time origin lead to:

$$D(\tau)=\overline{\mathbf{Y}}^{0\mathrm{T}}\left[\overline{\boldsymbol{\lambda}}(\tau)-\overline{\boldsymbol{\lambda}}(0)\right]=\overline{\mathbf{Y}}^{0\mathrm{T}}\overline{\boldsymbol{\lambda}}(\tau)\geq 0 \qquad \forall\tau\geq 0 \tag{52}$$

Another preliminary inequality is provided by the Drucker's postulate:

$$\left(\overline{\boldsymbol{\sigma}}-\overline{\boldsymbol{\sigma}}^{*}\right)^{\mathrm{T}}\dot{\overline{\boldsymbol{\varepsilon}}}^{p}+\left(\overline{\mathbf{p}}-\overline{\mathbf{p}}^{*}\right)^{\mathrm{T}}\dot{\overline{\boldsymbol{\zeta}}}^{p}\geq 0 \quad \forall\tau \tag{53}$$

Now the generalized total stresses and pressures, both actual and fictitious (the latter marked by stars and written for the SD factor $s$), are split into addends, namely into the poroelastic responses to: (*i*) live loads; (*ii*) plastic strains and permanent fluid content changes; (*iii*) dead loads. Then let the first and second addends be covered by a single symbol for convenience:

$$\left.\begin{aligned} \overline{\boldsymbol{\sigma}}&=\overline{\boldsymbol{\sigma}}'^{e}+\overline{\boldsymbol{\sigma}}^{p}+\overline{\boldsymbol{\sigma}}''^{e}=\overline{\boldsymbol{\sigma}}'+\overline{\boldsymbol{\sigma}}''^{e} \\ \overline{\mathbf{p}}&=\overline{\mathbf{p}}'^{e}+\overline{\mathbf{p}}^{p}+\overline{\mathbf{p}}''^{e}=\overline{\mathbf{p}}'+\overline{\mathbf{p}}''^{e} \end{aligned}\right\} \tag{54}$$

$$\left.\begin{aligned} \overline{\boldsymbol{\sigma}}^{*}&=s\,\overline{\boldsymbol{\sigma}}'^{e}+s\,\overline{\boldsymbol{\sigma}}^{p*}+\overline{\boldsymbol{\sigma}}''^{e}=s\,\overline{\boldsymbol{\sigma}}'^{*}+\boldsymbol{\sigma}''^{e} \\ \overline{\mathbf{p}}^{*}&=s\,\overline{\mathbf{p}}'^{e}+s\,\overline{\mathbf{p}}^{p*}+\overline{\mathbf{p}}''^{e}=s\,\overline{\mathbf{p}}'^{*}+\overline{\mathbf{p}}''^{e} \end{aligned}\right\} \tag{55}$$

Thus inequality (53) can be rewritten in the form:

$$\left(\overline{\boldsymbol{\sigma}}'-s\,\overline{\boldsymbol{\sigma}}'^{*}\right)^{\mathrm{T}}\dot{\overline{\boldsymbol{\varepsilon}}}^{p}+\left(\overline{\mathbf{p}}'-s\,\overline{\mathbf{p}}'^{*}\right)^{\mathrm{T}}\dot{\overline{\boldsymbol{\zeta}}}^{p}\geq 0 \quad \forall\tau \tag{56}$$

or, through some algebraic manipulations, in the alternative form:

$$\left(\overline{\boldsymbol{\sigma}}'-\overline{\boldsymbol{\sigma}}'^{*}\right)^{\mathrm{T}}\dot{\overline{\boldsymbol{\varepsilon}}}^{p}+\left(\overline{\mathbf{p}}'-\overline{\mathbf{p}}'^{*}\right)^{\mathrm{T}}\dot{\overline{\boldsymbol{\zeta}}}^{p}\geq\frac{s-1}{s}\left(\overline{\boldsymbol{\sigma}}'^{\mathrm{T}}\dot{\overline{\boldsymbol{\varepsilon}}}^{p}+\overline{\mathbf{p}}'^{\mathrm{T}}\dot{\overline{\boldsymbol{\zeta}}}^{p}\right)\quad\forall\tau \tag{57}$$

By substituting (54), (51), (19) and (38), successively, into the r.h.s. of (57), this inequality becomes:

$$\left(\overline{\boldsymbol{\sigma}}'-\overline{\boldsymbol{\sigma}}'^{*}\right)^{\mathrm{T}}\dot{\overline{\boldsymbol{\varepsilon}}}^{p}+\left(\overline{\mathbf{p}}'-\overline{\mathbf{p}}'^{*}\right)^{\mathrm{T}}\dot{\overline{\boldsymbol{\zeta}}}^{p}\geq\frac{s-1}{s}\left(\dot{D}+\overline{\boldsymbol{\lambda}}^{\mathrm{T}}\overline{\mathbf{H}}\dot{\overline{\boldsymbol{\lambda}}}-\overline{\mathbf{M}}''^{\mathrm{T}}\dot{\overline{\boldsymbol{\lambda}}}\right)\quad\forall\tau \tag{58}$$

whence, by integrating in time:

$$\begin{aligned}\int_0^{\tau}\left[\left(\overline{\boldsymbol{\sigma}}'-\overline{\boldsymbol{\sigma}}'^{*}\right)^{\mathrm{T}}\dot{\overline{\boldsymbol{\varepsilon}}}^{p}+\left(\overline{\mathbf{p}}'-\overline{\mathbf{p}}'^{*}\right)^{\mathrm{T}}\dot{\overline{\boldsymbol{\zeta}}}^{p}\right]\mathrm{d}\tau'\geq\frac{s-1}{s}\left[D(\tau)+\tfrac{1}{2}\overline{\boldsymbol{\lambda}}^{\mathrm{T}}\overline{\mathbf{H}}\overline{\boldsymbol{\lambda}}-\overline{\mathbf{M}}''^{\mathrm{T}}\overline{\boldsymbol{\lambda}}\right]=\\=\frac{s-1}{s}\left[\left(\overline{\mathbf{Y}}^{0}-\overline{\mathbf{M}}''\right)^{\mathrm{T}}\overline{\boldsymbol{\lambda}}(\tau)+\tfrac{1}{2}\overline{\boldsymbol{\lambda}}^{\mathrm{T}}(\tau)\overline{\mathbf{H}}\,\overline{\boldsymbol{\lambda}}(\tau)\right]\qquad\forall\tau\geq 0\end{aligned} \tag{59}$$

(*b*) Inequality (59) reflects Drucker postulate and SD under live loads amplified by $s$. Its l.h.s., which depends on the actual evolution of the system (its r.h.s. does not), will now be bounded from above by a quantity which does not depend. To this purpose, let us consider the poroelastic response to the actual inelastic variables $\overline{\boldsymbol{\varepsilon}}^{p}$ and $\overline{\boldsymbol{\zeta}}^{p}$ conceived as the only external actions. This response is governed by Eqs. (31)-(33) with all capped vectors set equal to zero and by homogeneous initial conditions.

Differentiating with respect to time the equilibrium equation (31), substituting it into the diffusion equation (32) and rearranging, the following first-order linear ordinary differential equation in $\overline{\mathbf{p}}(\tau)$ is arrived at:

$$\left[\overline{\mathbf{S}}+\overline{\mathbf{L}}^{\mathrm{T}}\overline{\mathbf{K}}^{-1}\overline{\mathbf{L}}\right]\dot{\overline{\mathbf{p}}}^{p}+\overline{\mathbf{V}}\,\overline{\mathbf{p}}^{p}=-\dot{\overline{\boldsymbol{\zeta}}}^{p}-\overline{\mathbf{P}}^{\mathrm{T}}\overline{\mathbf{Z}}\,\dot{\overline{\boldsymbol{\varepsilon}}}^{p} \tag{60}$$

Taking into account Eq. (19) and setting:

$$\overline{\mathbf{X}}=\overline{\mathbf{S}}+\overline{\mathbf{L}}^{\mathrm{T}}\overline{\mathbf{K}}^{-1}\overline{\mathbf{L}},\qquad \overline{\mathbf{Z}}_{p}=\overline{\mathbf{Z}}\,\overline{\mathbf{P}},\qquad \hat{\overline{\mathbf{N}}}=\overline{\mathbf{N}}^{p}+\overline{\mathbf{Z}}_{p}^{\mathrm{T}}\,\overline{\mathbf{N}}^{\sigma}, \tag{61}$$

equation (60) acquires the following compact form:

$$\overline{\mathbf{X}}\dot{\overline{\mathbf{p}}}^{p}+\overline{\mathbf{V}}\overline{\mathbf{p}}^{p}=-\overline{\widehat{\mathbf{N}}}\dot{\overline{\boldsymbol{\lambda}}} \tag{62}$$

The symmetric matrix $\overline{\mathbf{X}}$, Eq. (61a), is positive definite since its former addend $\overline{\mathbf{S}}$ is so, while the latter is positive definite or semidefinite depending on $\overline{\mathbf{L}}$. Then Eq. (62) can easily be shown to have the following solution:

$$\overline{\mathbf{p}}^{p}(\tau)=e^{-\tau\overline{\mathbf{X}}^{-1}\overline{\mathbf{V}}}\,\overline{\mathbf{p}}^{p}(0)-\int_{0}^{\tau}e^{-(\tau-\tau')\overline{\mathbf{X}}^{-1}\overline{\mathbf{V}}}\,\overline{\mathbf{X}}^{-1}\overline{\widehat{\mathbf{N}}}\dot{\overline{\boldsymbol{\lambda}}}(\tau')\,\mathrm{d}\tau' \tag{63}$$

Note that $\overline{\mathbf{p}}^{p}(0)=\mathbf{0}$, for the present purpose.

Making use of Eqs. (21), (23), (24) and (31), as well as in view of Eqs. (40) and (61b), the stresses due only to inelastic kinematic variables can be expressed as follows:

$$\overline{\boldsymbol{\sigma}}^{p}(\tau)=\overline{\mathbf{Z}}\,\overline{\boldsymbol{\varepsilon}}^{p}(\tau)+\overline{\mathbf{Z}}_{p}\,\overline{\mathbf{p}}^{p}(\tau)=\overline{\mathbf{Z}}\left[\overline{\boldsymbol{\varepsilon}}^{p}(\tau)+\overline{\mathbf{P}}\,\overline{\mathbf{p}}^{p}(\tau)\right] \tag{64}$$

where pressures are represented by Eq. (63).

(*c*) Focusing now on the integrand of Eq. (59), it can be given the following expressions by employing successively Eqs. (54) and (55) and Eq. (64), taking into account that $\dot{\overline{\boldsymbol{\varepsilon}}}^{p*}=\mathbf{0}$ and $\overline{\mathbf{p}}^{p*}=\mathbf{0}$:

$$\begin{aligned}\left(\overline{\boldsymbol{\sigma}}^{p}-\overline{\boldsymbol{\sigma}}^{p*}\right)^{\mathrm{T}}\dot{\overline{\boldsymbol{\varepsilon}}}^{p}+\left(\overline{\mathbf{p}}^{p}-\overline{\mathbf{p}}^{p*}\right)^{\mathrm{T}}\dot{\overline{\boldsymbol{\zeta}}}^{p}&=\left(\overline{\boldsymbol{\varepsilon}}^{p}-\overline{\boldsymbol{\varepsilon}}^{p*}\right)^{\mathrm{T}}\overline{\mathbf{Z}}\,\dot{\overline{\boldsymbol{\varepsilon}}}^{p}+\overline{\mathbf{p}}^{p\mathrm{T}}\left[\dot{\overline{\boldsymbol{\zeta}}}^{p}+\overline{\mathbf{Z}}_{p}^{\mathrm{T}}\dot{\overline{\boldsymbol{\varepsilon}}}^{p}\right]=\\&=\left(\overline{\boldsymbol{\varepsilon}}^{p}-\overline{\boldsymbol{\varepsilon}}^{p*}\right)^{\mathrm{T}}\overline{\mathbf{Z}}\left(\dot{\overline{\boldsymbol{\varepsilon}}}^{p}-\dot{\overline{\boldsymbol{\varepsilon}}}^{p*}\right)+\overline{\mathbf{p}}^{p\mathrm{T}}\left(\dot{\overline{\boldsymbol{\zeta}}}^{p}+\overline{\mathbf{Z}}_{p}^{\mathrm{T}}\dot{\overline{\boldsymbol{\varepsilon}}}^{p}\right)\end{aligned} \tag{65}$$

Then the integral (59) becomes:

$$\begin{aligned}&\int_{0}^{\tau}\left[\left(\overline{\boldsymbol{\sigma}}^{p}-\overline{\boldsymbol{\sigma}}^{p*}\right)^{\mathrm{T}}\dot{\overline{\boldsymbol{\varepsilon}}}^{p}+\left(\overline{\mathbf{p}}^{p}-\overline{\mathbf{p}}^{p*}\right)^{\mathrm{T}}\dot{\overline{\boldsymbol{\zeta}}}^{p}\right]\mathrm{d}\tau'=\\&=\frac{1}{2}\left[\left(\overline{\boldsymbol{\varepsilon}}^{p}-\overline{\boldsymbol{\varepsilon}}^{p*}\right)^{\mathrm{T}}\overline{\mathbf{Z}}\left(\overline{\boldsymbol{\varepsilon}}^{p}-\overline{\boldsymbol{\varepsilon}}^{p*}\right)\right]_{0}^{\tau}-\int_{0}^{\tau}\dot{\overline{\boldsymbol{\lambda}}}^{\mathrm{T}}(\tau')\int_{0}^{\tau'}\left[\overline{\widehat{\mathbf{N}}}^{\mathrm{T}}e^{-(\tau'-z)\overline{\mathbf{X}}^{-1}\overline{\mathbf{V}}}\,\overline{\mathbf{X}}^{-1}\overline{\widehat{\mathbf{N}}}\right]\dot{\overline{\boldsymbol{\lambda}}}(z)\,\mathrm{d}z\,\mathrm{d}\tau'\end{aligned} \tag{66}$$

The convolutive integral on the r.h.s. of Eq. (66) turns out to be always positive. In fact, it can be interpreted as the space- and time-cumulative energy in a fictitious poroelastic process due to a modelled field of permanent fluid content rate defined by

$\overline{\hat{\mathbf{N}}}\dot{\overline{\boldsymbol{\lambda}}}$ and conceived as the only external action operating on the system. This energy amounts to the sum of elastic strain energy stored in the fluid and of energy dissipated in the diffusion processes, while boundary and initial conditions, plastic deformations and displacements of the skeleton are kept zero.

Such an interpretation will be justified below by starting from the following equation which is a consequence of Eq. (63) derived in Subsection (*b*):

$$\int_0^\tau \dot{\overline{\boldsymbol{\zeta}}}^{p\mathrm{T}}\,\overline{\mathbf{p}}^p\,\mathrm{d}\tau' = -\int_0^\tau \dot{\overline{\boldsymbol{\zeta}}}^{p\mathrm{T}}(\tau')\int_0^{\tau'} e^{-(\tau'-z)\overline{\mathbf{X}}^{-1}\overline{\mathbf{V}}}\,\overline{\mathbf{X}}^{-1}\dot{\overline{\boldsymbol{\zeta}}}^p(z)\,\mathrm{d}z\,\mathrm{d}\tau' \quad \forall\tau\geq 0,\ \forall\dot{\overline{\boldsymbol{\zeta}}}^p \tag{67}$$

The integrand on the l.h.s. of Eq. (67) can be given alternative formulations. In fact, by using additivity (19b), mass conservation (26), Darcy law (27) and the constitutive equations (21)-(22) and by noting that $\overline{\mathbf{q}}_\Gamma^{\mathrm{T}}\,\overline{\mathbf{p}}^p = 0$ (in boundary nodes either the given flux or the given pressure is zero in the considered fictitious process) that integrand can be expressed as follows:

$$\begin{aligned}\dot{\overline{\boldsymbol{\zeta}}}^{p\mathrm{T}}\,\overline{\mathbf{p}}^p &= \dot{\overline{\boldsymbol{\zeta}}}^{\mathrm{T}}\,\overline{\mathbf{p}}^p - \dot{\overline{\boldsymbol{\zeta}}}^{e\mathrm{T}}\,\overline{\mathbf{p}}^p = \left(\overline{\mathbf{G}}^{\mathrm{T}}\overline{\mathbf{q}}^p - \overline{\mathbf{q}}_\Gamma\right)^{\mathrm{T}}\overline{\mathbf{p}}^p - \left(\overline{\mathbf{S}}\,\dot{\overline{\mathbf{p}}}^p\right)^{\mathrm{T}}\overline{\mathbf{p}}^p = \\ &= \overline{\mathbf{q}}^{p\mathrm{T}}\overline{\mathbf{G}}\overline{\mathbf{p}}^p - \overline{\mathbf{q}}_\Gamma^{\mathrm{T}}\,\overline{\mathbf{p}}^p - \dot{\overline{\mathbf{p}}}^{p\mathrm{T}}\,\overline{\mathbf{S}}\,\overline{\mathbf{p}}^p = -\overline{\boldsymbol{\pi}}^{p\mathrm{T}}\,\overline{\mathbf{k}}\,\overline{\boldsymbol{\pi}}^p - \dot{\overline{\mathbf{p}}}^{p\mathrm{T}}\,\overline{\mathbf{S}}\,\overline{\mathbf{p}}^p \qquad \forall\dot{\overline{\boldsymbol{\zeta}}}^p\end{aligned} \tag{68}$$

By integrating in time: $-\int_0^\tau \dot{\overline{\boldsymbol{\zeta}}}^{p\mathrm{T}}\,\overline{\mathbf{p}}^p\,\mathrm{d}\tau' = \int_0^\tau \overline{\boldsymbol{\pi}}^{p\mathrm{T}}\,\overline{\mathbf{k}}\,\overline{\boldsymbol{\pi}}^p + \dot{\overline{\mathbf{p}}}^{p\mathrm{T}}\,\overline{\mathbf{S}}\,\overline{\mathbf{p}}^p\,\mathrm{d}\tau' =$

$$= \int_0^\tau \overline{\mathbf{p}}^{p\mathrm{T}}\overline{\mathbf{G}}^{\mathrm{T}}\,\overline{\mathbf{k}}\,\overline{\mathbf{G}}\overline{\mathbf{p}}^p\,\mathrm{d}\tau' + \tfrac{1}{2}\overline{\mathbf{p}}^{p\mathrm{T}}\overline{\mathbf{S}}\,\overline{\mathbf{p}}^p > 0 \qquad \forall\tau>0,\ \forall\dot{\overline{\boldsymbol{\zeta}}}^p\neq\mathbf{0} \tag{69}$$

Since the material permeability matrix $\overline{\mathbf{k}}$ and the storage matrix $\overline{\mathbf{S}}$ are positive definite (the relevant quadratic forms represent dissipated energy by filtration and stored elastic energy in the compressible fluid, respectively), the positiveness stated in Eq. (67) is ascertained.

(*d*) As a consequence of Eq. (69), Eq. (66) yields the inequality:

$$\frac{1}{2}\left[\left(\overline{\boldsymbol{\varepsilon}}^p - \overline{\boldsymbol{\varepsilon}}^{p*}\right)^{\mathrm{T}}\overline{\mathbf{Z}}\left(\overline{\boldsymbol{\varepsilon}}^p - \overline{\boldsymbol{\varepsilon}}^{p*}\right)\right]_0^\tau \geq \int_0^\tau\left[\left(\overline{\boldsymbol{\sigma}}^p - \overline{\boldsymbol{\sigma}}^{p*}\right)^{\mathrm{T}}\dot{\overline{\boldsymbol{\varepsilon}}}^p + \left(\overline{\mathbf{p}}^p - \overline{\mathbf{p}}^{p*}\right)^{\mathrm{T}}\dot{\overline{\boldsymbol{\zeta}}}^p\right]\mathrm{d}\tau' \tag{70}$$

which, accounting for Eqs. (19a) and (43b) in its l.h.s., becomes:

$$\overline{\boldsymbol{\lambda}}^{*\mathrm{T}}\overline{\mathbf{A}}_0\,\overline{\boldsymbol{\lambda}}-\frac{1}{2}\overline{\boldsymbol{\lambda}}^{\mathrm{T}}\overline{\mathbf{A}}_0\,\overline{\boldsymbol{\lambda}} \geq \int_0^{\tau}\left[\left(\overline{\boldsymbol{\sigma}}^{p}-\overline{\boldsymbol{\sigma}}^{p*}\right)^{\mathrm{T}}\dot{\overline{\boldsymbol{\varepsilon}}}^{p}+\left(\overline{\mathbf{p}}^{p}-\overline{\mathbf{p}}^{p*}\right)^{\mathrm{T}}\dot{\overline{\boldsymbol{\zeta}}}^{p}\right]\mathrm{d}\tau' \tag{71}$$

Combining Eq. (71) with (59), inequality (50) is proven and, hence, is legitimated as a constraint in the *minmax* problem (47)-(50) leading to the optimal upper bound on quantity $Q=\overline{\mathbf{R}}^{\mathrm{T}}\,\overline{\boldsymbol{\lambda}}$.

## 5. Relaxed bounds

It may be fruitful to derive from Theorem 1 of Section 4.2, by successive relaxation of constraints, other bounds which are less stringent but easier to compute. They are gathered in the following statement.

*Theorem 2.*

The post-shakedown value $Q$ of a quantity linearly dependent (through the coefficient vector $\overline{\mathbf{R}}$) on the residual plastic multiplier vector $\overline{\boldsymbol{\lambda}}$ is bounded from above by the optimal values $\widetilde{Q}_2$, $\widetilde{Q}_3$, $\widetilde{Q}_4$ and $\widetilde{Q}_6$ of the following maximization problems:

$$\widetilde{Q}_2=\max_{\boldsymbol{\lambda}}\left\{\overline{\mathbf{R}}^{\mathrm{T}}\,\overline{\boldsymbol{\lambda}}\right\},\qquad \text{subject to:} \tag{72}$$

$$\overline{\mathbf{M}}'-\left(\overline{\mathbf{A}}_0+\overline{\mathbf{H}}\right)\overline{\boldsymbol{\lambda}}\leq\overline{\mathbf{Y}}^0-\overline{\mathbf{M}}'',\qquad \overline{\boldsymbol{\lambda}}\geq\mathbf{0} \tag{73}$$

$$\frac{s-1}{s}\left[\left(\overline{\mathbf{Y}}^0-\overline{\mathbf{M}}''\right)^{\mathrm{T}}\overline{\boldsymbol{\lambda}}+\frac{1}{2}\overline{\boldsymbol{\lambda}}^{\mathrm{T}}\overline{\mathbf{H}}\overline{\boldsymbol{\lambda}}\right]\leq\Omega \tag{74}$$

where:

$$\Omega=\min_{\boldsymbol{\lambda}^*}\left\{\frac{1}{2}\overline{\boldsymbol{\lambda}}^{*\mathrm{T}}\overline{\mathbf{A}}_0\,\overline{\boldsymbol{\lambda}}^*\right\},\qquad \text{subject to:} \tag{75}$$

$$s\,\overline{\mathbf{M}}'-s\left(\overline{\mathbf{A}}_0+\overline{\mathbf{H}}\right)\overline{\boldsymbol{\lambda}}^*\leq\overline{\mathbf{Y}}^0-\overline{\mathbf{M}}'',\qquad \overline{\boldsymbol{\lambda}}^*\geq\mathbf{0} \tag{76}$$

$$\widetilde{Q}_3=\max_{\boldsymbol{\lambda}}\left\{\overline{\mathbf{R}}^{\mathrm{T}}\,\overline{\boldsymbol{\lambda}}\right\},\qquad \text{subject to:} \tag{77}$$

$$\overline{\mathbf{M}}'-\left(\overline{\mathbf{A}}_0+\overline{\mathbf{H}}\right)\overline{\boldsymbol{\lambda}}\leq\overline{\mathbf{Y}}^0-\overline{\mathbf{M}}'',\qquad \overline{\boldsymbol{\lambda}}\geq\mathbf{0} \tag{78}$$

$$\frac{s-1}{s}\left(\overline{\mathbf{Y}}^0-\overline{\mathbf{M}}''\right)^{\mathrm{T}}\overline{\boldsymbol{\lambda}}\leq\Omega \tag{79}$$

$$\tilde{Q}_4 = \max_{\boldsymbol{\lambda}} \left\{ \overline{\mathbf{R}}^{\mathrm{T}} \overline{\boldsymbol{\lambda}} \right\}, \quad \text{subject to:} \tag{80}$$

$$\overline{\mathbf{M}}' - \left( \overline{\mathbf{A}}_0 + \overline{\mathbf{H}} \right) \overline{\boldsymbol{\lambda}} \leq \overline{\mathbf{Y}}^0 - \overline{\mathbf{M}}'', \qquad \overline{\boldsymbol{\lambda}} \geq \mathbf{0} \tag{81}$$

$$\frac{s-1}{s} \left[ \left( \overline{\mathbf{Y}}^0 - \overline{\mathbf{M}}'' \right)^{\mathrm{T}} \overline{\boldsymbol{\lambda}} + \frac{1}{2} \overline{\boldsymbol{\lambda}}^{\mathrm{T}} \overline{\mathbf{H}} \overline{\boldsymbol{\lambda}} \right] - \overline{\boldsymbol{\lambda}}_{\Omega}^{\mathrm{T}} \overline{\mathbf{A}}_0 \overline{\boldsymbol{\lambda}} + \frac{1}{2} \overline{\boldsymbol{\lambda}}^{\mathrm{T}} \overline{\mathbf{A}}_0 \overline{\boldsymbol{\lambda}} \leq 0 \tag{82}$$

$\overline{\boldsymbol{\lambda}}_{\Omega}$ being the optimal vector of minimization (75)-(76).

$$\tilde{Q}_6 = \max_{\boldsymbol{\lambda}} \left\{ \overline{\mathbf{R}}^{\mathrm{T}} \overline{\boldsymbol{\lambda}} \right\}, \quad \text{subject to:} \quad \frac{s-1}{s} \left( \overline{\mathbf{Y}}^0 - \overline{\mathbf{M}}'' \right)^{\mathrm{T}} \overline{\boldsymbol{\lambda}} \leq \Omega \tag{83}$$

and therefore:

$$\tilde{Q}_6 = \Omega \frac{s}{s-1} \max_k \left\{ \frac{\overline{R}_k}{\overline{Y}_k^0 - \overline{M}_k''} \right\} \tag{84}$$

*Proofs.*

The demonstration of the above statement is centered on the following inequality readily justified by means of simple algebra, account taken of the symmetry and positive semidefiniteness of matrix $\overline{\mathbf{A}}_0$:

$$\begin{aligned} &\overline{\boldsymbol{\lambda}}^{*\mathrm{T}} \overline{\mathbf{A}}_0 \overline{\boldsymbol{\lambda}} - \frac{1}{2} \overline{\boldsymbol{\lambda}}^{\mathrm{T}} \overline{\mathbf{A}}_0 \overline{\boldsymbol{\lambda}} = \\ &= \frac{1}{2} \overline{\boldsymbol{\lambda}}^{*\mathrm{T}} \overline{\mathbf{A}}_0 \overline{\boldsymbol{\lambda}}^* - \frac{1}{2} \left( \overline{\boldsymbol{\lambda}}^* - \overline{\boldsymbol{\lambda}} \right)^{\mathrm{T}} \overline{\mathbf{A}}_0 \left( \overline{\boldsymbol{\lambda}}^* - \overline{\boldsymbol{\lambda}} \right) \leq \frac{1}{2} \overline{\boldsymbol{\lambda}}^{*\mathrm{T}} \overline{\mathbf{A}}_0 \overline{\boldsymbol{\lambda}}^* \end{aligned} \tag{85}$$

Bound $\tilde{Q}_2$ is generated by relaxing the constraint (50), i.e. by dropping from the nonlinear constraint (50) the second quadratic form in the r.h.s. of Eq.(85a) in view of inequality (85b).

Bound $\tilde{Q}_3$ follows from $\tilde{Q}_2$ by eliminating the (nonnegative) quadratic form associated to the hardening matrix $\overline{\mathbf{H}}$.

Bound $\tilde{Q}_4$ arises from $\tilde{Q}_1$ simply by using the (feasible) vector $\overline{\boldsymbol{\lambda}}_{\Omega}$ obtained from optimization (75)-(76). Clearly, the above relaxation leading to $\tilde{Q}_3$ can be adopted to generate a (possibly worse) bound, say $\tilde{Q}_5$. In the light of this first study, the bound $\tilde{Q}_5$ seems to represent a good compromise between the conflicting requirements of computing economy and usefulness, and, hence, it will be focussed in the numerical tests of Sect. 6.

Finally, bound $\tilde{Q}_6$ is obtained from $\tilde{Q}_3$ by ignoring the constraint (78) which emanates from the necessary SD condition expressed by (41b).

Remarks.

(a) The above bounds, in view of their generations, fulfil the following inequalities:

$$\tilde{Q}_1 \le \tilde{Q}_4 \le \tilde{Q}_2 \le \tilde{Q}_3 \le \tilde{Q}_6; \qquad \tilde{Q}_1 \le \tilde{Q}_4 \le \tilde{Q}_5 \le \tilde{Q}_3 \le \tilde{Q}_6 \tag{86}$$

(b) The noteworthy mathematical features of the bounds are as follows.

The feasible domain of the *minmax* problem leading to $\tilde{Q}_1$ is nonconvex in the space $\{\bar{\boldsymbol{\lambda}}, \bar{\boldsymbol{\lambda}}^*\}$.

After the decoupling due to the first one of the relaxations involved in Theorem 2, bound $\tilde{Q}_2$ is generated by a sequence of two steps: (i) a convex quadratic program (QP) Eqs. (75)-(76); (ii) a convex nonlinear programming problem (NLP) Eqs. (72)-(74). By neglecting hardening, the NLP (ii) reduces to LP with expected drastic savings in computing $\tilde{Q}_3$.

Bound $\tilde{Q}_4$ emerges from a sequence of two decoupled phases, analogous to that of $\tilde{Q}_2$, while $\tilde{Q}_5$ preserves in the second phase the NLP formulation like $\tilde{Q}_4$ and unlike $\tilde{Q}_3$.

Finally, the most loose upper bound $\tilde{Q}_6$ can be computed merely by scanning a vector according to Eq. (84).

## 6. Examples

6.1 The preceding theoretical results have been numerically tested first with reference to the merely illustrative, simple two-dimensional plane-strain system shown in Fig. 1. The data and assumptions are as follows (the same as in [3]).

Material model (see Sec. 2): $E = 1$ GPa; $\nu = 0.2$; $\alpha = 0.98$; $M = 11.7$ GPa; isotropic permeability: $k = 4.10^{-5}$ m$^4$ MN$^{-1}$ s$^{-1}$; associative perfect plasticity (normality, no hardening); Drucker-Prager yield criterion characterized by compressive strength = 10 MPa and internal friction angle = 30°; zero bulk self-weight.

Boundary conditions: on the side AB: vanishing horizontal displacements and vertical tractions ($\hat{u}_1 = 0$, $\hat{t}_2 = 0$), pressure $\hat{p}''$ constant in time; on BC: $\hat{t}_1 = 0$, $\hat{t}_2 = \mu\, \hat{t}'$ with $\hat{t}' = \hat{q}\, sin(2\pi\tau/\mathrm{T})$ MPa and T = 2 s, no flux ($\hat{q} = 0$); on CD: $\hat{t}_1 = \hat{t}_2 = 0$, $\hat{p} = 0$; on AD: $\hat{u}_2 = 0$, $\hat{t}_1 = 0$, $\hat{q} = 0$. Initial conditions: $p^0 = p(x,0) = (1 - x/l)\,\hat{p}''$, with $l = 4$ m.

Space discretization by eight square, isoparametric FEs (Fig. 1): bi-quadratic shape functions (8 nodes) for displacements $u_1$, $u_2$; bi-linear interpolations (4 vertex nodes) for pressure $p$.

Two cases are considered, characterized by given pressure on AB, namely: (*i*) $\hat{p}''$=10 MPa; (*ii*) $\hat{p}''$=5 MPa.

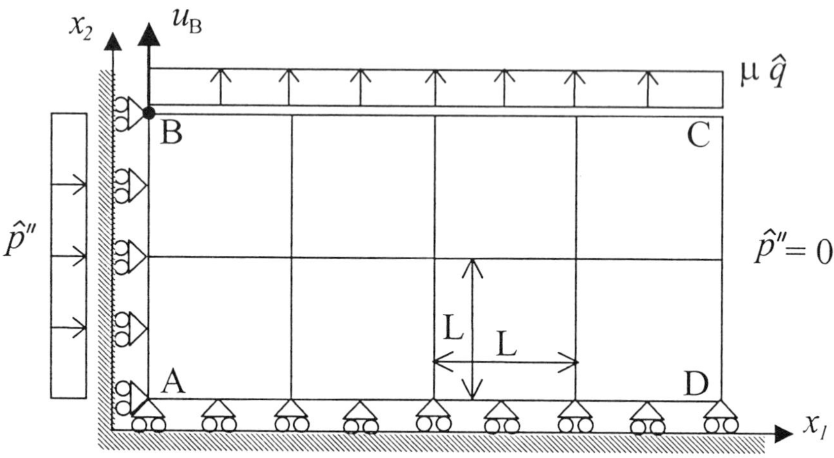


Fig. 1. Geometry and boundary conditions of the first illustrative example.

*Evolutionary (incremental) analyses.* First, the commercial nonlinear finite element code Abaqus [1] has been used to comparison purposes. Time-stepping poroplastic analyses of the above system have been carried out along a sequence of about 300 load cycles for various load amplifiers. Peculiar features of these computations are as follows: time-integration scheme with variable time-step; Newton-Rapson iterative solution of the step problem, with consistent tangent matrix predictor and backward difference corrector and default tolerances.

Fig. 2 visualizes some results of the evolutionary analyses for the case (*ii*) with load factors $\mu = 6$ and $\mu = 7$: the cumulative energy dissipation due to irreversible (plastic) deformative processes versus time; the fluctuation band of total displacement $u_2^{\mathrm{P}}$ of point B on the upper edge. For $\mu = 6$ the system is seen to shake down. For $\mu = 7$ it does not: dissipated energy keeps growing and incremental collapse ("ratchetting") occurs (Fig. 2b). The sequence of evolutionary analyses indicates that the critical threshold of SD limit $s$ is bracketed by $\mu = 6$ and $\mu = 7$ in the case (*ii*). For a number of load factors $\mu$ below $s$, the residual displacements of point B computed by Abaqus in the case (*i*) and (*ii*) are plotted in Fig. 3 and 4, respectively. It is worth noting that SD can be ascertained by time-marching incremental computations only if these are carried out over a large number of cycles. This circumstance, which is due to the diffusion processes (and, hence, is much less pronounced in plasticity) corroborates the importance of nonevolutionary methods in poroplasticity.

*Nonevolutionary direct analyses.* The direct methods proposed in this paper for SD analysis and bounding postSD quantities are now applied to the illustrative system of Fig. 1 with the following features: same FE discretization as above for displacements $u_1$, $u_2$ (eight square isoparametric bi-quadratic 8 nodes) and pressure $p$ (bi-linear, 4 vertex nodes); for stresses $\sigma_{11}$, $\sigma_{22}$, $\sigma_{33}$, $\sigma_{12}$ and fluxes $q_1$, $q_2$, bi-linear FEs with 4 nodes in the Gauss points; for plastic multipliers $\lambda_\alpha$, bi-linear FEs with 4 vertex nodes; the modelling of the variables conjugate to the above ones has been performed according to the concept of Prager's generalized variables (details in [5] [6] [17]). The Drucker-Prager yield criterion is approximated by 8 yield planes: these are tangent to the original cone in the principal stress space and are shifted axially so that their intersections with the deviatoric plane through the intersections of the cone with the axes define a hexagon whose area equals the one generated by the cone on that plane (see [3] for motivation and details). The other data are the same as for the evolutionary analyses.

On this basis, the static, primal LP problem, Eq. (41), has been solved to provide the SD limits and various upper bounds on the residual upward vertical displacement of the point B, for initial pressure on AB $\hat{p}'' = 10$ MPa (case *i*) and $\hat{p}'' = 5$ (case *ii*).

The results are gathered in Fig. 3 and 4 and give rise to the following comments.

(A) Reasonable agreement was found in both cases between the SD limits $s$ provided by direct LP computations ($s = 3.63$ in case *i*; $s = 7.10$ for case *ii*) and those attained by time-stepping applications of Abaqus with the original (non PWL) yield criterion ($s = 3.5$ for case *i*; $s = 6.5$ for case *ii*).

(B) Although the LP, QP and NLP solvers employed were parts of commercial software [18], the direct bounding techniques generally entail significant computational savings with respect to the determination of the postSD residual displacement by evolutionary analysis. For instance, the average CPU computing time (in seconds, on a workstation HP735) in case (*ii*) turned out to be as follows: 1200 for each one of the numerous step-by-step analyses; 980, 33 and 15 for bound $\tilde{Q}_5$, $\tilde{Q}_3$ and $\tilde{Q}_6$, respectively (including 15 $s$ for the common QP solution).

(a)

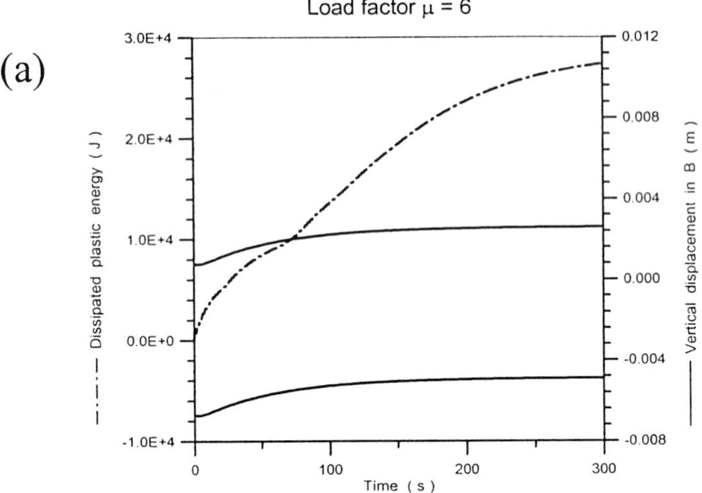


(b)

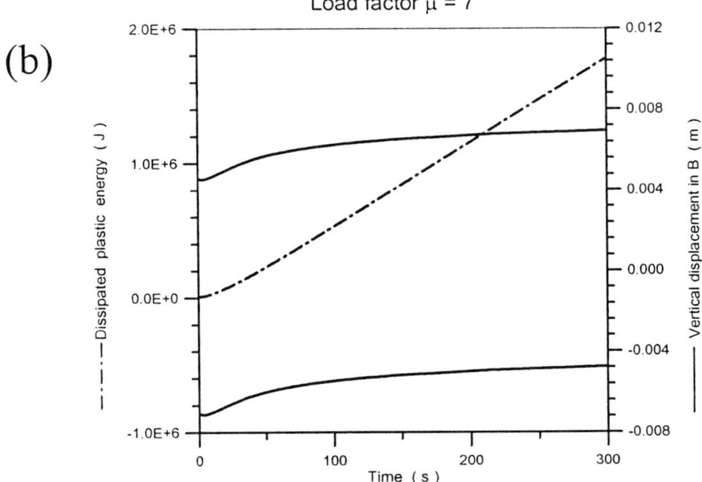


Fig. 2. Results of time-stepping analysis in case (*ii*) for μ = 6 (a) and μ = 7 (b): cumulative dissipated energy (dashed lines); fluctuation band of total vertical diplacement of point B (solid lines).

(C) In case ($i$) the two bounds $\tilde{Q}_3$ and $\tilde{Q}_5$ shown in Fig. 3 exceed the actual residual displacements by a factor of few units and, hence, may be useful for an integrity assessment of the system (e.g., for $\mu = 2$: $u_B = 5.40$ mm, $\tilde{Q}_5 = 16.7$ mm, $\tilde{Q}_3 = 31.1$ mm). In case ($ii$) with lower imposed pressure $\hat{p}'' = 5$ MPa (Fig. 4), the bounds are very loose and, hence, hardly useful (e.g., for $\mu = 6$: $u_B = 0.112$ mm, $\tilde{Q}_5 = 46$ mm, $\tilde{Q}_3 = 264$ mm, $\tilde{Q}_6 = 616$ mm) .

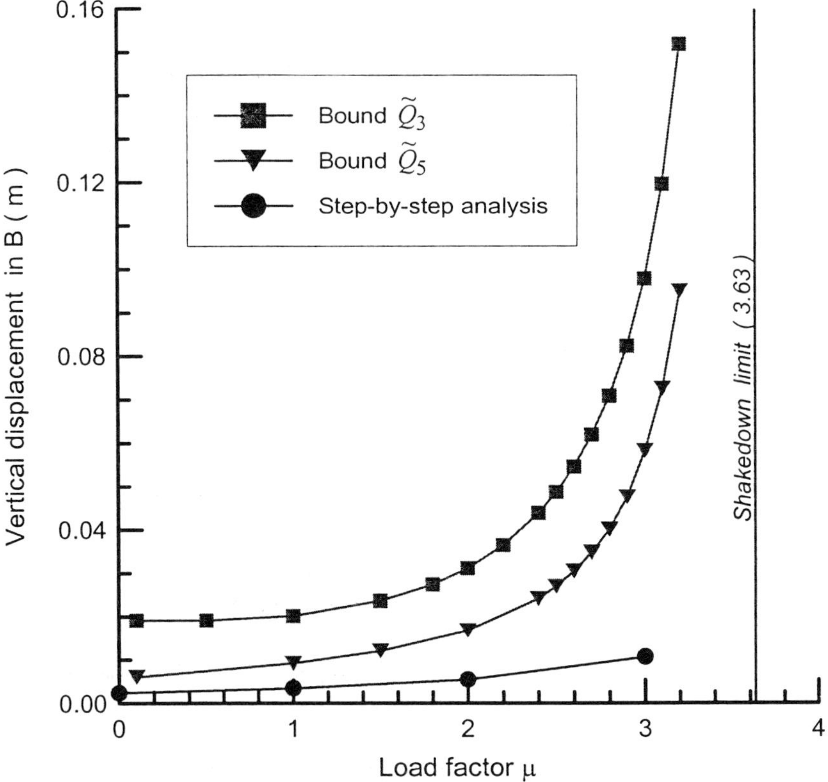


Fig. 3 Case ($i$) with $\hat{p}'' = 10$ MPa: upper bounds on the residual vertical displacement in B obtained from direct analyses and compared to the results of a time-stepping procedure for various load factors $\mu$.

(D) In the case ($i$) the simplest but coarsest bound $\widetilde{Q}_6$ could not be determined since some components of vector $\overline{\mathbf{Y}}^0 - \overline{\mathbf{M}}''$ (but not the corresponding ones in vector $\overline{\mathbf{R}}$) turn out to be negative for $\hat{p}'' = 10$ MPa and, hence, the LP problem (83) has an unbounded feasible domain and infinity as optimal value .

(E) When a bound has been computed for a given load factor μ, the determination of the same bound for another factor can exploit a substantial part of the computations already performed, with remarkable time savings .

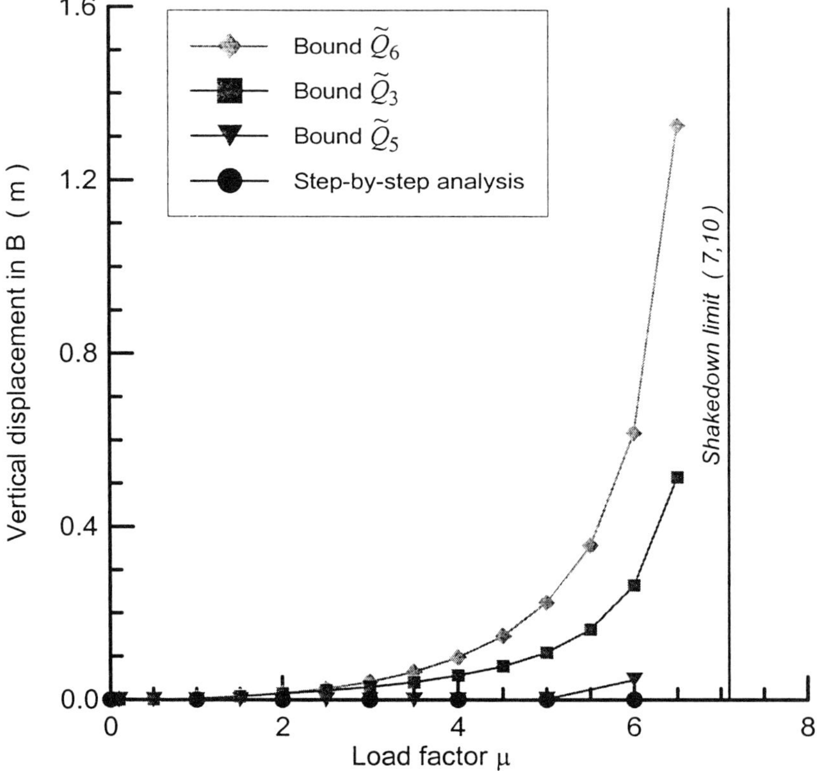


Fig. 4. Case ($ii$) with $\hat{p}'' = 5$ MPa: upper bounds on the residual vertical displacement in B obtained from direct analyses and compared to the results of a time-stepping procedure for various load factors μ.

6.2 The second example concerns the plane-strain FE model of a gravity dam shown in Fig. 5 and already considered in [3] as for SD analysis.

The essential features of the FE discretization and constitutive model (Drucker-Prager piecewiselinearized by secant planes) are the same as in the first example; however, the material model is now defined by fairly realistic parameters (partly suggested by the engineering situation dealt with in [8]), namely: E = 14 GPa; $\nu = 0.15$; $\alpha = 0.4$; $M = 11.8$ GPa; bulk self-weight 17 kN/m$^3$; horizontal and vertical permeability 0.1 and 0.01 m$^4$MN$^{-1}$s$^{-1}$, respectively. The live load is represented by the overtopping water pressure sinusoidally fluctuating in time with a period of 3 days; the dead load by the self-weight, by the full reservoir pressure upstream and by zero pressure downstream and on the dam top; zero flux and zero displacement are the boundary conditions along the foundation interface.

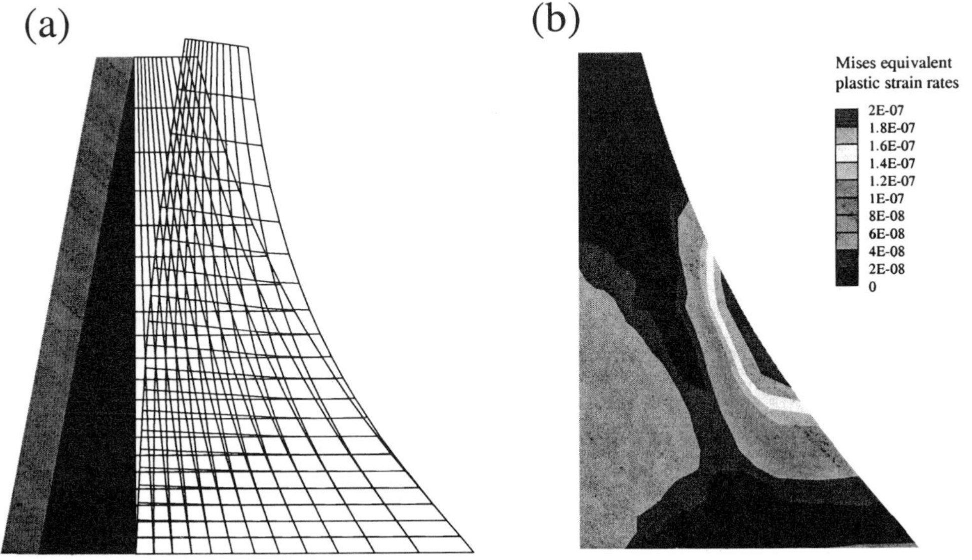


Fig. 5. Poroplastic model of a gravity dam: (a) geometry (40 m height, 27 m basis, 7 m crown), FE mesh (12x20), incremental collapse mechanism; (b) Mises equivalent plastic strain rates of collapse.

Using the mesh of 12x20 FEs depicted in Fig. 5a, the SD analysis led to: SD limit = 63.6 m of overtopping height; incremental collapse mechanism visualized by Fig. 5a in terms of nodal velocities and by Fig. 5b in terms of plastic strain rates.

For overtopping height (live load) of 25.0 m, i.e. one half of the SD limt, the upper bound $\tilde{Q}_5$ (see Sect. 5) on the horizontal residual displacement at the top turns out to be $\tilde{Q}_5 = 23$ cm, while the value provided by the evolutionary analysis through Abaqus is $u = 2.5$ cm. The bound in this test is very loose, but its strong dependence on the number of variables, visualized in Fig. 6, makes it susceptible to improvement by mesh refinement.

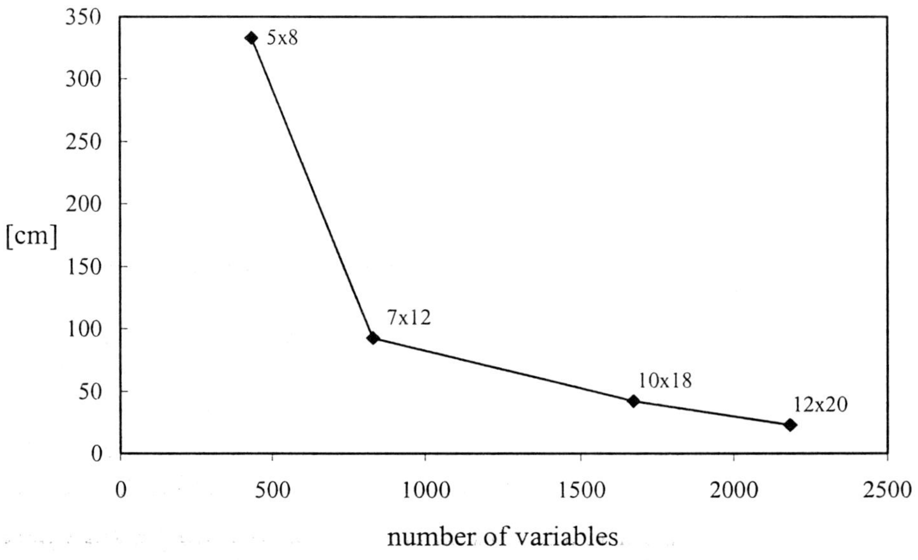


Fig. 6. Upper bound $\tilde{Q}_5$ on residual horizontal displacement versus the number of plastic multipliers $\lambda$ for the example of Fig. 5.

## 7. Conclusions

The post-shakedown values of history-dependent quantities have been bounded from above by history-independent bounds in the context of hardening poroplasticity on the basis of poroelastic shakedown theorems established in [2] and [3]. The theoretical developments presented exhibit well expected similarity with their counterparts in classical plasticity, but also novel features related to peculiar mechanical features of two-phase materials, primarily the dependence on physical time. Illustrative numerical tests have shown some potentialities and limitations of the present results. Quite restrictive hypotheses have been adopted in this (apparently first) contribution to a bounding theory in poroplasticity: no geometric effects; constant permeability; linear yield functions and hardening; periodic external actions; full saturation. Current research concerns alternative formulations of the shakedown and bounding theorems in poroplasticity and their generalizations in a number of directions (primarily partial saturation and the removal of the piecewiselinearization to avoid prohibitive numbers of variables for three-dimensional systems).

## Acknowledgements

The authors acknowledge the financial support by the Italian Ministry for University and Research (MURST) for a research project on "Integrity assessment of large dams". Valuable contributions of R. Ardito and L. Ghezzi, on the occasion of their master theses, are also gratefully acknowledged.

## References

[1] Abaqus/Standard (1999): Theory and User's manuals, rel. 5.8, HKS Inc., Providence, RI, USA.

[2] Cocchetti G., Maier *G. (1998): Static* shakedown theorems in piecewiselinearized poroplasticity, *Archive of Applied Mechanics,* **68,** 651-661.

[3] Cocchetti G., Maier G. (2000): Shakedown analysis in poroplasticity by linear programming, *Int. J. Num. Meth. Engng,* 47, 141-168.

[4] Cohn M. Z., Maier G. [Eds.] (1977): Engineering plasticity by mathematical programming, Pergamon Press, New York.

[5] Comi C., Maier G., Perego U. (1992): Generalized variables and extremum theorems in discretized elastic-plastic analysis, *Comp. Meth. Appl. Mech. Engng.,* **96,** 213-237.

[6] Comi C., Perego U. (1995): A unified approach for variationally consistent finite elements in elastoplasticity, *Comp. Meth. Appl. Mech. Engng.,* **121,** 323-344.

[7] Coussy O. (1995): Mechanics of porous continua, John Wiley & Sons, Chichester.

[8] Fauchet B., Coussy O., Carrère A., Tardieu B. (1992): Poroplastic analysis of concrete dams and their foundations, *Dam Engineering,* **2,** 165-192.

[9] Kamenjarzh J. A. (1996), Limit analysis of solids and structures, CRC Press, London.

[10] Koiter W. T. (1960): General theorems for elastic-plastic solids, in "Progress *in Solid Mechanics"*, Vol. 1, Sneddon J. N., Hill R. (Eds.), North-Holland Pub., Amsterdam, 167-221.

[11] König J. A. (1987): Shakedown of Elastic-Plastic Structures, Elsevier, Amsterdam.

[12] Kaliszky S. (1989): Plasticity: theory and engineering applications, Elsevier, Amsterdam, North Holland.

[13] Lewis R. W., Schrefler B. A. (1998): The finite element method in the static and dynamic deformation and consolidation of porous media, John Wiley & Sons, Chichester.

[14] Lubliner J. (1990): Plasticity theory, Mc Millan Publ., New York.

[15] Maier G. (1970): A matrix structural theory of piecewise-linear plasticity with interacting yield planes, *Meccanica*, **5**, 55-66.

[16] Maier G. (1973): Upper bounds on deformations of elastic-workhardening structures in the presence of dynamic and second-order effects, *Journal of Structural Mechanics*, **2**, 265-280.

[17] Maier G., Comi C. (1997): Variational finite element modelling in poroplasticity, in "Recent *developments in computational and applied mechanics"*, Reddy B. D. (Ed.), CIMNE, Barcelona, 180-199.

[18] Math Works Inc. (1996): Matlab User's Manual and Optimization Toolbox Manual, rel. 5.

[19] Neale B. G., Symonds P. S. (1951): A method for calculating the failure load for a framed structure subjected to fluctuating loads, *J. Instn. Civil Engrs.*, **35**, 186-197.

[20] Ponter A. R. S. (1972): An upper bound on the small displacements of elastic, perfectly plastic structures, *Journal of Applied Mechanics, Ser. E*, **39**, 959-963.

[21] Symonds P. S. (1951): Shakedown in continuous media, *Journal of Applied Mechanics*, **18**, 85-89.

# FATIGUE BEHAVIOR OF FIBER REINFORCED CONCRETE: COMPARISON BETWEEN MATERIAL AND STRUCTURAL RESPONSE

S. CANGIANO
*CTG Italcementi Group*
*via G. Camozzi 124, 24121 Bergamo, Italy*

AND

G.A. PLIZZARI
*Department of Civil Engineering, University of Brescia*
*via Branze 38, 25123 Brescia, Italy*

**Abstract.** In this paper, the role of steel and carbon fibers on fatigue behavior of both normal and high strength concrete is investigated. Experimental results from direct tension tests on cylindrical specimens and from four point bending tests on beam specimens allow for a comparison between material and structural behavior. Geometrically similar beams with different sizes are studied.

It is found that the envelope curves obtained from cyclic tests match the curves obtained from the static tests on cylinders of high-strength concrete and of fiber reinforced concrete quite well. This however does not always occur in beam specimens since the fatigue damage strongly depends on the fracture process zone development which is influenced by the specimen size and loading history. Finally, analytical models available in the literature for crack increment during inner loops and for cyclic behavior on the envelope curve are compared with the experimental results.

## 1. Introduction

Cracks are among the most common flaws in civil engineering structures and they are often present from the beginning of the life of a structure because of drying shrinkage or thermal gradients. As a consequence, inter-



*D. Weichert and G. Maier (eds.), Inelastic Analysis of Structures under Variable Loads,* 315–334.

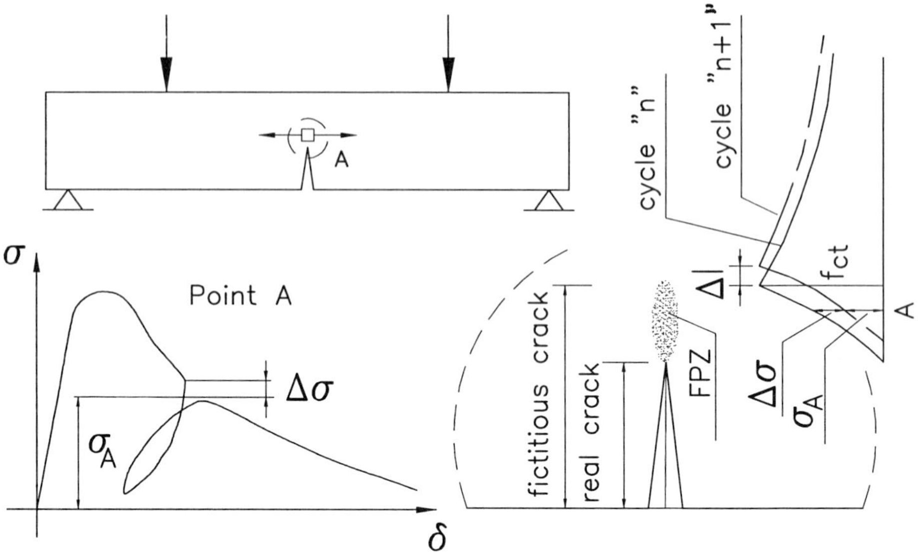


*Figure 1.* Stress distribution along the fracture process zone in concrete specimens under cyclic loading.

est in the tensile behavior of concrete has increased enormously in recent years, and the concept of fracture mechanics has been introduced in the field of concrete structures. Also, considering the cyclic nature of the variable actions, and the use of high-strength concrete that results in slender structures (in which the dead load forms a smaller part of the total load), it becomes evident that the evaluation of the structure safety factor should consider the behavior of the cracks under cyclic loading.

Slowik and co-workers [18] claimed that concrete damage under fatigue loading mainly occurs in the Fracture Process Zone (FPZ) present at the crack tip (Fig. 1). In fact, by applying cyclic loads with peaks every 4 or 5 cycles to wedge splitting specimens, the crack propagation rate increased remarkably after every peak. The authors attributed this rate increase to the presence of a larger FPZ after the application of the peak loads. The development of fatigue damage in the FPZ is confirmed by the known dependence on the loading history or the sequence effect found by some researchers who determined the fatigue strength of concrete structures by testing "structural specimens" such as beams or wedge splitting specimens [7,10]; in fact, in these specimens the FPZ size strongly depends on the loading history as well as on the specimen size [1,3]. This would also explain the inapplicability of the Palmgren-Miner rule [9,11] to determine the fatigue strength of concrete structures under tension or bending [7].

Fracture behavior of concrete structures under cyclic loads can be correctly studied only when the FPZ is present and the material behavior in the FPZ is known [13]. As a matter of fact, most of the experimental results from fatigue tests available in the literature are obtained from specimens, notched or unnotched, without a FPZ present at the beginning of the test [16].

The University of Brescia and the CTG Italcementi Group have recently undertaken a joint research project to study the interrelation between material and structural behavior in concrete fracture, both for static and cyclic loading. The complete research program is sketched in Fig. 2. In this paper, the results concerning the comparison between material and structural behavior under fatigue loading are presented. Material behavior is studied by means of uniaxial tensile tests on cylindrical specimens while the structural behavior is studied by performing four point bending tests on beams. Some results from the cyclic test, including the size effect, are reported in [15], while the results from static test are shown in [14].

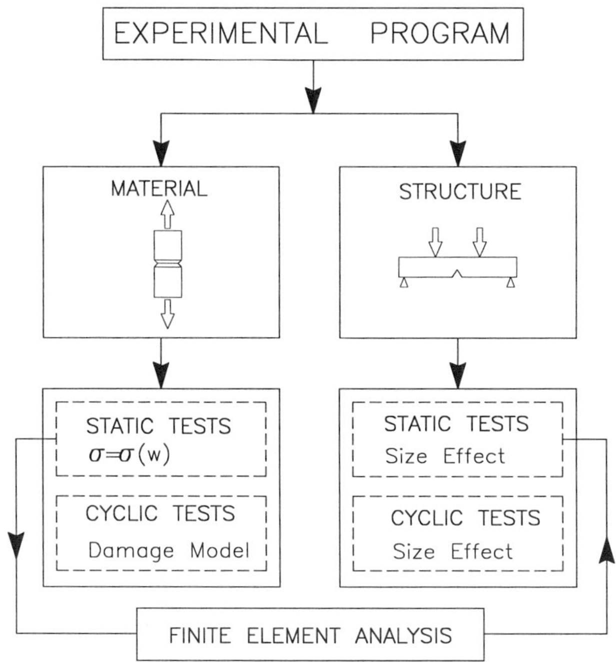


*Figure 2.* Scheme of the complete experimental program.

The experiments concern normal-strength concrete as well as high-strength concrete. The use of steel fibers in normal concrete and steel and carbon fibers in high-strength concrete allows to study the potential benefit of fibers on the behavior of concrete structures under cyclic loads. Tests are performed on pre-cracked specimens in which a fracture process zone is present. The experimental results are compared with damage models for fatigue loading available in the literature.

## 2. Experiments

As far as fatigue is concerned, a distinction should be made between low-cycle fatigue and high-cycle fatigue; the former involves few load cycles ($< 10^3 \div 10^4$) with high stresses (similar to those induced by an earthquake), while the latter is characterized by a much larger number of cycles with lower stresses (such as those induced by rotating machinery). Although a clear distinction between the two cases cannot be easily made, in this research-work, the term fatigue will refer to low-cycle fatigue and the tests mainly concern high stresses.

### 2.1. MATERIALS

Two types of cement were used (UNI-ENV 197): 1) class 32.5 R type CEM II/B-L for normal-strength concrete (NSC); 2) class 52.5 R type CEM I for high-strength concrete (HSC). The cement content was 370 $kg/m^3$ for the NSC matrices and 550 $kg/m^3$ for the HSC matrices; in the latter, 55 $kg/m^3$ of silica fume was also adopted (10% of the cement content).

Two different types of fibers were used: 1) the hooked steel fibers were low carbon, cold drawn, 30 mm long and had a diameter of 0.5 mm (aspect ratio = 60), an elastic modulus of 210.000 MPa and a tensile strength higher than 1100 MPa; 2) the carbon fibers were 20 mm long, had a diameter of 0.008 mm (aspect ratio = 2500), an elastic modulus ranging from 180.000 to 240.000 MPa and a tensile strength higher than 2000 MPa.

In the following, the NSC and HSC concretes reinforced with steel fibers will be identified by NSC-SFR and HSC-SFR respectively, while those (HSC) reinforced with carbon fibers will be identified by HSC-CFR. The content of steel fibers was limited to the minimum value suggested by the producers (30 $kg/m^3$). In order to make the comparison possible, the same volume fraction of carbon fibers (0.38%) was adopted.

All concrete mixes were made with siliceous aggregates, having a rounded shape and a maximum diameter of 15 mm. The grain size distribution was very close to the Bolomey curve [2].

Prior to the fatigue tests, some quasi-static tests were performed under monotonically increasing deformation to determine the fracture parameters of the plain and fiber-reinforced concrete.

Table 1 shows the concrete mix proportions, the properties of fresh concrete as well as the compression strength ($f_c$), the tensile strength ($f_{ct}$) and the elastic modulus of elasticity ($E_c$), all measured from cylinders.

TABLE 1. Composition and properties of the adopted concretes.

| Material | w/c | Superp. $[kg/m^3]$ | Density $[kg/m^3]$ | Slump $[mm]$ | $f_c$ [MPa] | $f_{ct}$ [MPa] | $E_c$ [MPa] | $G_F$ $[J/m^2]$ |
|---|---|---|---|---|---|---|---|---|
| NSC | 0.56 | 3.3 | 2374 | 130 | 45.4 | 4.01 | 31.600 | 151 |
| NSC-SFR | 0.59 | 4.4 | 2371 | 130 | 27.8 | 4.31 | 27.400 | 2137 |
| HSC | 0.29 | 12.7 | 2488 | 200 | 102.0 | 5.16 | 48.900 | 135 |
| HSC-SFR | 0.29 | 13.2 | 2508 | 170 | 106.4 | 5.57 | 49.400 | 830 |
| HSC-CFR | 0.29 | 13.2 | 2482 | 100 | 96.5 | 5.18 | 50.000 | 138 |

### 2.2. TESTING MACHINE

The tests were carried out by means of a very stiff servo-controlled testing machine. A servo-valve, with high dynamic response characteristics (400 Hz) was used to pilot the hydraulic actuator. The servo-valve was piloted by a current signal coming from a P.I.D. controller where the feed-back signal was compared with the reference signal generated by the software [13]. Furthermore, the software allowed for the acquisition of signals from transducers which were suitably amplified and converted from analog to digital by the A/D converter (Fig. 3).

### 2.3. SPECIMENS AND INSTRUMENTATION

The material behavior was studied by testing cylinders having a diameter of 80 mm and a height of 210 mm (Fig. 4). A triangularly shaped notch was made in the middle section of the cylinders, by mechanical turning.

In cylindrical specimens, the crack-opening data were acquired by three inductive transducers (LVDT; Linear Variable Differential Transformer) placed radially at $120^o$ on a base length of 30 mm (Fig. 4). In fact, the contribution of the elastic deformation to the measured displacement is negligible because of the small base length of the LVDTs. In addition, three resistance full-bridge displacement transducers (clip-gauges) were placed (equally spaced) between the three LVDTs. The average signal of the three clip-gauges was used as the feed-back signal. The rigid coupling of the specimen to the loading system was obtained by gluing it to the two loading

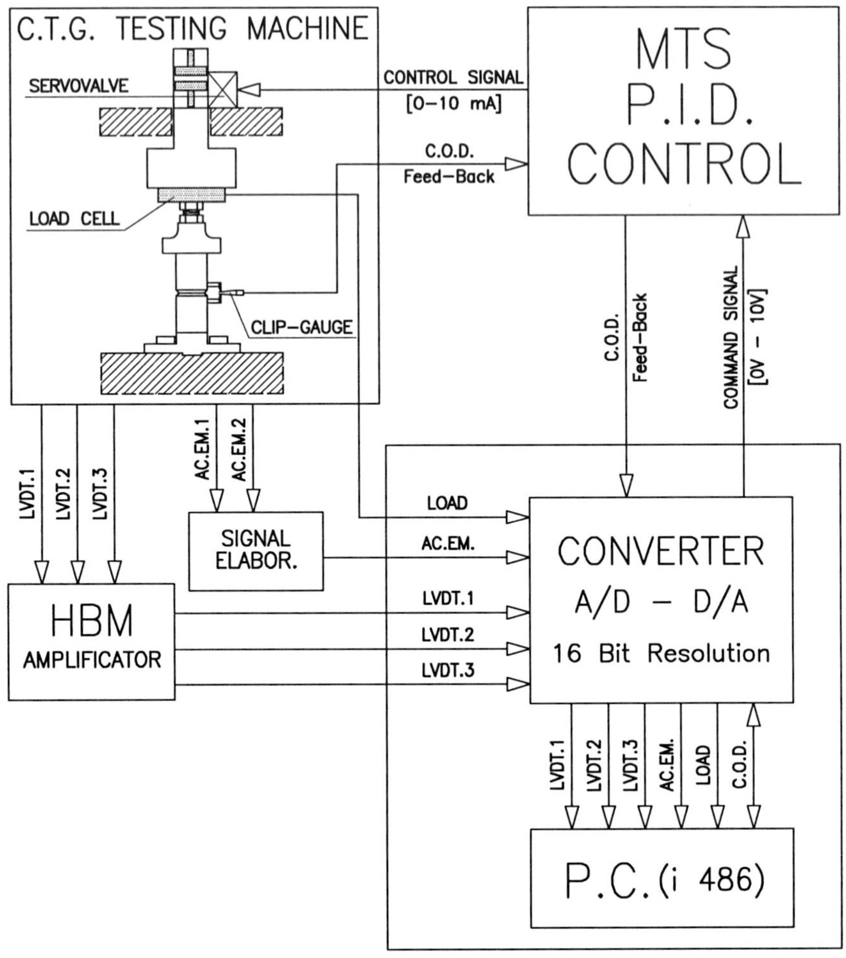


*Figure 3.* Scheme of the control of the cyclic tests on beam specimens.

plates by an epoxy resin. The upper plate was screwed to a bolt able to transfer the tension to a reversible load cell with a load capacity of 200 kN.

The study of structural behavior was performed on beams geometrically similar in two dimensions. These beams have overall lengths of 500, 1000 and 1500 mm (spans of 460, 920 and 1380 mm) and heights of 100, 200 and 300 mm respectively (Fig. 5). The width was equal to 100 mm for all the beams. The beams were notched in the midspan with a peak-shaped diamond circular saw. The ratio between the notch depth and height was kept constant (equal to 0.27) for all beams (Fig. 5). In order to limit the shear stresses in the mid-span zone, a four point loading scheme was adopted

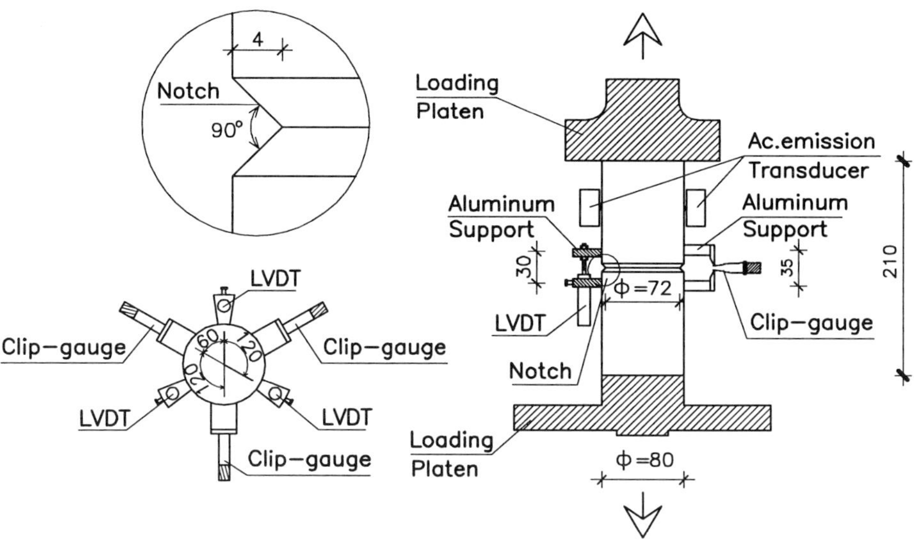


*Figure 4.* Geometrical characteristics and instrumentation of cylindrical specimens.

(Fig. 5). The ratio of the distance between the two upper loading points with the distance between the lower rollers was maintained constant and equal to 0.326 for all beam sizes. The Crack Mouth Opening Displacement (CMOD) was assumed as the feed-back quantity in the control loop of the test. The vertical displacement was measured at the upper loading points by means of two LVDTs fixed to an aluminum bar that was fixed at the beam ends (at mid-height; Fig. 5). The beams were supported by two roller bearings, one of which was fixed and the other was free to move horizontally. Steel plates were placed between the roller bearings and the beam to avoid local plastic deformation of concrete due to the high compression stresses.

Two pre-amplified transducers were placed on diametrically opposite points of the cylinders and on the upper face of the beam, near the notched section, to monitor the Acoustic Emission (A.E.) during the test; silicone grease was used as a coupling medium. The range of A.E. transducers was 0-50 MHz and their signal was amplified in order to obtain a linear output in the range 0-10 V. The A.E. activity was acquired in terms of a cumulative count of acoustic events which could be related to the development of the cracking process during the test [17].

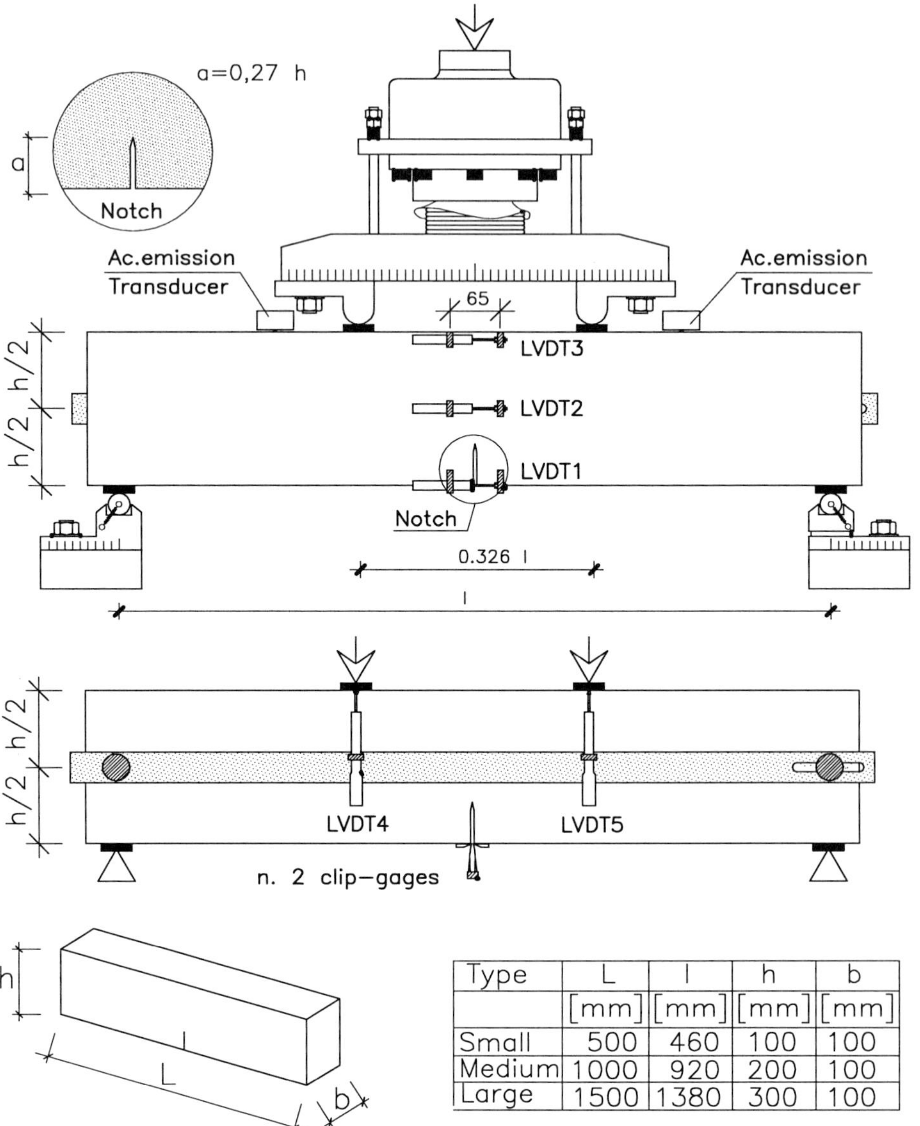


| Type | L | l | h | b |
|---|---|---|---|---|
| | [mm] | [mm] | [mm] | [mm] |
| Small | 500 | 460 | 100 | 100 |
| Medium | 1000 | 920 | 200 | 100 |
| Large | 1500 | 1380 | 300 | 100 |

*Figure 5.* Geometrical characteristics and instrumentation of beam specimens.

## 2.4. PLANNING AND CONTROL OF CYCLIC TESTS

In order to have a fracture process zone present at the beginning of the cyclic loading, both cylinders and beams were initially subjected to a quasi-static monotonic displacement up to the peak load. At the beginning of the descending branch of the load-displacement curve, when the load dropped to about 95% of the peak load, the specimens were unloaded (by the software) up to the chosen lower value of the following cyclic loading (curve OAB of Fig. 6). The experiment then continued with the inner loop phase (BC). In order to vary the load between an upper and a lower bound during inner loops, and to continue the cyclic test on the descending branch of the load-displacement curve (curve CD of Fig. 6), the experiments were performed under displacement control and the load levels were controlled by means of the software [13]. The test ended with a quasi-static monotonic deformation (curve DE of Fig. 6). The locus of broken curves joining the end of the reloading curve will be called "envelope curve" (curve CDE of Fig. 6).

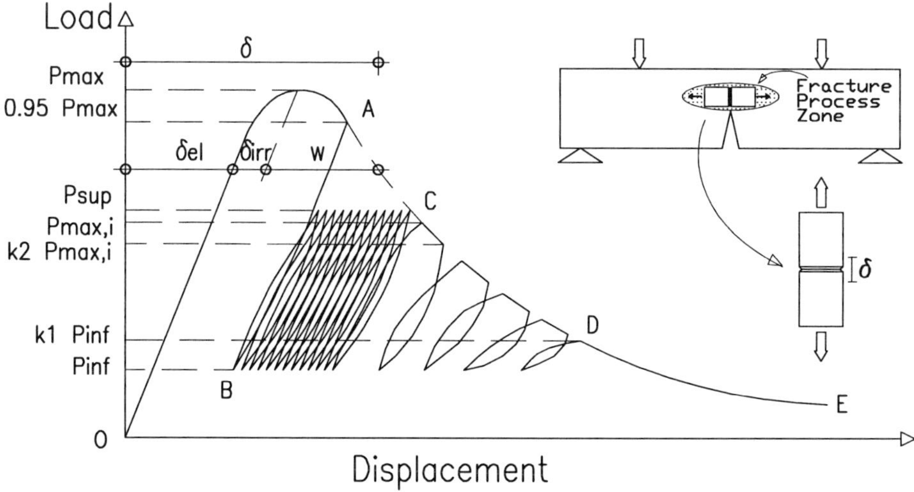


*Figure 6.* Scheme of the loading history for cyclic tests on cylinders and on beams.

## 3. Results and discussion

Typical load and cumulative A.E. curves plotted vs. C.M.O.D., as obtained on a medium NSC-SFR beam, are shown in Fig. 7. The A.E., which is normalized to the maximum value (related to the end of the test), remarkably increases during the inner loops and the cycles on the envelope curve, whereas the slope of the A.E. curve becomes softer in the final part of the test. Results from tensile tests of cylinders of the same material show that the A.E. activity is very small during the inner loops and remarkably increases during the cycles on the envelope curve (Fig. 8); this evidences that damage in the material mainly occurs during the cycles on the envelope curve. The A.E. measured during the inner loops on beam specimens demonstrates that, although the structure is subjected to inner loops, in some parts of the FPZ the material is subjected to cycles on the envelope curve (that provoke an increment of the A.E. measurement). This underlined the need to perform both inner loops and cycles on the envelope curve when studying the material behavior [13].

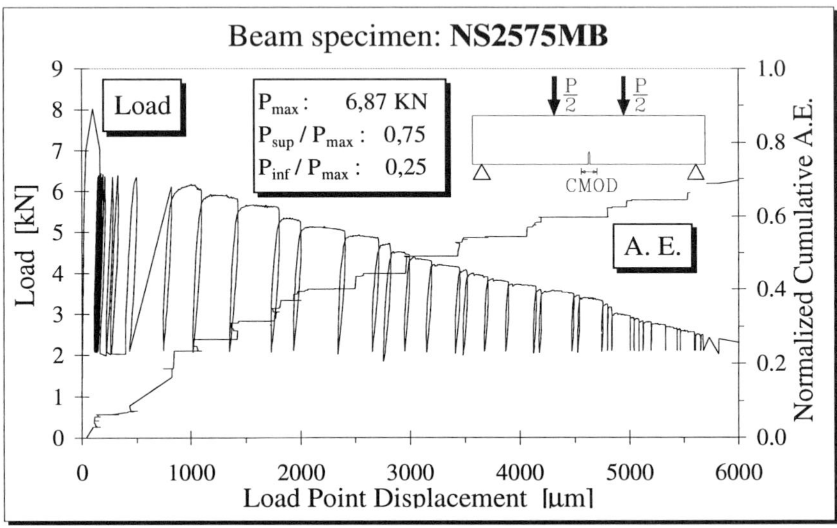


*Figure 7.* Typical results from cyclic tests as obtained from a medium beam of normal-strength concrete with steel fibers.

Fig. 9 exhibits the envelope curves obtained by cyclic tests and the curves of the post-peak tensile stress (assumed uniformly distributed on the cracked section) versus the average crack opening, as obtained by the static tests on cylinders of NSC-SFR. The crack opening was determined by subtracting the small elastic deformation and the irreversible displacement due to microcracking ($\delta_{irr}$ in Fig. 6) from the measured displacement [8]. It can be noticed that the envelope curves match the static curves quite well. Since the same good agreement was obtained for all the materials adopted [15], it can be accepted that, as already found for normal concrete without fibers [8,13], the envelope curve from direct tension tests on fiber-reinforced concrete cylinders can be approximated with the static monotonic curve. The comparison between envelope and static curves obtained from beams often did not evidence the same good fitting. A typical example is shown in Fig. 10 that exhibits a comparison between the envelope curves and the curves from the static tests obtained by the medium HSC-SFR beams. The difference between the material (from cylinders) and the structural behavior

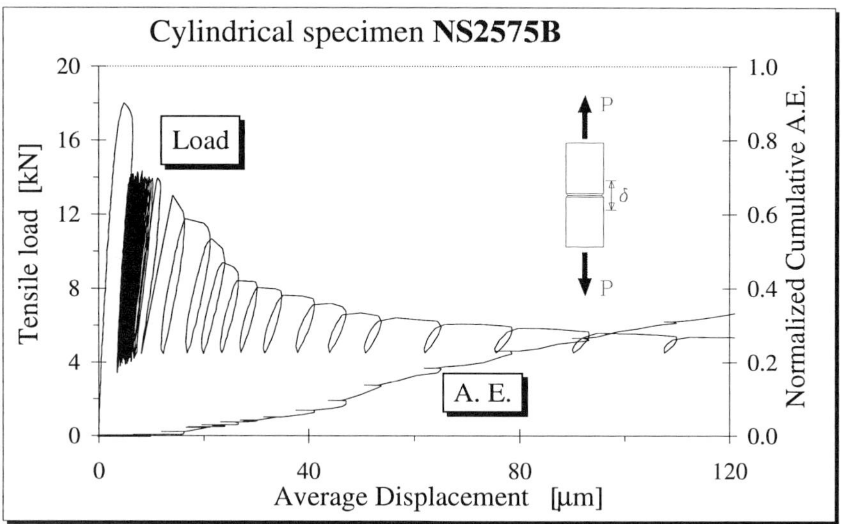


*Figure 8.* Typical results from cyclic tests as obtained from a cylinder of normal-strength concrete with steel fibers.

(from beams) mainly depends on the FPZ development whose dimension varies with the beam size.

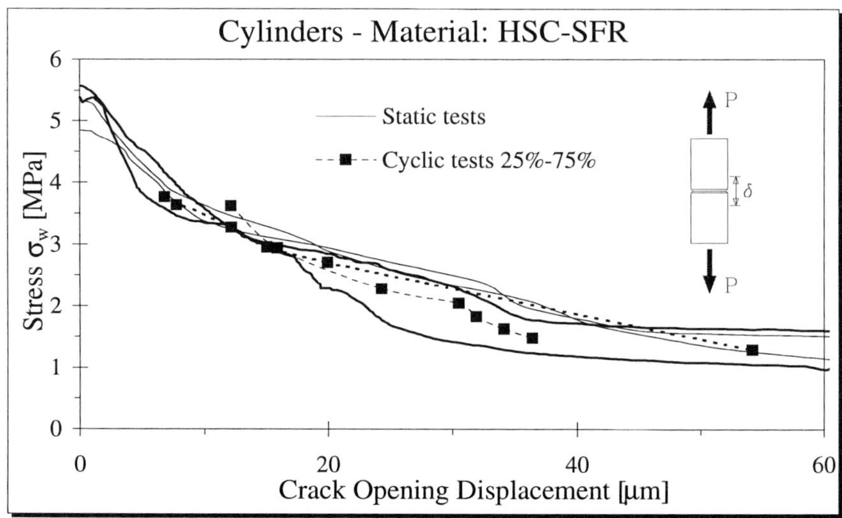


*Figure 9.* Comparison between the cyclic envelope curves and the quasi-static curves as obtained from cylinders of high-strength concrete with steel fibers.

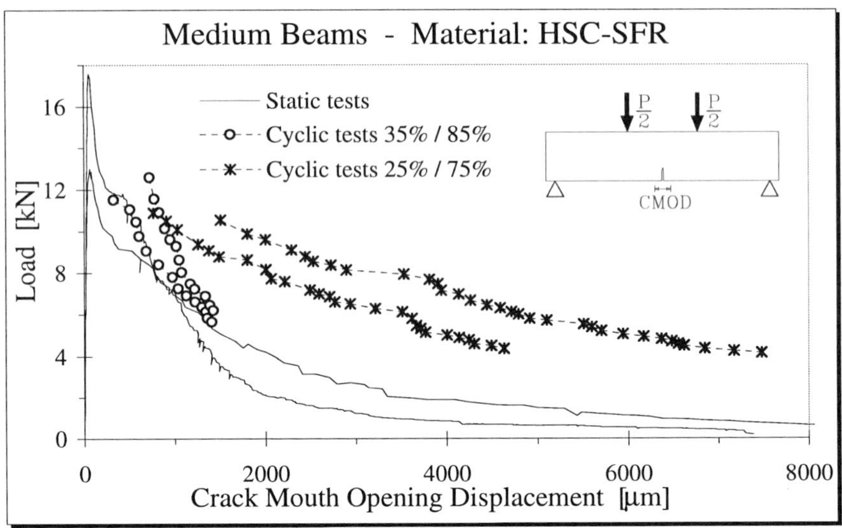


*Figure 10.* Comparison between the cyclic envelope curves and the quasi-static curves as obtained from medium beams of high-strength concrete with steel fibers [15].

The influence of the fracture process zone size is also evidenced in Fig. 11 that shows the number of inner loops ($N_{max}$) as a function of the ratio between the initial CMOD at the upper load of the first inner cycle ($CMOD_0$) and the CMOD at the last inner cycle (on the envelope curve, $CMOD_u$) in the medium beams. A high value of this ratio is probably due to a higher initial damage in the specimen that can be related to a larger FPZ size. In the legend, the number included in the parenthesis indicates the lower and the upper load levels as percentages of the peak load. The diagram evidences the decrease of the maximum number of inner loops with the increase of the $CMOD_0/CMOD_u$ ratio (i.e. the initial damage).

In summary, the results shown in this Section underline the need to perform uniaxial tensile tests on cracked cylinders to study fatigue behavior of concrete, since the results from beam specimens are strongly influenced by the fracture process zone present at the crack tip that depends on the specimen size. Although there are structural effects also in the uniaxial tensile tests, often related to the rotational stiffness of the loading platens [19], it is believed that the error made by using the average crack opening for modeling the crack cyclic behavior is small [8], also considering the errors associated with the determination of elastic and strength parameters $E_c$ and $f_c$.

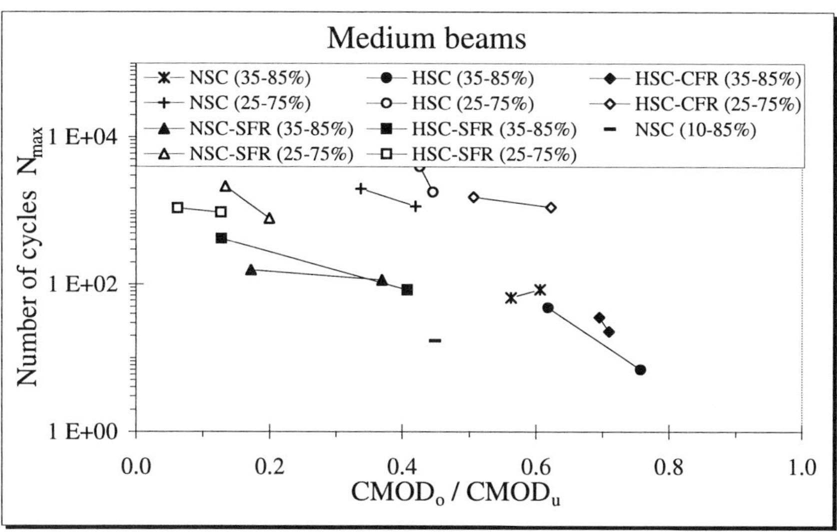


*Figure 11.* Maximum number of inner loops versus the ratio between CMOD at the upper load of the first and last cycles.

## 4. Comparison with analytical models

As mentioned above, tensile behavior of concrete under cyclic loading should be studied by means of direct tension tests and both inner loops and cycles on the envelope curve. In concrete in tension, after cracking occurs, the deformations tend to localize in the cracked section and damage has often been expressed in terms of crack-opening [4,8]. As far as the inner loops are concerned, the authors recently proposed to consider the crack opening at the upper stress level ($w_{sup}$) and to plot $dw_{sup}/dn$ versus $w_{sup}$ normalized to $w_u$ which is the crack opening on the envelope curve for the same upper stress level [13]. The fatigue life depends on the initial damage that can be related to deformation $w_0$ at the upper load after cracking (Fig. 12). Local failure occurs when the crack opening reaches the envelope curve ($w_{sup}/w_u = 1$). Since the envelope curve is very close to the quasi-static curve (Fig. 9), for a chosen maximum stress, $w_u$ can be determined either from static tests or from analytical models. On the basis of the observed experimental results, the following expression for the subcritical crack growth was proposed:

$$\frac{dw_{sup}}{dn} = \frac{Ae^{B\frac{w_{sup}}{w_u}}}{(1 - \frac{w_{sup}}{w_u})} \tag{1}$$

where $A$ and $B$ are parameters that depend on the material, the load levels and the load frequency. By integrating Eq. 1, the following relationship

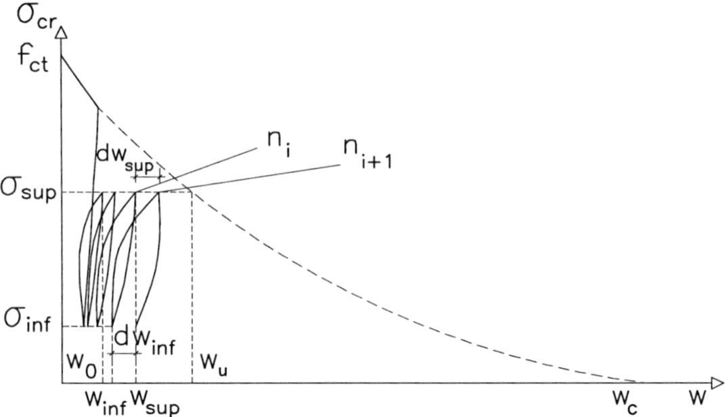


*Figure 12.* Scheme of the crack opening increase during inner loops [13].

between $N_{max}$, $w_0$ and $w_u$ was determined [13]:

$$N_{max} = \frac{1}{A}\left(\frac{w_u}{B^2 e^B} - \frac{(w_u - Bw_u + Bw_0)}{B^2 e^{B\frac{w_0}{w_u}}}\right) \tag{2}$$

Figure 13 exhibits typical experimental curves as obtained from NSC-SFR cylinders. In the same figure, the curve from Eq. 1 is also plotted by assuming the best-fit values of parameters $A$ and $B$ determined by the experimental values of $w_0$ and $w_u$. The best-fitting parameters determined for all the materials are shown in Table 2. Because of the very brittle behavior and the few cycles involved, parameters $A$ and $B$ for high-strength concrete without fibers were not determined. One can notice that the values of parameter $B$ for different types of concrete are similar while parameter $A$ for fiber-reinforced concrete (both normal and high-strength) is one order of magnitude lower than for concrete without fibers because of the slower crack growth. This result underlines the importance of fibers in concrete subjected to cyclic loads. Moreover, as mentioned above, fibers allow for residual post-cracking stresses remarkably greater than those of a similar concrete without fibers and thus require a longer distance to reach the envelope curve, where failure occurs. However, the increment of fatigue life also depends on fiber efficiency that is related to the fiber geometry, the concrete matrix and the maximum load level.

In Table 2, the logarithmic values of $N_{max}$, often used for fatigue results, are also reported. The proposed parameters are related to the load levels adopted in the present experimentation (25-75% of $P_{max}$) and different values should be obtained for different load levels. However, because of the few experimental results available, the study of the relationship between parameters $A$ and $B$, the material properties and the loading levels, needs further investigation. Once the relationship between these parameters and the stress levels is known, Eq. 2 could be used to determine the S-N (or Wöhler) curves for uniaxial tensile tests.

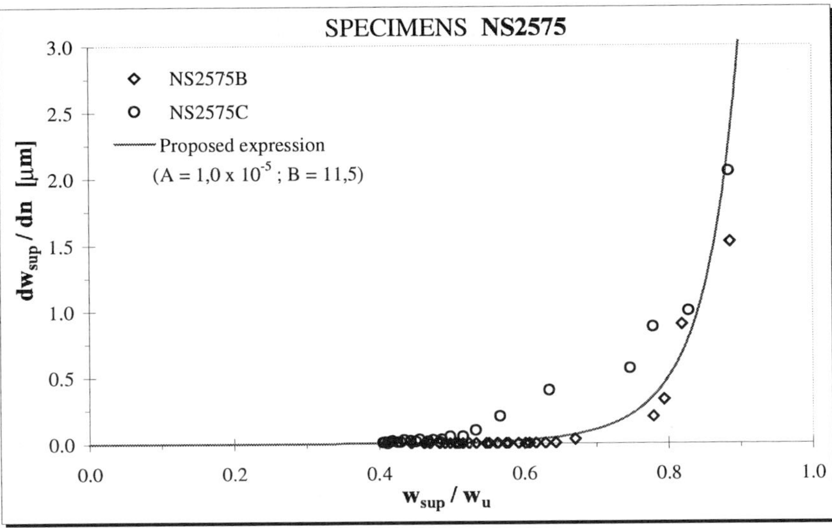


*Figure 13.* Crack opening rate at upper load in cylinders of normal-strength concrete with steel fibers [15].

TABLE 2. Comparison between the maximum number of cycles experimentally determined and calculated on the basis of the empirical damage model.

| Specimen | $w_0$ [$\mu$m] | $w_u$ [$\mu$m] | $A$ [$\mu$m] | $B$ | $N_{max}$ Exp. | $N_{max}$ Model | $\log(N_{max})$ Exp. | $\log(N_{max})$ Model |
|---|---|---|---|---|---|---|---|---|
| NC75/25A(*) | 5 | 20 | 1.0E-04 | 8 | 645 | 2116 | 2.81 | 3.33 |
| NC75/25B(*) | 5 | 12 | 1.0E-04 | 8 | 344 | 246 | 2.54 | 2.39 |
| NC75/25E(*) | 1 | 12 | 1.0E-04 | 8 | 5081 | 6097 | 3.71 | 3.79 |
| NS2575B | 5 | 14 | 1.0E-05 | 11.5 | 2450 | 1114 | 3.39 | 3.05 |
| NS2575C | 8 | 17 | 1.0E-05 | 11.5 | 136 | 292 | 2.13 | 2.47 |
| HS2575B | 4 | 11 | 1.0E-05 | 10 | 5131 | 1555 | 3.71 | 3.19 |
| HS2575C | 7 | 14 | 1.0E-05 | 10 | 238 | 378 | 2.38 | 2.58 |
| HC2575A | 8 | 15 | 1.0E-05 | 11 | 829 | 145 | 2.92 | 2.16 |
| HC2575B | 7 | 13 | 1.0E-05 | 11 | 1201 | 117 | 3.08 | 2.07 |

(*) The results from these specimens were presented in [13].

As far as cycles on the envelope curve are concerned, some models have been proposed [5,6]. Hordijk [8], on the basis of experimental results from normal-strength concrete specimens, proposed the *Continuous Function Model* that is characterized by three expressions: (I) Unloading branch (Eq. 3), (II) Crack increment (Eq. 4), (III) Reloading branch (Eq. 5; Fig. 14).

$$\frac{\sigma}{f_{ct}} = \frac{\sigma_{eu}}{f_{ct}} + \left(\frac{1}{3(w_{eu}/w_c) + 0.4}\right)\left\{0.014\left[\ln(\frac{w}{w_{eu}})\right]^5 - 0.57\left(1 - \frac{w}{w_{eu}}\right)^{0.5}\right\} \tag{3}$$

$$w_{inc} = 0.1 w_{eu}\left[\ln\left(1 + 3\frac{\sigma_{eu} - \sigma_L}{f_{ct}}\right)\right] \tag{4}$$

$$\frac{\sigma}{\sigma_L} = 1 + \left\{\frac{1}{c_3}\left(\frac{w - w_L}{w_{er} - w_L}\right)^{0.2c_3} + \left[1 - \left(1 - \frac{w - w_L}{w_{er} - w_L}\right)^2\right]^{c_4}\right\}\left(\frac{c_3}{c_3 + 1}\right)\left(\frac{\sigma_{er}}{\sigma_L} - \right. \tag{5}$$

Parameters $c_3$ and $c_4$ are given by the following expressions:

$$c_3 = 3\left(3 \cdot \frac{f_{ct} - \sigma_L}{f_{ct}}\right)^{\left(-1 - 0.5\frac{w_{eu}}{w_c}\right)}\left[1 - \left(\frac{w_{eu}}{w_c}\right)^{\left(\frac{0.71 \cdot f_{ct}}{f_{ct} - \sigma_L}\right)}\right] \tag{6}$$

$$c_4 = \left[2\left(3\frac{f_{ct} - \sigma_L}{f_{ct}}\right)^{-3} + 0.5\right]^{-1} \tag{7}$$

As envelope curve, Hordijk adopted the same expression he proposed for the static post-peak curves, characterized by the following equation:

$$\frac{\sigma}{f_{ct}} = \left[1 + \left(c_1\frac{w}{w_c}\right)^3\right]e^{\left(-c_2\frac{w}{w_c}\right)} - (1 - c_1^3)e^{(-c_2)} \tag{8}$$

where $c_1$=3, $c_2$=6.93 and the critical (stress-free) crack opening is given by:

$$w_c = 5.14 \cdot \frac{G_F}{f_{ct}} \tag{9}$$

The Hordijk model allowed a good fitting of the experimentally determined cycles on the envelope curve on cylinders of normal-strength concrete subjected to different load levels [12]. Figure 15 shows a typical example of a specimen subjected to a load variable between 0 and 75% of the maximum load. The same model matched the experimental results obtained

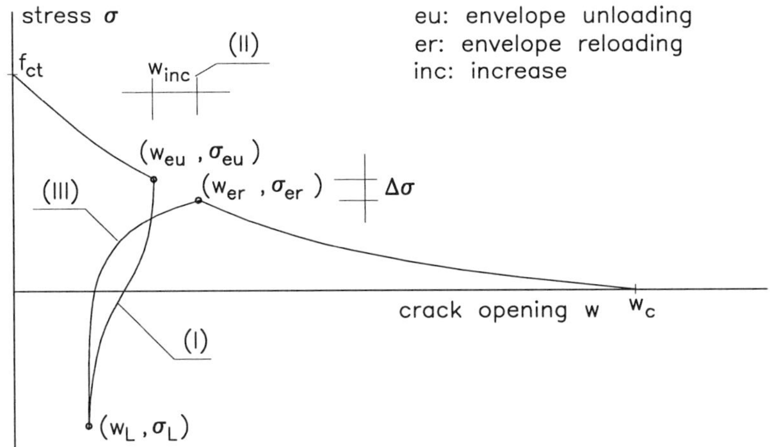


*Figure 14.* Continuous Function Model proposed by Hordijk [8].

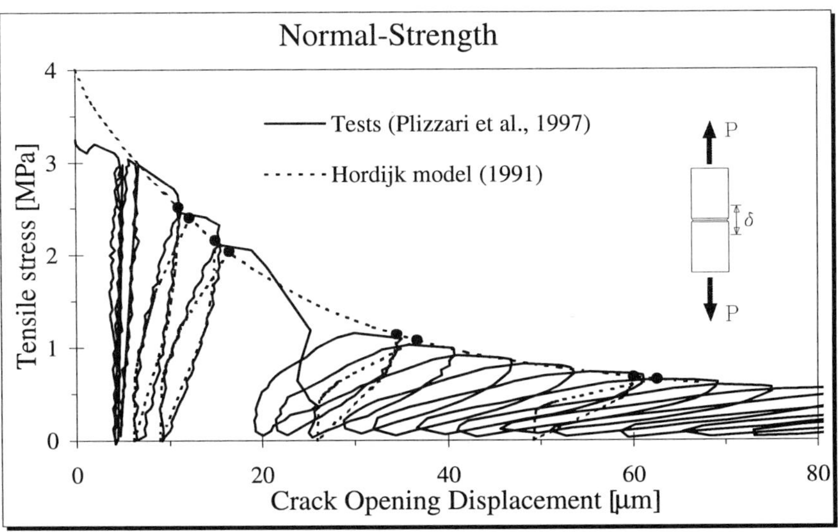


*Figure 15.* Comparison between the experimentally determined cycles on the envelope curve of normal-strength concrete [13] and the Continuous Function Model proposed by Hordijk [8].

here from specimens of high-strength concrete without fibers, as shown in Fig. 16. This model cannot be used for fiber-reinforced concrete that is characterized by a different envelope curve.

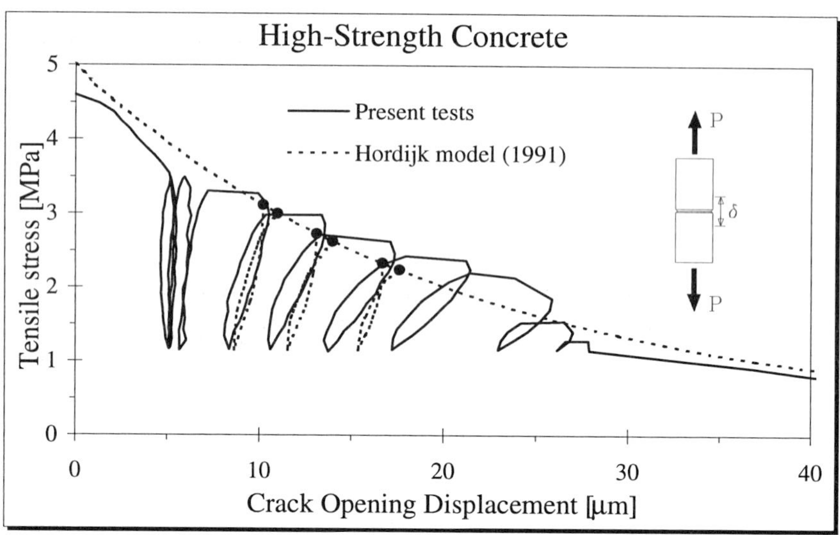


*Figure 16.* Comparison between the experimentally determined cycles on the envelope curve of high-strength concrete and the model proposed by Hordijk [8].

## 5. Concluding remarks

The behavior of concrete subjected to fatigue loading has been experimentally investigated by means of uniaxial tensile tests on cylindrical notched specimens and of four point bending tests on notched beams made of normal and high-strength concrete. The experimental results allowed for a comparison between the material (cylinders) and the "structural" (beams) behavior. The tests were performed on cracked specimens where a fracture process zone was present.

The main results can be summarized as follows:

1) The comparison between the A.E. measured on cylindrical and on beam specimens shows that in some parts of the fracture process zone of beams subjected to inner loops, the material behavior is subjected to cycles on the envelope curve (Figs. 7,8).

2) Fatigue behavior of beam specimens is strongly influenced by the fracture process zone size (and thus by the specimen size) so that fatigue tests on structural specimens are non helpful to determine the material behavior under fatigue loading. It should be determined by means of direct tension tests on cylinders that are less affected by structural behavior.

3) The envelope curves obtained from cyclic tests match the curves obtained from static tests on cylinders of fiber-reinforced concrete quite well, as already found for normal-strength concrete without fibers (Fig. 9).

This does not always occur in beam specimens since the fatigue damage depends on the fracture process zone development which is influenced by the loading history (Fig. 10).

4) The number of inner loops ($N_{max}$) obtained from cylindrical specimens remarkably diminishes in concrete with steel fibers. The effectiveness of hooked steel-fibers is greater in high-strength than in normal-strength cylindrical specimens. The influence of carbon fibers on $N_{max}$ is less marked than that of steel fibers.

5) The model for crack growth during inner loops previously proposed by the authors for normal-strength concrete, allowed for a good fitting of the experimental results obtained from high-strength concrete and fiber-reinforced concrete (Fig. 13).

6) Cycles on the envelope curve are well approximated by the model proposed by Hordijk [8], both from specimens of normal and high-strength concrete without fibers (Figs. 15,16).

## 6. Acknowledgements

The support from Dr Luigi Cassar, director of the CTG Research and Development Department, and from Dr Paolo Ursella, head of the CTG Laboratory for Testing Materials, is gratefully acknowledged. The authors would also like to thank Mr Nicola Cere for his assistance in carrying out the experiments. The financial support of the Italian National Council for Research (CNR, National project "Special materials for better structures") to the University of Brescia is also gratefully acknowledged.

## References

1. Bažant, Z.P.: 1995, 'Scaling Theories for Quasibrittle Fracture: Recent Advances and New Directions'. In: Wittmann F.H. (ed.): *Proceedings of Second International Conference on Fracture Mechanics of Concrete Structures.* Zurich, Switzerland, pp. 515–534.
2. Bolomey, J.: 1948, 'Granulation Continue ou Discontinue des Bétons'. *Revue des Matériaux de Constructions & Travaux Public* **48**, pp. 218–219.
3. Carpinteri, A. and Chiaia, B.: 1995, 'Multifractal Scaling Law for the Fracture Energy Variation of Concrete Structures'. In: Wittmann F.H. (ed.): *Proceedings of Second International Conference on Fracture Mechanics of Concrete Structures.* Zurich, Switzerland, pp. 581–596.
4. Cornelissen, H.A.W.: 1984, 'Fatigue Failure of Concrete in Tension'. *Heron*, Delft University of Technology, **29**(4).
5. Duda, H. and Konig, G.: 1991, 'Rheological Material Model for the Stress-Crack Width Relation of Concrete under Monotonic and Cyclic Tension'. *ACI Materials Journal* **88**(3), pp. 278–287.
6. Gylltoft, K.: 1984, 'A Fracture Mechanics Model for Fatigue in Concrete'. *Matériaux et Constructions* **17**, pp. 55–58.
7. Hilsdorf, H.K. and Kesler, C.E.: 1966, 'Fatigue Strength of Concrete Under Varying Flexural Stresses'. *Journal of the American Concrete Institute* **63**(10), pp. 1059–

1076.
8. Hordijk, D.A.: 1991, 'Local approach to Fatigue of Concrete'. Ph.D. Thesis, Delft University of Technology.
9. Miner, M.A.: 1945, 'Cumulative Damage in Fatigue'. *Transactions of the American Society of Mechanical Engineering* **67**, pp. A159–A164.
10. Oh, B.H.: 1991, 'Cumulative Damage Theory of Concrete under Variable-Amplitude Fatigue Loading'. *ACI Materials Journal* **88**(1), pp. 41–48.
11. Palmgren, A.: 1924, 'Die Lebensdauer von Kugellagern'. *Zeitschrift Verein Deutscher Ingenieure* **68**(14), pp. 339–341.
12. Plizzari, G.A., Cangiano, S., and Alleruzzo, S.: 1997, 'Sul comportamento del calcestruzzo fessurato sottoposto a carichi ciclici'. *Studi e Ricerche, Milan University of Technology* **18**, pp. 1–30 (in Italian).
13. Plizzari, G.A., Cangiano, S., and Alleruzzo, S.: 1997, 'The Fatigue behaviour of Cracked Concrete'. *Fatigue and Fracture of Engineering Materials and Structures* **20**(8), pp. 1195–1206.
14. Plizzari, G.A., Cangiano, S., and Cere, N.: 1996, 'Studio sperimentale sul comportamento fessurativo di calcestruzzi ad alta resistenza e fibrorinforzati sottoposti a carichi ciclici'. Technical Report 5.6/96, Department of Civil Engineering, University of Brescia, Brescia, Italy (in Italian).
15. Plizzari, G.A., Cangiano, S., and Cere, N.: 1999, 'Post-Peak Behavior of Fiber Reinforced Concrete Under Cyclic Tensile Loads'. *ACI Materials Journal.* 2000 (in print).
16. Reinhardt, H.W., and Cornelissen, H.A.W.: 1984, 'Post-Peak Cyclic behaviour of Concrete in Uniaxial Tensile and Alternating Tensile and Compressive Loading'. *Cement and Concrete Research* **14**, pp. 263–270.
17. Rossi, P., Robert, J.L., Gervais, J.P., and Bruhat, D.: 1989, 'Acoustic Emission Applied to study Crack Propagation in Concrete'. *RILEM Materials and Structures* **22**, pp. 574–584.
18. Slowik, V., Plizzari, G., and Saouma, V.E.: 1996, 'Fracture of Concrete Under Variable Amplitude Fatigue Loading'. *ACI Materials Journal* **93**(3), pp. 272–283.
19. van Mier, J.G.M., Vervuurt, A., and Schlangen, E.: 1994, 'Boundary and Size Effects in Uniaxial Tensile Tests: a Numerical and Experimental Study'. In: Z. Bažant (ed.): *US-Europe Workshop on Fracture and Damage of Quasi-Brittle Materials: Experiment, Modeling and Computer Analysis.* Prague, Czech Republic, pp. 289–302.
20. Yankelevsky, D., and Reinhardt, H.: 1987, 'Focal Point Model for Uniaxial Cyclic behaviour of Concrete'. In: *Computational Mechanics of Concrete Structures; Advances and Applications, IABSE Colloquium.* Delft, pp. 99–106.

# SHAKEDOWN ANALYSIS BY ELASTIC SIMULATION

C. POLIZZOTTO, G. BORINO and F. PARRINELLO
*Dipartimento di Ingegneria Strutturale e Geotecnica*
*Università di Palermo,*
*Viale delle Scienze, 90128 Palermo (Italy)*

P. FUSCHI
*Dipartimento Arte Scienza Tecnica del Costruire*
*Università di Reggio Calabria,*
*Via Melissari Feo di Vito, 89124 Reggio Calabria, (Italy)*

## 1. Introduction

Shakedown analysis of elastic plastic structures is widely credited as a valuable analytical/numerical tool for design purposes. For complex structures and loading conditions, e.g. for fast breeder nuclear reactor plants, full inelastic analysis is rarely performed, practically never within the early stages of the design advancement and the inherent decision process. The essential information therein needed can in fact be obtained, at moderate computational costs, by application of the shakedown methods and rules, at least within some limits related to the present developments of shakedown theory and its applicability to practical engineering problems, see e.g. Ponter *et al.* (1990), Carter *et al.* (1988), Ainsworth (1988), Goodall *et al.* (1991).

Nowadays, efforts to develop shakedown theory can be devided in two main research streams, one of which tends to generalize the theory as to account for more realistic constitutive models, e.g. nonlinear hardening (Maier, 1987; Comi and Corigliano, 1991; Polizzotto *et al.*, 1991; Stein *et al.*, 1992), damage (Hachemi and Weichert, 1992; Feng and Yu, 1995; Polizzotto *et al.*, 1996; Druyanov and Roman, 1998), creep (Ponter, 1972; Polizzotto, 1995), temperature dependent yield stress (Borino and Polizzotto, 1997a, 1997b; Borino, 1999), lack of normality rule (Pycko and Maier, 1995; Corigliano *et al.*, 1995); as well as for particular material and structural conditions, e.g. limited hardening (Weichert and Gross-Weege, 1988), large displacements (Weichert, 1986, 1990; Polizzotto and Borino, 1996),



*D. Weichert and G. Maier (eds.), Inelastic Analysis of Structures under Variable Loads,* 335–364.

unilateral contact boundary conditions (Polizzotto, 1997), cracks (Huang and Stein, 1996). Efforts in the other research stream tend to produce adequate computational methods with related finite element algorithms for the solution of practical engineering problems (see e.g. Stein *et al.*, 1992).

Among the latter methods, a new technique has emerged lately, referred to as "elastic compensation" by the first proposers (Hamilton *et al.*, 1996, and references therein quoted), or as "elastic simulation" by Ponter and Carter (1997a, 1997b) who gave rational bases to it with convergence criteria. This technique consists of a series of linear elastic analyses for a fictitious nonhomogeneous isotropic material with the elastic shear modulus to be updated at the beginning of every analysis. However, all these developments are concerned with the simple case of proportionally varying loadings. The present work is an attempt to generalize the above technique to the case of loadings arbitrarily varying within a given polyhedral multidimensional domain.

A compact notation is used. Boldface letters denote vectors or tensors. The 'dot' and 'colon' products between vectors and tensors denote the single and double index contraction operations, respectively; that is, $\mathbf{u} \cdot \mathbf{v} = u_i\, v_i$, $\boldsymbol{\sigma} : \boldsymbol{\varepsilon} = \sigma_{ij}\, \varepsilon_{ji}$, and $(\mathbf{C} : \boldsymbol{\varepsilon})_{ij} = \mathbf{C}_{ijhk}\, \varepsilon_{kh}$, where the indices denote orthogonal Cartesian components and the repeated index rule is applied. The time derivative is denoted by an upper dot, i.e. $\dot{a} = \partial a/\partial t$. $\nabla$ denotes the gradient operator, e.g. $\nabla\mathbf{u} = \{\partial u_i/\partial x_j\}$, $\nabla^s$ its symmetric part, e.g. $2\nabla^s\mathbf{u} = \{\partial u_i/\partial x_j + \partial u_j/\partial x_i\}$, 'div' is the divergence operator, e.g. $\text{div}\boldsymbol{\sigma} = \{\sigma_{ji,j}\}$. Other symbols will be defined in the text at their first appearence.

## 2. Shakedown limit state and governing equations

Let a structure of volume $V$ and boundary surface $S = \partial V$ be referred to an orthogonal Cartesian co-ordinate system $\mathbf{x} = (x_1, x_2, x_3)$ in its initial undeformed state. The body is constrained on the portion $S_D$ of its boundary $S$, where the displacements are prescribed. It is subjected to two categories of external actions, that is:

i) A permanent *mechanical* load, which in general consists of volume forces $\mathbf{b} = \alpha\,\bar{\mathbf{b}}(\mathbf{x})$ in $V$ and surface forces $\mathbf{p} = \alpha\,\bar{\mathbf{p}}(\mathbf{x})$ on $S_T = S - S_D$, where $\alpha$ is a (positive) scalar factor. This load (referred to as the *steady load* in the following) will be globally denoted with the symbol $P = \alpha\,\bar{P}$, $\bar{P}$ being the prescribed reference load, i.e. $\bar{P} = \{\bar{\mathbf{b}}(\mathbf{x}) \text{ in } V, \bar{\mathbf{p}}(\mathbf{x}) \text{ on } S_T\}$.

ii) Repeated or cyclic mechanical and/or kinematical loads, generally consisting of volume forces in $V$, surface forces on $S_T$, imposed (e.g. thermal) strains in $V$ and imposed displacements on $S_D$. These loads (referred to as the *unsteady loads* in the following) are linearly expressed in terms of a vector $\mathbf{Q}$ of independent parameters, which is allowed to range within a given (finite, closed) domain $\Pi$ of suitable dimensions. Without loss of generality, $\Pi$ can be assumed shaped as a convex hyperpolyhedron with $m \geq 2$ vertices, say $\mathbf{Q}_1$, $\mathbf{Q}_2$, ..., $\mathbf{Q}_m$, these vertices being referred to as the *dominant* (or *basic*) *loads*. Any path $\mathbf{Q}(t), t \geq 0$, in $\Pi$ is an *Admissible Load Path* (ALP). With no loss in generality, an ALP can be conceived as a cyclic (i.e. periodic) load history in which all the loads of some (either continuous, or discontinuous, or even discrete) load path in $\Pi$ are sequentially applied at every time interval $\Delta t$.

By assumption, the material is elastic perfectly plastic and associative, with a yield function as

$$f(\sigma) \equiv \phi(\sigma) - \sigma_y \leq 0 \tag{1}$$

where $\sigma_y$ is the yield stress and

$$\phi(\sigma) \equiv (\sigma : \mathbf{a} : \sigma)^{1/2} \tag{2}$$

where $\mathbf{a}$ is a positive definite (adimensional) fully symmetric fourth-rank tensor. The related dissipation function, $D(\dot{\varepsilon}^p)$, can be shown to read as:

$$D(\dot{\varepsilon}^p) = \sigma_y \, \psi\,(\dot{\varepsilon}^p) \tag{3}$$

where

$$\psi\,(\dot{\varepsilon}^p) \equiv \left(\dot{\varepsilon}^p : \mathbf{a}^{-1} : \dot{\varepsilon}^p\right)^{1/2}, \tag{4}$$

such that the stress $\boldsymbol{\sigma}$ corresponding to a given plastic strain rate, $\dot{\varepsilon}^p \neq \mathbf{0}$, is given by

$$\boldsymbol{\sigma} = \boldsymbol{\sigma}^*\,(\dot{\varepsilon}^p) = \frac{\partial D}{\partial \dot{\varepsilon}^p} = \frac{\sigma_y}{\psi\,(\dot{\varepsilon}^p)}\,\mathbf{a}^{-1} : \dot{\varepsilon}^p. \tag{5}$$

For a structure subjected to combined steady/unsteady loads, the shakedown problem can be formulated in various ways according to the particular design purpose. The most general way consists in specifying the unsteady load to within a scalar factor, i.e. $\mathbf{Q} = \beta\,\bar{\mathbf{Q}}$, $\bar{\mathbf{Q}} \in \bar{\Pi}$, and in determining the relevant *shakedown domain* in the $(\alpha, \beta)$-plane, i.e. the set of load points $(\alpha, \beta)$ for which shakedown occurs, and in particular the shakedown limit loads $(\alpha, \beta)$ on the *shakedown boundary*. In practice, it may be more convenient either to evaluate, at constant $\alpha$, the maximum value of $\beta$, say $\beta_{\rm sh} = g(\alpha)$, for which shakedown occurs; or to evaluate, at constant $\beta$, the maximum value of $\alpha$, say $\alpha_{\rm sh} = h(\beta)$, for which shakedown occurs. For the purposes of the present paper, the second way will be adopted with the proviso that $\beta = 1 < \beta_{\rm al}$, where $\beta_{\rm al}$ is the structure's *alternating plasticity safety factor*, given by $\beta_{\rm al} = \max \beta_{\rm sh}(\alpha)$ with respect to $\alpha$. (For $\beta = \beta_{\rm al}$ and

$\alpha$ within some range, the shakedown limit state is characterized by an impending alternating plasticity collapse mode; $\beta_{\rm al}$ can be considered known as it can be directly evaluated independently of the $\alpha_{\rm sh}$ values, (Polizzotto, 1993).) This choice will make shakedown limit analysis to exhibit features quite similar to those of plastic limit analysis. The following can in fact be observed:

a) In *plastic limit analysis*, the steady load $P = \alpha \bar{P}$ is applied upon the (unloaded) structure and the maximum value of $\alpha$, $\alpha_p$ say, is to be evaluated for which the structure is prone to an *instantaneous* plastic collapse mechanism (i.e. the compatible plastic strains occur simultaneously everywhere in $V$). Obviously, $\alpha_p$ depends on the (virgin) structure's resistance.

b) In *shakedown limit analysis*, the steady load $P = \alpha \bar{P}$ is applied upon the structure being pre-loaded by the unsteady loads $\mathbf{Q} \in \Pi$ and the maximum value of $\alpha$, $\alpha_{sh}$ say, is to be evaluated for which the structure is exposed to an *incremental* (or *ratchetting*) plastic collapse mechanism (indeed, a *noninstantaneous* plastic collapse mechanism, as the compatible plastic strains result from incompatible and anisochronous plastic strain fields respectively promoted by loads that differ from one another). Obviously, $\alpha_{\rm sh}$ depends on the marginal resistance of the pre-loaded structure; it coincides with $\alpha_p$ of point a) above if the pre-load is removed.

The evaluation of $\alpha_{\rm sh}$ can in principle be achieved making use of the well-known static (or lower bound) and kinematic (or upper bound) shakedown theorems, Koiter (1960), Gokhfeld and Cherniavsky (1980), König (1987). These theorems can be simplified by replacing the unsteady loads $\mathbf{Q} \in \Pi$ by the set of dominant loads $\mathbf{Q}_r$, $r \in I(m) \equiv \{1, 2, ..., m\}$. Thus, the following formulations hold for $\alpha_{\rm sh}$.

*Lower-bound theorem*

$$\alpha_{\rm sh} = \max \alpha \text{ s.t. :} \tag{6a}$$

$$\operatorname{div} \mathbf{s} + \alpha \bar{\mathbf{b}} = \mathbf{0} \quad \text{in } V, \qquad \mathbf{s} \cdot \mathbf{n} = \alpha \bar{\mathbf{p}} \quad \text{on } S_T \tag{6b}$$

$$f_r(\mathbf{s}) \equiv \phi\left(\mathbf{s} + \boldsymbol{\sigma}_r^E\right) - \sigma_y \leq 0 \quad \text{in } V, \quad \forall r \in I(m) \tag{6c}$$

where $\boldsymbol{\sigma}_r^E$ denotes the elastic stress response to $\mathbf{Q}_r$ and 's.t.' means 'subject to'.

*Upper-bound theorem*

$$\alpha_{\rm sh} = \min_{(\varepsilon_r^p, u)} \int_V \sum_{r=1}^{m} \left[D(\varepsilon_r^p) - \boldsymbol{\sigma}_r^E : \varepsilon_r^p\right] \mathrm{d}V \text{ s.t. :} \tag{7a}$$

$$\mathbf{e} \equiv \sum_{r=1}^{m} \varepsilon_r^p = \nabla^s \mathbf{u} \quad \text{in } V, \qquad \mathbf{u} = \mathbf{0} \quad \text{on } S_D \tag{7b}$$

$$(\bar{P}, \mathbf{u}) \equiv \int_V \bar{\mathbf{b}} \cdot \mathbf{u} \, \mathrm{d}V + \int_{S_T} \bar{\mathbf{p}} \cdot \mathbf{u} \, \mathrm{d}S = 1. \tag{7c}$$

Here, $\mathbf{n}$ is the unit external normal to $S$, such that eq. (6b) is the equilibrium requirement for the stress field, $\mathbf{s}$, with $P = \alpha \bar{P}$, eq. (7b) is the self-compatibility requirement for the *ratchet strain* $\mathbf{e}$ (i.e. compatible with vanishing distortions in $V$ and with displacements $\mathbf{u}$ vanishing on $S_D$). It is worth to remark that problems (6) and (7) are similar to the analogous problems of plastic limit analysis to which they reduce, respectively, if the pre-loading is removed, i.e. if $\boldsymbol{\sigma}_r^E \equiv \mathbf{0} \; \forall i \in I(m)$.

When the pre-loaded structure is subjected to the shakedown limit load, i.e. $P = P_{\rm sh} = \alpha_{\rm sh} \bar{P}$, the structure tends to find itself (and in practice it can be considered to be after a finite number of load cycles) in a *shakedown limit state*, characterized by an *impending* incremental plastic collapse mechanism, $\mathbf{e} = \mathbf{e}[\mathbf{u}]$. The latter mechanism is the result of $m$ incompatible plastic strain fields, $\boldsymbol{\varepsilon}_r^p$ say, respectively promoted by the loads $P_{\rm sh} \cup \mathbf{Q}_r$, with at least two such fields being nontrivial. In the framework of continuum solid mechanics, the equations governing the above limit state read as follows (Panzeca and Polizzotto, 1988; Polizzotto, 1994; Fuschi and Polizzotto, 1995):

$$\operatorname{div} \mathbf{s} + \alpha \, \bar{\mathbf{b}} = \mathbf{0} \quad \text{in } V, \qquad \mathbf{s} \cdot \mathbf{n} = \alpha \, \bar{\mathbf{p}} \quad \text{on } S_T \tag{8}$$

$$\mathbf{e} = \nabla^s \mathbf{u} \quad \text{in } V, \qquad \mathbf{u} = \mathbf{0} \quad \text{on } S_D \tag{9}$$

$$\mathbf{e} = \sum_{r=1}^{m} \boldsymbol{\varepsilon}_r^p \quad \text{in } V, \qquad \boldsymbol{\varepsilon}_r^p = \lambda_r \frac{\partial f_r}{\partial \mathbf{s}} (\mathbf{s}) \quad \text{in } V \; \forall \, r \in I(m) \tag{10a}$$

$$f_r(\mathbf{s}) \equiv \phi \left( \mathbf{s} + \boldsymbol{\sigma}_r^E \right) - \sigma_y \le 0, \quad \lambda_r \ge 0, \quad \lambda_r \, f_r = 0 \quad \text{in } V \; \forall \, r \in I(m) \tag{10b}$$

$$(\bar{P}, \mathbf{u}) = 1. \tag{11}$$

The latter equation set can be derived as the Euler-Lagrange equations pertaining to either problems (6) and (7). These equations do not describe an evolutive problem, but rather a state problem in which the plastic strain fields $\boldsymbol{\varepsilon}_r^p$ represent each (fictitious) incipient plastic strains promoted, respectively, by the loads $\mathbf{Q}_r \cup \alpha \bar{P}$ on increasing $\alpha$, these strains having no influence on the structure's stress state (Polizzotto *et al.*, 1991). The deformation process described by (8)-(11) is a *cyclically rigid-plastic one*, meaning that the structure undergoes a compatible plastic deformation after a complete load cycle (in which the pre-load $\mathbf{Q}$ jumps from one dominant load to another sequentially), which thus leaves unchanged the structure's stress and elastic strain states existing at the beginning of the cycle.

Equations (8)-(11) are usually considered altogether in the literature on shakedown only with the purpose to study the structure's features in the shakedown limit state, Panzeca and Polizzotto (1988), Polizzotto (1994), Fuschi and Polizzotto (1995), though they may as well be used to evaluate the shakedown limit load, $P_{\text{sh}}$, with the related impending incremental plastic collapse mechanism. These equations admit a unique solution for all, except for $\mathbf{s}$ which may be not uniquely determined in a region not exceeding that part of $V$ (if any) where $\mathbf{e} = \mathbf{0}$. These same equations can be used to study the plastic-limit-analysis approach to shakedown analysis, first proposed by Gokhfeld and Cherniavsky (1980). Some aspects of the method proposed by these authors will be briefly reviewed and suitably reinterpreted and integrated in the next Section.

## 3. Plastic-limit-analysis approach to shakedown

Gokhfeld and Cherniavsky (1980) transformed the shakedown limit analysis problem into a plastic analysis one by introducing, for the given material and structure/load system, a suitable "modified" elastic domain in the space of the time-independent stresses, $\mathbf{s}$. Writing $f(\mathbf{s} + \sigma_r^E) = f(\mathbf{s} - (-\sigma_r^E)) = 0$ suggests one to translate the yield surface, $f = 0$, centred at the stress origin, into one centred at the pre-stress point $\mathbf{s}_r^0 = -\sigma_r^E$. Making this operation for every $r \in I(m)$ produces a set of $m \geq 2$ translated yield surfaces, $f_r(\mathbf{s}) \equiv f(\mathbf{s} - \mathbf{s}_r^0) = 0$, which intersect with one another to form a multiple yield surface enclosing a domain $\mathcal{R}$ in the $\mathbf{s}$–stress space. This $\mathcal{R}$, generally changing with $\mathbf{x} \in V$, represents the *marginal* resistance domain of the structure pre-loaded sequentially by the dominant loads $\mathbf{Q}_r$, $r \in I(m)$. Thus, $\mathcal{R}$ is referred to as the structure's *marginal elastic domain.*

Figure 1 illustrates this conceptually simple operation. $\mathcal{R}$ collects the *transfer stresses*, that is the stress points $\mathbf{s}$ such that the *translated* elastic-stress-response domain $\widehat{\Pi}_\sigma = \Pi_\sigma + \mathbf{s}$, i.e. $\Pi_\sigma = \{\sigma^E(\mathbf{x}, \mathbf{Q}) : \mathbf{Q} \in \Pi\}$ translated by $\mathbf{s}$, does not exceed the (original) yield surface $f = 0$. For $\beta_{\text{al}} > 1$, as assumed, $\mathcal{R}$ always contains interior points, at least somewhere in $V$, (for $\beta_{\text{al}} = 1$, $\mathcal{R}$ would be still nonempty everywhere, but without interior points). There exists a one-to-one correspondence between the transfer stress points, $\mathbf{s} \in \mathcal{R}$, and the positions of the translated elastic-stress-response domain, $\widehat{\Pi}_\sigma$, inside the yield surface, $f = 0$. In particular: i) If $\mathbf{s}$ is an interior point of $\mathcal{R}$, then $\widehat{\Pi}_\sigma$ has an interior position in the yield surface; ii) If $\mathbf{s}$ lies on a regular portion of $\partial\mathcal{R}$, then $\widehat{\Pi}_\sigma$ is in a simple contact condition, i.e. has a single vertex in contact with $f = 0$; iii) If $\mathbf{s}$ coincides with a singular point of $\partial\mathcal{R}$, where two or more regular portions of $\partial\mathcal{R}$ intersect with one another, then $\widehat{\Pi}_\sigma$ is in a multiple contact condition, i.e. it has as many vertices in contact with $f = 0$. In other words, the multiplicity of the plastic activation condition for $\mathbf{s}$ (i.e. the subset of translated yield surfaces

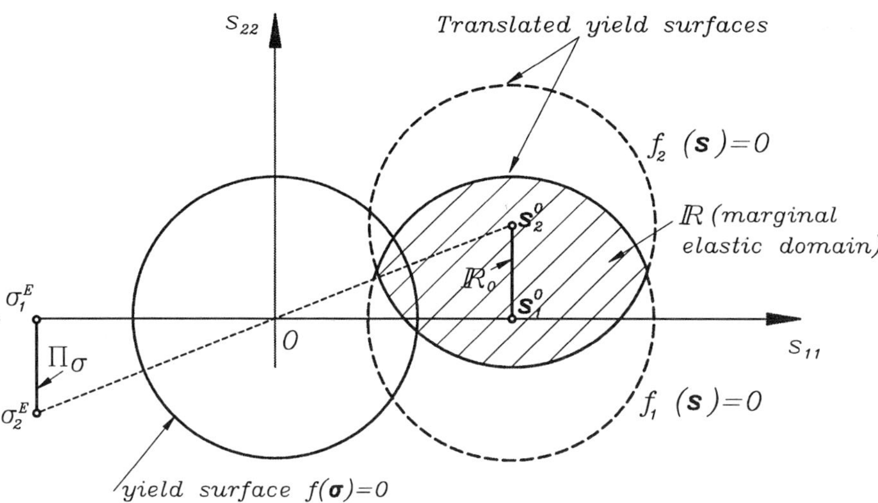


*Figure 1.* Sketch illustrating the generation of the marginal elastic domain $\mathcal{R}$ in the transfer stress space (s).

on which $\mathbf{s}$ lies, say $f_{r_i} = \phi(\mathbf{s} + \sigma^E_{r_i}) - \sigma_y = 0,\ i \in I(n),\ n \leq m)$ determines the multiplicity of the contact condition for $\widehat{\Pi}_\sigma$ with $f = 0$ (i.e. the subset of vertices of $\widehat{\Pi}_\sigma$, $\widehat{\boldsymbol{\sigma}}_{r_i} = \mathbf{s} + \boldsymbol{\sigma}^E_{r_i}$, $i \in I(n)$, $n \leq m$, lying on $f = 0$), and viceversa.

Equations (10a,b) provide the plasticity relationship between the transfer stress, $\mathbf{s}$, and the ratchet strain, $\mathbf{e}$, for assigned pre-stresses, $\mathbf{s}^0_r = -\boldsymbol{\sigma}^E_r$, $r \in I(m)$. At least in principle, these equations can be solved for $\mathbf{s}$ to obtain $\mathbf{s} = \mathbf{s}^*(\mathbf{e})$ and thus the marginal dissipation function, $D^{\mathrm{mg}} = D^{\mathrm{mg}}(\mathbf{e}) = \mathbf{s}^*(\mathbf{e}) : \mathbf{e}$, which corresponds to the marginal elastic domain $\mathcal{R}$.

Equations (10a,b) admit a particular form of the classical maximum dissipation theorem (Hill, 1950), that is:

$$D^{\mathrm{mg}}(\mathbf{e}) = \max_{(\mathbf{s})} \mathbf{s} : \mathbf{e} \quad \text{s.t. } f_r(\mathbf{s}) \leq 0\ \forall r \in I(m). \tag{12}$$

The latter theorem, herein referred to as the *Maximum marginal dissipation theorem*, can be phrased as: The stress $\mathbf{s} \in \partial\mathcal{R}$ corresponding to a given ratchet strain, $\mathbf{e}$, maximizes the marginal dissipation.

The optimal objective value of (12) is the marginal dissipation function, $D^{\mathrm{mg}}(\mathbf{e})$, that is:

$$D^{\mathrm{mg}}(\mathbf{e}) \;=\; \mathbf{s}^*(\mathbf{e}) : \mathbf{e} = \mathbf{s}^*(\mathbf{e}) : \sum_{r=1}^{m} \lambda_r \frac{\partial f_r}{\partial \mathbf{s}} = \sum_{r=1}^{m} \mathbf{s}^*(\varepsilon^p_r) : \varepsilon^p_r$$

$$= \sum_{r=1}^{m} \left[\mathbf{s}^*(\varepsilon_r^p) + \boldsymbol{\sigma}_r^E\right] : \varepsilon_r^p - \sum_{r=1}^{m} \boldsymbol{\sigma}_r^E : \varepsilon_r^p$$

$$= \sum_{r=1}^{m} \left[D(\varepsilon_r^p) - \boldsymbol{\sigma}_r^E : \varepsilon_r^p\right] \tag{13}$$

where the equality $\mathbf{s}^*(\mathbf{e}) = \mathbf{s}^*(\varepsilon_r^p)$ for $\varepsilon_r^p = \lambda_r \, \partial f_r/\partial \mathbf{s}$ has been used and $D(\varepsilon_r^p) = \sigma_y \, \psi(\varepsilon_r^p)$ by eqs. (3) and (4). $D^{\text{mg}}(\mathbf{e})$ is convex and one-degree homogeneous; it is also positive definite, provided $\mathcal{R}$ includes the origin $\mathbf{s} = \mathbf{0}$.

The solution to (12) includes the *optimal plastic strain path*, say $\varepsilon_{r_1}^p + \varepsilon_{r_2}^p + ... + \varepsilon_{r_n}^p = \mathbf{e}$, with $\varepsilon_r^p = \mathbf{0}$ for all $r \notin \{r_1, r_2, ...r_n\}$. The number $n \leq m$ of nonvanishing (nonisochronous) plastic strain increments, $\varepsilon_{r_i}^p$, $i \in I(n)$, in the ratchet strain $\mathbf{e}$ depends on whether the stress point $\mathbf{s}^*(\mathbf{e})$ is on a regular portion of $\partial\mathcal{R}$ (in which case $n = 1$), or coincides with a singular point of it (in which case $n > 1$, in general) —or, equivalently, on whether the translated elastic-stress-response domain, $\widehat{\Pi}_\sigma$, is in a simple, or multiple, contact condition with the yield surface, $f = 0$.

Problem (12) admits as dual the following:

$$D^{\text{mg}}(\mathbf{e}) = \min_{(\varepsilon_r^p)} \sum_{r=1}^{m} \left[D(\varepsilon_r^p) - \boldsymbol{\sigma}_r^E : \varepsilon_r^p\right] \qquad \text{s.t.} \quad \sum_{r=1}^{m} \varepsilon_r^p = \mathbf{e}, \tag{14}$$

which can be referred to as the *Minimum net plastic work theorem.* The latter theorem can be phrased as: The optimal plastic strain path corresponding to a given ratchet strain, $\mathbf{e}$, minimizes the net plastic work (i.e. the plastic work $\sum_{r=1}^{m} D(\varepsilon_r^p)$, less the work done by the pre-stresses, $\sum_{r=1}^{m} \boldsymbol{\sigma}_r^E : \varepsilon_r^p$).

With the above considerations in mind, eqs. (8)-(11) can be re-interpreted as the equations governing the (incremental, or noninstantaneous) plastic collapse limit state for the considered pre-loaded structure, this structure being endowed with the marginal elastic domain $\mathcal{R}$ in the s-stress space and the related marginal dissipation function $D^{\text{mg}}(\mathbf{e})$. Thus, problems (6) and (7) can be regarded, respectively, as lower-bound and upper-bound formulations within plastic limit analysis for the considered structure subjected to the steady load $\alpha \bar{P}$. These formulations are the basis of the plastic-limit-analysis approach to shakedown devised by Gokhfeld and Cherniavsky (1980) and pursued by others (Cocks, 1984). Though not all the potentialities of the latter approach have been explored yet, it is of interest here to study its interpretation and implementation via elastic simulation. This task will be accomplished in the next Sections.

## 4. A linear elastic composite material model

Let $\mathbf{s}$ and $\mathbf{e}$ be stress and strain states of an elastic composite material with $m \geq 2$ phases (or fractions) the complementary strain energy of which is

$$W = \frac{1}{2\mu} \sum_{r=1}^{m} \xi_r \, \phi^2(\mathbf{s} - \mathbf{s}_r^0). \tag{15}$$

Here, $\phi$ is the same plastic potential introduced in Section 2, eq. (2), $\mu > 0$ denotes the (scalar) *composite modulus*, the $\xi_r$, $r \in I(m)$, denote the *fraction coefficients* satisfying

$$\xi_r \geq 0 \; \forall r \in I(m); \; \sum_{r=1}^{m} \xi_r = 1, \tag{16$a, b$}$$

and finally $\mathbf{s}_r^0$, $r \in I(m)$, are assigned *pre-stress* tensors. In a more explicit form, $W$ of (15) reads

$$W = \frac{1}{2\mu} \sum_{r=1}^{m} \xi_r (\mathbf{s} - \mathbf{s}_r^0) : \mathbf{a} : (\mathbf{s} - \mathbf{s}_r^0). \tag{17}$$

The strain $\mathbf{e} = \mathbf{e}(\mathbf{s})$ can be derived from (17), i.e.

$$\mathbf{e} = \frac{\partial W}{\partial \mathbf{s}} = \frac{1}{\mu} \sum_{r=1}^{m} \xi_r \mathbf{a} : (\mathbf{s} - \mathbf{s}_r^0), \tag{18}$$

and it is the sum of contributions $\mathbf{e}_r$, i.e.

$$\mathbf{e}_r = \frac{\xi_r}{\mu} \mathbf{a} : (\mathbf{s} - \mathbf{s}_r^0), \quad r \in I(m) \tag{19}$$

from the $m$ fractions, respectively. A fraction is *active*, i.e. it actually contributes to deformation, if the related fraction coefficient, $\xi_r$, is nonvanishing. The notation

$$J = \{r_1, r_2, ..., r_n\} \subseteq I(m) \tag{20}$$

is used to specify the subset of active fractions. The composite becomes *perfectly rigid* for $\mu = \infty$.

An alternative expression for $\mathbf{e}(\mathbf{s})$ is derived from (18) taking into account (16b), i.e.

$$\mathbf{e} = \frac{1}{\mu} \mathbf{a} : (\mathbf{s} - \mathbf{s}_0) \tag{21}$$

where $\mathbf{s}_0$, defined by

$$\mathbf{s}_0 = \sum_{r=1}^{m} \xi_r \mathbf{s}_r^0, \tag{22}$$

is the *pre-stress centre*, i.e. a stress point enclosed by the (closed, convex) pre-stress hyperpolyhedron, $\mathcal{R}_0$, with vertices $\mathbf{s}_r^0$, $r \in I(m)$. The composite material as a whole has a complience $\mathbf{A} = (1/\mu)\mathbf{a}$ and a pre-stress $\mathbf{s}_0$. Correspondingly, the complemetary strain energy $W$ takes on the expression

$$W = \frac{1}{2\mu}\,\phi^2(\mathbf{s} - \mathbf{s}_0) + E_0 \geq E_0 \tag{23a}$$

$$E_0 = \frac{1}{2\mu}\left[\sum_{r=1}^{m} \xi_r \phi^2(\mathbf{s}_r^0) - \phi^2(\mathbf{s}_0)\right] \tag{23b}$$

where $E_0 \geq 0$ is the stored complementary strain energy of the *unstrained* material (i.e. for $\mathbf{s} = \mathbf{s}_0$, hence $\mathbf{e} = \mathbf{0}$). The (convex) equipotential surfaces $W(\mathbf{s})$ = const are homothetic with respect to the pre-stress centre $\mathbf{s}_0$, in which the surface $W(\mathbf{s}) = E_0$ degenerates. $\mathbf{e}(\mathbf{s})$ as given by (21) is normal to the equipontential surface $W(\mathbf{s})$ = const passing through the stress point $\mathbf{s}$, Figure 2.

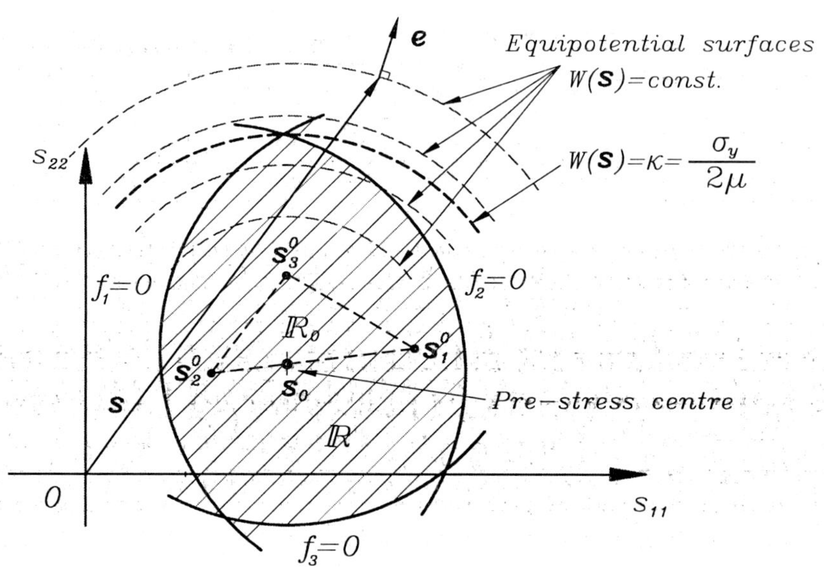


*Figure 2.* Sketch representing a two-dimensional marginal elastic domain ($\mathcal{R}$) and a system of equipotential surfaces $W(\mathbf{s})$ = const. with pre-stress centre $\mathbf{s}_0$.

Let eq. (21) be re-written by solving it for $\mathbf{s}$, that is:

$$\mathbf{s} = \mu\mathbf{a}^{-1} : \mathbf{e} + \mathbf{s}_0 = \mu\mathbf{a}^{-1} : (\mathbf{e} - \mathbf{e}_0) \tag{24}$$

where $\mathbf{B} = \mu \mathbf{a}^{-1}$ is the composite moduli fourth-rank tensor and

$$\mathbf{e}_0 = -\frac{1}{\mu}\mathbf{a} : \mathbf{s}_0 \tag{25}$$

is, by definition, the composite pre-strain tensor. The strain energy, $U(\mathbf{e})$, can be obtained as the Legendre transform of $W$, i.e.

$$\begin{aligned} U &= \mathbf{s} : \mathbf{e} - W \\ &= \frac{\mu}{2}\psi^2(\mathbf{e} - \mathbf{e}_0) + E_0 - \frac{\mu}{2}\psi^2(\mathbf{e}_0) \end{aligned} \tag{26}$$

where $\psi$ is the function in eq. (4). The following relations can be shown to hold:

$$U + W = (\mathbf{s} - \mathbf{s}_0) : \mathbf{e} + 2E_0, \tag{27}$$

$$\phi(\mathbf{s} - \mathbf{s}_0) = \mu\psi(\mathbf{e}), \tag{28}$$

$$\phi(\mathbf{s}) = \mu\psi(\mathbf{e} - \mathbf{e}_0). \tag{29}$$

## 5. Elastic simulation of the shakedown limit analysis problem

The cyclically rigid-plastic problem (8)-(11) can be transformed into a (fictitious) elastic one making use of the material model of Section 4. To this purpose, the material constants $\mu$, $\xi_r$, $\mathbf{s}_r^0$ introduced in Section 4 are given suitable values at every point $\mathbf{x} \in V$, that is:

$$\mu = \sigma_y / \sum_{j=1}^{m} \lambda_j \tag{30}$$

$$\xi_r = \lambda_r / \sum_{j=1}^{m} \lambda_j \qquad \forall r \in I(m) \tag{31}$$

$$\mathbf{s}_r^0 = -\boldsymbol{\sigma}_r^E \qquad \forall r \in I(m) \tag{32}$$

the first two of which hold provided $\mathbf{e} \neq \mathbf{0}$. Assuming for the moment $\mathbf{e} \neq \mathbf{0}$ everywhere in $V$, using (30) and (31) to eliminate the $\lambda$'s variables from (10a), and remarking that $\phi(\mathbf{s} - \mathbf{s}_r^0) = \sigma_y$ whenever $\xi_r > 0$, gives the stress-strain relation (21), or (24). Thus, replacing eq. (10a) with (21) and ignoring for the moment eq. (10b) and the normalization condition (11), it results that problem (8)-(11) can be cast as follows:

$$\operatorname{div} \mathbf{s} + \alpha\bar{\mathbf{b}} = \mathbf{0} \quad \text{in } V, \qquad \mathbf{s} \cdot \mathbf{n} = \alpha\bar{\mathbf{p}} \quad \text{on } S_T \tag{33a}$$

$$\mathbf{e} = \nabla^s \mathbf{u} \quad \text{in } V, \qquad \mathbf{u} = \mathbf{0} \quad \text{on } S_D \tag{33b}$$

$$\mathbf{e} = \frac{1}{\mu}\,\mathbf{a} : (\mathbf{s} - \mathbf{s}_0) \quad \text{in } V. \tag{33c}$$

As long as the material parameters $\mu$ and $\xi_r$ —the latter being carried in by $\mathbf{s}_0$, eq. (22)— are specified throughout $V$, eqs. (33a-c) represent a (nonhomogeneous) linear elastic problem for the load $\alpha\bar{P}$. In order that the solution to the latter problem be also a solution to problem (8)-(11), the plasticity conditions (10b) must be satisfied, i.e.

$$\phi(\mathbf{s} - \mathbf{s}_r^0) = \sigma_y \qquad \forall r \in J \quad \text{in } V \tag{34a}$$

$$\phi(\mathbf{s} - \mathbf{s}_r^0) < \sigma_y \qquad \forall r \notin J \quad \text{in } V \tag{34b}$$

where $J$ is the subset of active fractions at $\mathbf{x} \in V$, eq. (20), and $\mathbf{s}$ relates to the solution to the elastic problem.

If the plasticity conditions (34 a,b) are satisfied, the $\alpha$ value of problem (33 a-c) equals $\alpha_{\mathrm{sh}}$, i.e.

$$\alpha = \frac{\int_V \mathbf{s} : \mathbf{e}\, \mathrm{d}V}{(\bar{P}, \mathbf{u})} = \alpha_{\mathrm{sh}}, \tag{35}$$

where $\mathbf{s}$, $\mathbf{e}$, $\mathbf{u}$ pertain to the elastic problem solution. In fact, this $\alpha$ value is a lower bound to $\alpha_{\mathrm{sh}}$ because the pair $(\alpha, \mathbf{s})$ is a feasible solution to problem (6). But it is also an upper bound to $\alpha_{\mathrm{sh}}$ because the product $\mathbf{s} : \mathbf{e}$ in the numerator of the second member of (35) equals the net plastic work density corresponding to the set of plastic strain increments to which the strain $\mathbf{e}$ is equivalent, as in fact one can write:

$$\varepsilon_{r_i}^p = \frac{\xi_{r_i}}{\mu}\,\mathbf{a} : (\mathbf{s} - \mathbf{s}_{r_i}^0) \quad \forall r \in I(n) \text{ in } V, \tag{36a}$$

$$\mathbf{e} = \sum_{i=1}^{n} \varepsilon_{r_i}^p = \frac{1}{\mu}\,\mathbf{a} : (\mathbf{s} - \mathbf{s}_0) \quad \text{in } V, \tag{36b}$$

$$\mathbf{s} : \mathbf{e} = \sum_{i=1}^{n} \mathbf{s} : \varepsilon_{r_i}^p = \sum_{i=1}^{n} \left[ D(\varepsilon_{r_i}^p) - \boldsymbol{\sigma}_{r_i}^E : \varepsilon_{r_i}^p \right] \quad \text{in } V, \tag{36c}$$

such that $\mathbf{s} = \mathbf{s}^*(\mathbf{e})$ and the set $(\varepsilon_{r_i}^p,\ \mathbf{e},\ \mathbf{u})$ —to within an inessential scaling factor— is a feasible solution to problem (7). As a consequence $\alpha = \alpha_{\mathrm{sh}}$.

Note that expressing $\alpha$ as the ratio of eq. (35) is equivalent to taking into account the normalization condition (11). Also note that, at points $\mathbf{x} \in V$ where $\mathbf{e} = \mathbf{0}$ at a finite stress $\mathbf{s}$, it is $\mu = \infty$ by eq. (33c), i.e. the composite material is there perfectly rigid.

Always in the hypothesis that eqs. (34a,b) are satisfied, by (3) and (36a) one has

$$D(\varepsilon_r^p) = \sigma_y\, \psi(\varepsilon_r^p) = \frac{\xi_r}{\mu}\,\phi^2(\mathbf{s} - \mathbf{s}_r^0), \qquad \forall r \in J. \tag{37}$$

Furthermore, by the third in eq.(10b), it is $\xi_r\,\phi^2(\mathbf{s}-\mathbf{s}_r^0) = \xi_r\,\sigma_y^2$ for all $r \in I(m)$ and everywhere in $V$. Thus, at the shakedown limit solution, the complementary elastic strain energy $W$ of (15) takes on the optimal value

$$W = \kappa \equiv \frac{\sigma_y^2}{2\,\mu} \qquad \text{in } V \tag{38}$$

and the related plastic dissipation density equals twice the latter optimal value, i.e.

$$\sum_{r=1}^{m} D(\varepsilon_r^p) = 2\kappa = \frac{\sigma_y^2}{\mu} \qquad \text{in } V. \tag{39}$$

In other words, the plastic dissipation density equals twice the optimal complementary elastic strain energy density, i.e. $\sum_{i=1}^{m} D(\varepsilon_r^p)$ equals $2W$ at every $\mathbf{x} \in V$; additionally the value of $W$ turns to be inversely proportional to $\mu$, i.e. $W = \sigma_y^2/2\mu$, and is the sum of contributions $W_r = \xi_r\kappa$ from the $m$ phases, respectively. The optimal equipotential surface $W(\mathbf{s}) = \kappa$ encloses the marginal elastic domain $\mathcal{R}$; more precisely, it is tangent to $\mathcal{R}$ and the stress $\mathbf{s}$ pertaining to the solution to (33a-c) is the/a relevant contact point, Figure 2.

The elastic simulation problem presented in this Section is solvable by an attractive iteration procedure in which every iteration is a linear elastic analysis performed by eqs.(33a-c) with the material and load parameters, i.e. $\mu$, $\xi_r$ and $\alpha$, taken fixed at some (approximate) values. This procedure is explained in the next Section.

## 6. The iterative procedure

This procedure is aimed at providing a sequence of $\alpha$ values converging to $\alpha_{\rm sh}$. To this purpose, let the label LEAN($\xi_r$, $\mu$, $\alpha$) denote the (nonhomogeneous) linear elastic problem obtained from (33 a-c) by taking the fields $\xi_r$, $\mu$ and the load parameter $\alpha$ fixed at some values. LEAN($\xi_r$, $\mu$, $\alpha$) provides the linear elastic response, to the (fixed) load $P = \alpha\bar{P}$, of an elastic body made with a fictitious pre-stressed composite material characterized by the material parameters $\xi_r(\mathbf{x})$ and $\mu(\mathbf{x})$ throughout $V$.

Let $\xi_r^{(N-1)}(\mathbf{x})$, $\mu^{(N-1)}(\mathbf{x})$, $\alpha^{(N-1)}$ denote approximate values of $\xi_r(\mathbf{x})$, $\mu(\mathbf{x})$, $\alpha$ and assume $\mu^{(N-1)}(\mathbf{x})$ finite. Solving LEAN($\xi_r^{(N-1)}$, $\mu^{(N-1)}$, $\alpha^{(N-1)}$) generates the fields $\mathbf{s}^{(N)}$, $\mathbf{e}^{(N)}$, $\mathbf{u}^{(N)}$, which are then utilized to generate new values of the parameters, i.e. $\xi_r^{(N)}(\mathbf{x})$, $\mu^{(N)}(\mathbf{x})$, $\alpha^{(N)}$; then, LEAN($\xi_r^{(N)}$, $\mu^{(N)}$, $\alpha^{(N)}$) is solved, to obtain $\mathbf{s}^{(N+1)}$, $\mathbf{e}^{(N+1)}$, $\mathbf{u}^{(N+1)}$, and so forth.

In order to avoid computational troubles that may arise for $\mu = \infty$, it is convenient to introduce an upper bound to $\mu$, i.e. $\mu \le \mu^{\infty}$, where $\mu^{\infty}$ is a suitably large scalar. The consequence of the latter provision is that the material is deformable everywhere in the body and that possible region(s) of $V$ with zero ratchet strain in the shakedown limit state can be represented only approximately with $\mu = \mu^{\infty}$ and thus with a suitably small ratchet strain. Whenever the regions with vanishing ratchet strain are known in advance, e.g. in case of flexional plastic-hinge models, the fictitious elastic material can be taken as perfectly rigid within these regions.

At the $N$-th iteration, once LEAN$(\xi_r^{(N-1)}, \mu^{(N-1)}, \alpha^{(N-1)})$ has been solved, and thus $\mathbf{s}^{(N)}$, $\mathbf{e}^{(N)}$, $\mathbf{u}^{(N)}$ computed, the following operations must be accomplished to generate the updated parameters, $(\xi_r^{(N)}, \mu^{(N)}, \alpha^{(N)})$.

### 6.1. UPDATING THE SUBSET OF ACTIVE FRACTIONS AT $\mathbf{x} \in V$

This operation is accomplished by considering the material in a specified strain state, $\mathbf{e} = \mathbf{e}^{(N)}$, and finding the stress point $\mathbf{s}^{*(N)} = \mathbf{s}^*(\mathbf{e}^{(N)})$ on $\partial\mathcal{R}$, from where $\mathbf{e}^{(N)}$ departs along the external normal. The updated set of active fractions, $J^{(N)}$, specifies the $n \leq m$ translated yield surfaces on which $\mathbf{s}^{*(N)}$ lies. Mathematically, this problem can be solved by addressing either (12), or (14). But a simpler procedure consists in successive steps in which, starting from a single active fraction, another fraction is recognized as active at every new step till the right number ($n$) of active fractions.

In the first step, by hypothesis, $\mathbf{s}^{*(N)}$ lies on a regular portion of $\partial\mathcal{R}$, say $f_k(\mathbf{s}^{*(N)}) = 0$, where $k \in I(m)$. The related optimal plastic strain path is $\varepsilon_k^p = \mathbf{e}^{(N)}$, $\varepsilon_r^p = \mathbf{0}\ \forall r \in I(m)$ and $\neq k$; also, by eq.(13), the marginal dissipation $D^{\mathrm{mg}}(\mathbf{e}^{(N)}) = D(\mathbf{e}^{(N)}) + \mathbf{s}_k^0 : \mathbf{e}^{(N)}$ turns out to coincide with the optimal objective value of (14). That is, the inequality holds:

$$D(\mathbf{e}^{(N)}) + \mathbf{s}_k^0 : \mathbf{e}^{(N)} < D(\mathbf{e}^{(N)}) + \mathbf{s}_r^0 : \mathbf{e}^{(N)} \quad \forall r \in I(m),\ r \neq k \tag{40}$$

and thus

$$\mathbf{s}_k^0 : \mathbf{e}^{(N)} < \mathbf{s}_r^0 : \mathbf{e}^{(N)} \quad \forall r \in I(m),\ r \neq k, \tag{41}$$

which enables $k$ to be identified. Then, using (5), the stress $\mathbf{s}^{*(N)}$ is obtained as:

$$\mathbf{s}^{*(N)} = \mathbf{s}_{(1)}^{*(N)} = \frac{\partial D(\mathbf{e}^{(N)})}{\partial \mathbf{e}^{(N)}} + \mathbf{s}_k^0 = \frac{\sigma_y}{\psi(\mathbf{e}^{(N)})}\, \mathbf{a}^{-1} : \mathbf{e}^{(N)} + \mathbf{s}_k^0. \tag{42}$$

If, then, $f_r(\mathbf{s}_{(1)}^{*(N)}) \leq 0\ \forall r \in I(m)$ and $\neq k$, it is $n^{(N)} = 1$ and $J^{(N)} = \{r_1 = k\}$. Otherwise, a second step must be undertaken.

In the latter case, let $h \in I(m)$ be evaluated such that:

$$f_h\left(\mathbf{s}_{(1)}^{*(N)}\right) > f_r\left(\mathbf{s}_{(1)}^{*(N)}\right) \quad \forall r \in I(m) : f_r\left(\mathbf{s}_{(1)}^{*(N)}\right) > 0 \tag{43}$$

and let one set $n^{(N)} = 2$ and $J^{(N)} = \{r_1 = k, r_2 = h\}$. Assume that —through a procedure to be considered later on— the pertinent stress

$\mathbf{s}^{*(N)} = \mathbf{s}^{*(N)}_{(2)}$ has been computed. Obviously, $f_k\left(\mathbf{s}^{*(N)}_{(2)}\right) = f_h\left(\mathbf{s}^{*(N)}_{(2)}\right) = 0$. The latest values assumed for $n^{(N)}$ and $J^{(N)}$ are the right values if $f_r\left(\mathbf{s}^{*(N)}_{(2)}\right) \leq 0\ \forall r \in I(m)$ and $\neq (k,h)$. Otherwise, a third step must be accomplished following a procedure similar to the second one. In this way, the final values of $n^{(N)}$ and $J^{(N)}$ can be determined.

### 6.2. UPDATING THE MATERIAL PARAMETERS $\mu$ AND $\xi_r$ AT $\mathbf{x} \in V$

Let $J = \{r_1, r_2, ..., r_n\}$ be known. Then eqs. (16b) and (34a) can be utilized to determine $\mu$ and $\xi_r$. By eq.(33c) rewritten as

$$\mathbf{s} = \mu\,\mathbf{a}^{-1} : \mathbf{e} + \mathbf{s}_0 \tag{44}$$

with $\mathbf{s}_0$ given by (22) and with $\mu$ and $\xi_{r_i}$ considered unknown, but $\mathbf{e} = \mathbf{e}^{(N)}$ known, eq.(34a) gives:

$$\phi^2\left(\mu\,\mathbf{a}^{-1} : \mathbf{e} + \mathbf{s}_0 - \mathbf{s}^0_{r_i}\right) = \sigma_y^2, \quad \forall i \in I(n) \tag{45}$$

which is equivalent to

$$\begin{aligned} \mu^2\,\psi^2(\mathbf{e}) \quad &+ \quad 2\mu\,\mathbf{e} : \left(\mathbf{s}_0 - \mathbf{s}^0_{r_i}\right) \\ &+ \quad \phi^2\left(\mathbf{s}_0\right) - 2\mathbf{s}_0 : \mathbf{a} : \mathbf{s}^0_{r_i} + \phi^2\left(\mathbf{s}^0_{r_i}\right) = \sigma_y^2, \quad \forall i \in I(n). \end{aligned} \tag{46}$$

Writing the latter also for $i+1$ and then making the difference gives

$$\begin{aligned} \mathbf{s}_0 : \mathbf{a} : \left(\mathbf{s}^0_{r_i} - \mathbf{s}^0_{r_{i+1}}\right) \quad = \quad &\frac{1}{2}\left[\phi^2\left(\mathbf{s}^0_{r_i}\right) - \phi^2\left(\mathbf{s}^0_{r_{i+1}}\right)\right] - \mu\,\mathbf{e} : \left(\mathbf{s}^0_{r_i} - \mathbf{s}^0_{r_{i+1}}\right), \\ &\forall i \in I(n-1). \end{aligned} \tag{47}$$

Remembering (22), eq. (47) together with (16b) constitute a system of $n \leq m$ linear algebraic equations with the $\xi_{r_i}$ as unknowns, i.e.

$$\sum_{j=1}^{n} M_{ij}\,\xi_{r_j} = d_i \qquad \forall i \in I(n) \tag{48}$$

where, by definition,

$$M_{ij} = \left(\mathbf{s}^0_{r_i} - \mathbf{s}^0_{r_{i+1}}\right) : \mathbf{a} : \mathbf{s}^0_{r_j} \qquad \forall i \in I(n-1),\ \forall j \in I(n) \tag{49a}$$

$$M_{nj} = 1 \qquad \forall j \in I(n) \tag{49b}$$

$$d_i = \tfrac{1}{2}\left[\phi^2\left(\mathbf{s}^0_{r_i}\right) - \phi^2\left(\mathbf{s}^0_{r_{i+1}}\right)\right] - \mu\left(\mathbf{s}^0_{r_i} - \mathbf{s}^0_{r_{i+1}}\right) : \mathbf{e} \quad \forall i \in I(n-1) \tag{50a}$$

$$d_n = 1. \tag{50b}$$

For $n = 1$, eq.(48) reduces to $\xi_{r_1} = 1$ ($M_{11} = 1$, $d_1 = 1$). Since the coefficients $M_{ij}$ of eqs. (49a,b) do not depend on $\mathbf{e}$, it is convenient to invert (48) by writing:

$$\xi_{r_i} = \sum_{j=1}^{n} Z_{ij}\, d_j = \bar{\xi}_{r_i} + \mu\, \mathbf{T}_i^0 : \mathbf{e} \tag{51}$$

where the $Z_{ij}$ are elements of the inverse matrix $\mathbf{M}^{-1}$ and, by definition:

$$\bar{\xi}_{r_i} = \tfrac{1}{2} \textstyle\sum_{j=1}^{n-1} Z_{ij} \left[\phi^2\left(\mathbf{s}_{r_j}^0\right) - \phi^2\left(\mathbf{s}_{r_{j+1}}^0\right)\right] + Z_{in} \quad \forall i \in I(n) \tag{52a}$$

$$\mathbf{T}_i^0 = -\textstyle\sum_{j=1}^{n-1} Z_{ij}\left(\mathbf{s}_{r_j}^0 - \mathbf{s}_{r_{j+1}}^0\right) \quad \forall i \in I(n). \tag{52b}$$

It can be verified that the following identities hold, i.e.

$$\textstyle\sum_{i=1}^{n} Z_{ij} = 0 \quad \forall j \in I(n-1), \qquad \sum_{i=1}^{n} Z_{in} = 1 \tag{53a,b}$$

$$\textstyle\sum_{i=1}^{n} \bar{\xi}_{r_i} = 1, \qquad \sum_{i=1}^{n} \mathbf{T}_i^0 = \mathbf{0}. \tag{53c,d}$$

Another equation can be derived by multiplying (46) by $\xi_{r_i}$ and then summing with respect to $i \in I(n)$. Thus, one can write the equation:

$$\mu = \frac{1}{\psi(\mathbf{e})}\left\{\sigma_y^2 - \left[\sum_{i=1}^{n} \xi_{r_i}\, \phi^2\left(\mathbf{s}_{r_i}^0\right) - \phi^2\left(\mathbf{s}_0\right)\right]\right\}^{1/2} \tag{54}$$

where $\mathbf{s}_0$ is the pre-stress tensor, eq.(22). By topological reasons, the pre-stress centre finds itself on the boundary of $\mathcal{R}_0$, i.e. $\mathbf{s}_0 \in \partial\mathcal{R}_0$. For $n = 1$, eq. (54) simplifies as

$$\mu = \sigma_y / \psi(\mathbf{e}) \tag{55}$$

which coincides with a result given by Ponter and Carter (1997a, 1997b).

Substituting the $\xi_{r_i}$ from (51) into (54), the latter equation can be written as

$$\begin{aligned}
\psi^2(\mathbf{e})\,\mu^2 &= \sigma_y^2 - \sum_{i=1}^{n}\left(\bar{\xi}_{r_i} + \mu\,\mathbf{T}_i^0 : \mathbf{e}\right)\phi^2\left(\mathbf{s}_{r_i}^0\right) \\
&+ \phi^2\left(\sum_{i=1}^{n}\left[\left(\bar{\xi}_{r_i} + \mu\mathbf{T}_i^0 : \mathbf{e}\right)\otimes \mathbf{s}_{r_i}^0\right]\right) \\
&= \sigma_y^2 - \sum_{i=1}^{n}\bar{\xi}_{r_i}\,\phi^2\left(\mathbf{s}_{r_i}^0\right) - \mu\left(\sum_{i=1}^{n}\phi^2\left(\mathbf{s}_{r_i}^0\right)\,\mathbf{T}_i^0\right) : \mathbf{e}
\end{aligned}$$

$$+ \quad \phi^2 \left( \sum_{i=1}^{n} \bar{\xi}_{r_i} \mathbf{s}^0_{r_i} \right) + 2\,\mu \left( \sum_{i=1}^{n} \bar{\xi}_{r_i} \mathbf{s}^0_{r_i} \right) : \mathbf{a} : \left( \sum_{j=1}^{n} \mathbf{s}^0_{r_j} \otimes \mathbf{T}^0_j \right) : \mathbf{e}$$

$$+ \quad \mu^2 \mathbf{e} : \left[ \left( \sum_{i=1}^{n} \mathbf{T}^0_i \otimes \mathbf{s}^0_{r_i} \right) : \mathbf{a} : \left( \sum_{j=1}^{n} \mathbf{s}^0_{r_j} \otimes \mathbf{T}^0_j \right) \right] : \mathbf{e}. \tag{56}$$

This is a second-degree algebraic equation with $\mu$ as unknown, i.e.

$$(\mathbf{e} : \mathbf{C}_2 : \mathbf{e})\, \mu^2 - 2\, (\mathbf{C}_1 : \mathbf{e})\, \mu - C_0 = 0 \tag{57}$$

where $\mathbf{C}_2$, $\mathbf{C}_1$ and $C_0$ turn out to be tensors of second, first and zero-th orders, respectively, expressed as

$$\mathbf{C}_2 = \mathbf{a}^{-1} - \left( \sum_{i=1}^{n} \mathbf{T}^0_i \otimes \mathbf{s}^0_{r_i} \right) : \mathbf{a} : \left( \sum_{j=1}^{n} \mathbf{s}^0_{r_j} \otimes \mathbf{T}^0_j \right) \tag{58$a$}$$

$$\mathbf{C}_1 = \left( \sum_{i=1}^{n} \bar{\xi}_{r_i} \mathbf{s}^0_{r_i} \right) : \mathbf{a} : \left( \sum_{j=1}^{n} \mathbf{s}^0_{r_j} \otimes \mathbf{T}^0_j \right) - \frac{1}{2} \left( \sum_{i=1}^{n} \phi^2 \left( \mathbf{s}^0_{r_i} \right) \mathbf{T}^0_i \right) \tag{58$b$}$$

$$C_0 = \sigma_y^2 - \sum_{i=1}^{n} \bar{\xi}_{r_i}\, \phi^2 \left( \mathbf{s}^0_{r_i} \right) + \phi^2 \left( \sum_{i=1}^{n} \bar{\xi}_{r_i} \mathbf{s}^0_{r_i} \right). \tag{58$c$}$$

Equation (57) likely has a positive root that reads

$$\mu = \frac{\mathbf{C}_1 : \mathbf{e} + \sqrt{\mathbf{e} : [\mathbf{C}_1 \otimes \mathbf{C}_1 + C_0\, \mathbf{C}_2] : \mathbf{e}}}{\mathbf{e} : \mathbf{C}_2 : \mathbf{e}}, \tag{59}$$

which for $n = 1$, since correspondingly $\mathbf{T}^0_1 = \mathbf{0}$ hence $\mathbf{C}_2 = \mathbf{a}^{-1}$, $\mathbf{C}_1 = \mathbf{0}$ and $C_0 = \sigma_y^2$, will coincide with (55). By (59), $\mu$ turns out to depend on $\mathbf{e}$ explicitly, as well as on the set of active fractions, $J$, besides the pre-stresses, $\mathbf{s}^0_r$, through the coefficient tensors $C_0$, $\mathbf{C}_1$, $\mathbf{C}_2$, but not on the fraction coefficients, $\xi_r$.

For $\mathbf{e} = \mathbf{e}^{(N)}$ and $J = J^{(N)}$, eq. (59) gives $\mu^{(N)}$, whereas eq.(51) gives the $\xi_r^{(N)}$. Then, the stress $\mathbf{s}^{*(N)}$ can be computed by eq. (44), i.e.

$$\mathbf{s}^{*(N)} = \mu^{(N)} \mathbf{a}^{-1} : \mathbf{e}^{(N)} + \mathbf{s}_0^{(N)} \tag{60}$$

where

$$\mathbf{s}_0^{(N)} = \sum_{i=1}^{n} \xi_{r_i}^{(N)} \mathbf{s}^0_{r_i}, \tag{61}$$

whereas the optimal plastic strain path can be determined by (19), i.e.

$$\varepsilon^{p(N)}_{r_i} = \frac{\xi^{(N)}_{r_i}}{\mu^{(N)}} \mathbf{a} : \left(\mathbf{s}^{*(N)} - \mathbf{s}^0_{r_i}\right), \qquad \forall i \in I(n). \tag{62}$$

The above updating procedure for $\mu$ and $\xi_r$ must be integrated with that for $J$ described in Subsection 6.1. The resulting combined procedure can be described as follows.

i) Start with $n = 1$. Thus, $J^{(N)} = \{r_1 = k\}$ with $k$ given by (41), $\xi^{(N)}_{r_1} = 1$ (all other $\xi^{(N)}_r$ being vanishing) and, by (54) or (55), $\mu^{(N)} = \sigma_y/\psi(\mathbf{e}^{(N)})$. The stress $\mathbf{s}^{*(N)}$ obtained from (60) and (61) will coincide with that given by (42). No further step is required if $f_r\left(\mathbf{s}^{*(N)}\right) \leq 0 \; \forall r \in I(m)$; otherwise go to next step.

ii) Set $n = 2$. Thus, $J^{(N)} = \{r_1 = k, \, r_2 = h\}$, where $k$ is the same as at the first step, $h$ is provided by (43). Then compute the quantities $\bar{\xi}^{(N)}_{r_i}$, $\mathbf{T}^{0(N)}_i$, $C^{(N)}_0$, $\mathbf{C}^{(N)}_1$, $\mathbf{C}^{(N)}_2$ using eqs. (52), (53), (58a-c), and subsequently $\mu^{(N)}$ by eq. (59) and $\xi^{(N)}_r$ by eq.(51). No further step is needed, provided the computed $\xi^{(N)}_{r_i}$ are each positive and, moreover, the stress $\mathbf{s}^{*(N)}$ obtained by eqs. (60) and (61) satisfies the yield conditions, i.e. $f_r(\mathbf{s}^{*(N)}) \leq 0 \; \forall r \in I(m)$; otherwise, a new step must be started. And so forth till the final values of $\mu^{(N)}$ and $\xi^{(N)}_r$. Should the computed value of $\mu$ be greater than $\mu^\infty$, one has to set $\mu^{(N)} = \mu^\infty$.

### 6.3. UPDATING THE LOAD PARAMETER $\alpha$

This operation can be accomplished either by a static approach, or a kinematic one, as in the following.

— *Static approach*, (based on the static theorem of limit analysis), in which the updated load parameter $\alpha^{(N)} \equiv \alpha^{(N)}_{\mathrm{ST}}$ is obtained by solving the following problem at every point $\mathbf{x} \in V$, i.e.

$$\gamma(\mathbf{x}, N) = \max \gamma \quad \text{s.t.} \; \phi\left(\gamma \mathbf{s}^{(N)} - \mathbf{s}^0_r\right) - \sigma_y \leq 0 \; \forall r \in I(m), \tag{63}$$

where $\mathbf{s}^{(N)} = \mathbf{s}^{(N)}(\mathbf{x})$ and $\mathbf{s}^0_r = \mathbf{s}^0_r(\mathbf{x})$, and then

$$\gamma^{(N)} = \min_{x \in V} \gamma(\mathbf{x}, N) \tag{64a}$$

$$\alpha^{(N)}_{\mathrm{ST}} = \alpha^{(N-1)}_{\mathrm{ST}} \gamma^{(N)}. \tag{64b}$$

Problem (63), which amounts to scaling the stress vector $\mathbf{s}^{(N)}(\mathbf{x})$ such as to find the farthest intersection point with $\partial\mathcal{R}$, generally admits a solution.

If this is not the case (i.e. the stress vector $\mathbf{s}^{(N)}$ does not intersect $\partial\mathcal{R}$) at points $\mathbf{x}$ of some region $\bar{V} \subset V$, then the minimum operation in (64a) must be achieved for $\mathbf{x} \in V_1 = V - \bar{V}$. (It is reasonable to conjecture that such occurrence likely presents itself only at the beginning of the iteration procedure —provided the latter procedure is convergent.)

A simple way to solve (63) consists in solving the $m$ equations $f_r(\gamma \mathbf{s}^{(N)}) = 0$ for $\gamma$, that is the quadratic equations

$$\gamma^2 \phi^2 \left(\mathbf{s}^{(N)}\right) - 2\gamma \left(\mathbf{s}^{(N)} : \mathbf{a} : \mathbf{s}_r^0\right) + \phi^2 \left(\mathbf{s}_r^0\right) - \sigma_y^2 = 0, \; r \in I(m) \tag{65}$$

each of which in general has real roots. Considering the largest root, one can write:

$$\gamma_r(\mathbf{x}, N) = \left\{ \mathbf{s}^{(N)} : \mathbf{a} : \mathbf{s}_r^0 + \left[ \left(\mathbf{s}^{(N)} : \mathbf{a} : \mathbf{s}_r^0\right)^2 - \left(\phi^2 \left(\mathbf{s}_r^0\right) - \sigma_y^2\right) \phi^2 (\mathbf{s}^{(N)}) \right]^{\frac{1}{2}} \right\} / \phi^2(\mathbf{s}^{(N)}) \tag{66a}$$

$$\gamma(\mathbf{x}, N) = \min_{r \in I(m)} \gamma_r(\mathbf{x}, N). \tag{66b}$$

The discriminant within the square brackets in (66a) is negative at points $\mathbf{x} \in \bar{V}$, if any, where problem (63) has no solution.

Considering that the stresses $\gamma^{(N)} \mathbf{s}^{(N)}$ are in equilibrium with $\alpha_{\text{ST}}^{(N)} \bar{P}$ and that the stresses $\mathbf{s}^{(N)}$ are in equilibrium with $\alpha_{\text{ST}}^{(N-1)} \bar{P}$, follows that the equality (64b) holds good; also, as the stresses $\gamma^{(N)} \mathbf{s}^{(N)}$ —provided (63) admits solution everywhere in $V$— are plastically admissible in the whole body, then (64b) provides a lower bound to $\alpha_{\text{sh}}$, i.e. $\alpha_{\text{ST}}^{(N)} \leq \alpha_{\text{sh}}$.

— *Kinematic approach*, (based on the upper bound theorem of limit analysis), in which the updated load parameter $\alpha^{(N)} \equiv \alpha_{\text{KIN}}^{(N)}$ is given by

$$\alpha_{\text{KIN}}^{(N)} = \frac{\int_V \mathbf{s}^{*(N)} : \mathbf{e}^{(N)} \, \mathrm{d}V}{(\bar{P}, \mathbf{u}^{(N)})} \tag{67}$$

where $\mathbf{s}^{*(N)}$ is taken from (60). As the set $(\varepsilon_{ri}^{p(N)}, \mathbf{e}^{(N)}, \mathbf{u}^{(N)})$ —where the plastic strain increments $\varepsilon_{ri}^{p(N)}$ are given by (62)— is a feasible solution to problem (7), it can be concluded that (67) provides an upper bound to $\alpha_{\text{sh}}$, that is $\alpha_{\text{KIN}}^{(N)} \geq \alpha_{\text{sh}}$.

The updating procedure is so complete. What is still needed in order to start the iterative procedure is a suitable initialization solution. A choice may be the (true) elastic response to $\bar{P}$, i.e. $\mathbf{s}^{(0)} = \bar{\mathbf{s}}^E$, $\mathbf{e}^{(0)} = \bar{\mathbf{e}}^E$, $\mathbf{u}^{(0)} = \bar{\mathbf{u}}^E$. Another (perhaps better) choice is the elastic limit solution for the prestressed structure, that is, the set $\mathbf{s}^{(0)} = \mathbf{s}^{\text{el}}$, $\mathbf{e}^{(0)} = \mathbf{e}^{\text{el}}$, $\mathbf{u}^{(0)} = \mathbf{u}^{\text{el}}$ where $\mathbf{s}^{\text{el}} = \alpha^{\text{el}} \, \bar{\mathbf{s}}^E$ and $\alpha^{\text{el}}$ is the related elastic limit load multiplier. Whatever the choice made, the initialization solution ($\mathbf{s}^{(0)}$, $\mathbf{e}^{(0)}$, $\mathbf{u}^{(0)}$) can be used with the updating procedure previously described to obtain the initial values ($\xi_r^{(0)}$, $\mu^{(0)}$, $\alpha^{(0)}$). The iteration procedure can thus be started for $N = 1, 2, ...$ etc. and continued till convergence.

## 7. About convergence

Two convergence criteria for the iteration procedure of Section 6 can be envisaged, one of which relates to the static approach, the other to the kinematic approach.

For the *static approach*, the convergence criterion requires:

- Problem (63) always admits solution at all points $\mathbf{x} \in V$, or equivalently, the discriminant of (66a) is nonnegative, i.e. the condition

$$\Delta = \sigma_y^2 - \phi^2\left(\mathbf{s}_r^0\right) + \frac{\left(\mathbf{s}^{(N)} : \mathbf{a} : \mathbf{s}_r^0\right)^2}{\phi^2\left(\mathbf{s}^{(N)}\right)} \geq 0 \qquad \forall r \in I(m) \text{ in } V \tag{68}$$

  is satisfied at every iteration, $N = 1, 2, ....$

- At every iteration, the stress $\mathbf{s}^{(N)}$ provided by LEAN($\xi_r^{(N-1)}$, $\mu^{(N-1)}$, $\alpha^{(N-1)}$) is plastically admissible, i.e. $f_r\left(\mathbf{s}^{(N)}\right) = \phi\left(\mathbf{s}^{(N)} - \mathbf{s}_r^0\right) - \sigma_y \leq 0$ $\forall r \in I(m)$ and $\forall \mathbf{x} \in V$.

Then, if these requisites are satisfied, obviously it is $\gamma(\mathbf{x}, N) \geq 1$ by (63) and thus $\gamma^{(N)} \geq 1$ by (64a). It thus results, by (64b), $\alpha_{\text{ST}}^{(N)} \geq \alpha_{\text{ST}}^{(N-1)}$. Since $\alpha_{\text{ST}}^{(N)} \leq \alpha_{\text{sh}}$, it can be concluded that, under the above restrictions, the static load multiplier, $\alpha_{\text{ST}}^{(N)}$, converges monotonically to $\alpha_{\text{sh}}$.

For the *kinematic approach*, the convergence criterion requires:

- At every iteration $N = 1, 2, ...$, the solution provided by LEAN($\xi_r^{(N-1)}$, $\mu^{(N-1)}$, $\alpha^{(N-1)}$) is characterized by a complementary potential energy, i.e.

$$W^{(N-1)}(\mathbf{s}) = \frac{1}{2\mu^{(N-1)}} \sum_{r=1}^{m} \xi_r^{(N-1)} \phi^2\left(\mathbf{s} - \mathbf{s}_r^0\right), \tag{69}$$

  such that, at every $\mathbf{x} \in V$, the equipotential surface $W^{(N-1)}(\mathbf{s}) = \text{const}$ passing through $\mathbf{s}^{(N)}$ encompasses the marginal elastic domain $\mathcal{R}$, that is, the inequality

$$W^{(N-1)}\left(\mathbf{s}^{(N)}\right) \geq \sigma_y^2 / 2\mu^{(N-1)} \qquad \text{in } V \tag{70}$$

  is satisfied.

Then, if this requisite is complied with, the inequality

$$\left(\mathbf{s}^{(N)} - \mathbf{s}^{*(N)}\right) : \mathbf{e}^{(N)} \geq 0 \qquad \text{in } V \tag{71}$$

is also satisfied as a consequence of the fact that the stress point $\mathbf{s}^{*(N)}$ lies on $\partial\mathcal{R}$ and is thus not external to the equipotential surface $W^{(N-1)}(\mathbf{s}) =$ const passing through $\mathbf{s}^{(N)}$. Thus, considering the integral of (71), one can write, by the virtual work principle,

$$\int_V \mathbf{s}^{*(N)} : \mathbf{e}^{(N)} \, \mathrm{d}V \leq \int_V \mathbf{s}^{(N)} : \mathbf{e}^{(N)} \, \mathrm{d}V = \alpha_{\mathrm{KIN}}^{(N-1)} \left(\bar{P}, \mathbf{u}^{(N)}\right) \tag{72}$$

and, inserting the latter result in (67), one has

$$\alpha_{\mathrm{KIN}}^{(N)} \leq \alpha_{\mathrm{KIN}}^{(N-1)}. \tag{73}$$

As $\alpha_{\mathrm{KIN}}^{(N)} \geq \alpha_{\mathrm{sh}}$, one can conclude that, under the above restriction, the kinematic load multiplier, $\alpha_{\mathrm{KIN}}^{(N)}$, converges monotonically to $\alpha_{\mathrm{sh}}$.

The above convergence criteria, particularly that relating to the static approach, are rather restrictive in nature. It seems quite improbable, except perhaps for a category of simple problems, that the fictitious elastic solution computed by LEAN( $\xi_r$, $\mu$, $\alpha$) may comply with the pertinent requisite during the entire iterative procedure. However, the above criteria provide sufficient conditions for convergence. Convergence may actually occur in a larger range of structural conditions and elastic solution characteristics, which can only be addressed through suitable numerical experiments at this time.

## 8. Numerical application

A simple illustrative example is presented. To this purpose the elastic simulation technique formulated in the previous Sections has been implemented into the finite element code FEAP (Zienkiewicz and Taylor, 1988), where a new macro-procedure able to solve sequences of fictitious linear elastic problems has been introduced. The structural shakedown problem of a square plate with a central circular hole and subjected to a combined steady/unsteady mechanical loading acting on the plate edges has been numerically analyzed. The material is elastic-perfectly plastic with associative flow rule and a Von Mises yield function. In order to account for the plain stress state of the plate the yield function of eq.(2) has been taken, with the vector notation, as

$$\phi(\sigma) = \left(\sigma'^{T} \mathbf{a}\, \sigma'\right)^{1/2} \tag{74}$$

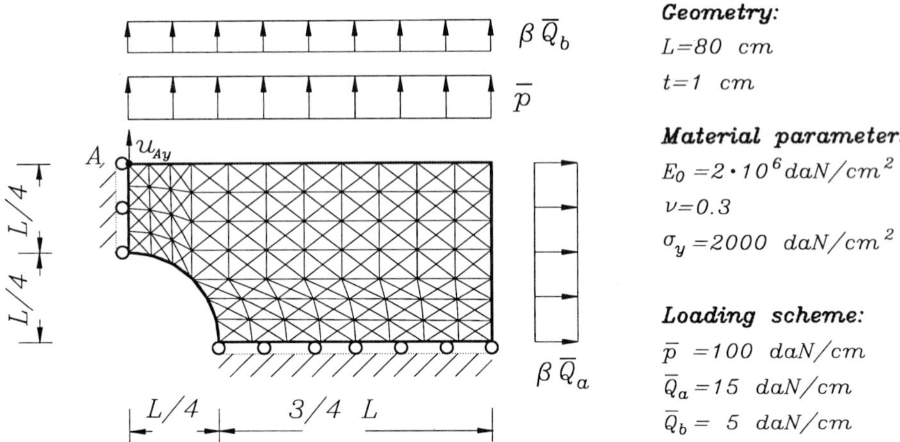


*Figure 3.* Plate subjected to combined loads and discretization mesh by constant stress-constant strain triangular finite elements.

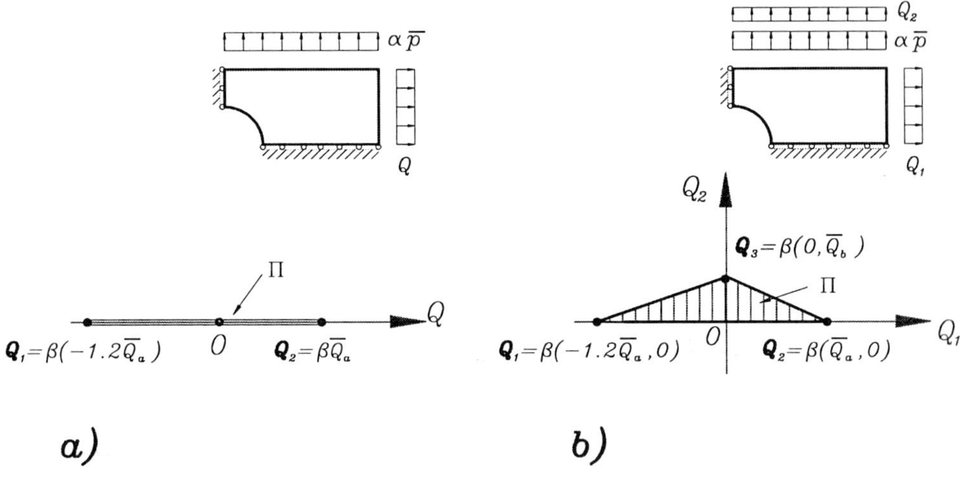


*Figure 4.* Unsteady load domains and loading schemes of the plate: a) One-dimensional case with $\mathbf{Q}_1 \leq \mathbf{Q} \leq \mathbf{Q}_2$, $\mathbf{Q}_1 = \beta(-1.2\bar{Q}_a)$, $\mathbf{Q}_2 = \beta\bar{Q}_a$; b) Two-dimensional case with $(\mathbf{Q}_1, \mathbf{Q}_2) \in \Pi$ and $\Pi$ specified by the vertices $\mathbf{Q}_1 = \beta(-1.2\bar{Q}_a, 0)$, $\mathbf{Q}_2 = \beta(\bar{Q}_a, 0)$, $\mathbf{Q}_3 = \beta(0, \bar{Q}_b)$.

where $\boldsymbol{\sigma}'^T = [\sigma'_x, \sigma'_y, \sigma'_{xy}]$ is the vector of deviatoric stress components and $\mathbf{a}$ is a positive definite constant matrix, i.e.

$$\mathbf{a} = \frac{3}{2}\begin{bmatrix} 1 & 0 & 0 \\ 0 & 1 & 0 \\ 0 & 0 & 2 \end{bmatrix} \tag{75}$$

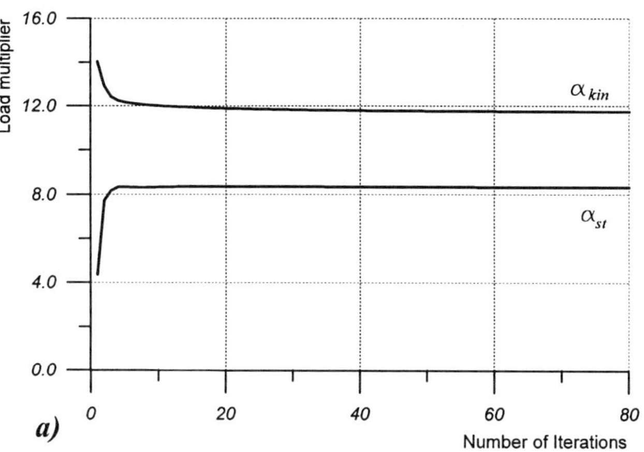


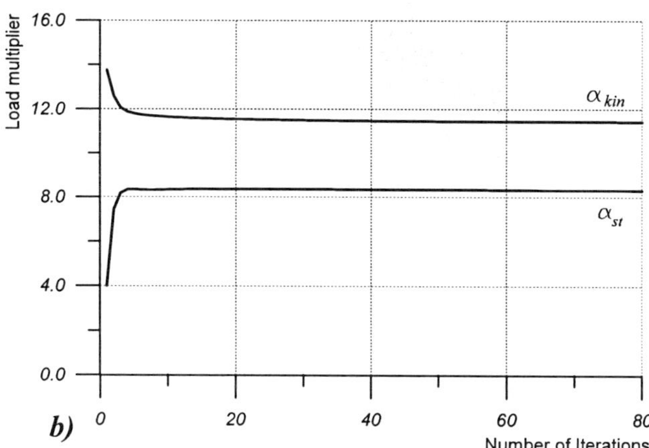


*Figure 5.* Static and kinematic (steady) load parameters resulting from the numerical iterative procedure reported as functions of the iteration number in the first load case (a) and the second load case (b).

Figure 3 shows the structural geometry, the loading scheme, the F.E. discretization adopted and the relevant material data. Structural symmetries allow one to analyze a quarter of plate.

The discretization details are as follows: i) a mesh with 272 constant stress/constant strain triangular elements has been employed; ii) two different load domains $\Pi$ have been considered for the unsteady loads, namely a) $\Pi$ is the linear segment $\beta(-1.2\bar{Q}_a) \leq Q \leq \beta\bar{Q}_a$ depicted in Figure 4(a), and b) $\Pi$ is the triangle of vertices $\mathbf{Q}_1 = \beta(-1.2\bar{Q}_a, 0)$, $\mathbf{Q}_2 = \beta(\bar{Q}_a, 0)$, $\mathbf{Q}_3 = \beta(0, \bar{Q}_b)$ depicted in Figure 4(b). A uniform steady load $\alpha\bar{p}$ is acting on the longer edge of the plate, as shown in Figures 3 and 4(a,b). The marginal resistance domain correspondent to the load cases of Figures 4 (a,b) are respectively a two- and a three-surface domain in the $\mathbf{s} = (s_x, s_y, s_{xy})$ transfer stress space.

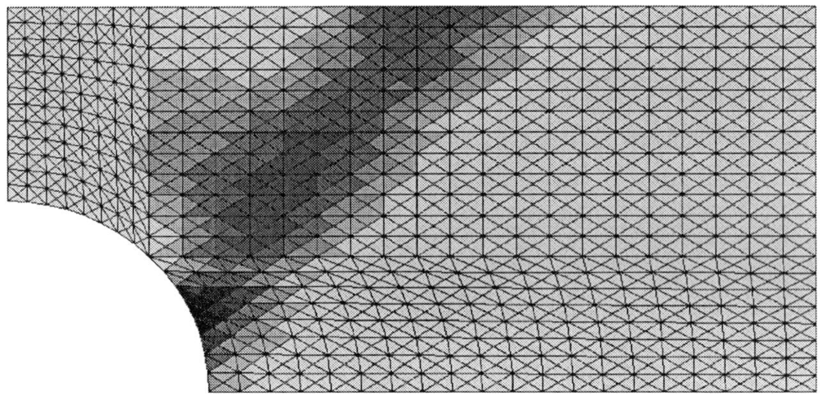

*Figure 6.* F.E. discretized plate and element plastic dissipation distribution for the load case of Figure 5(a) at convergence attained with $\alpha = \alpha_{\text{KIN}} = 11.73$.

The iterative procedure of Section 6 is aimed to determine, for a fixed amplitude of the unsteady loads (i.e. for an assigned value of $\beta < \beta_{\text{al}}$), the maximum value of the steady load multiplier, $\alpha_{\text{sh}}$, for which shakedown occurs. The obtained numerical results are depicted in Figures 5(a,b) where the static and kinematic load multipliers are plotted versus the number of iterations with reference to the loading cases of Figures 4(a,b), respectively. Figure 6 shows the finite element plastic dissipation distribution for the first loading condition when convergence is attained with $\alpha = \alpha_{\text{KIN}} = 11.73$.

In order to check the solution accuracy, evolutive elastic-plastic analyses have been also carried out, always for fixed amplitude ($\beta$) of the unsteady loads. As a result of these analyses, the displacement history $u_{Ay}^{(+)}$ at point $A$ (specified in Figure 3), has been computed with the steady load value obtained by the elastic simulation procedure, i.e. with $\alpha = \alpha_{\mathrm{ST}} = 8.33$ and $\alpha = \alpha_{\mathrm{KIN}} = 11.73$ and reported in Figure 7. These analyses confirm that, for $\alpha = \alpha_{\mathrm{KIN}} = 11.73$, shakedown does not occur as actually the structure exhibits an incremental collapse mode, whereas for $\alpha = \alpha_{\mathrm{ST}} = 8.33$, after some cycles, the structural response appears to be fully elastic so that shakedown takes place. Adopting a number of trial analyses, the real shakedown load multiplier resulted $\alpha = \alpha_{\mathrm{sh}} = 11.5$, which shows to be closer to the kinematic bound.

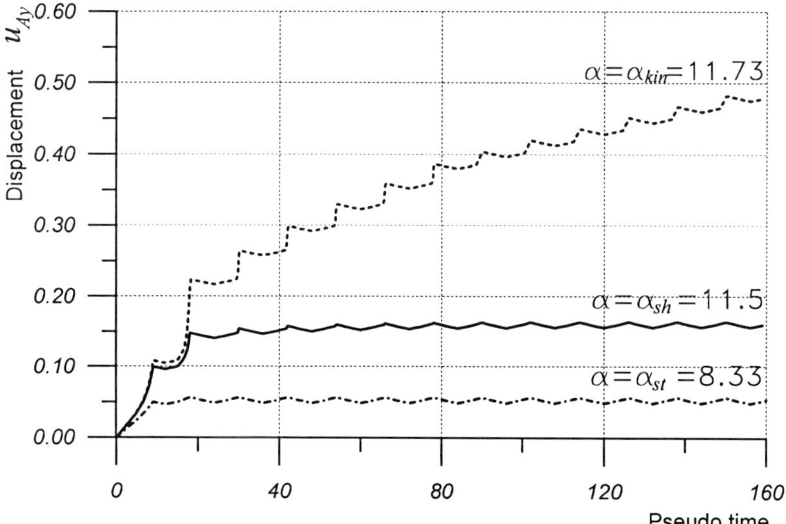


*Figure 7.* Displacement histories $u_{Ay}$ resulting from evolutive elastic-plastic analyses with fixed $\beta$ and $\alpha = 11.73$ (dot line), $\alpha = 8.33$ (dash-dot line) and $\alpha = \alpha_{\mathrm{sh}} = 11.5$ (solid line).

Finally, in order to investigate the mesh sensitivity, the same problem was solved using four different meshes. The results, shown in Figure 8, prove that for tha analyzed problem a not extremely refined mesh can produce sufficiently accurate results.

## 9. Summary and conclusions

We have presented a computationally oriented approach to the classical problem of shakedown limit analysis for elastic perfectly plastic structures. This approach substantiates in an analysis procedure referred to as

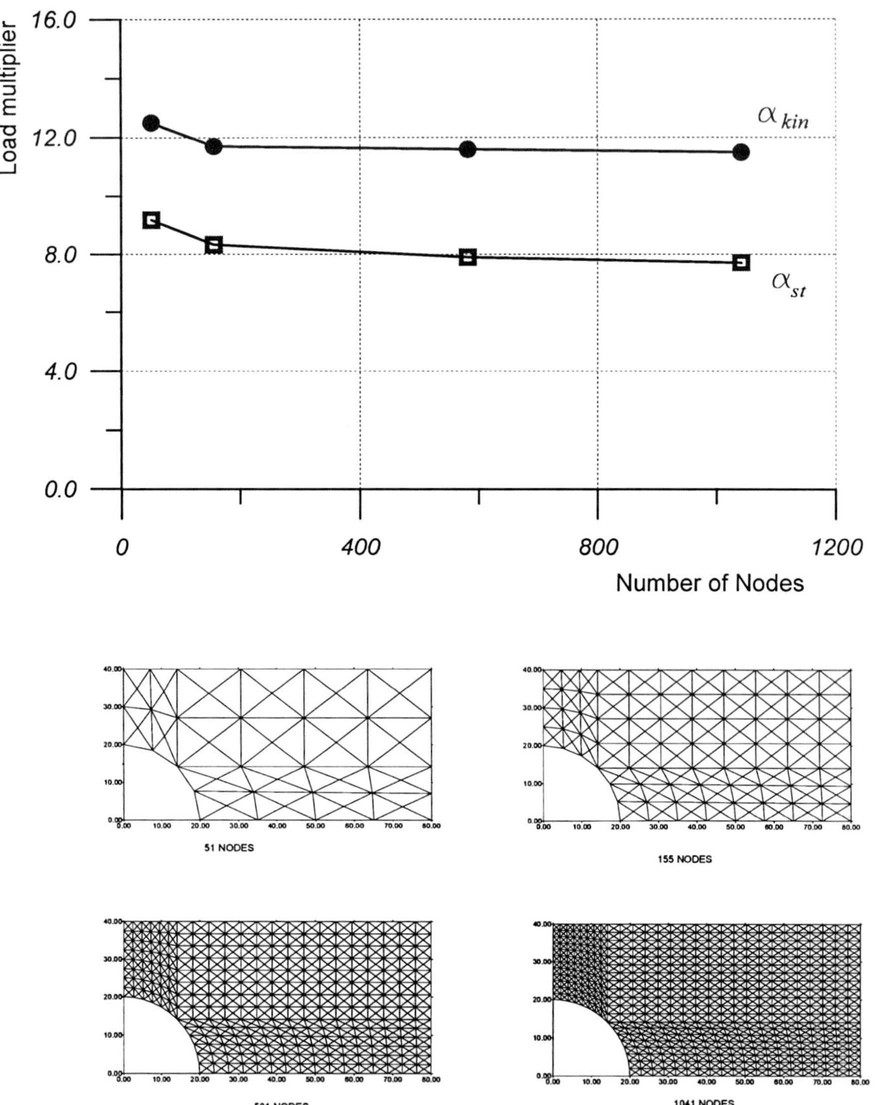


*Figure 8.* Diagrams of the static and kinematic load multipliers vs. mesh nodes, obtained by the elastic simulation technique for four different FE meshes.

*elastic simulation technique*, generalization of known elasticity-based methods of shakedown analysis under one-dimensional variable loads (Ponter and Carter, 1997a, 1997b) to the more general case of multidimensional variable loads. The generalization, by no means a straightforward task, has been achieved with the aid of the concept of *linear elastic pre-stressed composite material model* which, through its phases and fractions, is able to simulate the complex plastic behaviour of the material at the shakedown limit state. This fictitious elastic material is characterized by a scalar modulus ($\mu$) and $m$ fractions coefficients ($\xi_r$, $r = 1, 2, ..., m$), with $m$ being the number of vertices of the polyhedral load domain ($\Pi$). These material constants are a-priori unknown, but their correct values, that make the elastic composite material to be computationally equivalent to the actual cyclic rigid-plastic material of the shakedown limit analysis problem, are determined through an iterative procedure. At every iteration, a (nonhomogeneous) linear elastic analysis problem is solved with approximate (fixed) values of the above material constants, which can thus be improved for use in the next iteration. Energy statements enable the elastic simulation effectiveness to be characterized.

The loading scheme considered is a combined steady/unsteady load, in which some unsteady (i.e. repeated, or cyclic) loads are allowed to vary arbitrarily within a polyhedral multi-dimensional domain ($\Pi$) maintained at fixed amplitude, and a steady mechanical load is specified within a scalar factor ($\alpha$). In the elastic simulation problem, the unsteady load is accounted for through the material pre-stressing system, whereas the shakedown value of $\alpha$, i.e. $\alpha_{\text{sh}}$, is obtained by the iterative procedure of the elastic simulation technique.

Two sufficient convergence criteria have been given, which however turn out to be quite restrictive. An illustrative example herein presented shows that convergence of the proposed iterative procedure is satisfactory, particularly for the kinematic approach, but more extensive numerical experiments are necessary in order to express a safe judgement on the proposed method. This task will be accomplished in a subsequent more extensive paper.

The proposed method applies —at least in principle— with any number of dimensions and vertices of $\Pi$. For one-dimensional load domains $\Pi$, it substantially coincides with the method proposed by Ponter and Carter (1997a,b). The proposed method also applies to the particular case of plastic limit analysis.

The plastic material model considered in this paper is one with a Mises yield function. Generalizations to other types of yield functions can be achieved without great complications, as e.g. the Druker-Prager cap model already employed by Ponter *et al.* (to appear) for the plastic limit analysis problem of soil structures via elastic simulation technique with a one-dimensional loading. These generalizations will be addressed elsewhere.

The elastic simulation technique previously proposed constitutes a promising computational method for plastic and shakedown limit analysis, due to the reason that the numerical analyses require the use of elasticity FE computer programs that are included in the existing commercial computer software packages. Special numerical algorithms are only requested for the updating process of the iterative procedure. Further research efforts are being made on this side.

## References

1. Ainsworth, R.A. (1988) Assessment procedures for high temperature crack growth, in *Recent Advances for high temperature plant*, Inc. Mech. Engrg., London, pp. 103–107.
2. Borino, G. (1999) Consistent shakedown theorems for materials with temperature dependent yield functions, *J. Solids Struct.*, (in press).
3. Borino, G., Polizzotto, C. (1997a) Shakedown theorems for a class of materials with temperature-dependent yield stress, in D.R.J. Owen, E. Oñate, E. Hinton (eds.), *Computational Plasticity, Fundamentals and Applications*, CIMNE, Barcelona, Spain, pp.475–480.
4. Borino, G., Polizzotto, C. (1997b) Reformulation of shakedown theorems for materials with temperature-dependent yield stress, *Proc. 2nd Int. Symp. on thermal stresses and related topics*, Rochester Institute of Technology, pp.89–92.
5. Carter, K.F., Ponter, A.R.S., Ellington, J.P., Ponter, A.G. (1988) Interaction diagrams for use in the construction of design codes, in *Recent Advances for high temperature plant*, Inc. Mech. Engrg., London, pp. 7–12.
6. Cocks, A.C.F. (1984) Lower-bound shakedown analysis of a simply supported plate carrying a uniformly distributed load and subjected to cyclic thermal loading, *Int. J. Mech. Sci.*, **26**, 471–475.
7. Comi, C., Corigliano, A. (1991) Dynamic shakedown in elastoplastic structures with general internal variable constitutive law, *Int. J. Plasticity*, **7**, 679–692.
8. Corigliano, A., Maier, G., Pycko, S. (1995) Dynamic shakedown analysis and bounds for elastoplastic structures with nonassociative internal variable constitutive laws, *Int. J. Solids Structures*, **32**, 3145–3166.
9. Druyanov, B., Roman, I. (1998) On adaptation (shakedown) of a class of damage elastic plastic bodies to cyclic loading, *Eur. J. Mech., A/Solids*, **17**, 71–78.
10. Feng, X., Yu, S. (1995) Damage and shakedown analysis of structures with strain-hardening, *Int. J. Plasticity*, **11**, 237–249.
11. Frederick, C.O., Armstrong, P.J. (1966) Convergent internal stresses and steady cyclic state of stress, *J. Strain Analysis*, **1**, 154–162.
12. Fuschi, P., Polizzotto, C. (1995) The shakedown load boundary of an elastic-plastic structure under combined cyclic/steady loads, *Meccanica*, **30**, 155-174.
13. Gokhfeld, D.A., Cherniavsky, D.F. (1980) *Limit Analysis of Structures at Thermal Cycling*, Sijthoff & Noordhoff, Alphen an Der Rijn, The Netherlands.
14. Goodall, I.W., Goodman, A.M., Chell, G.C., Ainsworth, R.A., Williams, J.A. (1991) An assessment procedure for high temperature response of structures, Nuclear Electric PLC Report, Berkeley Nuclear Laboratories, Gloucester, U.K.
15. Hachemi, A., Weichert, D. (1992) An extension of the static shakedown theorem to a certain class of inelastic materials with damage, *Arch. Mech.*, **44**, 491-498.
16. Hamilton, R., Boyle, J.T., Shi, J., Mackenzie, D. (1996) Shakedown load bounds by elastic finite element analysis, in R.F. Sammataro, D.J. Ammerman (eds.), *Development, validation and application of inelastic methods for structural analysis and design*, The American Society of Mechanical Engineers, New York, N.Y., PVP - Vol.343, pp. 205–209.

17. Hill, R. (1950) *The Mathematical Theory of Plasticity*, Clarendon Press, Oxford, U.K.
18. Huang, Y., Stein, E. (1996) Shakedown of a cracked body consisting of kinematic hardening material, *Eng. Frac. Mech.*, **1**, 107–112.
19. Kamenjarzh, J., Weichert, D. (1992) On kinematic upper bounds for the safety factor in shakedown theory, *Int. J. Plasticity*, **8**, 827–837.
20. Kamenjarzh, J., Merzljakov, A. (1994) On kinematic method in shakedown theory: I. Duality of extremum problems, and II. Modified kinematic method, *Int. J. Plasticity*, **10**, 363–380, 381–392.
21. Koiter, W.T. (1960) General theorems of elastic-plastic solids, in J.N. Sneddon, R. Hill (eds.), *Progress in Solid Mechanics*, North Holland, Amsterdam, Vol. 1, pp.167-221.
22. König, J.A., Maier, G. (1981) Shakedown analysis of elastoplastic structures: A review of recent developments, *Jour. of Nuclear Engng. Design*, **66**, 81–95.
23. König, J.A. (1982a) On upper bounds to shakedown loads, *ZAMM*, **52**, 421–428.
24. König, J.A. (1982b) On some recent developments in the shakedown theory, *Advances in Mechanics*, **5**, 237–258.
25. König, J.A. (1987) *Shakedown of Elastic-Plastic Structures*, PWN-Polish Scientific Publishers, Warsaw and Elsevier, Amsterdam.
26. Maier, G. (1987) A generalization to nonlinear hardening of the first shakedown theorem for discrete elastic-plastic models, Atti Acc. Lincei Rend. Fis. **8**, LXXX, 161-174.
27. Martin, J.B. (1975) *Plasticity: Fundamentals and General Results*, MIT Press, Cambridge, Ma.
28. Panzeca, T., Polizzotto, C. (1988) On shakedown of elastic-plastic solids, *Meccanica*, **23**, 94–101.
29. Pham, D.C., Stumpf, H. (1994) Kinematic approach to shakedown analysis of some structures, *Quarterly of Appl. Math.*, Vol. LII, No.4, 707–719.
30. Polizzotto, C., Borino, G., Fuschi, P. (1990) On the steady-state response of elastic-perfectly plastic solids to cyclic loads, in M. Kleiber and J.A. König (eds.), *Inelastic Solids and Structures*, Pineridge Press, Swansea, UK, pp. 473-488.
31. Polizzotto, C., Borino, G., Caddemi, S., Fuschi, P. (1991) Shakedown problems for material models with internal variables, *Eur. J. Mech. A/Solids*, **10**, 621–639.
32. Polizzotto, C. (1993) On the conditions to prevent plastic shakedown of structures, Parts I and II, *ASME J. Appl. Mech.*, **60**, 15-19 and 20-25.
33. Polizzotto, C. (1993b) A study on plastic shakedown of structures, Parts I and II, *ASME J. Appl. Mech.*, **60**, 318-323 and 324-330.
34. Polizzotto, C. (1994) On elastic plastic structures under cyclic loads, *Eur. J. Mech. A/Solids*, **13**, 149-173.
35. Polizzotto, C. (1995) Elastic-viscoplastic solids subjected to thermal and loading cycles, in Z. Mróz, D. Weichert, S. Dorosz (eds.), *Inelastic behaviour of structures under variable loads*, Kluwer Academic Publishers, Dordrecht, The Netherlands, pp. 95-128.
36. Polizzotto, C., Borino, G. (1996) Shakedown and steady-state responses of elastic-plastic solids in large displacements, *Int. J. Solids Structures*, **23**, 3415–3437.
37. Polizzotto, C., Borino, G., Fuschi, P. (1996) An extended shakedown theory for elastic-plastic-damage material models, *Eur. J. Mech. A/Solids*, **15**, 825–858.
38. Polizzotto, C. (1997) Shakedown of elastic-plastic solids with frictionless unilateral boundary conditions, *Int. J. Mech. Sci.*, **7**, 819-828.
39. Ponter, A.R.S. (1972) Deformation, displacement and work bounds for structures in a state of creep and subjected to variable loads, *J. Appl. Mech.*, **39**, 953–958.
40. Ponter, A.R.S. (1983) *Shakedown and Ratchetting Below the Creep Range*, Report EUR, 8702 EN, Commission of the European Communities, Brussels.
41. Ponter, A.R.S., Karadeniz, S. (1985) An extended shakedown theory for structures that suffer cyclic thermal loadings, Parts 1 and 2, *ASME J. Appl. Mech.*, **52**, 877–

882 and 883–889.

42. Ponter, A.R.S., Karadeniz, S., Carter, K.F. (1990) The computation of shakedown limits for structural components subjected to variable thermal loading, Report EUR 12686, Directorate General Science, Research Development, Commission of the European Communities, Brussels.
43. Ponter, A.R.S., Karadeniz, S., Carter, K.F. (1990) Extended upper bound shakedown theory and finite element method for axisymmetric thin shells, in M. Kleiber, J.A. König (eds.), *Inelastic Solids and Structures*, Pineridge Press, Swansea, UK, pp. 433-449.
44. Ponter, A.R.S., Carter, K.F. (1997a) Limit state solutions based upon linear elastic solutions with a spatially varying elastic modulus, *Comput. Methods Appl. Mech. Engrg.*, **140**, 237–258.
45. Ponter, A.R.S., Carter, K.F. (1997b) Shakedown state simulation techniques based on linear elastic solutions, *Comput. Methods Appl. Mech. Engrg.*, **140**, 259–279.
46. Ponter, A.R.S., Fuschi, P., Engelhardt, M. (1999) Limit analysis for a general class of yield conditions, (to appear).
47. Pycko, S., Maier, G. (1995) Shakedown theorems for some classes of nonassociated hardening elastic-plastic material models, *Int. J. Plasticity*, **11**, 367–395.
48. Sewell, M.J. (1987) *Maximum and Minimum Principles*, Cambridge University Press, Cambridge, UK.
49. Stein, E., Zhang, G., König, J.A. (1992) Shakedown with nonlinear strain hardening including structural computation using finite element method, *Int. J. Plasticity*, **8**, 1–31.
50. Weichert, D. (1986) On the influence of geometrical nonlinearities on the shakedown of elastic-plastic structures, *Int. J. Plasticity*, **2**, 135–148.
51. Weichert, D., Gross-Weege, J. (1988) The numerical assessment of elastic-plastic sheets under variable mechanical and thermal loads using a simplified two-surface yield condition, *Int. J. Mech. Sci.*, **30**, 757–767.
52. Weichert, D. (1990) Advances in geometrical nonlinear shakedown theory, in M. Kleiber, J.A. König (eds.), *Inelastic Solids and Structures*, Pineridge Press, Swansea, UK, pp. 489-502.
53. Zarka, J., Casier, J. (1979) Cyclic loading on elastic-plastic structures, in S. Nemat-Nasser (eds.), *Mechanics Today*, Pergamon Press, Oxford, UK, pp. 93-198.
54. Zienkiewicz, O.C., Taylor, R.L. (1988) *The finite element method*, McGraw-Hill, Maidenhead, 4-th ed.

# APPLICATION OF THE KINEMATIC SHAKEDOWN THEOREM TO PAVEMENTS DESIGN

M. BOULBIBANE and I.F. COLLINS
*School of Engineering, University of Auckland, New Zealand.*

Abstract. The concepts and methods of shakedown theory, which describes the ultimate response of an elastic/plastic structure to cyclic loads, are used to analyse the response of unbound pavements. The pavements are modelled as multilayered structures of Mohr-Coulomb material, characterised by their elastic moduli, cohesion and friction angle. The upper bound theorem of classical shakedown theory is used to estimate the critical shakedown load, and employed to predict the relative importance of various failure mechanisms such as subsurface slip and rut formation.

## 1. Introduction

The techniques of analysis and design of pavements (roads and runways) are not so developed as for many other types of structures. This is partly due to the inherent complex material responses of the various constituent layers in a pavement, but also due to the need to model these responses to a sequence of repeated loads over a long time frame. Furthermore, most current procedures ignore the results of much laboratory testing on roading materials and fail to use any of the theories and models of modern soil mechanics. Various types of pavements can be classified according to the nature of the material and thickness of the top surface layer. Many involve a substantial asphalt layer. Since asphalt possesses significant viscous properties it is not readily modelled as a rate-independent elastic/plastic material. In many parts of the world, however, including Australasia, U.S.A and South Africa, where traffic densities are low the pavements are "unbound". In such a pavement the top asphalt/bitumen layer is very thin, serving simply as weatherproofing, and does not contribute to the structural strength of the pavement. This strength is provided by the basecourse layers (made of various type of aggregate) and the original subgrade. Since the viscous bituminous layer in an unbounded pavement can be ignored, such a pavement can be modelled by layers of rate-independent elastic/plastic materials. The response of such a structure to a repeated loading history can be described by shakedown



*D. Weichert and G. Maier (eds.), Inelastic Analysis of Structures under Variable Loads,* 365–380.

theory, which provides a rational model for explaining the observed failure mechanisms such as rut formation, subsurface slip and fatigue. It has been recognised for some years now that shakedown theory may well provide the key to a more rationally based pavement design procedure. In particular it is suggested that the concept of the "shakedown load" should be the basic design parameter. When the operating load is less than this critical value, the pavement may suffer some permanent deformation and damage for a finite number of load applications, but beyond this, further pavement traffic will not produce any further deterioration to the pavement structure. For higher operating loads the pavement will "fail" in the sense that permanent deformations of some sort will continue to build up indefinitely. Whilst most early applications of shakedown theory were to frame structures and pressure vessels, more recently the theory has been used very successfully to model various aspects of the wear of metallic surfaces subject to repeated sliding and rolling loads (e.g. Ponter *et al.* [23], Johnson [17]. The possible application of shakedown theory to pavement analysis seems to be first suggested by Sharp and Booker [26]. Further studies of increasing realism and complexity have been made by Raad *et al.* [24], Collins and Cliffe [7], Collins *et al.* [8, 9], Collins and Boulbibane [10, 11], Raad and Weichert [25] and Yu and Hossain [27].

The present approach to pavement design involves the computation of the critical shakedown load associated with various proposed failure mechanisms. Early work on this problem concentrated on the static approach to calculating the shakedown load, via Melan's theorem [21] and assuming two-dimensional pavement models in which the tyre is replaced by an infinitely long cylinder. The calculation of the shakedown load can either be calculated using the "method of conics" introduced by Sharp and Booker [26] or by using finite element methods together with linear [27] or non-linear [1, 24] programming procedures. A discussion of the relative merits of these methods can be found in [4]. In all these calculations the pavement layers are modelled as elastic/plastic structures, with linear or non-linear elastic moduli, and perfectly plastic yielding according to the Mohr-Coulomb criterion. Whilst these methods work well for the two-dimensional model they do not generalise easily to three-dimensions, as the number of constraints in the mathematical programming formulation becomes excessive.

The authors have pioneered the dual kinematic approach to calculating the shakedown load using Koiter's upper bound theorem [19]. This approach has a number of advantages. Mathematically it replaces a very large linear programming problem with a much smaller non-linear optimisation problem, which enables three-dimensional problems to be solved with relative ease. Early research in this area [8, 9] concentrated on failure mechanisms

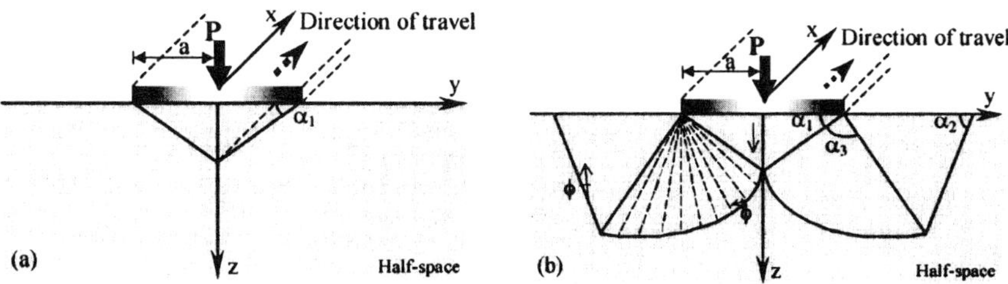


*Figure 1.* Two proposed failure mechanisms for pavements.

in which subsurfaces or surface slip occurred in the travel direction, paralleling the metal surface wear study of Ponter *et al.* [23]. More recently calculations have been made for the complementary mode of failure, where the subgrade motion is perpendicular to the travel direction and the loaded tyre penetrates into the pavement so forming a rut [3, 10].

All the above studies assume a normal or associated flow rule. However, it is well established experimentally that in many situations the flow rule appropriate to granular materials, such as roading aggregates, is non-associated. Some plane strain, lower bound calculations for non-associated materials have been presented by Boulbibane [1] and Boulbibane and Weichert [2]. More recently we have performed three- dimensional upper bound calculation for non -associated materials making use of the ideas of Drescher and Detournay [13] and are discussed here in section 4.1.4

In this paper we will present an overview of the research to-date, including descriptions of the various failure modes, optimisation methods and material models, and discuss aspects of basecourse and subgrade modelling.

## 2. Slip and rut formation models

According to the classical theorem given by Koiter, the kinematic method provides an upper bound of the true limit load inducing collapse. This upper bound is obtained by assuming a failure mechanism (see Fig. 1) and calculating the load factor, which is defined to be the ratio of plastic to elastic work rate in the assumed mechanism:

$$\lambda = \frac{\int_v \sigma_{ij}^p e_{ij}^p dV}{\int_v \sigma_{ij}^e e_{ij}^p dV} \tag{1}$$

where $e_{ij}^p$ and $\sigma_{ij}^p$ are the time-independent strain-rate and plastic stress fields associated with the assumed kinematically admissible velocity field and $\sigma_{ij}^e$ is the elastic stress field produced by the applied load. In this class of problem the co-ordinate in the travel direction (i.e. $x$ in Fig. 1) plays the role of the time variable. Each point on a line $y = constant$ $z = constant$ undergoes the same stress history. Hence each mechanism and associated residual stress field must be independent of $x$. In essence the estimates are obtained by postulating some form of failure mechanism and equating the elastic and plastic rates of working in this mechanisms. The resulting value of the load is then an upper bound to the actual shakedown load, so that if a number of possible mechanisms are studied, the lowest gives the best estimate. We use block sliding mechanisms so that the plastic strain-rate $e_{ij}^p$ is only non zero on the velocity discontinuities and the plastic work rate can be computed as in the well known limit analysis calculation for monotonic loading (e.g. Chen [6]). The elastic work rate is simply calculate by integrating the virtual elastic work-rate along discontinuities, the elastic stresses being computed using a standard elastic layer program such as BISAR. On a given discontinuity the plastic work-rate per unit length is simply $c[v_t]$, the cohesion times the jump in the tangential velocity component. The corresponding expression for the elastic work-rate is $c^e[v_t]$ where:

$$c^e = |\sigma_{nt}^e| - \sigma_{nn}^e tan\phi \tag{2}$$

and $\sigma_{nn}$, $\sigma_{nt}$ are the normal and tangential stress components, respectively and $\phi$ is the internal angle of friction. In this study two broad classes of mechanisms, shown in Fig. 1(a) and (b) for a one layer pavement, are considered. In the first, failure occurs by slipping in the travel direction in a V-shaped channel. Such failure would be observed by the formation of surface shear cracks at the edges of the channel. Calculations demonstrate that in such failures the channel edge occurs very close to the edge of the loaded zone. Figure 1(b) shows one of many possible rut formation mechanisms in which the aggregate material moves downwards and sideways. A number of such mechanisms have been studied, c.f. [3, 12] and the best solution found so far is that illustrated here. This mechanism bears some similarity to the well known bearing capacity solution for a footing [5], but is different in that the displaced material disturbs the surface very close to the loaded area, rather than over a region 2-4 times the radius of the loaded area as is the case in the foundation problem. It is to be emphasised that this model only predicts the onset of rutting and does not attempt to model the complete development of the rut. (N.B. An attempt to model the incremental development of a groove in the corresponding metal contact problem has been made by Kapoor and Johnson [18]. The results are most conveniently presented in terms of a non-dimensional load factor:

$$\mu = \frac{\lambda P}{\pi c a^2} \tag{3}$$

where $\lambda P$ is the applied normal shakedown load, $a$ is the radius of the circular loaded region and $c$ is the cohesion of the basecourse layer.

## 3. Optimisation procedure

The mechanism shown in Fig. 1(b) is defined by the 3 angles ($\alpha_1$, $\alpha_2$, $\alpha_3$). The minimum value of $\lambda$ for this class of mechanisms is found by numerical optimisation. Initial attempts at optimisation using the standard Newton procedure proved difficult because of the presence of a large number of local minima. It has been necessary therefore to use more robust global optimisation routine, and the simulated annealing (SA) procedure devised by Goffe *et al.* [15] has been found to be very effective. In this method the ratio of plastic work to elastic rate work (Eq. 1) is the function to be minimised and the control parameter in the algorithm, which is initially set to high value, is decreased in steps until the optimal solution is reached. The procedure of using the SA algorithm to solve an optimisation problem can be summarised as follows:

(a) arbitrarily assign a configuration and calculate the corresponding function value;

(b) initialise the control parameter and the number of perturbations at each step;

(c) a configuration is randomly selected in the vicinity of the current configuration and the change in function is calculated;

(d) the current configuration is replaced by the new configuration according to the Metropolis criterion which is based on the Monte Carlo method;

(e) repeat steps (c) and (d) N times;

(f) reduce the control parameter and repeat form (c) until a specified criterion for stopping the algorithm is satisfied.

With the Metropolis criterion there is a non-zero probability for transition to a configuration with a higher function value than the current configuration. This is the property that allows the process to escape configurations corresponding to local minima in its search for the configuration corresponding to the global minimum.

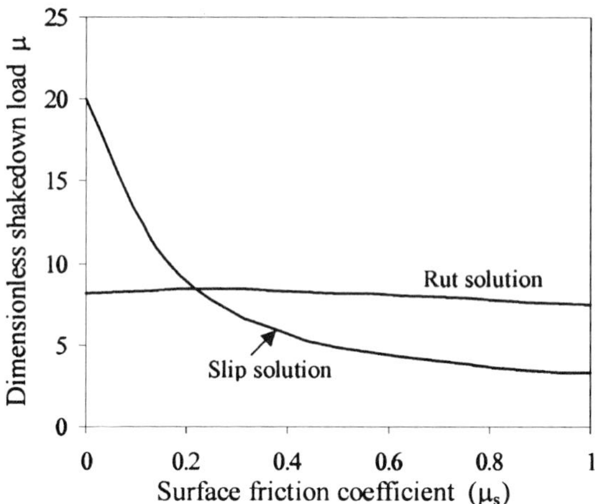


*Figure 2.* Variation of load factor with surface friction for slip and rut solutions.

## 4. Applications

### 4.1. SHAKEDOWN IN A HALF-SPACE

#### 4.1.1. *Effect of surface friction*

In a uniform homogeneous pavement model, $\mu$ will be a function of the surface loading friction coefficient $\mu_s = Q/P$ (the ratio of the surface tangential and normal loads), $\phi$ the internal angle of friction of the basecourse material and the form of the loading distribution, here assumed uniform. However, it will not depend on the stiffness of the material. This is because the elastic stress field is proportional to the applied loads and independent of the material's stiffness. In a layered model, the dimensionless load factor will depend additionally on the ratios of the layer cohesions and stiffnesses, the values of the angle of internal friction and of Poisson's ratio in each layer and the ratio of the layer depth to the radius of the loaded area.

The variation of $\mu$ with $\mu_s$ for the two types of mechanism is shown in Fig. 2 for a single layer with $\phi = 35°$. It is seen that for $\mu_s < 0.2$ the rutting mode is the preferred failure mechanism, whilst for larger values of $\mu_s$ the slip mode is predicted to occur.

#### 4.1.2. *Effect of load distribution*

It is well known that the uniform load distribution is not a very realistic approximation to the tyre load on the pavement, and also some numerical difficulties are introduced by the singularity caused by the discontinuity in the loading distribution on the stresses in the pavement. Here, an ef-

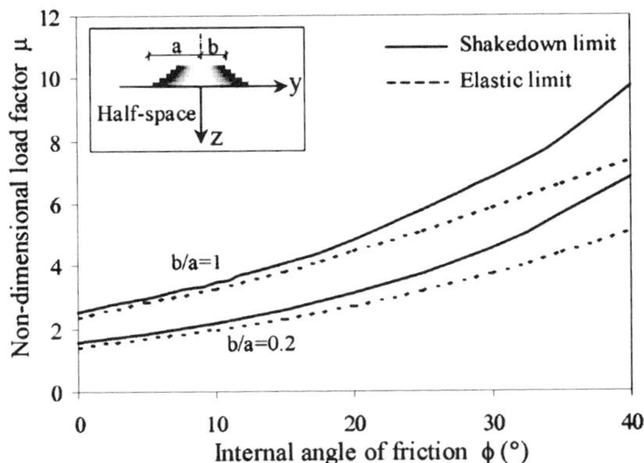


*Figure 3.* The dependence of shakedown load on the shape of the load distribution.

fort is made to simulate a more realistic load distribution by using of a superposition of 10 uniform load distributions.

The dependence of the non-dimensional load factor for a uniform half-space, on the assumed shape of the loaded distribution is illustrated in Fig. 3. The pressure distribution takes the form of a cone frustrum with upper radius $b$ and lower radius $a$. It is seen that both the elastic limit and shakedown load decrease when the uniform pressure is replaced by a graded pressure distribution, and that the difference in value between the shakedown limit and the elastic limit is most pronounced at higher friction angles.

#### 4.1.3. *Effect of dual loads*

In practice loads are frequently applied through dual wheels and the question arises as to whether the value of the shakedown load on one wheel is significantly effected by the presence of the adjacent load. Two possible types of failure mechanism are considered here. In the first the two wheels act separately, whilst in the second the two wheels effectively act as a single loaded area. Since in the optimal solutions the deforming region is confined to the edge of the loaded area, it is found that in fact there is very little interference between the deformation modes and the mechanism in Fig. 4(a) is always the more critical. Nevertheless the presence of a neighbouring loaded patch does effect the value of the shakedown load for a given tyre, since the elastic stress field beneath the tyre is now modified. This effect is illustrated in Fig. 5(a), where the shakedown loads for a single tyre with load $P$ is compared with that for a dual system with a load $P$ on each tyre. The presence of the neighbouring load is seen to increase the shakedown

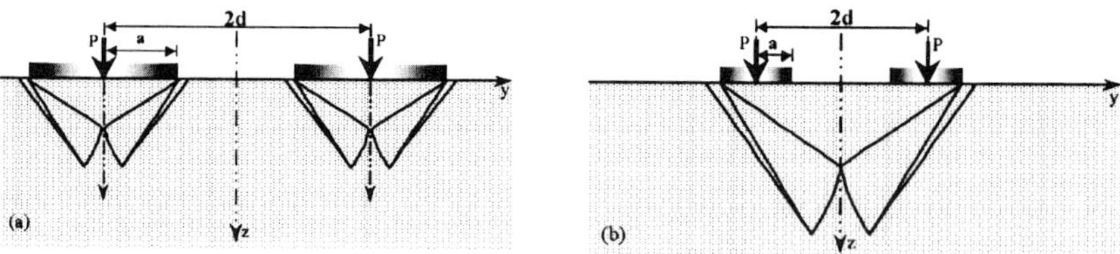


*Figure 4.* Two possible mechanisms for failure under dual loads.

load slightly, an effect more noticeable at higher friction angles.

It is also of interest to compare the shakedown load corresponding to a load $P$ on a single tyre with that of $P/2$ on each wheel of a dual wheel system. This is illustrated in Fig. 5(b), where now $\widetilde{\mu} = \lambda P_t/\pi c a^2$, $P_t$ is being the total load on the system, is plotted against $\phi$. (Note that $\widetilde{\mu} = \mu$ for a single tyre, but $\widetilde{\mu} = 2\mu$ for the dual wheel system). It is seen that, as expected, the value of the shakedown load is significantly increased (more than doubled in fact) by splitting the load between the two tyres.

#### 4.1.4. *Effect of a non-associated flow rule*

In the above calculation the shakedown load is calculated assuming that the material obeys the normality rule. In reality soils and granular materials frequently exhibit non-associated behaviour Lade [20]. In this section an extension of the kinematic method of the shakedown analysis for non-associated materials is given for translational failure mechanisms.

The straightforward generalisation of limit analysis techniques to non-associated materials does not work, since the expression for the plastic dissipation rate depends on the individual stress components, which of course are unknown. However, Drescher and Detournay [13] have shown that for plane strain block sliding modes, estimates of failure loads can still be obtained by using a "comparison" associated material with cohesion $\widetilde{c}$ and friction angle $\widetilde{\phi}$. The equation (2) can be written for a non-associated material as:

$$c^e = \mid \sigma^e_{nt} \mid - \sigma^e_{nn} tan\widetilde{\phi} \tag{4}$$

where

$$\widetilde{c} = \omega c \qquad \text{and} \qquad tan\widetilde{\phi} = \omega tan\phi \tag{5}$$

and

$$\omega = \frac{cos\phi cos\psi}{1 - sin\phi sin\psi} \tag{6}$$

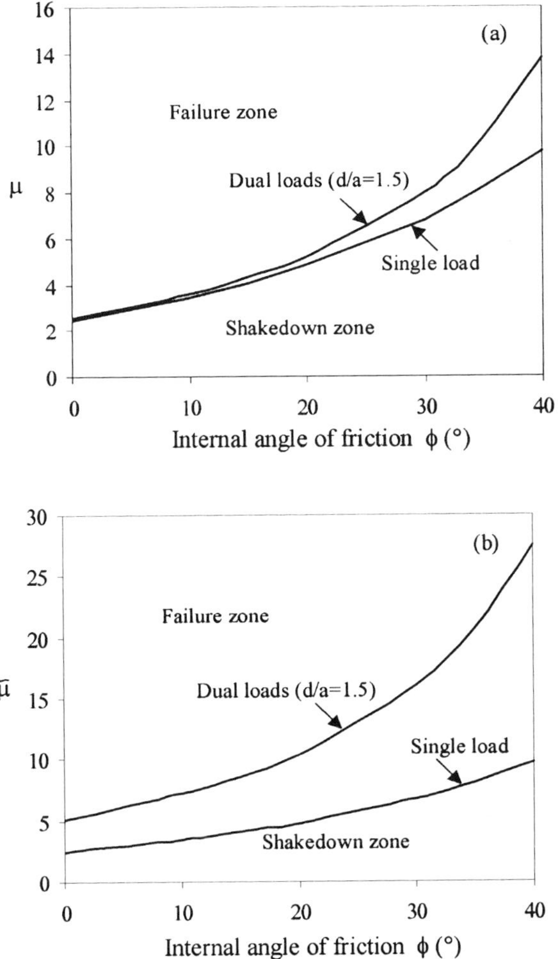


*Figure 5.* Comparison of values of dimensionless load factors for single and dual loads.

and the jump in normal velocity is $[v_n] = [v_t]tan\psi$, where $\psi$ and $\phi$ are the dilation and internal friction angles respectively. For more details (see Drescher and Detournay [13].

To illustrate the applicability of the procedure, a number of calculations are presented. The first example, shown in Fig. 6, shows the variation of dimensionless shakedown load with dilation angle for four given friction angles.

Figure 7 shows the difference in the values of the shakedown load calculated by assuming firstly the associated flow rule and secondly assuming

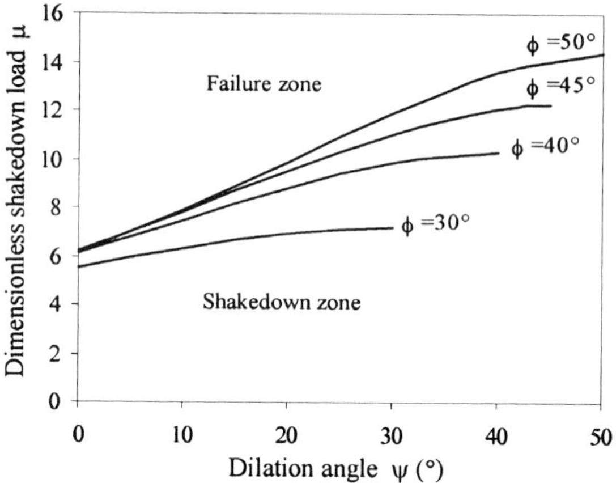


*Figure 6.* Variation of $\mu$ against dilation angle for homogeneous semi-infinite pavement.

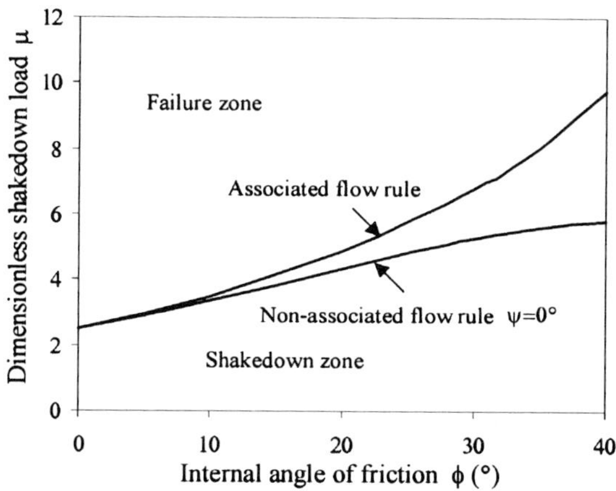


*Figure 7.* Comparison of associated and non-associated shakedown bahaviour.

that the material is incompressible ($\psi = 0°$). The results obtained for $\psi = 0°$ can be considered as applying to a fully saturated pavement under undrained conditions. For small friction angles the shakedown load is relatively unaffected by the flow rule, but the drop in the value of the predicted shakedown load for friction angles in the range 30°- 40° is highly significant.

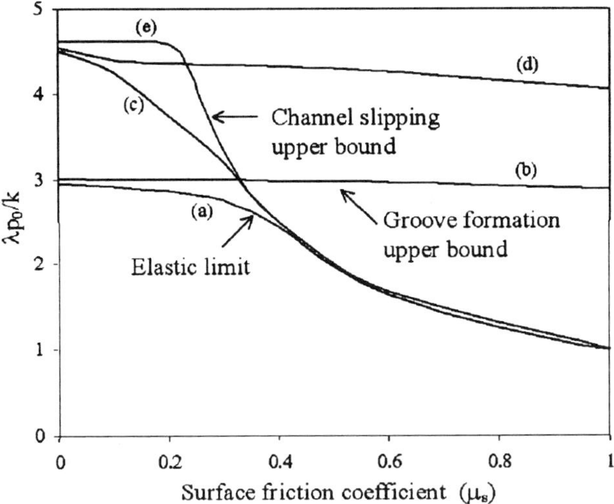


*Figure 8.* Variation of various critical dimensionless load factors with surface friction coefficient for Hertzian pressure distribution on a circular loading patch.

4.1.5. *Shakedown loads for purely cohesive materials*

For zero friction angles, the solutions apply to purely cohesive materials and also to metals. The present results can hence be compared with the extant studies on metal deformations. A detailed account of this comparison is given in Boulbibane and Collins [4]. Some of these are summarised in Fig. 8 for a Hertzian pressure distribution (with maximum pressure $p_0$) applied over a circular contact patch. The dimensionless load factor is plotted against the surface friction coefficient for (a) the elastic limit, (b) the optimal rut formation mode in Fig 1(b), (c) the groove formation mode discussed by Kapoor and Johnson [18] who assumed that $\alpha_3 + 2\alpha_1 = \pi$ and used a lower bound analyse, (d) the sub optimal rut formation mode in Fig. 1(b) in which $\alpha_1$ and $\alpha_3$ are restricted to satisfy the Kapoor and Johnson condition and (e) the upper bound solution given by Ponter *et al.* [23] assuming a channel slipping failure mode. It is concluded that the groove or rut formation mode is the likely failure mechanism for value of $\mu_s$ less than about 0.3, but that for higher frictional coefficients failure is more likely to be due to slipping in the travel direction. However, in both cases the shakedown load is very close to the elastic limit load. This result is important for some proposed plastic road design procedures, many of which are based on plastic collapse load (i.e. foundation load) calculations (cf. Burd and Frydman [5], Fredlund *et al.* [14], Houlsby and Burd [16]).

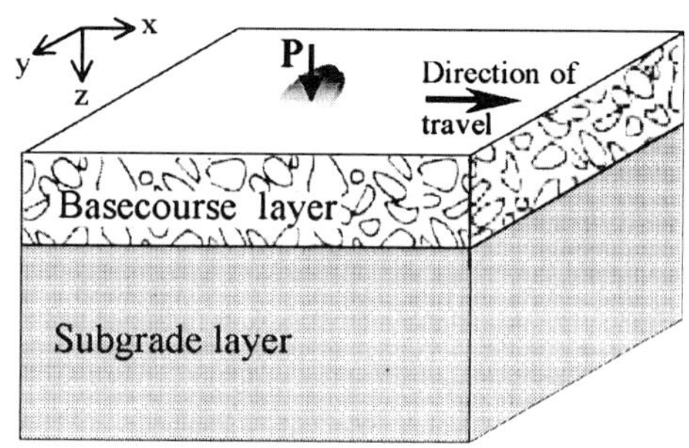


*Figure 9.* Multi-layered pavement.

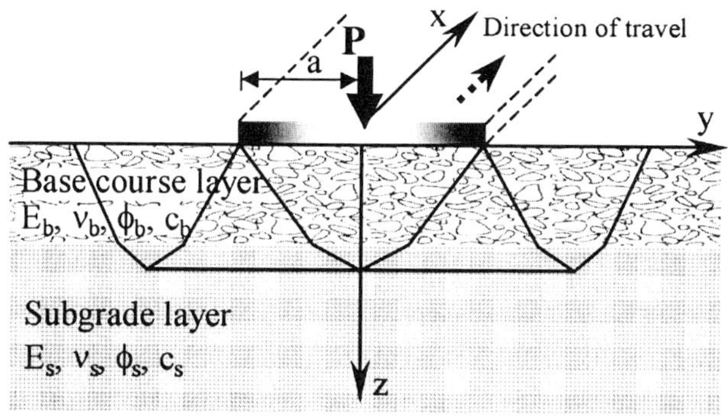


*Figure 10.* Plastic collapse mechanism for multi-layered pavement structure.

### 4.2. SHAKEDOWN IN A MULTI-LAYER SYSTEM

In reality pavements are multilayer systems and rutting failures frequently occur when the pavement has a relatively weak subgrade. Detailed calculations for two layered pavements have also been performed, and some of the results have already been presented in Collins and Boulbibane [10]. An example of a possible mechanisms for a pavement with a single basecourse of depth $h$ is shown in Fig. 10. The discontinuity lines must have a change in slope as they cross the basecourse/subgrade boundary whenever the value of the internal friction angle is different in the two layers. This is also a feature of the limit analysis solution of Michalowski and Shi [22] for the failure of footings on layered soils.

For the two-layer pavement model the dimensionless shakedown load parameter $\mu$ defined to be $\lambda P/\pi c_b a^2$, depends on a number of dimensionless parameters, viz.:

$$\mu = \mu(\phi_b,\ \phi_s,\ \mu_s,\ \nu_b,\ \nu_s,\ E_b/E_s,\ c_b/c_s,\ h/a) \tag{7}$$

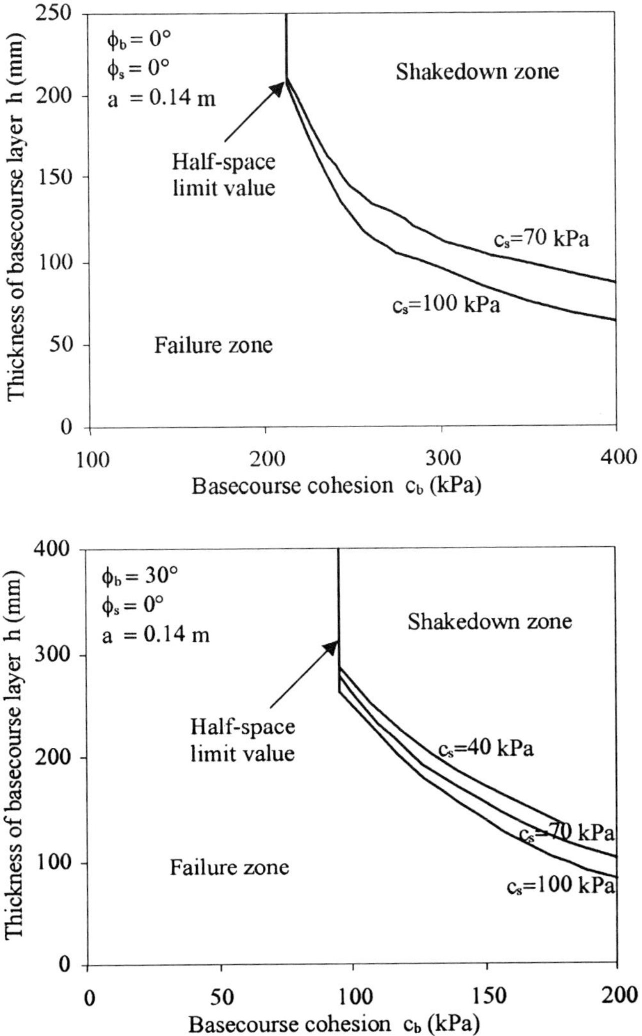


*Figure 11.* Variation of basecourse thickness with basecourse cohesion.

In all the calculations presented here $\phi_s = 0°$, so the subgrade is assumed to be a saturated clay and its cohesion $c_s$ can be interpreted as its undrained shear strength, whilst $\nu_b = 0.35$, $\nu_s = 0.4$ and $E_b/E_s = 3$. We hence concentrate on the effect of the major design parameters, $\phi_b$ the internal angle of friction of the basecourse material, $c_b$ and $c_s$ the cohesions of the two layers and $h$ the thickness of the basecourse. Some preliminary results of calculations for this situation are shown in Fig. 11 in a form suitable for design purposes. The variation of $\mu$ with $h/a$ for various basecourse friction angles for a fixed cohesion ratio is shown in Fig. 11. Graphs such as these enable the critical basecourse thickness at which shakedown just occurs to be determined for given values of applied load, radius of loaded area, and basecourse and subgrade strength parameters. The graphs exhibit "cut offs" at points where the cohesion of the basecourse is sufficiently low for failure to occur in the basecourse alone as predicted by the half-space solutions of the previous section.

## 5. Conclusions

There is widespread appreciation that existing design procedures for unbound pavements are unsatisfactory, and it is necessary to improve the basic model. The shakedown model offers a number of advantages:

(a) The strength as well as the stiffness properties are included, so that realistic failure mechanisms, such as rutting can be modelled.

(b) The repeated nature of the loading is automatically accounted for, and so this procedure is superior to an adapted "foundation" method, which presumes monotonic loading.

(c) The basic model can be extended to include a number of design issues, such as multilayering, non-linear elastic response, moisture content, use of stabilisation layers, or geotextiles.

The main shortcoming, which applies to all pavement shakedown analyses presented to date, is the use of a perfectly plastic Coulomb material model. Since our aim is to study the ultimate behaviour of a pavement, it could be argued that the model parameters used should be those corresponding to the "critical" or "steady state" model of the aggregate and subgrade materials. However, a truly satisfactory model will not eventuate until the concepts of the shakedown theory have been united with a realistic model of the material response of the various pavement constituent materials to cyclic loading, deduced from laboratory experiments.

## Acknowledgements

Much of this research was funded by a grant from the New Zealand Foundation for Research in Science and Technology.

## References


1. Boulbibane, M. (1995) *Application of the shakedown theory to non-associative elastic-plastic media: case of geomaterials*, Ph.D. Thesis, No. 1496, University of Science and Technology of Lille, France, (in French).
2. Boulbibane, M. and Weichert, D. (1997) Application of shakedown theory to soils with non-associated flow rules, *Mechanics Research Communications*,**5**, 516-519.
3. Boulbibane, M. and Collins, I.F. (1999) Geotechnical Models of Failure in Unbound Pavements, *Proceeding of the 8th Australia New Zealand Conference on Geomechanics*, Hobart, pp. 733-738.
4. Boulbibane, M. and Collins, I.F. (2000) The calculation of shakedown loads for contact problems, AEPA 2000 Hong-Kong (to be published).
5. Burd, H.J. and Frydman, S. (1997) Bearing capacity of plane-strain footing on layered soils, *Canadian Geotech. J.*, **34**, 241-253.
6. Chen, W.F. (1975) *Limit analysis and soil plasticity*, Elsevier, Amsterdam.
7. Collins, I.F. and Cliffe, P.F. (1987) Shakedown in frictional materials under moving surface loads, *Int. J. Num. Anal. Methods Geomechanics*, **11**,409-420.
8. Collins, I.F., Wang, A.P. and Sanders, L.R. (1993) Shakedown theory and the design of unbound pavements, *Road and Transport Research*, **2**, 29-38.
9. Collins, I.F., Wang, A.P. and Sanders, L.R. (1993) Shakedown in layered pavements under moving surface loads, *Int. J. Num. and Anal. Methods in Geomechanics*, **17**, 165-174.
10. Collins, I.F. and Boulbibane, M. (1997) Pavements as structures subjected to repeated loadings, *The Mechanics of Structures and Materials*, Grzebieter, Al-Mahaidi and Wilson (eds) Balkema, Rotterdam, pp. 511-516.
11. Collins, I.F. and Boulbibane, M. (1998) The application of shakedown theory to pavement design, *Metals and Materials*, **4**, 832-837.
12. Collins, I.F. and Boulbibane, M. (in press) A geotechnical analysis of unbound pavements based upon shakedown theory, *ASCE J. of Geotechnical and Geoenvironmental Engineering.*
13. Drescher, A. and Detournay, E. (1993) Limit load in translational failure mechanisms for associative and non-associative materials, *Géotechnique*, **43**, 443-456.
14. Fredlund, D.G., Morgenstern, N.R. and Widger, R.A. (1978) The shear strength of unsaturated soils, *Canadian Geotech. J.*, **15**, 313-321.
15. Goffe, B., Ferri, G.D. and Rogers, J. (1994) Global optimization of statical functions with simulated annealing, *Journal of Econometrics*, **60**, 65-99.
16. Houlsby, G.T. and Burd, H.J. (1999) Understanding the behaviour of unpaved roads on soft clay, *Proceedings of the 12th European Conference on Soil Mechanics and Geotechnical Engineering - Geotechnical Engineering for Transportation Infrastructure*, Amsterdam, **1**, pp. 31-42.
17. Johnson, K.L. (1992) The application of shakedown principles in rolling and sliding contact, *Eur. J. Mech., A/Solids*, **11**, 155-172.
18. Kapoor, A. and Johnson, K.L. (1992) Effect of changes in contact geometry on shakedown of surfaces in rolling/sliding contact, *Int. J. Mech. Sci.*, **3**, 223-239.
19. Koiter, W.T. (1960) General theorems for elastic-plastic solids, *Progress in Solid Mechanics* (Eds. I.N. Sneddon and R. Hill), North Holland, Amsterdam, pp. 165-221.
20. Lade, P.V. (1984) Failure criterion for frictional materials, *Mechanics of Eng. Materials*, ed. By C.S. Desai and R.H. Gallagher, John Wiley and Sons, 385-402.
21. Melan, E. (1938) Zur Plastizität des räumlichen Kontinuums, *Ing. Arch.*, **9** 116-126.

22. Michalowski, R.L. and Shi, L. (1995) Bearing capacity of footings over two-layer foundation soils, *J. of Geotechnical Engineering*, 421-428.
23. Ponter, A.R., Hearle, A.D. and Johnson, K.L. (1985) Application of the kinematic shakedown theorem to rolling and sliding contact points, *J. Mech. Phys. Solids*, **33**, 339-362.
24. Raad, L., Weichert, D. and Haidar A. (1989) Shakedown and fatigue of pavements with granular bases, *Transportation Research Record*, **1227**, 159-172.
25. Raad, L. and Weichert, D. (1995) Stability of pavement structures under long term repeated loading, Mroz *et al.* (eds.), *inelastic Behaviour of Structures under Variable Loads*, Kluwer Academic Publishers, pp. 473-496.
26. Sharp, R.W. and Booker, J.R.(1984) Shakedown of pavements under moving surface loads, *J. of Transport Eng.*, **110**, 1-14.
27. Yu, H.S. and Hossain, M.Z. (1998) Lower bound shakedown analysis of layered pavements using discontinuious stress fields, *Computer Methods in Applied Mechanics and Engineering*, **167**, 209-222.

# Mechanics

## *SOLID* MECHANICS AND ITS APPLICATIONS

*Series Editor*: G.M.L. Gladwell

49. J.R. Willis (ed.): *IUTAM Symposium on Nonlinear Analysis of Fracture.* Proceedings of the IUTAM Symposium held in Cambridge, U.K. 1997 ISBN 0-7923-4378-6
50. A. Preumont: *Vibration Control of Active Structures.* An Introduction. 1997 ISBN 0-7923-4392-1
51. G.P. Cherepanov: *Methods of Fracture Mechanics: Solid Matter Physics.* 1997 ISBN 0-7923-4408-1
52. D.H. van Campen (ed.): *IUTAM Symposium on Interaction between Dynamics and Control in Advanced Mechanical Systems.* Proceedings of the IUTAM Symposium held in Eindhoven, The Netherlands. 1997 ISBN 0-7923-4429-4
53. N.A. Fleck and A.C.F. Cocks (eds.): *IUTAM Symposium on Mechanics of Granular and Porous Materials.* Proceedings of the IUTAM Symposium held in Cambridge, U.K. 1997 ISBN 0-7923-4553-3
54. J. Roorda and N.K. Srivastava (eds.): *Trends in Structural Mechanics.* Theory, Practice, Education. 1997 ISBN 0-7923-4603-3
55. Yu.A. Mitropolskii and N. Van Dao: *Applied Asymptotic Methods in Nonlinear Oscillations.* 1997 ISBN 0-7923-4605-X
56. C. Guedes Soares (ed.): *Probabilistic Methods for Structural Design.* 1997 ISBN 0-7923-4670-X
57. D. François, A. Pineau and A. Zaoui: *Mechanical Behaviour of Materials.* Volume I: Elasticity and Plasticity. 1998 ISBN 0-7923-4894-X
58. D. François, A. Pineau and A. Zaoui: *Mechanical Behaviour of Materials.* Volume II: Viscoplasticity, Damage, Fracture and Contact Mechanics. 1998 ISBN 0-7923-4895-8
59. L.T. Tenek and J. Argyris: *Finite Element Analysis for Composite Structures.* 1998 ISBN 0-7923-4899-0
60. Y.A. Bahei-El-Din and G.J. Dvorak (eds.): *IUTAM Symposium on Transformation Problems in Composite and Active Materials.* Proceedings of the IUTAM Symposium held in Cairo, Egypt. 1998 ISBN 0-7923-5122-3
61. I.G. Goryacheva: *Contact Mechanics in Tribology.* 1998 ISBN 0-7923-5257-2
62. O.T. Bruhns and E. Stein (eds.): *IUTAM Symposium on Micro- and Macrostructural Aspects of Thermoplasticity.* Proceedings of the IUTAM Symposium held in Bochum, Germany. 1999 ISBN 0-7923-5265-3
63. F.C. Moon: *IUTAM Symposium on New Applications of Nonlinear and Chaotic Dynamics in Mechanics.* Proceedings of the IUTAM Symposium held in Ithaca, NY, USA. 1998 ISBN 0-7923-5276-9
64. R. Wang: *IUTAM Symposium on Rheology of Bodies with Defects.* Proceedings of the IUTAM Symposium held in Beijing, China. 1999 ISBN 0-7923-5297-1
65. Yu.I. Dimitrienko: *Thermomechanics of Composites under High Temperatures.* 1999 ISBN 0-7923-4899-0
66. P. Argoul, M. Frémond and Q.S. Nguyen (eds.): *IUTAM Symposium on Variations of Domains and Free-Boundary Problems in Solid Mechanics.* Proceedings of the IUTAM Symposium held in Paris, France. 1999 ISBN 0-7923-5450-8
67. F.J. Fahy and W.G. Price (eds.): *IUTAM Symposium on Statistical Energy Analysis.* Proceedings of the IUTAM Symposium held in Southampton, U.K. 1999 ISBN 0-7923-5457-5
68. H.A. Mang and F.G. Rammerstorfer (eds.): *IUTAM Symposium on Discretization Methods in Structural Mechanics.* Proceedings of the IUTAM Symposium held in Vienna, Austria. 1999 ISBN 0-7923-5591-1

# Mechanics

## *SOLID* MECHANICS AND ITS APPLICATIONS

*Series Editor*: G.M.L. Gladwell

69. P. Pedersen and M.P. Bendsøe (eds.): *IUTAM Symposium on Synthesis in Bio Solid Mechanics.* Proceedings of the IUTAM Symposium held in Copenhagen, Denmark. 1999 ISBN 0-7923-5615-2
70. S.K. Agrawal and B.C. Fabien: *Optimization of Dynamic Systems.* 1999 ISBN 0-7923-5681-0
71. A. Carpinteri: *Nonlinear Crack Models for Nonmetallic Materials.* 1999 ISBN 0-7923-5750-7
72. F. Pfeifer (ed.): *IUTAM Symposium on Unilateral Multibody Contacts.* Proceedings of the IUTAM Symposium held in Munich, Germany. 1999 ISBN 0-7923-6030-3
73. E. Lavendelis and M. Zakrzhevsky (eds.): *IUTAM/IFToMM Symposium on Synthesis of Nonlinear Dynamical Systems.* Proceedings of the IUTAM/IFToMM Symposium held in Riga, Latvia. 2000 ISBN 0-7923-6106-7
74. J.-P. Merlet: *Parallel Robots.* 2000 ISBN 0-7923-6308-6
75. J.T. Pindera: *Techniques of Tomographic Isodyne Stress Analysis.* 2000 ISBN 0-7923-6388-4
76. G.A. Maugin, R. Drouot and F. Sidoroff (eds.): *Continuum Thermomechanics.* The Art and Science of Modelling Material Behaviour. 2000 ISBN 0-7923-6407-4
77. N. Van Dao and E.J. Kreuzer (eds.): *IUTAM Symposium on Recent Developments in Non-linear Oscillations of Mechanical Systems.* 2000 ISBN 0-7923-6470-8
78. S.D. Akbarov and A.N. Guz: *Mechanics of Curved Composites.* 2000 ISBN 0-7923-6477-5
79. M.B. Rubin: *Cosserat Theories: Shells, Rods and Points.* 2000 ISBN 0-7923-6489-9
80. S. Pellegrino and S.D. Guest (eds.): *IUTAM-IASS Symposium on Deployable Structures: Theory and Applications.* Proceedings of the IUTAM-IASS Symposium held in Cambridge, U.K., 6–9 September 1998. 2000 ISBN 0-7923-6516-X
81. A.D. Rosato and D.L. Blackmore (eds.): *IUTAM Symposium on Segregation in Granular Flows.* Proceedings of the IUTAM Symposium held in Cape May, NJ, U.S.A., June 5–10, 1999. 2000 ISBN 0-7923-6547-X
82. A. Lagarde (ed.): *IUTAM Symposium on Advanced Optical Methods and Applications in Solid Mechanics.* 2000 ISBN 0-7923-6604-2
83. D. Weichert and G. Maier (eds.): *Inelastic Analysis of Structures under Variable Loads.* Theory and Engineering Applications. 2000 ISBN 0-7923-6645-X

Kluwer Academic Publishers – Dordrecht / Boston / London

## ICASE/LaRC Interdisciplinary Series in Science and Engineering

1. J. Buckmaster, T.L. Jackson and A. Kumar (eds.): *Combustion in High-Speed Flows*. 1994 ISBN 0-7923-2086-X
2. M.Y. Hussaini, T.B. Gatski and T.L. Jackson (eds.): *Transition, Turbulence and Combustion*. Volume I: Transition. 1994 ISBN 0-7923-3084-6; set 0-7923-3086-2
3. M.Y. Hussaini, T.B. Gatski and T.L. Jackson (eds.): *Transition, Turbulence and Combustion*. Volume II: Turbulence and Combustion. 1994 ISBN 0-7923-3085-4; set 0-7923-3086-2
4. D.E. Keyes, A. Sameh and V. Venkatakrishnan (eds): *Parallel Numerical Algorithms*. 1997 ISBN 0-7923-4282-8
5. T.G. Campbell, R.A. Nicolaides and M.D. Salas (eds.): *Computational Electromagnetics and Its Applications*. 1997 ISBN 0-7923-4733-1
6. V. Venkatakrishnan, M.D. Salas and S.R. Chakravarthy (eds.): *Barriers and Challenges in Computational Fluid Dynamics*. 1998 ISBN 0-7923-4855-9
7. M.D. Salas, J.N. Hefner and L. Sakell (eds.): *Modeling Complex Turbulent Flows*. 1999 ISBN 0-7923-5590-3

KLUWER ACADEMIC PUBLISHERS – DORDRECHT / BOSTON / LONDON

## *ERCOFTAC* SERIES

1. A. Gyr and F.-S. Rys (eds.): *Diffusion and Transport of Pollutants in Atmospheric Mesoscale Flow Fields.* 1995 ISBN 0-7923-3260-1
2. M. Hallbäck, D.S. Henningson, A.V. Johansson and P.H. Alfredsson (eds.): *Turbulence and Transition Modelling.* Lecture Notes from the ERCOFTAC/IUTAM Summerschool held in Stockholm. 1996 ISBN 0-7923-4060-4
3. P. Wesseling (ed.): *High Performance Computing in Fluid Dynamics.* Proceedings of the Summerschool held in Delft, The Netherlands. 1996 ISBN 0-7923-4063-9
4. Th. Dracos (ed.): *Three-Dimensional Velocity and Vorticity Measuring and Image Analysis Techniques.* Lecture Notes from the Short Course held in Zürich, Switzerland. 1996 ISBN 0-7923-4256-9
5. J.-P. Chollet, P.R. Voke and L. Kleiser (eds.): *Direct and Large-Eddy Simulation II.* Proceedings of the ERCOFTAC Workshop held in Grenoble, France. 1997 ISBN 0-7923-4687-4

KLUWER ACADEMIC PUBLISHERS – DORDRECHT / BOSTON / LONDON